**The
Pion-Nucleon
System**

The Pion-Nucleon System

B. H. Bransden and
R. G. Moorhouse

Princeton
University Press

Copyright © 1973
by Princeton
University Press
All rights reserved.

L.C. Card: 73-39795
ISBN: 0-691-08115-8
(hardcover edition)
0-691-08129-8
(paperback edition)

Printed in the
United States
of America
By Princeton
University Press,
Princeton, New Jersey

PREFACE

This book gives an account, suitable for first or second year graduate students, of the hadronic interactions of pions and nucleons and the structure of nucleons. Because of the hadronic SU3 symmetry the book has necessarily also the larger theme of the structure of baryons in general, so we include much material and ideas from outside the field of pion-nucleon interactions. The reader could regard the subject of the book as the hadronic structure of baryons as illustrated particularly by the pion-nucleon interaction. Since very many of our ideas come from pion-nucleon experiments, we have not felt unduly restricted at using this approach; at the same time its degree of narrowness has enabled us to give a reasonably full treatment of many topics within the format of a moderate-sized book.

In writing we have necessarily had in mind the modern history of the subject, comprised perhaps within the last 12 years. Experimentally we have seen a steady accretion of increasingly accurate data showing, for instance, the existence of a very large number of resonances, but with no single experiment making a watershed of thought. Theoretically our period begins with the eightfold way, and the general acceptance of the metastable baryons and the unstable baryonic resonances as belonging to one class of objects. There has been a movement away from interest in such dynamical explanations of particle spectra as the dispersion relation bootstrap and towards a more model-building approach, including composite models of baryon structure as well as other types. All these models have the task of explaining in some sense the great number of (unstable) baryons. There has also arisen a great body of data-fitting based on the construction of amplitudes using Regge pole exchange and the phenomenological idea of duality as the possible equivalence of, on the one hand, the exchanged Regge pole structure and, on the other, the resonance dominant structure of certain amplitudes.

We hope that the double authorship has enabled us to deal even-handedly with the various approaches, with the exception that we have not found it possible within a book of reasonable compass to give an extended account of quantum field theory. This, however, is not a serious limitation, and otherwise the book is sufficiently self-contained for the graduate student to use it as a text. Although no attempt has been made to provide a bibliography that is complete in any sense, we have given at the end of each chapter rather more references to the original literature than is usual in a text, so that the reader should have no difficulty in following up any topic of particular interest to him.

Finally, it is a pleasure to acknowledge many helpful conversations and suggestions from our colleagues in the Universities of Durham and Glasgow and our particular debt to Dr. F. D. Gault for his critical reading of the draft manuscript.

B. H. B.
R. G. M.

APRIL, 1972

CONTENTS

PREFACE	v
A Note on Units and Conventions	3
CHAPTER 1 – Introduction	7
1.1 The Field of Study	7
1.2 Isospin	11
CHAPTER 2 – Low Energy Pion Scattering by Nucleons and the Photoproduction of Pions	21
2.1 The Scattering Matrix and Cross Sections	21
2.2 Low Energy π-Meson-Nucleon Scattering	43
2.3 The Low Energy Scattering of K^+ and K^- Mesons by Nucleons	65
2.4 Photoproduction of π Mesons from Nucleons	68
CHAPTER 3 – Forward and Fixed Momentum Transfer Dispersion Relations	89
3.1 Introduction	89
3.2 Forward Scattering of Neutral Particles	92
3.3 Fixed Momentum Transfer Dispersion Relations	102
3.4 Dispersion Relations for Pion-Nucleon Elastic Scattering	105
3.5 Forward Dispersion Relations for $K^{\pm}N$ Scattering	121
CHAPTER 4 – Analytic Properties of Scattering Amplitudes	127
4.1 Potential Scattering	128
4.2 The Relativistic S-Matrix	142
4.3. Partial Wave Dispersion Relations for πN Scattering	153
CHAPTER 5 – Formation of Nucleon Resonances	167
5.1 Baryon Resonances, Formation and Production	167
5.2 Resonance Theory	169
5.3 Partial Wave Analysis of Elastic Scattering Experiments	195
5.4 Resonances: Assessment of Results of Partial Wave Analysis	203

CHAPTER 6 – Symmetries and Classification of Particles
and Resonances 217
  6.1 Isospin, an SU_2 Symmetry 217
  6.2 Young Tableaux and the Permutation Symmetry of
      Wave Functions 224
  6.3 The Eightfold Way, an SU_3 Symmetry 230
  6.4 Spinning Quarks and SU_6 244
  6.5 The L-Excitation Quark Model 250
  6.6 SU_3 Couplings and N^* Decays 258
  6.7 Electromagnetic Interactions and U-Spin 269
  6.8 Electromagnetic Decays of Baryon Resonances 271

CHAPTER 7 – Current Algebra and Sum Rules 285
  7.1 Born Approximation 286
  7.2 Mesons and Currents 294
  7.3 Soft Pions 299
  7.4 Current Algebra 306
  7.5 S-Wave Pion-Nucleon Scattering Lengths 311

CHAPTER 8 – Scattering at Higher Energies 319
  8.1 Elastic Scattering at Laboratory Momenta $\gtrsim 1$ GeV ... 319
  8.2 Total Cross Sections and the Pomeranchuk Theorem .. 336
  8.3 Particle Exchange and Regge Exchange 340
  8.4 t-Channel Regge Poles 344
  8.5 The ρ-Regge Pole and Charge Exchange Scattering .. 347
  8.6 Factorization and Zeros of Regge Pole Residues ... 351
  8.7 Elastic Scattering, Regge Poles, and Absorption ... 354
  8.8 The s-Channel, Duality, and Peripherality 361
  8.9 Backward Pion-Nucleon Scattering, u-Channel Exchange
      and Nucleonic Regge Trajectories 369
  8.10 u-Channel Regge Poles and Daughter Trajectories .. 375

CHAPTER 9 – Pion-Nucleon Dynamics 381
  9.1 Introduction 381
  9.2 The Analysis of Low Energy πN Scattering by Partial
      Wave Dispersion Relations 384
  9.3 Duality, Exchange-Degeneracy, and Dual Models .. 407
  9.4 Padé Approximates 419

CHAPTER 10 – Pion-Nucleon Inelastic Scattering 425
  10.1 Analysis of Three Particle Final States 428
  10.2 Single Pion Production Below 1 GeV and the
       Isobar Model 434
  10.3 The Peripheral Model 440

10.4	Resonance Production at High Energies and the Quark Model	447
10.5	Multiparticle Production at High Energies	451

CHAPTER 11 – The Form Factors of the Nucleon and Pion — 459
  11.1 Introduction — 459
  11.2 The Elastic Scattering of Electrons by Protons — 463
  11.3 The Neutron Form Factors and the Isotopic Spin Decomposition — 474
  11.4 The Theory of Form Factors — 478
  11.5 The Pion Form Factor — 486
  11.6 Deep Inelastic Scattering and the Parton Model — 498

APPENDIX A – Angular Momentum — 509
  A.1 The Legendre Polynomials and Spherical Harmonics — 509
  A.2 The Clebsch-Gordan Coefficients — 511
  A.3 The Rotation Matrices — 514

APPENDIX B – Formalism for Pion Photoproduction — 517

The
Pion-Nucleon
System

A NOTE ON UNITS AND CONVENTIONS

Except where otherwise indicated natural units are used throughout the text for which $\hbar = c = 1$. Unit length is then equal to $(\frac{1}{M})$, where M is a mass that can be chosen conveniently. If M is in units of Mev/c^2, the unit of length is equal to $\frac{1.97314}{M} \times 10^{-11}$ cm. When M is equal to the pion mass, the unit of length is 1.41×10^{-13} cm and the unit of time becomes 0.47×10^{-23} sec.

When quoting the results of experiments we also use as a unit of length the Fermi, which is equal to 10^{-13} cm, and cross sections are quoted in millibarns (mb) or microbarns (μb), which are equal to $10^{-27} cm^2$ or $10^{-30} cm^2$ respectively.

COVARIANT QUANTITIES

The metric adopted is the following:

$$g_{\mu\nu} = \begin{pmatrix} 1 & 0 & 0 & 0 \\ 0 & -1 & 0 & 0 \\ 0 & 0 & -1 & 0 \\ 0 & 0 & 0 & -1 \end{pmatrix}$$

so that a vector A^μ is equivalent to $(A^0, \vec{A})$ where $\vec{A}$ is a three-dimensional vector, and A_μ has components $(A^0, -\vec{A})$. The convention is used that we sum over repeated indices and:

$$A_\mu = g_{\mu\nu} A^\nu .$$

The scalar product of A and B is given by:

$$A \cdot B = A^0 B^0 - \vec{A} \cdot \vec{B} .$$

We also use the notation

$$\Box = \frac{\partial}{\partial x_\mu} \frac{\partial}{\partial x^\mu} = -\nabla^2 + \frac{\partial^2}{\partial x_0^2}$$

DIRAC SPINORS AND MATRICES

For the Dirac γ matrices we use the representation for which:

$$\vec{\gamma} = \beta \vec{\alpha}, \gamma^0 = \beta$$

and:

$$\vec{\alpha} = \begin{pmatrix} 0 & \vec{\sigma} \\ \vec{\sigma} & 0 \end{pmatrix} \qquad \beta = \begin{pmatrix} 1 & 0 \\ 0 & -1 \end{pmatrix}$$

where $\sigma_1, \sigma_2, \sigma_3$ are the usual 2×2 Pauli matrices and $\mathbf{1}$ denotes the 2×2 unit matrix.

We then use the notation that:

$$\gamma^\mu p_\mu \equiv \gamma \cdot p = \gamma^0 p^0 - \vec{\gamma} \cdot \vec{p} = \not{p} .$$

The positive and negative frequency spinors $u^{(r)}(p), v^{(r)}(p)$ satisfying the Dirac equations:

$$(\not{p} - m) u^{(r)}(p) = 0$$
$$(\not{p} + m) v^{(r)}(p) = 0$$

DIRAC SPINORS AND MATRICES

are normalized so that:

$$\bar{u}^{(r)}(p)\, u^{(s)}(p) = \delta_{rs}$$
$$\bar{v}^{(r)}(p)\, v^{(s)}(p) = -\delta_{rs}$$

where:

$$\bar{u}^{(r)}(p) = u^{\dagger(r)}(p)\gamma^0, \quad \bar{v}^{(r)}(p) = v^{\dagger(r)}(p)\gamma^0.$$

The auxiliary quantities $\sigma^{\mu\nu}$ and γ^5 are defined as:

$$\sigma^{\mu\nu} = \frac{i}{2}[\gamma^\mu, \gamma^\nu]$$

$$\gamma^5 = \gamma_5 = i\gamma^0\gamma^1\gamma^2\gamma^3.$$

Consistently with the preceding definitions we adopt the Bjorken-Drell field operators but we take an invariant normalization for state vectors (cf. Chapter 2). For example for spinor fields:

$$\psi(x) = (2\pi)^{-\frac{3}{2}} \sum_r \int d^3\vec{p}\, \sqrt{\frac{M_N}{E_p}}\, [e^{-ipx}\, u^r(p)\, a^r(p) + e^{ipx}\, v^r(p)\, b^{r*}(p)]$$

where, if $|0\rangle$ is the vacuum state, normalized to unity:

$$a^r(p)\,|p', s\rangle = \sqrt{E_p}\, \delta_{rs}\, \delta(\vec{p}-\vec{p}\,')\,|0\rangle$$

$$\{a^r(p), a^s(p')\} = \delta_{rs}\, \delta(\vec{p}-\vec{p}\,')$$

CHAPTER 1

INTRODUCTION

1.1. THE FIELD OF STUDY

The proton, p, is a positively charged spin $\frac{1}{2}$ particle of mass 938.3 MeV/c^2 and the neutron, n, is a neutral spin $\frac{1}{2}$ particle of mass 939.6 MeV/c^2. They each have, by conventional definition, positive parity, $P = +1$ (Sakurai, 1964). The positively and negatively charged pions, $\pi^\pm$, are spin 0 particles of mass 139.6 and the neutral charge pion, π^0, is a spin 0 particle of mass 135.0. The parity of all three pions is found to be negative, $P = -1$ (Sakurai, 1964; Nishijima, 1963; Källen, 1964). Pions and nucleons interact strongly, and this interaction is responsible for at any rate the longer range part of the nucleon-nucleon force.

It is the investigation and understanding of these interactions, and of the concomitant structure of the pions and nucleons themselves, together with the broader implications for elementary particles as a whole, that is the field of study of this book. The typical experiment is the collision of a pion with a nucleon target, with the detection of the reaction products. These may be a single pion and a nucleon (elastic or charge-exchange scattering) or many pions and a nucleon, or more rarely it may involve strange particles (strangeness exchange reactions) or baryon-anti-baryon pairs. In all cases the experiments carry implications for the interactions

TABLE 1-1. The particles of the lowest-mass baryon octet, and the particles of the lowest-mass baryon decuplet, displayed in hypercharge-isospin (Y, I) multiplets, together with their masses and widths, or lifetimes where more appropriate, and principal decay modes. For further details see Particle Data Group (1972).

Name	Y	I	Mass (MeV/c^2)	Width (MeV)	Mean Life (sec)	Principal Decay Modes
p	1	$\frac{1}{2}$	938.26		Stable	
n			939.55		$(.932\pm.014)10^3$	$pe^-\nu$
Λ	0	0	1115.6		$(2.51\pm.03)10^{-10}$	πN
Σ^+	0	1	1189.4		$(.810\pm.013)10^{-10}$	πN
Σ^0			1192.5		10^{-14}	$\Lambda\gamma$
Σ^-			1197.3		$(1.64\pm.06)10^{-10}$	πN
Ξ^0	-1	$\frac{1}{2}$	1315		$(3.03\pm.18)10^{-10}$	$\pi^0\Lambda$
Ξ^-			1321		$(1.66\pm.04)10^{-10}$	$\pi^-\Lambda$
Δ^{++}	1	1	1236	120		πN
Δ^+						
Δ^0						
Δ^-						
$\Sigma^+(1385)$	0	1	1385	36 ± 3		$\pi\Lambda, \pi\Sigma$
$\Sigma^0(1385)$						
$\Sigma^-(1385)$						
$\Xi^0(1530)$	-1	$\frac{1}{2}$	1530	7.3 ± 1.7		$\pi\Xi$
$\Xi^-(1530)$						
Ω^-	-2	0	1672		$(1.3^{+.4}_{-.3})10^{-10}$	$K^-\Lambda, \pi\Xi$

1.1. THE FIELD OF STUDY

TABLE 1-2. The (lowest-mass) nonet of pseudoscalar mesons ($J^P = 0^-$) and the (lowest-mass) nonet of vector mesons ($J^P = 1^-$), together with masses and widths, or lifetimes where more appropriate, and principal decay modes. The nonets are displayed in isospin-hypercharge multiplets (I, Y) and charge conjugation, C, or G-parity, G, are given where they are good quantum numbers. For further details see Particle Data Group (1972).

Name, Y	I^G	C	Mass (MeV/c^2)	Width (MeV)	Mean Life (sec)	Principal Decay Modes
$\pi^\pm$, 0	1^-		139.58		$(2.6024 \pm .0024) \times 10^{-8}$	$\mu\nu$
π^0, 0		+	134.97		$(0.84 \pm .10) \times 10^{-16}$	$\gamma\gamma$
K^+, K^-	$\frac{1}{2}$		493.84 ± .10		$(1.2371 \pm .0026) \times 10^{-8}$	$\mu^\pm\nu, \pi^\pm\pi^0$
K^0, 1; $\bar{K}^0$, -1			497.79 ± .15		$\begin{cases} 50\% K^0_{\text{short}} (.862 \pm .006) \times 10^{-10} \\ 50\% K^0_{\text{long}} (5.172 \pm .042) \times 10^{-8} \end{cases}$	$\pi\pi$ $\pi\pi\pi, \pi e\nu, \pi\mu\nu$
η, 0	0^+	+	548.8 ± 0.6	$(2.63 \pm .58) \times 10^{-3}$		$\gamma\gamma, 3\pi^0, \pi^+\pi^-\gamma$
X^0, 0	0^+	+	957.1 ± 0.6	< 4		$\eta\pi\pi, \pi^+\pi^-\gamma, \gamma\gamma$
$\rho^\pm$, 0	1^+		765 ± 10	135 ± 20		$\pi\pi$
ρ^0, 0		$-$				
K^{+*}, K^{-*}	$\frac{1}{2}$		891.7 ± 0.5	50.1 ± 1.1		$K\pi$
K^{0*}, 1; $\bar{K}^{0*}$, -1			897.8 ± 2.0			
ω, 0	0^-	$-$	783.9 ± 0.3	10.0 ± 0.6		$\pi^+\pi^-\pi^0, \pi^0\gamma$
ϕ, 0	0^-	$-$	1019.1 ± 0.5	4.4 ± 0.3		$K^+K^-, K^0_s K^0_L, \pi^+\pi^-\pi^0$

in the collision. Generally, the smaller the number of final particles (except in the special case of the measurement of total cross sections), the less unobvious the implications are. For example, the charge exchange reaction $\pi^- + p \to \pi^0 + n$ lends itself to an interpretation in terms of particle (Reggeized ρ^-meson) exchange between the pions and the nucleons, while the behavior of the diffraction peak in high energy elastic scattering gives indications of particle structure and the forces between particles or their possible component parts. In pion-nucleon collisions at lower energies one can detect the formation of long-lived intermediate states or resonances, whose width is of the order of 100 MeV, and whose spin and parity is determined by partial wave analysis of elastic and charge exchange scattering or, possibly, through other reactions. These resonances are granted the status of elementary particles and at the same time may be regarded in some sense as excited states of the nucleons so that determination of their energy levels, widths, spins, parities, and other quantum numbers may give important indications on the classification and symmetries of elementary particles and even on their structure. As extended to include strange particles, the study of these energy levels is rather naturally known as *baryon spectroscopy*.

The two nucleons belong to a set of eight $J^P = \frac{1}{2}^+$ baryons (shown in Table 1-1) classified as an octet of the SU3 group. The three pions belong to a set of nine $J^P = 0^-$ mesons (shown in Table 1-2) classified as an octet-singlet mixture under the SU3 symmetry. The SU3 symmetry though broken, as evinced by the large mass differences within multiplets shown in Table 1-1, implies certain relationships between the various pseudoscalar meson-baryon interactions, which though not holding by any means exactly, are observed to exist to a certain approximate degree. Also the structures of the various baryons, as revealed for example in their masses or electromagnetic interactions as well as in meson-baryon interactions, are related, approximately, through SU3, as are the structures of the mesons.

Consequently any conclusions about the pion-nucleon interaction will have certain consequences for, and relations with, other pseudoscalar meson-baryon interactions. Where we discuss the classification of resonances we will naturally make the relations very explicit and complete. In other cases, for example high energy scattering, the detailed discussion will be confined to pion-nucleon scattering, and the application of the methods to processes such as kaon-nucleon scattering will be left implicit. The pion-nucleon interactions form a convenient experimental and theoretical subsystem in which many features of strong interactions can be studied, and techniques developed and expounded with a minimum of discursive interruption.

Among the pions and nucleons only the proton is stable. The neutron undergoes β-decay, $n \to p + e + \bar{\nu}_e$, with a lifetime of $(1.01 \pm 0.03) \times 10^3$ sec; the charged pions decay by $\pi^\pm \to \mu^\pm + \bar{\nu}_\mu$ with a lifetime of $(2.55 \pm 0.03) \times 10^{-8}$ sec; the principal decay mode (98.8%) of the π^0 is $\pi^0 \to 2\gamma$, with the subsidiary mode (1.2%) $\pi^0 \to \gamma + e^+ + e^-$, and the lifetime is approximately 10^{-16} sec. The time of strong interactions is less than 10^{-22} sec; thus, so far as strong interactions are concerned, all pions and nucleons are effectively stable particles, and their decays play no role whatever.

1.2. ISOSPIN

Historically, isotopic spin has its origin in low energy nuclear physics, where it was observed that the neutron-neutron, neutron-proton, and proton-proton forces were approximately equal. In these circumstances it was natural to regard the neutron and proton as two states of the same particle, the nucleon. In many low energy nuclear physics applications the charge state of the nucleon is an irrelevant internal quantum number. This of course is only approximately true since there are perturbations arising from electromagnetic interactions, in particular the neutron-proton mass difference, itself believed to be due to electromagnetic interactions, and the proton-proton electromagnetic interaction, responsible for the tendency to neutron excess in high-mass nuclei.

The isotopic spin formalism for the nucleons was built on a simple analogy with the nonrelativistic theory of a spin $\frac{1}{2}$ particle, which also has two possible internal states. (As we shall see, especially in Chapter 6, the analogy is mathematically exact since both the nucleon and the spin $\frac{1}{2}$ particle are bases of a 2×2 representation of the group SU2). On this analogy the proton and neutron are described by the spinors:

$$p = \begin{pmatrix} 1 \\ 0 \end{pmatrix}, \quad n = \begin{pmatrix} 0 \\ 1 \end{pmatrix} \tag{1-1}$$

Analogously to the Pauli spin matrices we write the isospin matrices:

$$\tau_1 = \begin{pmatrix} 0 & 1 \\ 1 & 0 \end{pmatrix}, \quad \tau_2 = \begin{pmatrix} 0 & -i \\ i & 0 \end{pmatrix}, \quad \tau_3 = \begin{pmatrix} 1 & 0 \\ 0 & -1 \end{pmatrix} \tag{1-2}$$

which act on the spinors in eq. (1-1). The proton state is an eigenstate of τ_3 of eigenvalue $+1$, and the neutron state is an eigenstate of τ_3 of eigenvalue -1. The vector $\frac{1}{2}\vec{\tau}$ is the analogue of the Pauli spin operator $\frac{1}{2}\vec{\sigma}$ and the proton and neutron spinors in eq. (1-1) correspond to spin states quantized along the 3-direction. The terms τ_1, τ_2, τ_3 are generators of rotations about the 1, 2, 3 axes respectively in isospin space; rotations in isospin space lead to transformations between protons and neutrons, and thus the equivalence of protons and neutrons is expressed by the hypothesis of invariance of the nucleon properties under rotations in isospin space. Also τ_1, τ_2, τ_3 obey the usual angular momentum commutation relations (corresponding to the Lie algebra of SU2):

$$[\tau_i, \tau_j] = 2i\, \epsilon_{ijk}\, \tau_k \tag{1-3}$$

1.2. ISOSPIN

where $\epsilon_{ijk} = \pm 1$ if i, j, k are an even or odd permutation respectively of 1, 2, 3 and $\epsilon_{ijk} = 0$ if any two of i, j, k are equal. It is also easily verified that:

$$\tau_i \tau_j + \tau_j \tau_i = 2\delta_{ij} . \qquad (1\text{-}4)$$

If we have a state of 2 nucleons, we may denote the isospin operator for the first nucleon by $\frac{1}{2}\vec{\tau}^{(1)}$ and the isospin operator for the second nucleon by $\frac{1}{2}\vec{\tau}^{(2)}$ so that the total isospin operator for the system is as follows:

$$\vec{I} = \frac{1}{2}\vec{\tau}^{(1)} + \frac{1}{2}\vec{\tau}^{(2)} . \qquad (1\text{-}5)$$

Two nucleons, which each have isospin $\frac{1}{2}$, can form eigenstates of total isospin of eigenvalue either 0 or 1. We denote the first and second nucleon by superscripts (1) and (2) respectively, and then the normalized eigenstates are:

$$I = 0, I_3 = 0 : \frac{1}{\sqrt{2}} (p^{(1)} n^{(2)} - p^{(2)} n^{(1)}) \qquad (1\text{-}6)$$

$$I = 0 \begin{cases} I_3 = 1 & p^{(1)} p^{(2)} \\ I_3 = 0 & \frac{1}{\sqrt{2}}(p^{(1)} n^{(2)} + p^{(2)} n^{(1)}) \\ I_3 = -1 & n^{(1)} n^{(2)} \end{cases} \qquad (1\text{-}7)$$

We have remarked that the interaction of two nucleons is charge-independent; rotations in isospin space are the transformations which bring about changes in the charge states, and charge-independence may be restated as the hypothesis that *nuclear forces are invariant under rotations in isospin space*. It is this *isospin invariance* that makes isospin an essential physical concept. The Hamiltonian that describes a 2-nucleon system contains isospin operators; the general isospin invariant form is:

$$H = H' \vec{\tau}^{(1)} \cdot \vec{\tau}^{(2)} + H'' \qquad (1\text{-}8)$$

where H' and H'' do not involve $\vec{\tau}^{(1)}$ or $\vec{\tau}^{(2)}$. It is immediately evident, on using the fact (shown in Chapter 6) that the $\vec{\tau}^{(i)}$ sandwich between isospinors corresponding to nucleon i transforms as a vector under isospin rotations, that eq. (1-8) is invariant under rotations in isospin space and, from eqs. (1-3) and (1-4), it is the only such form.

Historically also, isospin of the pions originated in low energy nuclear physics. Yukawa (1935) predicted the existence of a boson whose exchange between nucleons would give rise to the internucleon force and from the relation between the range of the force and the mass of the exchanged boson, the latter was predicted to be in the region of 100-200 MeV/c^2. With the discovery (Anderson and Neddermeyer, 1937) of the muon in cosmic rays of mass ~ 100 MeV/c^2, Yukawa's particle was thought to have been discovered, since the fact that the muon spin is $\frac{1}{2}$ was not at that time known. Since the muons are charged, the problem was to find a meson-nucleon interaction involving charged mesons that yet gave rise to charge-independent nuclear forces. The answer (Kemmer, 1938) was to postulate a triplet of mesons forming a vector in isospin space. Anticipating the discovery of the pion in 1947 (Lattes et al, 1947) these are the π^-, π^0, π^+ having charges $-1, 0, +1$ respectively, corresponding to third component of isospin being $-1, 0, +1$ respectively. The internucleon forces were postulated to arise from the virtual emission of a pion at one nucleon and its absorption at the other nucleon. The emission or absorption of pions from nucleons is the basic process, and it is clear that if this basic πNN interaction is invariant under rotations in isospin space, so will be the internucleon force. The form of this interaction in isospin space can be deduced as follows.

A pion is represented by a vector in isospin space, so that the nature of the pion, that is, whether it is positive, negative, or neutral, is denoted by a vector $\vec{\phi}$ or equivalently three numbers: $\vec{\phi} \equiv (\phi_1, \phi_2, \phi_3)$. Now we consider the basic πNN interaction, $N \rightarrow N + \pi$, and let the initial nucleon have isospinor N_1, the final nucleon have isospinor N_2 and the pion have isovector $\vec{\phi}$. Then the form:

1.2. ISOSPIN

$$N_1^\dagger \vec{\tau} N_2 \cdot \vec{\phi} \tag{1-9}$$

($\dagger$ = Hermitian conjugate) is invariant under rotations in isospin space. Expression (1-9) is the unique isospin invariant form of the πNN interaction.

Since for example a proton can only transform to a proton with emission of a π, it is easily seen that the conventions of eqs. (1-1) and (1-2) (together with the further convention that the pion triplet transforms like the angular momentum triplet $Y_{1m}(\theta\phi)$) imply:

$$\pi^- = \frac{1}{\sqrt{2}}\begin{pmatrix}1\\-i\\0\end{pmatrix}, \pi^0 = \begin{pmatrix}0\\0\\1\end{pmatrix}, \pi^+ = -\frac{1}{\sqrt{2}}\begin{pmatrix}1\\i\\0\end{pmatrix} \tag{1-10}$$

with corresponding isospin operators for the pion (see also Chapter 6) being:

$$\tfrac{1}{2}T_1 = \begin{pmatrix}0&0&0\\0&0&-i\\0&i&0\end{pmatrix}, \tfrac{1}{2}T_2 = \begin{pmatrix}0&0&+i\\0&0&0\\-i&0&0\end{pmatrix}, \tfrac{1}{2}T_3 = \begin{pmatrix}0&-i&0\\i&0&0\\0&0&0\end{pmatrix} \tag{1-11}$$

In a pion-nucleon scattering experiment the initial state consists of one pion with isospin 1 and one nucleon with isospin $\frac{1}{2}$, so that any initial state can be expressed as a superposition of a state of total isospin $\frac{1}{2}$ and total isospin $\frac{3}{2}$. These states are as follows:

$$I = \tfrac{3}{2} \begin{cases} I_3 = +\tfrac{3}{2} : |p\pi^+\rangle \\ I_3 = \tfrac{1}{2} : \sqrt{\tfrac{2}{3}}|p\pi^0\rangle + \sqrt{\tfrac{1}{3}}|n\pi^+\rangle \\ I_3 = -\tfrac{1}{2} : \sqrt{\tfrac{1}{3}}|p\pi^-\rangle + \sqrt{\tfrac{2}{3}}|n\pi^0\rangle \\ I_3 = -\tfrac{3}{2} : |n\pi^-\rangle \end{cases} \tag{1-12}$$

$$I = \tfrac{1}{2} \begin{cases} I_3 = \tfrac{1}{2} : \sqrt{\tfrac{1}{3}}|p\pi^0\rangle - \sqrt{\tfrac{2}{3}}|n\pi^+\rangle \\ I_3 = -\tfrac{1}{2} : \sqrt{\tfrac{2}{3}}|p\pi^-\rangle - \sqrt{\tfrac{1}{3}}|n\pi^0\rangle \ . \end{cases} \tag{1-13}$$

There is good evidence that all the strong interactions are invariant under isospin rotations so that for example initial states which have total isospin $\frac{3}{2}$ only make transitions to final states with total isospin $\frac{3}{2}$, except for possible small perturbations, usually unimportant in the study of strong interactions, caused by electromagnetic and weaker interactions. Consequently eigenstates of total isospin such as that shown in eqs. (1-12) or (1-13) are the appropriate objects of study in pion-nucleon interactions.

As an explicit test of isospin invariance we mention the reaction:

$$d + d \to He^4 + \pi^0 \tag{1-14}$$

a forbidden reaction since the deuteron and He^4 have isospin zero and the pion has isospin 1. Akimov et al. (1960) and Poirier and Pripstein (1963) have shown that the cross section for this reaction is less than $\frac{1}{100}$ of the expected rate if isospin were not conserved.

Charge Conjugation and G-parity

The operator of charge conjugation, C, transforms particle into antiparticle so that:

$$C | p > = | \bar{p} >$$
$$C | n > = | \bar{n} > . \tag{1-15}$$

Both strong and electromagnetic interactions are invariant under charge conjugation (Sakurai, 1964) and the pion properties are:

$$C | \pi^\pm > = - | \pi^\mp >$$
$$C | \pi^0 > = + | \pi^0 > . \tag{1-16}$$

We will show that eq. (1-16) follows from the invariance of strong interactions under P, C and isospin transformations, by noting that the pion can transform, through the fundamental πNN transformation into a nucleon-anti-nucleon system whose properties under C can be deduced:

1.2. ISOTOPIC SPIN

$$[I_3, I_1 \pm i I_2] = \pm(I_1 \pm i I_2)$$

so that $(I_1 \pm i I_2)$ applied to a state of an isospin multiplet with eigenvalue of I_3 equal to I_3' give a state of the same isospin multiplet with eigenvalue of I_3 equal to $I_3' \pm 1$. Applying these to the proton-neutron multiplet:

$$|p\rangle = (I_1 + i I_2)|n\rangle, \quad |n\rangle = (I_1 - i I_2)|p\rangle \quad (1\text{-}17a)$$

To maintain the linear relation:

$$Q/e = I_3 + \frac{1}{2} B \quad (1\text{-}18)$$

between charge Q isospin I_3 and baryon number B of nonstrange[1] particles, the third component of isospin of $\bar{n}$ must be $+\frac{1}{2}$ and of $\bar{p}$ must be $-\frac{1}{2}$. Consequently, with a certain phase convention:

$$|\bar{n}\rangle = -(I_1 + i I_2)|\bar{p}\rangle, \quad |\bar{p}\rangle = -(I_1 - i I_2)|\bar{n}\rangle \quad (1\text{-}17b)$$

C evidently anti-commutes with I_3, from eq. (1-18), and from eqs. (1-17) and (1-15) it anti-commutes with I_1 and commutes with I_2:

$$C I_1 = -I_1 C, \quad C I_2 = I_2 C, \quad C I_3 = -I_3 C. \quad (1\text{-}19)$$

With our phase conventions $(-\bar{n}, \bar{p})$ transforms as (p, n) under isospin transformation, and the nucleon-anti-nucleon isospin wave function corresponding to π^0 is thus $\frac{1}{\sqrt{2}}(\bar{p}p - \bar{n}n)$. Also by parity and angular momentum conservation it must be in a singlet spin state and orbital S-state. Consequently its wave function, anti-symmetric from the Pauli principle, is as follows:

$$\frac{1}{2\sqrt{2}}(\bar{p}(1) p(2) + \bar{p}(2) p(1) - \bar{n}(1) n(2) - \bar{n}(2) n(1))(\alpha(1)\beta(2)$$
$$- \alpha(2)\beta(1)) R(1, 2) \quad (1\text{-}20)$$

---

[1] For strange particles, eq. (1-18) becomes $Q/e = I_3 + \frac{1}{2}(B+S)$.

where α, β are the two possible spin states and $R(1,2)$ is a symmetrical spatial wave function. Obviously equation (1-20) is an eigenstate of C of eigenvalue +1, which gives the π^0 result of eq. (1-16). The $\pi^\pm$ results can be obtained similarly.

Since C is not a good quantum number for charged particles (in general terms it anti-commutes with I_3), it is useful to define the G-parity, which commutes with $\vec{I}$:

$$G = C e^{i\pi I_2} \tag{1-21}$$

as a product of charge conjugation and a rotation of 180° about the 2-axis in isospin-space. Such a rotation converts the pion isovectors as follows:

$$|\pi^+\rangle = -\frac{1}{\sqrt{2}}\begin{pmatrix}1\\i\\0\end{pmatrix} \rightarrow \frac{1}{\sqrt{2}}\begin{pmatrix}1\\-i\\0\end{pmatrix} = |\pi^-\rangle$$

$$|\pi^-\rangle = \frac{1}{\sqrt{2}}\begin{pmatrix}1\\-i\\0\end{pmatrix} \rightarrow \frac{1}{\sqrt{2}}\begin{pmatrix}-1\\-i\\0\end{pmatrix} = |\pi^+\rangle$$

$$|\pi^0\rangle = \begin{pmatrix}0\\0\\1\end{pmatrix} \rightarrow \begin{pmatrix}0\\0\\-1\end{pmatrix} = -|\pi^0\rangle \ .$$

Combined with eq. (1-16) these give:

$$G|\pi^\pm\rangle = -|\pi^\pm\rangle$$
$$G|\pi^0\rangle = -|\pi^0\rangle \tag{1-22}$$

so that the pion has G-parity -1, and obviously any odd number of pions has G-parity -1 and any even number G-parity $+1$. G-parity is conserved in strong interactions, but not in electromagnetic interactions, and since it commutes with I, it has the same value for any member of the same isospin multiplet.

Since G is a good quantum number for pions, it is also a good quantum number for any system which can transform by strong interactions into pions only. In particular this holds for the $N\bar{N}$ system, and to find the G-parity for such a system it is sufficient to consider the neutral member of an isospin multiplet. The $N\bar{N}$ system has total isospin, $I = 0$ or 1. For the neutral member the rotation $e^{i\pi I_2}$ results in $(-1)^I$ so that $G = C(-1)^I$. By an argument similar to that preceding equation (1-20), we find $C = (-1)^{L+S}$ where L is the orbital and S the spin angular momentum of the system. This gives, for the $N\bar{N}$ system, the result:

$$G = (-1)^{I+L+S} \tag{1-23}$$

REFERENCES

Yu. K. Akimov, O. V. Savchenko, and L. M. Sovoko, Z.h. eksp. Geom. Fiz. 38, 304 (1960).

C. D. Anderson and S. H. Neddermeyer, Phys. Rev. 51, 884, (1937).

Particle Data Group, "Review of Particle Properties," Phys. Letters, April 1972.

G. Kallen, *Elementary Particle Physics* (Addison-Wesley, Reading, Mass., (1964).

N. Kemmer, Proc. Cambridge Phil. Soc. 34, 354 (1938).

C. M. G. Lattes, H. Muirhead, G. P. S. Occhialini, and C. F. Powell, Nature 159, 694 (1947).

K. Nishijima, *Fundamental Particles* (Benjamin, New York, 1963).

J. A. Poirier, and M. Pripstein, Phys. Rev. 130, 1171 (1963).

J. J. Sakurai, *Invariance Principles and Elementary Particles* (Princeton University Press, Princeton, 1964).

H. Yukawa, Proc. Phys. Math. Soc. of Japan 17, 48 (1935).

CHAPTER 2

LOW ENERGY PION SCATTERING BY NUCLEONS AND THE PHOTOPRODUCTION OF PIONS

In this chapter, we discuss the low energy scattering of pions by nucleons and the associated process of photoproduction of pions from nucleons. The low energy region, for this purpose, is the energy region in which inelastic scattering is inappreciable. The low energy experiments are analyzed with the help of the quantum theory of scattering, making use, in particular, of partial wave expansions. It is assumed that the reader has some acquaintance with this theory,[1] but to establish our notation and conventions a summary of some of the more important results that are required is given below.

2.1. THE SCATTERING MATRIX AND CROSS SECTIONS

The probability P_{ba} of finding a system in free particle state b after scattering from an initial free particle state a, will be written as:

---

[1] Detailed accounts of scattering theory can be found in the books by Mott and Massey (1965), Newton (1966), and Goldberger and Watson (1964).

$$P_{ba} = |S_{ba}|^2 \qquad (2\text{-}1)$$

where the probability amplitudes S_{ba} form the elements of the scattering or S-matrix. Sometimes it is convenient to write S_{ba} as:

$$S_{ba} = <b|S|a> \qquad (2\text{-}2)$$

where S is an operator that generates the final free particle state from the initial free particle state.

The free particle states will be chosen with an invariant normalization, and in this case, the probability of transitions between the initial and final states is an intrinsic property of the system, so that P_{ba} and S_{ba} are Lorentz invariant. The particular normalization adopted for a state vector $|\vec{p}; m>$ describing a single particle of mass M momentum $\vec{p}$ and z-component of spin m, is that:

$$<\vec{p}'; m'|\vec{p}; m> = \delta_{mm'}\delta(\vec{p}-\vec{p}')E(p) \qquad (2\text{-}3)$$

where $E(p)$ is the energy of the particle (in natural units with $\hbar = c = 1$):

$$E(p) = \sqrt{p^2 + M^2} \qquad (2\text{-}4)$$

Noninteracting many-particle state vectors are built as direct products of the single-particle states and are normalized accordingly. An arbitrary state vector $|\psi>$, describing a single particle of mass M, can be expanded in terms of the free particle state vectors:

$$|\psi> = \sum_m \int \frac{d\vec{p}}{E(p)} |\vec{p}; m><\vec{p}; m|\psi> \qquad (2\text{-}5)$$

where the closure relation for the plane wave states is:

$$\sum_m \int \frac{d\vec{p}}{E(p)} |\vec{p}, m><\vec{p}, m| = 1 \quad . \qquad (2\text{-}6)$$

2.1. THE SCATTERING MATRIX AND CROSS SECTIONS

Unitarity

Making use of the fact that the probability that *some* final state of the system is reached is unity, it is seen that:

$$\sum_b P_{ba} = 1 \tag{2-7}$$

where the sum runs over all possible final states. From this and using the fact that the free particle states form a complete set, it follows that the S-matrix is unitary:

$$S^\dagger S = SS^\dagger = 1 \tag{2-8}$$

or

$$\sum_c \langle b|S^\dagger|c\rangle \langle c|S|a\rangle = \delta_{ab} \, .$$

The sum over c, and the delta function δ_{ab}, must be interpreted as shorthand for the closure and normalization conditions, eqs. (2-3) and (2-5), extended to cover the many-particle case. In any physical situation, the system in the final state must have the same total momentum and energy as in the initial state, and δ_{ba} must contain a four-dimensional delta function $\delta^4(P_a - P_b) = \delta(E_a - E_b)\delta(\vec{P}_a - \vec{P}_b)$ as a factor, where $P_a(E_a, \vec{P}_a)$ and $P_b(E_b, \vec{P}_b)$ are total energy-momentum vectors of the states a and b.

The Transition Rate

If there is no interaction between the particles of the system, the initial and final states cannot differ and S must reduce to the unit operator. From this, we see that the probability that a transition caused by an interaction has occurred will be determined by the operator $(S-1)$, and accordingly a transition operator T is defined as:

$$T = \frac{1}{2i}(S - 1) \, . \tag{2-9}$$

The factor $\frac{1}{2i}$ has been introduced purely for convenience later on. In terms of T, the unitarity relation of eq. (2-8) becomes:

$$T^\dagger T = \tfrac{1}{2} i(T^\dagger - T) \ . \tag{2-10}$$

The matrix elements of T with respect to the free particle state vectors $|a\rangle$ and $|b\rangle$, $\langle b|T|a\rangle$ contain as factors delta functions expressing the conservation of energy and momentum, and we may write:

$$\langle b|T|a\rangle = \delta^4(P_a - P_b) \, T_{ba} \ . \tag{2-11}$$

The reduced transition matrix elements T_{ba} are defined only for states b and a which have the same total energy and momentum.

Because of the presence of the four-dimensional delta function, the probability that a transition has occurred, which is equal to $|2i\langle b|T|a\rangle|^2$, is singular, even after integrating over the energy and momentum of the group of final states observed. The transition probability must be proportional to the overlap of the wave-packets representing the initial and final system. In the limiting case of plane waves, this overlap is infinite and is represented by the factor $\delta^4(P_a - P_b)$ which can be expressed as:

$$\delta^4(P_a - P_b) = (2\pi)^{-4} \int d^4 x \, e^{-i(P_a - P_b) \cdot x}$$

$$= (2\pi)^{-4} \int_{-\infty}^{\infty} dt' \, e^{-i(E_a - E_b)t'} \int d\vec{x}' \, e^{i(\vec{P}_a - \vec{P}_b) \cdot \vec{x}'} . \tag{2-12}$$

We can now take into account the fact that, however long the experiment, the wave-trains representing the initial and final states must be of finite duration and at some time in the past the overlap must vanish. To do this we may add a small positive imaginary part $i\epsilon$ to E_a and replace the infinite integral over t' by an integral up to a finite time t:

$$\int_{-\infty}^{\infty} dt' \, e^{-i(E_a - E_b)t'} \to \int_{-\infty}^{t} dt' \, e^{-i(E_a - E_b + i\epsilon)t'} . \tag{2-13}$$

2.1. THE SCATTERING MATRIX AND CROSS SECTIONS

The overlap of the wave-packets up to the time t, is then of the order $(\frac{1}{\epsilon})$. [For details see Goldberger and Watson (1964).] Similarly the integral over the space coordinates can be restricted to a large, but finite, volume V. The probability that the transition has occurred will now be a function of ϵ, t, and V; however the quantity that is measured is the transition rate per unit volume, which is defined as:

$$W_{ba} = \frac{1}{V}\frac{d}{dt}|2i<b|T|a>|^2 \qquad (2\text{-}14)$$

and this is finite in the limit $\epsilon \to 0$, $V \to \infty$ and is independent of t. We find, after a little algebra, that:

$$W_{ba} = (2\pi)^{-4}\,\delta(E_a - E_b)\,\delta(\vec{P}_a - \vec{P}_b)\,4\,|T_{ba}|^2 \,. \qquad (2\text{-}15)$$

The reduced transition matrix T_{ba} and the transition rate W_{ba} are invariants provided that the state vectors $|a>$ and $|b>$ are direct products of single particle state vectors with invariant normalization, defined by eq. (2-3).

The Cross Section

When the initial state of the system consists of two particles moving colinearly, a scattering cross section can be defined as:

$$\sigma = \sum_b \frac{1}{F} W_{ba} \qquad (2\text{-}16)$$

where the sum is over the final states observed in the experiment and F is the incident flux, which is defined as the number of particles crossing a unit cross-sectional area normal to the beam per unit time. This number is invariant under Lorentz transformations along the direction of the beam.

To evaluate F we have to know the particle density represented by the plane wave states with the normalization of eq. (2-3). The wave-function in the position representation, corresponding to the single particle state $|\vec{p}>$, is given by (ignoring spin variables):

$$\psi(\vec{x}) \equiv \langle \vec{x} | \vec{p} \rangle = (2\pi)^{-3/2} \sqrt{E(|\vec{p}|)} \exp(i\vec{p}\cdot\vec{x}) .$$

This represents a particle density $|\psi(\vec{x})|^2$ of $(2\pi)^{-3} E(|\vec{p}|)$ particles per unit volume. The particle density of the two-particle state a, in which the momenta of the two particles are p_1 and p_2 with energies E_1 and E_2 is then $(2\pi)^{-6} E_1(|\vec{p}_1|) E_2(|\vec{p}_2|)$. In the center of mass frame, $\vec{p}_1 = -\vec{p}_2 = \vec{q}$, so that the incident flux is given by:

$$F = (2\pi)^{-6} E_1(q) E_2(q) v$$

where v is the relative velocity of the colliding particles. The velocity of each particle is (p_i/E_i) and the relative velocity is $q\left(\dfrac{1}{E_1} + \dfrac{1}{E_2}\right)$ from which F can be written as:

$$F = (2\pi)^{-6} qW \tag{2-17}$$

where W denotes the total center of mass energy:

$$W = E_1(q) + E_2(q) .$$

If the final state b consists of n particles with masses M_3, M_4, ..., $M_{(2+n)}$ with momenta $\vec{p}_3, \vec{p}_4, ..., \vec{p}_{(2+n)}$, the total cross section for scattering from the two-particle state a into b is:

$$\sigma_{ba}(W) = \frac{1}{(2\pi)^4 F} \int \delta\left(\vec{p}_1 + \vec{p}_2 - \sum_{j=3}^{(n+2)} \vec{p}_j\right) \delta(E_a - E_b) 4|T_{ba}|^2 \prod_{j=3}^{(n+2)} \frac{d\vec{p}_j}{E_j(p_j)} . \tag{2-18}$$

The integration over the final states has been carried out with the weighting factors required by the closure relation, eq. (2-6). A summation over spin and other variables is understood.

The four-dimensional delta function, $|T_{ba}|^2$ and the integration factors $d\vec{p}_j/E_j$ are all invariants. We shall now show that an invariant expression for F can be given, so that the total cross section is also an invariant.

2.1. THE SCATTERING MATRIX AND CROSS SECTIONS

The four-vector P_a that represents the total energy and momentum of the two-particle initial state has components in the center of mass system $P_0 = W$ and $\vec{P}_a = \vec{p}_1 + \vec{p}_2 = 0$ so that:

$$W^2 = + P_a \cdot P_a = \sum_{\mu=0}^{3} (P_a)_\mu (P_a)^\mu . \tag{2-19}$$

This expression can be taken as the *definition* of W in any frame and as $(P_a)^2$ is an invariant, W is also invariant. An invariant expression for q is then defined by writing q in terms of W:

$$q^2 = [W^2 - (M_1 + M_2)^2][W^2 - (M_1 - M_2)^2] / 4W^2 . \tag{2-20}$$

With W and q defined for arbitrary frames in this way, the flux F given by eq. (2-17) also transforms as an invariant, and eq. (2-18) can be taken to be the definition of an invariant total cross section in an arbitrary frame, even when the particles in the incident state are not moving co-linearly.

Collisions with Two-Particle Final States

A most important particular case is when both the initial and final states a and b contain two particles. In the center of mass system, we set $\vec{p}_1 = -\vec{p}_2 = \vec{q}_a$, $\vec{p}_3 = -\vec{p}_4 = \vec{q}_b$ and:

$$\sigma_{ba}(W) = \frac{16\pi^2}{q_a W} \int \delta(W - E_3 - E_4) |T_{ba}|^2 \frac{d\vec{q}_b}{E_3 E_4} = \frac{16\pi^2}{W^2} \left(\frac{q_b}{q_a}\right) \int d\Omega_b |T_{ba}|^2 \tag{2-21}$$

where $d\Omega_b$ is an element of solid angle in the direction of $\vec{q}_b$ specified by the polar angles (θ_b, ϕ_b). From conservation of energy, q_a and q_b are related by:

$$W = \sqrt{M_1^2 + q_a^2} + \sqrt{M_2^2 + q_a^2} = \sqrt{M_3^2 + q_b^2} + \sqrt{M_4^2 + q_b^2} . \tag{2-22}$$

The differential cross section for scattering into a particular direction is defined as:

$$\frac{d\sigma_{ba}}{d\Omega_b} = \frac{16\pi^2}{W^2} \frac{q_b}{q_a} |T_{ba}|^2 . \qquad (2\text{-}23)$$

When the particles are spinless, rotational invariance requires that T_{ba} must depend on a three-dimensional scalar quantity, and the only scalar available is $(\vec{q}_b \cdot \vec{q}_a) = q_a q_b \cos\theta_b$ where we have taken $\vec{q}_a$ as z axis so that θ_b is the scattering angle. It follows that T_{ba} is a function of W and θ_b only. Here b and a indicate a dependence on the particular channels concerned, defined by the masses of the particles. The differential cross section is not invariant under a Lorentz transformation, since the scattering angle θ_b is not a (four-dimensional) scalar.

It is customary to define a new quantity, f_{ba}, known as the scattering amplitude, in place of T_{ba}. It is defined as:

$$f_{ba}(\theta) = \frac{4\pi}{W} \sqrt{\frac{\omega_b}{\omega_a}} T_{ba} \qquad (2\text{-}24)$$

where ω_a, ω_b are the reduced energies in the initial and final states:

$$\frac{1}{\omega_a} = \frac{1}{E_1(q_a)} + \frac{1}{E_2(q_a)}; \quad \frac{1}{\omega_b} = \frac{1}{E_3(q_b)} + \frac{1}{E_4(q_b)} . \qquad (2\text{-}25)$$

In terms of the scattering amplitude the differential cross section is:

$$\frac{d\sigma_{ba}}{d\Omega_b} = \frac{v_b}{v_a} |f_{ba}(\theta)|^2 \qquad (2\text{-}26)$$

where:

$$v_b = q_b/\omega_b, \quad v_a = q_a/\omega_a$$

The velocities v_b and v_a are relative velocities of the colliding particles in the final and initial states, and in a configuration space representation f_{ba} is the amplitude of the outgoing spherical waves in channel b, when the incident wave is a plane wave of unit amplitude in channel a.

2.1. THE SCATTERING MATRIX AND CROSS SECTIONS

The Optical Theorem

From the unitarity relation of eq. (2-10), by introducing a complete set of states b, and a two-particle state a, we obtain:

$$\sum_b <a|T^\dagger|b><b|T|a> = \text{Im}<a|T|a> . \tag{2-27}$$

The amplitude on the right-hand side refers to scattering, in which the final and initial states are the same; that is, it is an amplitude for elastic scattering in the forward direction. The expression on the left-hand side is proportional to the total cross section σ for all processes in which the initial state is the state a. Using eq. (2-18), we can write:

$$\sigma = \sum_b \sigma_{ba} = \frac{4}{(2\pi)^4 F} \sum_b |<a|T|b>|^2 = \frac{16\pi^2}{W q_a} \text{Im } T_{aa}(\theta = 0) \tag{2-28}$$

or in terms of the scattering amplitude f_{ba}:

$$\sigma = \frac{}{q_a} \text{Im } f_{aa}(\theta_a = 0) . \tag{2-29}$$

This is the relation known as the optical theorem and expresses the fact that the probability flux lost from the incident beam is to be equated with the total probability flux for scattering into all open channels.[2]

The Reduction of the S-matrix

The S-matrix can be reduced to diagonal or block diagonal form by making an expansion of the initial and final state vectors in eigenvectors of angular momentum. We shall consider a system consisting of a set of two-particle channels. If the momenta of particles 1 and 2 are $\vec{p}_1$ and $\vec{p}_2$, the state vectors will be normalized in conformity with eq. (2-3), that is:

---
[2] The term "channel" will be used to denote a particular set of reacting particles with a specified set of quantum numbers. An "open" channel is one that is a possible final configuration of the system at the energy concerned; a "closed" channel is a configuration of the system that cannot be reached from the initial state, usually because of energetic considerations.

$$\langle \vec{p}_1, \vec{p}_2; a | \vec{p}_1', \vec{p}_2'; b \rangle = \delta_{ab} \delta(\vec{p}_1 - \vec{p}_1') \delta(\vec{p}_2 - \vec{p}_2') E(p_1) E(p_2) \quad (2\text{-}30)$$

where a and b denote the spin and other intrinsic quantum numbers of the particles. It is most convenient to work in the center of mass system in which:

$$\vec{p}_1 = -\vec{p}_2 = \vec{q} \quad (2\text{-}31)$$

and this case omitting a factor $\delta(\vec{p}_1 + \vec{p}_2) = \delta(0)$, which expresses the overall conservation of momentum, we can label the two-particle states as $|\vec{q}; a\rangle$, $|\vec{q}'; b\rangle$ and normalize so that:

$$\langle \vec{q}; a | \vec{q}'; b \rangle = \delta_{ba} \frac{E_1(q) E_2(q)}{q^2} \delta(q' - q) \delta(\Omega' - \Omega)$$

where we have used polar coordinates (q, θ, ϕ) and $\delta(\Omega - \Omega) \equiv \delta(\cos\theta - \cos\theta') \delta(\phi - \phi')$. In terms of the total center of mass energy $W = E_1 + E_2$, the factor $\delta(q - q')$ can be written as $\frac{dW}{dq} \delta(W' - W)$, so that:

$$\langle \vec{q}; a | \vec{q}'; b \rangle = \delta_{ba} \frac{W}{q_a} \delta(\Omega - \Omega') \delta(W - W') . \quad (2\text{-}32)$$

We shall normally omit the factor $\delta(W - W')$, since we are only concerned with transitions in which the initial and final states have the same total energy.

Scattering of Spinless Particles

For spinless particles, the total angular momentum $\vec{J}$ is equal to the orbital angular momentum $\vec{L}$ and eigenvectors can be formed, providing a representation in which J^2 and J_3 are diagonal, by projecting the plane wave states with the spherical harmonics $Y_{\ell, m}(\theta, \phi)$:

$$|\ell, m; a\rangle = N_a \int d\Omega_a Y_{\ell, m}(\theta_a, \phi_a) |\vec{q}_a; a\rangle . \quad (2\text{-}33)$$

If the states $|\ell, m; a\rangle$ are normalized by:

2.1. THE SCATTERING MATRIX AND CROSS SECTIONS

$$<\ell',m';a'|\ell,m;a> = \delta_{\ell\ell'}\cdot\delta_{mm'}\cdot\delta_{aa'} \qquad (2\text{-}34)$$

then the coefficient $N_a = \sqrt{\dfrac{q_a}{W}}$ and:

$$<\vec{q}_a;a'|\ell,m;a> = \delta_{aa'}\sqrt{\dfrac{W}{q_a}}\,Y_{\ell,m}(\theta_a,\phi_a)\ . \qquad (2\text{-}35)$$

The S-matrix then has the expansion, using the completeness of the states $|\ell,m;a>$:

$$<\vec{q}_b;b|S|\vec{q}_a;a>$$

$$= \sum_{\ell,\ell'}\sum_{m,m'}\sum_{c,c'} <\vec{q}_b;b|\ell',m';c'><\ell',m';c'|S|\ell,m;c><\ell,m;c|\vec{q}_a;a> \qquad (2\text{-}36)$$

The rotational invariance of S shows that S is diagonal in both ℓ and m, and, further, since the probability of a transition cannot depend on the orientation of the complete system, the S-matrix will be independent of the value of m. From this and using eq. (2-35):

$$<\vec{q}_b;b|S|\vec{q}_a;a> = \sum_{\ell,m} S^\ell_{ba}\,Y_{\ell,m}(\theta_b,\phi_b)\,Y^*_{\ell,m}(\theta_a,\phi_a)\dfrac{W}{\sqrt{q_a q_b}}$$

$$= \sum_\ell \dfrac{1}{4\pi}(2\ell+1)\dfrac{W}{\sqrt{q_a q_b}}\,S^\ell_{ba}\,P_\ell(\cos\theta) \qquad (2\text{-}37)$$

where $S^\ell_{ba} \equiv <\ell,m;b|S|\ell,m;a>$ is a function of W, the center of mass energy, θ is the angle between $\vec{q}_b$ and $\vec{q}_a$, and the addition theorem for the spherical harmonics has been used (Appendix A). If only elastic scattering is possible, that is, if only one channel is open at the particular energy concerned, then the matrix S^ℓ_{ba} consists only of one element. The unitarity condition then becomes:

$$|S^\ell_{aa}|^2 = 1\ .$$

This is satisfied by setting:

$$S^\ell_{aa} = \exp(2i\delta_\ell) \qquad (2\text{-}38)$$

where δ_ℓ is a real phase shift. As $T = \frac{1}{2i}(S-1)$, the corresponding T-matrix expansion is:

$$T_{aa} = \sum_\ell \frac{1}{4\pi}(2\ell+1)\frac{W}{q_a} T^\ell_{aa} P_\ell(\cos\theta) \qquad (2\text{-}39a)$$

where:

$$T^\ell_{aa} = \frac{1}{2i}(\exp(2i\delta_\ell) - 1) \qquad (2\text{-}39b)$$

and the scattering amplitude $f_{aa}(\theta)$ is:

$$f_{aa}(\theta) = \frac{4\pi}{W} T_{aa} = \sum_\ell \frac{1}{2iq_a}(2\ell+1)\{\exp(2i\delta_\ell) - 1\} P_\ell(\cos\theta). \qquad (2\text{-}40)$$

The usefulness of this parametrization of the scattering amplitude arises because if R is the range of interaction, only particles with impact parameters b (see Chapter 8) $\lesssim R$ will be scattered. As $b \approx \frac{\{\ell(\ell+1)\}^{\frac{1}{2}}}{q_a} \approx \frac{(\ell+\frac{1}{2})}{q_a}$, at each energy only a finite number of phase shifts will be significantly different from zero, the number increasing with increasing energy. In the usual spectroscopic notation the partial waves with $\ell = 0, 1, 2, 3 \ldots$ are called $s, p, d, f, g \ldots$ waves.

Provided the effective interaction decreases faster than an inverse power of the interparticle separation, the phase shifts can be shown to have the following expansion near threshold, known as an effective range expansion:

$$q^{2\ell+1} \cot\delta_\ell = \frac{1}{a_\ell} + \frac{1}{2} r_\ell q^2 + \ldots \qquad (2\text{-}41)$$

where the constants a_ℓ and r_ℓ are known as the scattering length and effective ranges.

The partial cross sections for a given orbital angular momentum take the form:

$$\sigma_\ell(\text{elastic}) = \frac{4\pi}{q_a^2}(2\ell+1)\sin^2\delta_\ell \qquad (2\text{-}42)$$

2.1. THE SCATTERING MATRIX AND CROSS SECTIONS

so that $\sigma_\ell(\text{elastic}) \leq \dfrac{4\pi(2\ell+1)}{q_a^2}$ — a consequence of unitarity in this single-channel case. When more than one channel is open, the diagonal elements of S^ℓ must be of magnitude less than one, so we write:

$$S^\ell_{aa} = \eta_\ell \exp(2i\delta_\ell) \quad \text{and} \quad T^\ell_{aa} = \frac{1}{2i}(\eta_\ell \exp(2i\delta_\ell) - 1) \tag{2-43a}$$

from which:

$$f_{aa}(\theta) = \sum_{\ell=0}^{\infty} \frac{1}{2iq_a}(2\ell+1)(\eta_\ell \exp(2i\delta_\ell) - 1) P_\ell(\cos\theta) \tag{2-43b}$$

where η_ℓ are real parameters known as the inelasticity or absorption parameters and $0 \leq \eta_\ell \leq 1$. The partial elastic cross section is then:

$$\sigma_\ell(\text{elastic}) = \frac{\pi(2\ell+1)}{q_a^2}\{1 + \eta_\ell^2 - 2\eta_\ell \cos 2\delta_\ell\}. \tag{2-43c}$$

The total cross section for a given partial wave is obtained from the optical theorem which shows that:

$$\sigma_\ell(\text{total}) = \frac{2\pi}{q_a^2}(2\ell+1)(1 - \eta_\ell \cos 2\delta_\ell) \tag{2-44}$$

and we have:

$$\sigma_\ell(\text{total}) \leq \frac{4\pi(2\ell+1)}{q_a^2}. \tag{2-45}$$

The inelastic cross section, $\sigma_\ell(\text{inel}) = \sigma_\ell(\text{total}) - \sigma_\ell(\text{elastic})$, is:

$$\sigma_\ell(\text{inel}) = \frac{\pi}{q_a^2}(2\ell+1)(1 - \eta_\ell^2) \tag{2-46}$$

so that:

$$\sigma_\ell(\text{inel}) \leq \frac{\pi}{q_a^2}(2\ell+1). \tag{2-47}$$

Collisions of Spinless Particles with Particles of Spin S

As our next example, consider the case of scattering of a spinless particle by a particle of spin S. In this case, the eigenstates in the representation where J^2 and J_3 are diagonal, will be found by projecting with $\mathcal{Y}_{j,\ell,s}^m$ where:

$$\mathcal{Y}_{j,\ell,s}^m = \sum_{m_\ell, m_s} C_{\ell s}(j, m; m_\ell, m_s) X_{m_s}^S Y_{\ell, m_\ell}(\theta, \phi) . \quad (2\text{-}48)$$

In this equation $C_{\ell s}(j, m; m_\ell, m_s)$ is a Clebsch-Gordon coefficient, $X_{m_s}^S$ is the spin wave function of one particle, and $Y_{\ell,m}$ is spherical harmonic describing the orbital angular momentum (see Appendix A).

Then corresponding to eq. (2-37), we have that:

$$<\vec{q}_b; b |S| \vec{q}_a; a> = \sum_{j,\ell,\ell'} \sum_{m_\ell, m_s} \frac{W}{\sqrt{q_a q_b}} \sqrt{\frac{(2\ell+1)}{4\pi}} C_{\ell s}(j, m; 0, m_s)$$

$$C_{\ell' s'}(j, m; m'_\ell, m'_s) Y_{\ell', m'_\ell}(\theta_b, \phi_b) S_{\ell' s' b; \ell s a}^j$$

(2-49)

where the primed quantities refer to the final state b and the unprimed to the state a and where the direction of the momentum of the initial state, $\vec{q}_a$, has been taken as the polar axis. It should be noted that as in the previous example, S^j does not depend on m because of rotational invariance. The more general case where both particles are of nonzero spin can be discussed by combining the spin functions of each particle into eigenstates of the total spin. In the present case, the scattering amplitude f_{ba} becomes

$$f_{ba} = \sum_{j,\ell',\ell} \sum_{m'_\ell, m'_s} \frac{2\pi}{i\sqrt{q_a q_b}} \sqrt{\frac{\omega_b}{\omega_a}} \sqrt{\frac{2\ell+1}{4\pi}} C_{\ell s}(j, m; 0, m_s)$$

$$C_{\ell' s'}(j, m; m'_\ell, m'_s) Y_{\ell', m'_\ell}(\theta, \phi) (S_{\ell' s' b; \ell s a}^j - \delta_{\ell' s' b, \ell s a}) . \quad (2\text{-}50)$$

2.1. THE SCATTERING MATRIX AND CROSS SECTIONS

A particular case of great interest, is the elastic scattering of a spin 0 particle by a spin $\frac{1}{2}$ particle, as in π-meson-nucleon scattering. In this case, the rules for the addition of angular momenta confine the values of j, for a given ℓ, to $j = \ell \pm \frac{1}{2}$. Conservation of parity (which is always conserved in strong interactions) requires $(-1)^\ell = (-1)^{\ell'}$, which shows that $\ell = \ell'$. When only a single channel is open, the scattering matrix is in diagonal form, with $S^{\ell \pm \frac{1}{2}} = \exp(2i\delta_{\ell\pm})$, and there are two partial wave amplitudes for each ℓ (except for $\ell = 0$) defined by:

$$f_{\ell\pm} = \frac{1}{2iq_a} \{\exp(2i\delta_{\ell\pm}) - 1\}. \tag{2-51}$$

If inelastic scattering is possible the scattering amplitude for elastic scattering is modified as in eq. (2-43) by the introduction of an inelasticity parameter $\eta_{\ell\pm}$.

The expression eq. (2-50) for the scattering amplitude can now be simplified. The states a and b are only distinguished (apart from the different directions of $\vec{q}_a$ and $\vec{q}_b$) by the components of the spin along the Z direction denoted by m_s and m_s' where $m_s, m_s' = \pm \frac{1}{2}$. The amplitude for scattering without change in spin direction, $f = f_{m_s, m_s}$ and the amplitude for scattering with "spin flip," $\bar{g} = f_{m_s', m_s}$, $m_s \neq m_s'$, may be introduced. On using the appropriate Clebsch-Gordon coefficients (Appendix A), we find that:

$$f(\theta) = \sum_\ell \{(\ell+1)f_{\ell+} + \ell f_{\ell-}\} \sqrt{\frac{4\pi}{2\ell+1}} Y_{\ell,0}(\theta,\phi)$$

$$\bar{g}(\theta,\phi) = \sum_\ell \{f_{\ell+} - f_{\ell-}\} \sqrt{\frac{4\pi}{2\ell+1}} \sqrt{\ell(\ell+1)} Y_{\ell,2m_s}(\theta,\phi)$$

or in terms of Legendre polynomials:

$$f(\theta) = \sum_{\ell} \{(\ell + 1) f_{\ell+} + \ell f_{\ell-}\} P_\ell(\cos \theta) \quad \text{(a)}$$

$$\bar{g}(\theta,\phi) = \sum_{\ell} \{f_{\ell+} - f_{\ell-}\} \sin \theta \, P'_\ell(\cos \theta) (-1)^{\frac{1}{2}+m_s} \exp(2im_s\phi) \quad \text{(b)}$$

(2-52)

where:

$$P'_\ell(\cos \theta) \equiv \frac{d}{d \cos \theta} P_\ell(\cos \theta) \, .$$

The differential cross section for an unpolarized beam is found by summing over the final, and averaging over the initial, spin states:

$$\frac{d\sigma}{d\Omega} = \sum_{m_s',m_s} \frac{1}{(2s+1)} |f_{m_s',m_s}(\theta,\phi)|^2 = |f(\theta)|^2 + |\bar{g}(\theta,\phi)|^2. \quad (2\text{-}53)$$

The scattering amplitude $f_{m_s',m_s}(\theta,\phi)$ can also be written in the form:

$$f_{m_s',m_s}(\theta,\phi) = f(\theta)\delta_{m_sm_s'} + ig(\theta) <m_s'|\vec{\sigma}|m_s> \cdot \vec{n} \quad (2\text{-}54)$$

$$g(\theta) = \sum_{\ell} (f_{\ell+} - f_{\ell-}) \sin \theta \, P'_\ell(\cos \theta).$$

The $\vec{\sigma}$ are the usual Pauli spin matrices and $\vec{n}$ is the unit vector:

$$\vec{n} = \frac{\vec{q}_a \times \vec{q}_b}{|\vec{q}_a \times \vec{q}_b|} \, . \quad (2\text{-}55)$$

That $f_{m_s',m_s}(\theta,\phi)$ must be of this form, could have been seen immediately by appealing to rotational invariance. The only vectors available from which f_{m_s',m_s} can be constructed are $\vec{q}_a$, $\vec{q}_b$ and $\vec{\sigma}$; from these we can construct the scalar products $(\vec{q}_a, \vec{q}_b)$, $(\vec{\sigma} \cdot \vec{q}_a \times \vec{q}_b)$, $\vec{\sigma} \cdot \vec{q}_a$, and $\vec{\sigma} \cdot \vec{q}_b$; and, of these, only the first two conserve parity. Considered as a 2×2 matrix in spin space, f_{m_s',m_s} can be expanded as eq. (2-54) where f and g are functions of $(\vec{q}_a \cdot \vec{q}_b) = q^2 \cos \theta$ and of W. This form of the scattering amplitude is useful when we discuss polarization phenomena in the next section.

2.1. THE SCATTERING MATRIX AND CROSS SECTIONS

Polarization and the Density Matrix

The plane wave functions $|p; m\rangle$, on which the discussion of the scattering of spin $\frac{1}{2}$ particles by spin 0 particles was based, were defined so that in the rest frame of the fermion the states $|0; \pm\frac{1}{2}\rangle$, abbreviated as $|\pm\frac{1}{2}\rangle$, were eigenstates of the component of the angular momentum along the Z axis,

$$J_Z |\pm\frac{1}{2}\rangle = \pm\frac{1}{2}|\pm\frac{1}{2}\rangle . \qquad (2\text{-}56)$$

In the rest frame $S_Z = J_Z$ where S_Z is the component of the spin, and in this section, all operations involving spin are carried out in the rest frame of the particle concerned. By performing a rotation, new states $|\psi_\pm\rangle$ can be constructed from the states $|\pm\frac{1}{2}\rangle$ and it can be shown that:

$$|\psi_+\rangle = a|\tfrac{1}{2}\rangle + b|-\tfrac{1}{2}\rangle \qquad (2\text{-}57a)$$

$$|\psi_-\rangle = -b^*|\tfrac{1}{2}\rangle + a^*|-\tfrac{1}{2}\rangle \qquad (2\text{-}57b)$$

where $a = \cos(\theta/2)$ and $b = e^{i\phi}\sin(\theta/2)$. In the rest frame, the states $|\psi_\pm\rangle$ are quantized parallel or anti-parallel to the direction specified by a unit vector $\vec{u}$ with polar angles (θ,ϕ). In other words, if S_u is the component of the spin in the direction of $\vec{u}$, then:

$$S_u |\psi_\pm\rangle = \pm\tfrac{1}{2}|\psi_\pm\rangle . \qquad (2\text{-}58)$$

The states $|\psi_\pm\rangle$ are the most general that can be formed. These general states can be expressed (in the basis of the states $|\pm\frac{1}{2}\rangle$) by a two-dimensional column matrix $\psi = \binom{a}{b}$. The differential cross section for scattering from one such state $\psi_i = \binom{a_i}{b_i}$ to another $\psi_f = \binom{a_f}{b_f}$ can be calculated from the scattering amplitude, eq. (2-54), where the quantities f_{m_s',m_s} form the elements of a 2×2 matrix f. This differential cross section is:

$$\left.\frac{d\sigma}{d\Omega}\right|_{i \to f} = |(\psi_f, f\psi_i)|^2 \ . \tag{2-59}$$

In practice the beams or targets employed in experiments do not consist of particles which are all in the same pure spin state. In fact a certain fraction of the beam (or target) particles will be found in each of the pure states ψ_a.

The Density Matrix

In order to describe such an incoherent mixture of states the concept of the density matrix may be introduced. If the expectation value of an observable A in the normalized pure state ψ_a is $<\psi_a|A|\psi_a>$, then the average of A for an incoherent mixture of states, such that a fraction p_a of the system is in the pure state ψ_a, is:

$$\text{Av } A = \sum_a p_a <\psi_a|A|\psi_a> \tag{2-60}$$

where the sum runs over a complete set of pure states ψ_a and:

$$\sum_a p_a = 1, <\psi_a|\psi_a> = 1 \ .$$

The density operator ρ is defined as:

$$\rho = \sum_a p_a |\psi_a><\psi_a| \tag{2-61}$$

in terms of which the average of A given by eq. (2-60) becomes:

$$\text{Av } A = \text{Tr}(\rho A) \ . \tag{2-62}$$

The matrix elements of ρ and the trace may be evaluated with respect to some complete set of orthonormal functions $|\omega_j>$ as follows:

$$\rho_{ij} = \sum_a p_a <\omega_i|\psi_a><\psi_a|\omega_j> \ . \tag{2-63}$$

2.1. THE SCATTERING MATRIX AND CROSS SECTIONS

The matrix ρ_{ij} is the density matrix, and:

$$\text{Av } A = \text{Tr}(\rho A) = \sum_{ij} \rho_{ij} \langle \omega_j | A | \omega_i \rangle . \qquad (2\text{-}64)$$

From eq. (2-61) ρ is self-adjoint and the trace of ρ is unity:

$$\rho^+ = \rho, \text{ Tr } \rho = 1 .$$

A mixed system is completely specified by the density matrix, and this specification has the advantage of treating mixed and pure states on the same footing. For example, if the system is entirely in the pure state $|\psi_b\rangle$ so that $p_a = \delta_{ab}$, the density operator reduces to a projection operator on the state $|\psi_b\rangle$:

$$\rho = |\psi_b\rangle\langle\psi_b| ; \rho^2 = \rho . \qquad (2\text{-}65)$$

Scattering of Particles with Spin $\frac{1}{2}$

In our particular example of a beam of spin $\frac{1}{2}$ particles, the pure states $|\psi_a\rangle$ for which we have defined a differential cross section [see eq. (2-59)] correspond to the states $|\psi_+\rangle$ and $|\psi_-\rangle$, while the set of orthonormal states $|\omega_j\rangle$ correspond to the states $|\pm\frac{1}{2}\rangle$. The density matrix is explicitly:

$$\rho_{ij} = p_+ \begin{pmatrix} aa^* & ab^* \\ ba^* & bb^* \end{pmatrix} + p_- \begin{pmatrix} bb^* & -ab^* \\ -ba^* & aa^* \end{pmatrix}. \qquad (2\text{-}66)$$

The most general 2×2 matrix with unit trace can be written in terms of the three Pauli matrices σ_i and the unit matrix 1, so that we may write:

$$\rho = \frac{1}{2}(1 + \vec{P} \cdot \vec{\sigma}) \qquad (2\text{-}67)$$

where the three coefficients P_i form the polarization vector $\vec{P}$. To determine the significance of $\vec{P}$ we calculate the average of the spin $\vec{S} = \frac{1}{2}\vec{\sigma}$ using eq. (2-62) and eq. (2-67):

$$\text{Av } \vec{S} = \text{Tr}(\rho \vec{S}) = \frac{1}{2} \vec{P} . \tag{2-68}$$

The polarization vector is accordingly twice the average of the spin.

Like the spin vector, the polarization is only defined in the rest frame of the particle concerned.

The density matrix ρ_f for beam of particles scattered into the direction $\vec{p}_b$ from an incident beam described by ρ_i can be found with the help of the matrix f. The amplitude of the spin state in the scattered beam arising from a pure state ψ_i is $Nf\psi_i$, where N is the factor required to normalize the final state wave-function. From eq. (2-61) the density matrix for the scattered beam is:

$$\begin{aligned}\rho_f &= \sum_i p_i |N|^2 \, (f\psi_i)(f\psi_i)^\dagger \\ &= |N|^2 \, f \left\{ \sum_i p_i \psi_i \psi_i^\dagger \right\} f^\dagger \\ &= |N|^2 \, f \rho_i f^\dagger . \end{aligned} \tag{2-69}$$

The condition $\text{Tr}\,\rho_f = 1$ determines N and we find that:

$$\rho_f = \frac{f \rho_i f^\dagger}{\text{Tr}(f \rho_i f^\dagger)} . \tag{2-70}$$

The form of this relation is quite general, and for the particular case of scattering of spin 0 by spin $\frac{1}{2}$ particles, it may be expressed in terms of $\vec{P}_f$ and $\vec{P}_i$, the polarization vectors of the scattered and incident beams. From eq. (2-68) and eq. (2-63) we have [in conjunction with eq. (2-67) and eq. (2-70)]:

$$\begin{aligned}\vec{P}_f &= \text{Av } \vec{\sigma} = \text{Tr}(\vec{\sigma} \rho_f) \\ &= \frac{\text{Tr}(\vec{\sigma} f f^\dagger) + \text{Tr}(\vec{\sigma} f (\vec{\sigma} \cdot \vec{P}_i) f^\dagger)}{\text{Tr}(f f^\dagger) + \text{Tr}(f \vec{\sigma} \cdot \vec{P}_i f^\dagger)} . \end{aligned} \tag{2-71}$$

2.1. THE SCATTERING MATRIX AND CROSS SECTIONS

The differential cross section for an unpolarized beam with $\vec{P}_i = 0$ can be written, using eq. (2-59):

$$\frac{d\sigma}{d\Omega} = \frac{1}{2} \operatorname{Tr} (ff^\dagger) \qquad (2\text{-}72)$$

and the polarization produced from an unpolarized beam is:

$$\vec{P}_f = \frac{\operatorname{Tr} (\sigma f^\dagger f)}{2(d\sigma/d\Omega)} . \qquad (2\text{-}73)$$

Using the explicit form of f given by eq. (2-54), $\vec{P}_f$ becomes:

$$\vec{P}_f = \vec{n} 2 \operatorname{Im} \{f(\theta) g^*(\theta)\} / (d\sigma/d\Omega) . \qquad (2\text{-}74)$$

The polarization is therefore in the direction $\vec{n}$, which is at right angles to the plane of scattering.

The polarization produced by a collision may be detected by allowing the scattered beam of spin $\frac{1}{2}$ particles to be again scattered by a target of spin 0.

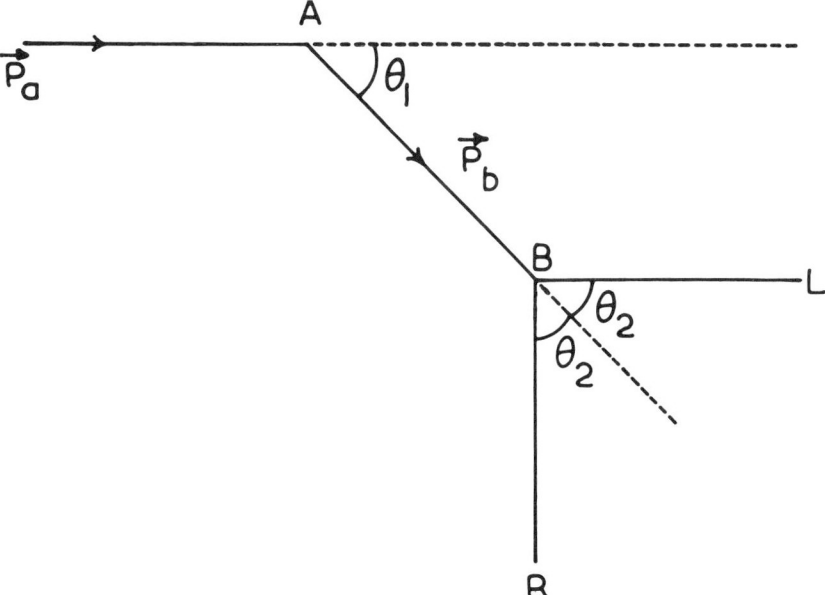

Fig. 2-1. Illustration of the scattering angles involved in a double scattering experiment.

The double scattering is illustrated in Fig. 2-1. After the collision at A, the beam will be polarized in a direction pointing out of the paper. This direction may be taken as the X axis, and for the second collision at B, AB may be taken as the Z axis. The differential cross section, summed over all final spin states, arising from a particular pure spin state ψ_a is, from eq. (2-59):

$$\frac{d\bar{\sigma}}{d\Omega} = (\psi_a, f^\dagger f \psi_a) . \qquad (2\text{-}75)$$

If the incident beam is an incoherent mixture of states defined by a density matrix ρ_i, the average cross section summed over all final spin states becomes:

$$\frac{d\bar{\sigma}}{d\Omega} = \sum_a p_a (\psi_a, f^\dagger f \psi_a)$$

$$= \text{Tr} (f \rho_i f^\dagger) \qquad (2\text{-}76)$$

where p_a is the fraction of the beam in state ψ_a as before. Applying eq. (2-76) to the second scattering at B, and using eq. (2-67):

$$\frac{d\bar{\sigma}}{d\Omega} = |f(\theta_2)|^2 + |g(\theta_2)|^2 + 2 \, [\text{Im} \, \{f(\theta_2) \, g^*(\theta_2)\}] \, n_x \, P_A(\theta_1) \qquad (2\text{-}77)$$

where $P_A(\theta_1)$ is the polarization produced at the first collision. If the plane of the second scattering is the same as that of the first, then for scattering to the right (Fig. 2-1) $n_x = +1$, and for scattering to the left $n_x = -1$. The asymmetry ϵ is then defined (where the angle of scattering θ_2 is the same to the right and left) as:

$$\epsilon = \frac{(d\bar{\sigma}/d\Omega)|_R - (d\bar{\sigma}/d\Omega)|_L}{(d\bar{\sigma}/d\Omega)|_R + (d\bar{\sigma}/d\Omega)|_L} \qquad (2\text{-}78)$$

where $\frac{d\bar{\sigma}}{d\Omega}\bigg|_{R,L}$ stands for the differential cross section in the right- and left-hand directions. From eq. (2-77):

2.2. LOW ENERGY π-MESON-NUCLEON SCATTERING

$$\epsilon = 2 P_A(\theta_1) \, \text{Im} \, \{f(\theta_2) g^*(\theta_2)\} / \{|f(\theta_2)|^2 + |g(\theta_2)|^2\}$$
$$= P_B(\theta_2) \, P_A(\theta_1) \, . \quad (2\text{-}79)$$

If the polarization P_B produced in a collision between an unpolarized beam and the second target is known, then this expression determines ϵ. If the second scattering is between particles of the same type as the first, and if $\theta_1 = \theta_2$, then:

$$\epsilon = \{P_A(\theta_1)\}^2 \quad (2\text{-}80)$$

so that the magnitude but not the sign of P_A can be measured.

2.2. LOW ENERGY π-MESON-NUCLEON SCATTERING

At the lowest energies, the angular distributions for the elastic scattering of π^+ and π^- mesons by protons have been measured, together with total cross sections and the polarization of the recoil proton:

$$\pi^+ + p \rightarrow \pi^+ + p \quad (2\text{-}81)$$

$$\pi^- + p \rightarrow \pi^- + p \, . \quad (2\text{-}82)$$

Angular distributions and total cross sections for the charge exchange reaction:

$$\pi^- + p \rightarrow \pi^0 + n \quad (2\text{-}83)$$

have also been measured, but the corresponding reactions involving neutrons rather than protons cannot be observed directly, although an indirect study is possible through the interaction of π mesons with deuterons.

It was explained in Chapter 1 that the π mesons form an isotopic triplet with $I = 1$, the π^+, π^- and π^0 mesons being in eigenstates of the third component of the isotopic spin I_3, with $I_3 = +1$, 0, and -1

respectively. In the same way, the proton and neutron form an isotopic doublet with $I = \frac{1}{2}$, and $I_3 = +\frac{1}{2}$ and $-\frac{1}{2}$. The isotopic properties of a system of one π meson and one nucleon can then be described in terms of state vectors $|I, I_3 \rangle$ that are eigenstates of the total isotopic spin operator and its third component. The possible values of the total isotopic spin I are $\frac{1}{2}$ and $\frac{3}{2}$, and, with the appropriate Clebsch-Gordon coefficients, the isotopic state vectors for the physical channels may be written in terms of $|I, I_3 \rangle$ as:

$$|\pi^+ + p \rangle = \sum_I C_{1\frac{1}{2}}(I, \tfrac{3}{2}; 1, \tfrac{1}{2}) | I, I_3 \rangle = |\tfrac{3}{2}, \tfrac{3}{2} \rangle$$

$$|\pi^- + p \rangle = \sum_I C_{1\frac{1}{2}}(I, -\tfrac{1}{2}; -1, \tfrac{1}{2}) | I, I_3 \rangle = \tfrac{1}{\sqrt{3}} [|\tfrac{3}{2}, -\tfrac{1}{2} \rangle + \sqrt{2} |\tfrac{1}{2}, -\tfrac{1}{2} \rangle]$$

$$|\pi^0 + n \rangle = \sum_I C_{1\frac{1}{2}}(I, -\tfrac{1}{2}; 0, -\tfrac{1}{2}) | I, I_3 \rangle = \tfrac{1}{\sqrt{3}} [\sqrt{2} |\tfrac{3}{2}, -\tfrac{1}{2} \rangle - |\tfrac{1}{2}, -\tfrac{1}{2} \rangle]$$

(2-84)

Complete charge independence requires that the transition matrix be diagonal in the basis of the states $|I, I_3 \rangle$ and that the matrix elements depend on the value of I but not on that of I_3. This is reasonable only to the extent that the Coulomb scattering of the charged particles and the mass differences, between the π^0 and the $\pi^\pm$ mesons, and between the neutron and proton, can be neglected. Coulomb scattering is very important, particularly at low energies, but it is possible to take this into account and to calculate the charge-independent nuclear amplitudes to a sufficient approximation. How this is done will be described later. The mass differences, amounting, in the extreme case, to 5 MeV, are important only at kinetic energies near threshold. In the charge-independent model, the physical amplitudes are related to those for scattering in an eigenstate of isotopic spin, T^I, by using eq. (2-84):

2.2. LOW ENERGY π-MESON-NUCLEON SCATTERING

$$\langle \pi^+ + p |T| \pi^+ + p \rangle = T^{\frac{3}{2}}$$

$$\langle \pi^- + p |T| \pi^- + p \rangle = \frac{1}{3} T^{\frac{3}{2}} + \frac{2}{3} T^{\frac{1}{2}}$$

$$\langle \pi^- + p |T| \pi^0 + n \rangle = \frac{\sqrt{2}}{3} (T^{\frac{3}{2}} - T^{\frac{1}{2}}) . \qquad (2\text{-}85)$$

The amplitudes for π meson-neutron scattering may be found in terms of T^I, in the same way.

The total cross section for $\pi^- p$ scattering below the pion production threshold, $\sigma_T(\pi^- p)$, is the sum of the cross sections for the reactions of eqs. (2-82) and (2-83) — elastic and charge-exchange scattering. If the amplitudes, eqs. (2-85), are squared, a relationship is found between this cross section, the cross sections for scattering in a pure isotopic spin state, $\sigma(I)$, and the cross section for elastic $\pi^+ p$ scattering, $\sigma(\pi^+ p)$:

$$\sigma\left(\frac{3}{2}\right) = \sigma(\pi^+ p)$$

$$\sigma\left(\frac{1}{2}\right) = \frac{1}{2} [3\sigma_T(\pi^- p) - \sigma(\pi^+ p)] . \qquad (2\text{-}86a)$$

It is easy to see that this relation holds generally for the total cross sections $\sigma_T(\pi^+ p)$, $\sigma_T(I)$, even above the pion production and other inelastic thresholds:

$$\sigma_T\left(\frac{3}{2}\right) = \sigma_T(\pi^+ p)$$

$$\sigma_T\left(\frac{1}{2}\right) = \frac{1}{2} [3\sigma_T(\pi^- p) - \sigma_T(\pi^+ p)] . \qquad (2\text{-}86b)$$

Each of the transition matrices T^I can be expanded in terms of partial waves; the general formulae obtained earlier in eqs. (2-51) and (2-52) for the scattering of a spin 0 by a spin $\frac{1}{2}$ particle will hold, if we make the replacements:

$$f_{\ell\pm} \to f_{\ell\pm}^{\frac{3}{2}} \qquad \text{for } \pi^+ p \text{ scattering}$$

$$f_{\ell\pm} \to \frac{1}{3}\left(f_{\ell\pm}^{\frac{3}{2}} + 2 f_{\ell\pm}^{\frac{1}{2}}\right) \qquad \text{for } \pi^- p \text{ scattering}$$

$$f_{\ell\pm} = \frac{\sqrt{2}}{3}\left(f_{\ell\pm}^{\frac{3}{2}} - f_{\ell\pm}^{\frac{1}{2}}\right) \qquad \text{for charge exchange scattering} \qquad (2\text{-}87)$$

where the partial wave amplitudes $f_{\ell\pm}^I$ are those for scattering in a channel with isotopic spin I. To complete our notation, the phase shifts and inelasticity parameters corresponding to each of the partial waves will be written as $\delta_{\ell\pm}^I$ and $\eta_{\ell\pm}^I$ respectively.

Coulomb Corrections to πp Scattering Amplitudes

At nonrelativistic velocities the scattering amplitude $f_c(\theta)$ for scattering by the Coulomb potential between particles of charge z_1 and z_2 may be obtained by solving the Schrodinger equation (Messiah, 1964, p. 421; Mott and Massey, 1965, p. 53). In the center of mass system, it is found that:

$$f_c(\theta) = \frac{-a}{2q \sin^2(\theta/2)} \exp\left[-i\alpha \log(\sin^2(\theta/2)) + 2i\sigma_0\right] \qquad (2\text{-}88)$$

where $\alpha = \dfrac{z_1 z_2 \omega e^2}{q}$ and $\sigma_0 = \arg \Gamma(1 + i\alpha)$. The charges z_1, z_2 are in units of e, the charge on the electron, ω is the reduced energy:

$$\omega = \frac{\sqrt{M_1^2 + q^2} \; \sqrt{M_2^2 + q^2}}{\sqrt{M_1^2 + q^2} + \sqrt{M_2^2 + q^2}}$$

and q is the center of mass momentum. As the Coulomb interaction is independent of spin, there is no spin flip amplitude corresponding to eq. (2-88). The amplitude $f_c(\theta)$ can be expanded into partial waves and the corresponding Coulomb phase shifts of order ℓ are σ_ℓ where:

$$\sigma_\ell = \arg \Gamma(1 + i\alpha + \ell) . \qquad (2\text{-}89)$$

2.2. LOW ENERGY π-MESON-NUCLEON SCATTERING

When the Coulomb interaction is modified by a short-range nuclear interaction, the partial waves of lower order are not determined by the σ_ℓ alone. Each amplitude is now determined by a phase shift $\epsilon_{\ell\pm}$ which can be written as the sum of a nuclear phase shift $\delta_{\ell\pm}$ and the Coulomb phase shift σ_ℓ:

$$\epsilon_{\ell\pm} = \sigma_\ell + \delta_{\ell\pm} . \tag{2-90}$$

If the first $(N+1)$ partial waves are so modified, we may write:

$$f(\theta) = f_c(\theta) + \sum_{\ell=0}^{N} \{(\ell+1)\bar{f}_{\ell+} + \ell \bar{f}_{\ell-}\} P_\ell(\cos\theta) - \sum_{\ell=0}^{N} \frac{(2\ell+1)}{2iq}(e^{2i\sigma_\ell} - 1) P_\ell(\cos\theta)$$

$$g(\theta) = \sum_{\ell=0}^{\infty} \{\bar{f}_{\ell+} - \bar{f}_{\ell-}\} \sin\theta \, P'_\ell(\cos\theta) \tag{2-91}$$

where the first $(N+1)$ Coulomb partial waves have been subtracted from $f(\theta)$ and replaced by the modified waves,[3] with $\bar{f}_{\ell\pm} = \frac{1}{2iq}(e^{2i\epsilon_{\ell\pm}} - 1)$. In terms of the nuclear amplitudes $f_{\ell\pm} = \frac{1}{2iq}(e^{2i\delta_{\ell\pm}} - 1)$, the expressions for f and g become:

$$f(\theta) = f_c(\theta) + \sum_{\ell=0}^{N} e^{2i\sigma_\ell} \{(\ell+1) f_{\ell+} + \ell f_{\ell-}\} P_\ell(\cos\theta)$$

$$g(\theta) = \sum_{\ell=0}^{N} e^{2i\sigma_\ell} \{f_{\ell+} - f_{\ell-}\} \sin\theta \, P'_\ell(\cos\theta) . \tag{2-92}$$

For π^+ and π^- scattering the nuclear amplitudes $f_{\ell\pm}$ must be replaced by the correct combination of amplitudes for scattering in the isotopic spin states with $I = \frac{3}{2}$, $I = \frac{1}{2}$, as in eq. (2-87).

---

[3] It should be noticed that the partial wave series for Coulomb scattering does not converge; this is the reason that $f_c(\theta)$ appears in eq. (2-91) in an unexpanded form. The convergence of the remaining terms in eq. (2-91) depends only on the nuclear scattering and is rapid.

The nuclear phase shifts $\delta_{\ell\pm}$ are still not the true charge-independent phase shifts which would describe the scattering if the electromagnetic interaction could be switched off. These phase shifts may be calculated approximately by noting that the Coulomb and nuclear interactions overlap only very slightly. In fact the parameter that characterizes the range of the Coulomb interaction may be taken to be the Bohr radius of a π^--p atom, $\sim 10^{-11}$ cm, which is much greater than the range of the nuclear π-p interaction, $\sim 10^{-13}$ cm. It is then a good approximation to neglect the Coulomb interaction within a sphere of radius R, where R is of the order 10^{-13} cm. If this is done, the wave function inside the sphere is exactly defined by the charge-independent phase shifts, while outside the sphere the wave function can be expressed in terms of Coulomb wave functions. For example, in π^+p scattering, the $\ell = 0$ partial wave function just inside the sphere is:

$$\phi(r) = \sin(qr) + \tan\bar{\delta}_0 \cos(qr)$$

while in the exterior region it is:

$$\phi_e(r) = F_0(qr) + \tan\delta_0 \, G_0(qr)$$

where F_0 and G_0 are the regular and irregular solutions of the radial Schrodinger equation for the Coulomb potential. For $\ell = 0$, (Mott and Massey, 1965), F_0 and G_0 have asymptotic form:

$$F_0(qr) \sim \sin(qr - \alpha \log(2qr) + \sigma_0)$$

$$G_0(qr) \sim \cos(qr - \alpha \log(2qr) + \sigma_0) \ .$$

The phase shifts $\bar{\delta}_0$ and δ_0 are related by smoothly joining the interior and exterior wave functions at $r = R$:

$$\frac{d}{dr}(\log\phi(r)) = \frac{d}{dr}(\log\phi_e(r)), \ r = R \ .$$

The result to first order in α is as follows:

2.2. LOW ENERGY π-MESON-NUCLEON SCATTERING

$$\delta_0 = \bar{\delta}_0 + a[0.5772 + \log(2qR) - \text{Ci}(2qR)\cos(2\bar{\delta}_0) + \text{Si}(2qR)\sin(2\bar{\delta}_0)]. \tag{2-93}$$

The correction for the $\ell = 1$ phase shifts and the corrections for the $\pi^- p$ scattering (which involves both isospin states) have been given (Van Hove, 1952). The difference between $\bar{\delta}_0$ and δ_0 is very small except at small momenta, but must be taken into account in the determination of the scattering lengths. If the highest accuracy is required near the threshold, the effects of the Coulomb potential within the sphere of radius R can be allowed for by using some suitable model for the charge distribution within the proton (Hamilton and Woolcock, 1958).

For momenta much in excess of 10 to 20 MeV/c, the velocity of the meson, (q/ω), is no longer very small compared with c, and the Coulomb amplitudes must be corrected to allow for relativistic effects (Solmitz, 1954; Foote et al., 1961). To first order in a this amounts to replacing f_c by $\bar{f}_c$ and adding to the spin-flip amplitude $g(\theta)$ the term $\bar{g}_c$ where:

$$\bar{f}_c(\theta) = f_c(\theta)[1 + \tfrac{1}{2}\nu_\pi \nu_p(1+\cos\theta) - \tfrac{1}{4}\nu_p^2(2\mu_p^2-1)(1-\cos\theta)]$$

$$\bar{g}_c(\theta) = \frac{a\sin\theta}{2q\sin^2(\theta/2)} \frac{\tfrac{1}{2}\mu_p \nu_p \nu_\pi + \tfrac{1}{4}(2\mu_p-1)\nu_p^2}{1+\nu_\pi \nu_p} \tag{2-94}$$

where μ_p is the magnetic moment of the proton in nuclear magnetons and ν_π, ν_p are the center of mass velocities of the π meson and proton. In addition, the Coulomb phase shifts σ_ℓ appearing in eqs. (2-91) should be modified (Foote et al., 1961), but the bulk of the correction is accounted for by using the amplitudes $\bar{f}_c$ and $\bar{g}_c$ and in practice the corrections to the σ_ℓ may be ignored.

The Analysis of the Experimental Cross Sections

The measured total cross sections for $\pi^+ p$ and $\pi^- p$ scattering are shown in Figs. 2-2 and 2-3 plotted against the kinetic energy of the π meson, T_π, in the laboratory system in which the target proton is at rest.

50   LOW ENERGY PION SCATTERING BY NUCLEONS

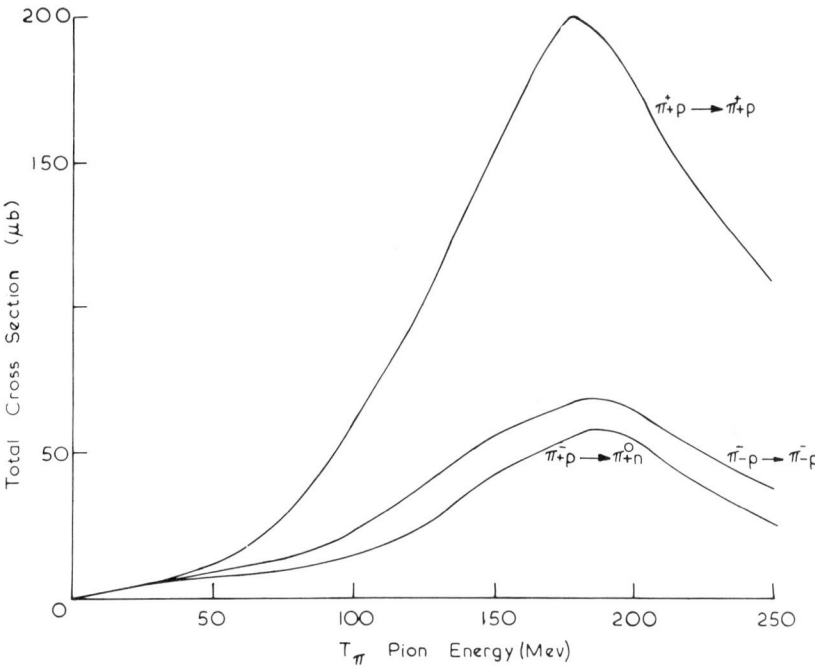

Fig. 2-2. Total cross sections for π^+ and π^- scattering by protons for laboratory pion energies T_π in the range 0 to 250 MeV.

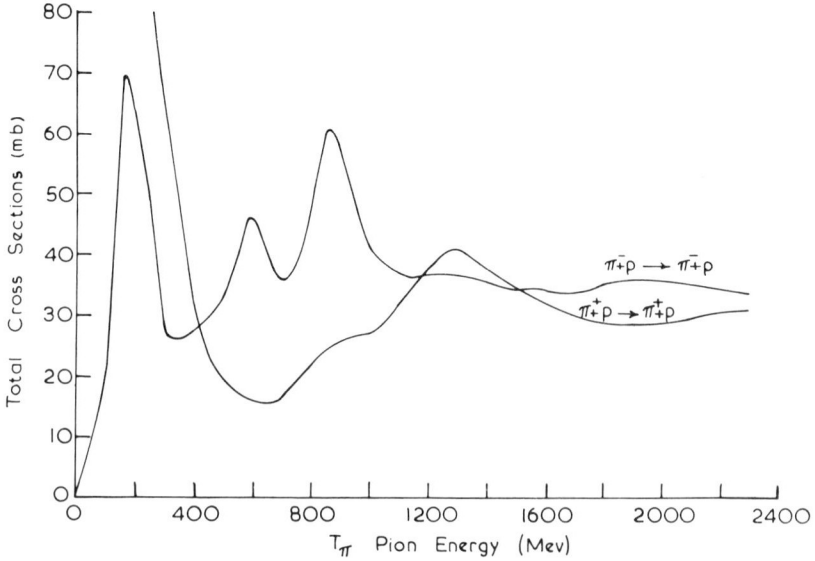

Fig. 2-3. Total cross sections for π^+ and π^- scattering by protons for laboratory pion energies T_π in the range 250 MeV to 2.4 GeV.

2.2. LOW ENERGY π-MESON-NUCLEON SCATTERING

$$T_\pi = \{\sqrt{q_L^2 + M_\pi^2} - M_\pi\}$$

where $\vec{q}_L$ is the laboratory momentum and M_π the mass of the π mesons. T_π is connected with the total center of mass energy W, by:[4]

$$T_\pi = \frac{1}{2M_N}[W^2 - (M_\pi + M_N)^2]$$

$$\vec{q}_L = \frac{\vec{q}W}{M_N} \qquad (2\text{-}95)$$

where M_N is the nucleon mass and $\vec{q}$ is the center of mass momentum.

The threshold for the production of a single π meson in the reaction:

$$\pi + N \rightarrow \pi + \pi + N$$

is $T_\pi = 170$ MeV; but, experimentally, inelastic scattering is unimportant below 300 MeV and this energy will be taken as the upper limit of the low energy region considered in this chapter.

The most noticeable feature of the total cross sections is the large peak centered at $T_\pi = 198$ MeV in both the π^+ and π^- scattering. From eq. (2-87) it is seen that if scattering occurs predominantly in the isotopic spin state $I = \frac{3}{2}$, the cross sections for elastic $\pi^+ p$, elastic $\pi^- p$, and charge-exchange scattering, are in the ratio:

$$\sigma(\pi^+ p) : \sigma_T(\pi^- p) : \sigma(\text{C.E.}) = 9 : 1 : 2 \ . \qquad (2\text{-}96)$$

At energies close to the peak in the cross sections, this ratio is approximately satisfied by the experimental data, suggesting that the strong scattering in this region arises from the $I = \frac{3}{2}$ state and that scattering in the $I = \frac{1}{2}$ state is comparatively unimportant.

---

[4] Note that W^2 is the norm of the total energy momentum vector of the system. In terms of the laboratory coordinates this vector is $(q_L^0, \vec{q}_L)$ with $q_L^0 = M_\pi + M_N + T_\pi$.

The shape and magnitude of the cross section can be explained, if one of the phase shifts $\delta_{\ell\pm}$ rapidly increases through the angle $(\pi/2)$, at an energy near the observed peak. Such a behavior is associated with the formation of a resonance or meta-stable state, and this topic will be discussed in detail in Chapter 5. Now, near the energy at which the resonance occurs, the behavior of a resonant phase shift can be represented by:

$$\tan \delta = \left(\frac{\Gamma/2}{W_R - W}\right) + A(W) \tag{2-97}$$

where Γ and A are slowly varying functions of W. The corresponding partial wave scattering amplitude, in the case that the background $A(W)$ is small, is given by:

$$qf_\ell = \frac{\Gamma/2}{(W_R - W) - i\Gamma/2}$$

giving rise to a partial cross section, with what is known as a Breit-Wigner shape:

$$\sigma = \frac{\pi(2j+1)}{q^2} \left[\frac{\Gamma^2}{(W_R - W)^2 + \Gamma^2/4}\right] \tag{2-98}$$

Assuming that the observed peak in the π-N cross section is due to the formation of a resonance, several clues to the partial wave responsible can be found in the total cross section data. For a start, the $\pi^+ + p$ total cross sections rise rapidly above threshold, being proportional at q^4 for small q. This suggests that the important partial wave is a p wave, because the effective range formula [eq. (2-41)] shows that the partial cross section of order ℓ is expected to behave like $q^{4\ell}$ at threshold. If a single partial wave is resonant and the background is small, the maximum cross section can be found from eqs. (2-52) and (2-53). It is seen that for scattering of unpolarized particles, the partial cross section for total angular momentum j is:

$$\sigma_j = 8\pi(j + \tfrac{1}{2}) |f_{\ell\pm}|^2 \tag{2-99}$$

where $j = \ell \pm \tfrac{1}{2}$ so that $\sigma_j \leq 4\pi(2j+1)/q^2$.

2.2. LOW ENERGY π-MESON-NUCLEON SCATTERING

At the resonance energy, the phase shift of the resonant partial wave is $(\pi/2)$ and σ_j will attain its maximum value. The measured total cross section for $\pi^+ p$ scattering is about 200 mb at the maximum, and this is consistent with $j = \frac{3}{2}$. The resonance state is therefore tentatively identified with the P_{33} state.[5] This identification is completely confirmed by polarization measurements and by the dispersion relation analysis to be described in Chapter 3.

A resonance occurs when a meta-stable state is formed, and the view will be developed that these states are not essentially different from the states stable under the strong interactions (such as the nucleon or the hyperon states), in the same way that the ground state of an atom is not different in nature from an excited state. Accordingly the resonant states will often be referred to as particles, in the same way as the stable states. The particles of baryon number $B = 1$, hypercharge $Y = 1$ will be denoted by $N^*(W)$, where W is the total center of mass energy of the resonance, which we shall normally give in MeV. In this notation, the P_{33} resonance is due to the formation of the $N^*(1236)$.

To determine the width of the resonance, the total cross section may be fitted by a Breit-Wigner formula. For $\pi^+ p$ scattering:

$$\sigma(\pi^+ p) = \frac{2\pi}{q^2} \left[\frac{\Gamma^2}{(W - W_R)^2 + \Gamma^2/4} \right] + B(q) \tag{2-100}$$

where W is the center of mass energy, W_R is the resonance energy, and the nonresonant background scattering is represented by $B(q)$. As B is small, it is sufficient to treat it as an energy-independent parameter when fitting the data near the resonance. To provide the correct threshold behavior, the resonance width Γ must behave like q^3 for small q; this is achieved by writing (Blatt and Weisskopf, 1952):

---

[5] We use the conventional notation $\ell_{2I,2j}$ where ℓ, j are the orbital and total angular momenta, and I is the isotopic spin of the state.

$$\Gamma(q) = 2(qR)^3 \gamma / (1 + (qR)^2) \tag{2-101}$$

where R is a constant defining the interaction radius, and γ is a constant called the reduced width. The parameters found by fitting the experimental cross section are (Deans and Holliday, 1966):

$$W_R = 1236 \text{ MeV}, \quad \gamma = 71 \text{ MeV}, \quad R = 0.81 \, (\hbar/M_\pi c)$$

$$B = 4 \text{ mb} \qquad \Gamma(q_R) = 121 \text{ MeV} .$$

The position and width obtained depends slightly on the specific parametrization employed, but the latest analyses (Rittenberg et al., 1971), based on new, very accurate, and definitive measurements of the total elastic and charge exchange cross sections (Carter et al., 1971; Bugg et al., 1971) are in essential agreement with these figures. The new values are $1231 \leq W_R \leq 1234$ Mev and $110 \leq \Gamma \leq 120$ Mev.

Phase Shift Analysis

To confirm the existence of the resonance, and to discover the detailed behavior of the scattering matrix, a phase shift analysis of the differential cross sections must be made. This is achieved by minimizing the expression:

$$F = \sum_i \left[\frac{I_i^E - I_i^P}{\epsilon_i} \right]^2 \tag{2-102}$$

with respect to the phase shifts as parameters,[6] where I_i^E is a measured cross section with statistical error ϵ_i and I_i^P is the corresponding parametrized cross section. The sum ranges over all the different experimental measurements at the energy concerned. Provided that the errors are independent and normally distributed, the quantity F will follow a "χ^2" distribution. The expected value of F will be equal to the number

---

[6] At energies above the inelastic threshold, the inelasticities are no longer equal to unity and should be included in the minimization.

2.2. LOW ENERGY π-MESON-NUCLEON SCATTERING

of degrees of freedom (the difference between the number of pieces of data and the number of parameters), and the "goodness of fit" may be obtained from standard statistical tables, (for example, Pearson and Hartley, 1954). Near the minimum, F will be of the form:

$$F = F_{MIN} + \sum_{i,j} G_{ij}(x_i - x_i^m)(x_j - x_j^m) \qquad (2\text{-}103)$$

where x_i are the varied parameters x_i^m is the value of x_i at the minimum and G is a real symmetric matrix. The inverse matrix G^{-1} is known as the error matrix, and the diagonal elements $(G^{-1})_{ii}$ are equal to the standard deviation of each parameter x_i. The off-diagonal elements of (G^{-1}) are associated with the correlations between the parameters. When fitting scattering data in this way, it is possible to reduce F to the expected value, and to obtain a reasonable confidence level. A good example of such a fit with a full analysis of errors using the error matrix can be consulted in the paper by Anderson et al., (1955), where the elastic scattering of π mesons by protons at 89 MeV is investigated.

In many cases, the absolute normalization of a set of angular distribution or other data may be in error. The error is a correlated error applying to the whole data set and is distinct from the statistical error associated with each member of the set. At any given energy, several sets of data may be used, perhaps originating from different laboratories with different normalization errors. This situation can be met by introducing a set of normalization parameters $\lambda_{n(i)}$, which would take the value unity if the normalization of a set were correct. Then in place of eq. (2-102), we construct the expression (Arndt and MacGregor (1967)):

$$F = \sum_{i,n} \left[\frac{\lambda_n l^E_{i,n} - l^P_{i,n}}{\epsilon_{i,n}} \right]^2 + \sum_n \left[\frac{\lambda_n - 1}{\Delta_n} \right]^2 \qquad (2\text{-}104)$$

where Δ_n is the normalization error in the nth set of data. The sum over n is over all the data sets employed, while the sum over i is a sum over each member of a data set. The function F is then to be minimized with respect both to the phase shift parameters and also to the normalization parameters.

The relationship between the cross section and the phase shifts is highly nonlinear, and in general a number of different minima will be found, even if only a small number of phase shifts are involved. To take a particular example, consider $\pi^+ p$ scattering at 100 MeV. It is expected from the argument following eq. (2-40), that only s and p waves will be important, so that the measured differential cross sections are of the form (apart from Coulomb scattering):

$$\frac{d\sigma(\pi^+ p)}{d\Omega} = |f_{0+}^{\frac{3}{2}} + \left(2f_{1+}^{\frac{3}{2}} + f_{1-}^{\frac{3}{2}}\right)\cos\theta|^2 + |f_{1+}^{\frac{3}{2}} - f_{1-}^{\frac{3}{2}}|^2 \sin^2\theta . \quad (2\text{-}105)$$

The term $|f_{1+}^{\frac{3}{2}} - f_{1-}^{\frac{3}{2}}|$ depends only on the difference $\left(\delta_{1+}^{\frac{3}{2}} - \delta_{1-}^{\frac{3}{2}}\right)$, while the coefficient of $\cos\theta$, in the amplitude, is:

$$\left(2f_{1+}^{\frac{3}{2}} + f_{1-}^{\frac{3}{2}}\right) = \frac{1}{2iq}\left(2\exp\left(2i\delta_{1+}^{\frac{3}{2}}\right) + \exp\left(i\delta_{1-}^{\frac{3}{2}}\right) - 3\right) . \quad (2\text{-}106)$$

It is now possible to find two sets of phase shifts $\delta_{1\pm}^{\frac{3}{2}}$ and $\Delta_{1\pm}^{\frac{3}{2}}$ such that:

$$\delta_{1+}^{\frac{3}{2}} - \delta_{1-}^{\frac{3}{2}} = \Delta_{1+}^{\frac{3}{2}} - \Delta_{1-}^{\frac{3}{2}}$$

and

$$2\exp\left(2i\delta_{1+}^{\frac{3}{2}}\right) + \exp\left(i\delta_{1-}^{\frac{3}{2}}\right) = 2\exp\left(2i\Delta_{1+}^{\frac{3}{2}}\right) + \exp\left(i\Delta_{1-}^{\frac{3}{2}}\right) \quad (2\text{-}107)$$

and both sets will fit the same experimental cross section. One set could be consistent with the idea of a resonance in the p_{33} state with $\delta_{1+}^{\frac{3}{2}}$ large and $\delta_{1-}^{\frac{3}{2}} \approx 0$, and in the other set neither $\Delta_{1+}^{\frac{3}{2}}$ nor $\Delta_{1-}^{\frac{3}{2}}$ need be

2.2. LOW ENERGY π-MESON-NUCLEON SCATTERING

resonant. This ambiguity, known as the Fermi-Yang ambiguity, caused some difficulty in early studies of πN scattering. Another difficulty arises because the differential cross section is invariant under the interchange of all the odd and even parity partial waves for the same I and J. That is, two sets of phase shifts $\delta^I_{\ell\pm}$ and $\Delta^I_{\ell\pm}$ provide the same differential cross section if:

$$\delta^I_{\ell+} = \Delta^I_{(\ell+1)-} \qquad \delta^I_{\ell-} = \Delta^I_{(\ell-1)+} . \qquad (2\text{-}108)$$

This is known as the Minami ambiguity. It is removed by measuring the polarization of the recoil proton, or by using the extra information provided by dispersion relations which are discussed in a later chapter. It is also easily seen that the signs of the phase shifts are not determined by eq. (2-105), the cross section being invariant under the simultaneous change in sign of all the phase shifts. This ambiguity is not important in practice, because it is resolved by examining whether the interference with the Coulomb interaction is constructive or destructive. In general the ambiguities of a phase shift analysis increase rapidly with energy, partly because of the increasing numbers of partial waves, but also because of the possibility of inelastic scattering, so that the inelasticity parameters $\eta^I_{\ell\pm}$ have to be introduced, in addition to the phase shifts $\delta^I_{\ell\pm}$.

In principle, the ambiguities at each energy may be removed by performing a complete set of polarization experiments. In the present example, these would include measurements of the polarization produced by scattering from an unpolarized beam, given by eq. (2-74) and the spin rotation coefficients A and R, which are defined in the following manner. If, as usual $\vec{q}_a$ and $\vec{q}_b$ denote the incident and final momenta in the center of mass system and the nucleons in the target are characterized by a polarization vector $\vec{P}^{(1)}$ and the recoil nucleons by a polarization vector $\vec{P}^{(2)}$, then from eq. (2-71):

$$\left(\frac{d\bar{\sigma}}{d\Omega}\right)\vec{P}^{(2)} = \{|f|^2 - |g|^2\}\vec{P}^{(1)} - 2\,\text{Re}\,(fg^*)\,\vec{n}\times\vec{P}^{(1)}$$
$$+ 2\,|g|^2\,(\vec{n}\cdot\vec{P}^{(1)})\,\vec{n} + 2\,\text{Im}\,(fg^*)\,\vec{n} . \qquad (2\text{-}109)$$

The differential cross section $(d\bar{\sigma}/d\Omega)$ for scattering from the polarized target, without regard to the final spin direction, is given by:

$$\left(\frac{d\bar{\sigma}}{d\Omega}\right) = |f|^2 + |g|^2 + 2 \operatorname{Im}(g^*f)\, \vec{P}^{(1)} \cdot \vec{n}\,. \tag{2-110}$$

The unit vectors $\vec{m}$ and $\vec{s}$ may now be defined as:

$$\vec{m} \equiv \frac{\vec{q}_a \times \vec{n}}{|q_a|} \quad \text{and} \quad \vec{s} \equiv \frac{\vec{q}_b \times \vec{n}}{|q_b|}\,. \tag{2-111}$$

The component of $\vec{P}^{(2)}$ in the direction $\vec{s}$ is, from eq. (2-109):

$$\vec{P}^{(2)} \cdot \vec{s} = \left(\frac{d\sigma/d\Omega}{d\bar{\sigma}/d\Omega}\right)\left(\frac{A\vec{q}_a}{|q_a|} + R\vec{m}\right) \cdot \vec{P}^{(1)}\,. \tag{2-112}$$

where $(d\sigma/d\Omega)$ is the differential scattering cross section for an unpolarized target, and where A and R are defined by:

$$A\left(\frac{d\sigma}{d\Omega}\right) = (|f|^2 - |g|^2)\sin\theta + 2\operatorname{Re}(fg^*)\cos\theta$$

$$R\left(\frac{d\sigma}{d\Omega}\right) = (|f|^2 - |g|^2)\cos\theta - 2\operatorname{Re}(fg^*)\sin\theta\,. \tag{2-113}$$

The amplitudes f and g must be replaced by the appropriate combinations of the amplitudes for isotopic spin $\frac{1}{2}$ and $\frac{3}{2}$ to obtain formulae for π^+, π^- and charge exchange scattering. It should be noted that while a measurement of the recoil polarization determines $\operatorname{Im}(fg^*)$, a measurement of A and R determines $\operatorname{Re}(fg^*)$.

The coefficients A and R may be found by measuring the polarization $\vec{P}^{(2)}$ in a second scattering. Suppose the up-down asymmetry (with respect to the plane of scattering of the *first* collision) produced in the *second* collision is ϵ_A when the target is polarized parallel to $\vec{q}_a$, and ϵ_R when the target is polarized parallel to $\vec{m}$. Then proceeding as in the discussion on page 41 we have that:

2.2. LOW ENERGY π-MESON-NUCLEON SCATTERING

$$\epsilon_A = P^{(3)} A P^{(1)}$$

$$\epsilon_R = P^{(3)} R P^{(1)} \qquad (2\text{-}114)$$

where $P^{(3)}$ is the analyzing power of the second scattering, which is equal to the polarization that would be produced at the second scattering from an unpolarized beam. The measurement of $d\sigma/d\Omega$, P, R, and A for both π^+ and π^- scattering determines the four complex elements (f and g for each isotopic spin state) of the scattering matrix at each angle and energy. In fact, no complete set of measurements exists at any energy in the low energy region, and in any case no measurements are perfectly accurate, with the result that unique values of the scattering matrix cannot be obtained straightforwardly.

The elastic and charge exchange cross sections have been measured[7] at close energy intervals from threshold upwards. The polarization of the recoil proton in elastic scattering has been measured at energies above 200 MeV, but for charge exchange scattering, polarization measurements

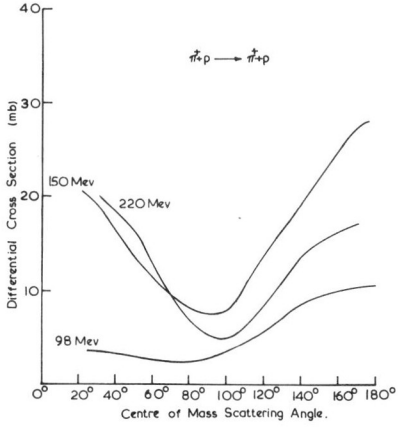

Fig. 2-4. Differential cross sections for the elastic scattering of π^+ mesons by protons.

---

[7] Bibliographies of the data from 0 to 300 MeV are given by Hull and Lin (1965) and by Roper, Wright, and Feld (1965).

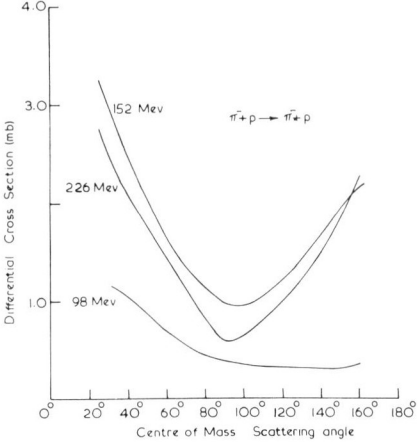

Fig. 2-5. Differential cross sections for the elastic scattering of π^- mesons by protons.

are scarce. No measurements of A and R have been made as yet in this energy region. Typical data at 98, 150, and 220 MeV are shown in Figs. 2-4 and 2-5.

Many phase shift analyses have been carried out at different energies, and using the polarization data it is possible to confirm the existence of the p_{33} resonance. The large phase shifts are well determined; but, for the reasons we have touched on, it is impossible to pick a unique set of phase shifts at each energy, and the small phase shifts are not completely determined by this procedure. To overcome this difficulty, two general methods or a combination of both have been suggested.

First some appeal to theory may be made to determine some or all of the phase shifts, or to constrain them in some way, and examples of this technique will be discussed in Chapter 5. Alternatively, the smooth variation of the phase shifts and inelasticity parameters with energy may be used to connect the low energy region, where the ambiguities are unimportant, to higher energy regions, where the ambiguities are serious. To do this the phase shifts must be expressed as smooth functions of the

2.2. LOW ENERGY π-MESON-NUCLEON SCATTERING

energy, depending on a number of parameters λ_i, which are determined by minimizing F in eq. (2-102) or eq. (2-104), where the sum is extended over all suitable experiments in a chosen energy interval. In an analysis of all the data in the energy range 0 to 350 MeV, Roper, Wright, and Feld (1965) and Roper and Wright (1965) parametrized the nonresonant phase shifts by polynomials in the momenta, writing:

$$\tan \delta_{\ell\pm}^I = q^{2\ell+1} \sum_{n=0}^{N} \lambda_n q^n \qquad (2\text{-}115)$$

where the λ_i are the parameters to be varied in the minimization. This expression guarantees the correct threshold behavior of the partial wave amplitudes as $q \to 0$. The inelasticity parameters $\eta_{\ell\pm}^I$ were all set equal to unity, which is a good approximation below 300 MeV. The resonant partial wave amplitude for the p_{33} state, $f_{1+}^{\frac{3}{2}}$, was represented as the sum of a Breit-Wigner resonance and a background:

$$q f_{1+}^{\frac{3}{2}}(q) = -\frac{1}{2}i \left[\frac{\Gamma(q)}{(E-E_R) + \tfrac{1}{2}i\Gamma(q)} \right] + \frac{1}{2} i \, (\exp(2i\delta_B) - 1) \qquad (2\text{-}116)$$

where E is the center of mass energy of the π meson, $E = \sqrt{q^2 + M_\pi^2}$, and E_R is the value of E at resonance. The width $\Gamma(q)$ was allowed an energy variation suggested by Layson (1963), which ensured that $\Gamma(q) \propto q^3$ for small q:

$$\Gamma(q) = \frac{4\gamma^2 M_N}{(E+E_R)} \left[(qR) j_1^2(qR) + n_1^2(qR) \right] \qquad (2\text{-}117)$$

where γ and R are constants. The form of eq. (2-116) is not automatically unitary for arbitrary values of the parameters, but the parameter ranges may be restricted so that unitarity is preserved.

Below 200 MeV, only s and p waves are large, but near this energy d waves become important, and to obtain the best fits to the data

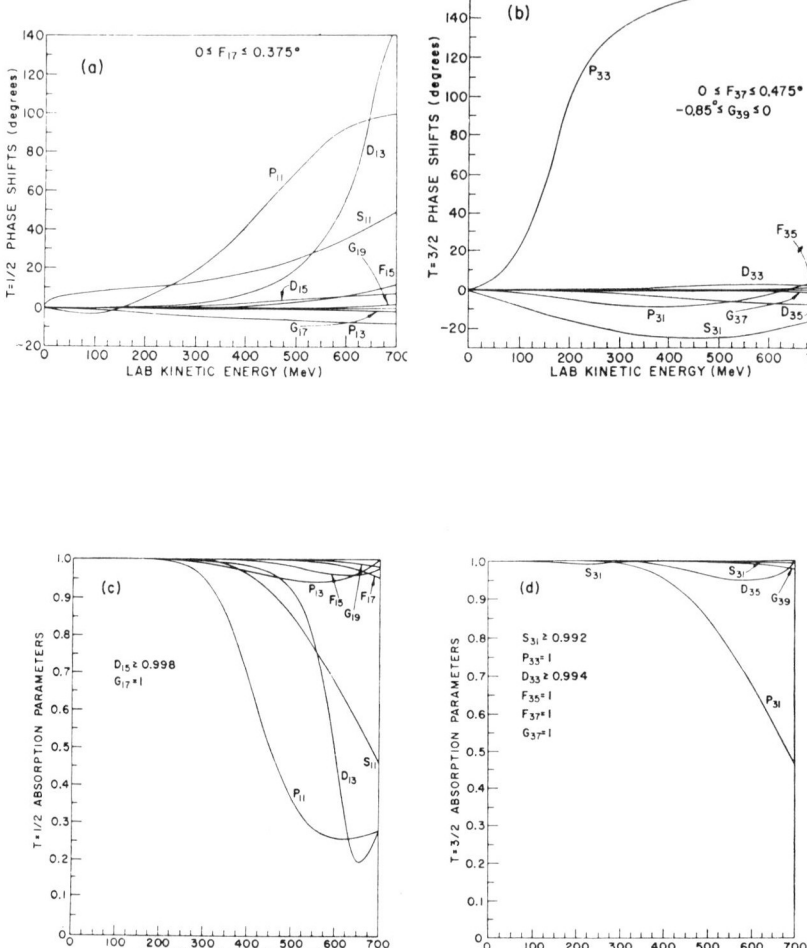

Fig. 2-6. The phase shifts and inelasticity parameters for pion-nucleon scattering below 700 MeV (from the work of Roper, Wright, and Field, 1965).

2.2. LOW ENERGY π-MESON-NUCLEON SCATTERING

it is necessary to take into account f waves as well. Even when all the data are included, a number of solutions can be found, between which it is difficult to distinguish. However these differ only in fine details[8] and all are close to the particular solution shown in Fig. 2-6.

Apart from the resonant p_{33} phase shift, one of the most interesting features of the solution is that the p_{11} phase shift is negative at low energies, but passes through zero near 170 MeV to become positive and rapidly increasing at the end of our energy range. This is particularly important because many dynamical models, constructed to account for the nucleon state (which has the same quantum numbers as the p_{11} partial wave), are inconsistent with this behavior and must be rejected.

When one fits data at a single energy, the standard χ^2 test can be employed to assess the accuracy of the fit. The normalized χ^2, defined as:

$$\chi^2 = \frac{F}{(M-N)}$$

where M is the number of pieces of data and N the number of parameters, can be reduced to less than unity in the best fits, giving a reasonable confidence level. Unfortunately, when one fits data at a number of energies simultaneously, particularly when the data arise from different laboratories, it must be recognized that some of the data will be inconsistent with the rest because of unknown *absolute* errors. Under these conditions, no precise meaning can be given to the values of χ^2 at the minimum, which, in fact, turns out to be two or three times the value normally obtained in fits at single energies. Despite this it is clear that, apart from fine details, the phase shifts shown in Fig. 2-6 are essentially correct because of the general agreement between results obtained with differing parametrizations. The continuation of this analysis to higher energies, with the identification of higher resonant states, will be left until Chapter 5.

---

[8] Another parametrization based on dispersion relations used by Hull and Lin (1965) gives results in essential agreement with those described in this paragraph.

Scattering Lengths

At low energies (less than 100 MeV), the phase shifts are roughly represented by the formulae:

$$q^{2\ell+1} \cot \delta_{\ell\pm}^I = \frac{1}{a_{\ell\pm}^I} . \qquad (2\text{-}118)$$

However to obtain accurate values for the scattering lengths, defined by:

$$a_{\ell\pm}^I = \ell t_{q\to 0}\left(\delta_{\ell\pm}^I / q^{2\ell+1}\right)$$

the low energy data must be extrapolated to zero momentum by a more elaborate method (see Hamilton, 1966). In doing this, as well as in making complete Coulomb corrections, including the inner Coulomb correction, it is necessary to allow for the mass differences between the $\pi^\pm$ and π^0 mesons and the neutron and the proton, which make the charge exchange reaction exothermic to the extent of 3 MeV. This may be done by writing the low energy s wave charge exchange cross section as [see eq. (2-26)]:

$$\sigma(\pi^- + p \to \pi^0 + n) = 4\pi \frac{v_b}{v_a} \frac{2}{9} \left| f_{0+}^{\frac{3}{2}}(q) - f_{0+}^{\frac{1}{2}}(q) \right|^2 \qquad (2\text{-}119)$$

where v_b and v_a are the final and initial relative velocities in the center of mass system.

An analysis of experiments at energies below 45 MeV has provided the following results (Hamilton, 1966). The $\pi^+ + p$ cross section defines $a_{0+}^{\frac{3}{2}}$, which has the value:

$$a_{0+}^{\frac{3}{2}} = -0.091 \pm 0.005 \left(\frac{\hbar}{M_\pi c}\right) \simeq -0.13 \text{ fermi} . \qquad (2\text{-}120)$$

The $\pi^- + p$ cross section defines the combination $\left(2a_{0+}^{\frac{1}{2}} + a_{0+}^{\frac{3}{2}}\right)$ which is found to have the value:

$$\left(2a_{0+}^{\frac{1}{2}} + a_{0+}^{\frac{3}{2}}\right) = 0.270 \pm 0.008 \left(\frac{\hbar}{M_\pi c}\right) \simeq 0.38 \text{ fermi} . \quad (2\text{-}121)$$

The charge exchange scattering is determined by the combination $\left(a_{0+}^{\frac{1}{2}} - a_{0+}^{\frac{3}{2}}\right)$, but the extrapolation of the experiments to zero energy in this case is open to some doubt and results have been quoted in the following range (Hamilton, 1966; Donald et al., 1966):

$$\left(a_{0+}^{\frac{1}{2}} - a_{0+}^{\frac{3}{2}}\right) = 0.266 \text{ to } 0.291 \left(\frac{\hbar}{M_\pi c}\right) . \quad (2\text{-}122)$$

The corresponding figure calculated from eqs. (2-120) and (2-121) is 0.271 ± 0.008. The agreement between these figures checks the validity of charge-independence in the low energy $\pi + N$ system. This test, although convincing, is not as accurate as that given by studying the reaction $d + d \to \text{He} + \pi^0$. This is forbidden by isotopic spin conservation. No events were found in the experiments of Poirier and Pripstein (1963), who showed that the isotopic spin is conserved at least to 99.74%.

2.3. THE LOW ENERGY SCATTERING OF K^+ AND K^- MESONS BY NUCLEONS

At this point it is of interest to consider briefly the analysis of the parallel system of K-meson scattering by nucleons at low energies.[9]

$K^\pm p$ Scattering

The analysis of low energy K^+-proton scattering can be developed in exactly the same way as for pion-nucleon scattering. In Chapter 1, we

---

[9] A detailed review has been given by Bransden (1968).

saw that the K mesons have isotopic spin $\frac{1}{2}$. As $I_3 = +\frac{1}{2}$ for both the K^+ meson and the proton, the K^+p system is in an eigenstate of isotopic spin with $I = 1$. At low energies no other system having the same quantum numbers ($B = 1$, $Y = 2$, $Q = 2$, $I = 1$) exists, so that the scattering is purely elastic. The lowest inelastic threshold is for the production of a single π meson, and this becomes important above a laboratory K-meson momentum of $p_K = 900$ MeV/c, through the production and decay of the $N^*(1236)$ resonance:

$$K^+ + p \to K^+ + N^* \to K^+ + N + \pi . \qquad (2\text{-}123)$$

Below 900 MeV/c, where the scattering is elastic, the angular distribution is isotropic and the data are fitted by assuming that only s-waves are important. The phase shift δ_0 is well represented by an effective range formula:

$$q \cot \delta_{0+}^1 = \frac{1}{a_{0+}^1} + \frac{1}{2} r_{0+}^1 q^2 \qquad (2\text{-}124)$$

with $a_{0+}^1 = (-0.29 \pm 0.02)$ fermi and $r_0 = 0.6 \pm 0.6$ fermi (Goldhaber et al., 1962; see also Martin and Perrin, 1969).

At higher momenta provision must be made both for the inclusion of further phase shifts and for the inclusion of inelasticity parameters. Some progress has been made with a phase shift analysis, in which the phase shifts and inelasticity parameters have been given energy-dependent parametric forms.

$K^- + p$ Scattering

Unlike $\pi + N$ and $K^+ + N$ scattering, the $K^- + p$ system is complicated by the fact that a number of two-body channels are "open" at low energies. The total hypercharge of the $K^- + p$ system is $Y = 0$, and the baryon number is $B = 1$. It follows that hyperon-π meson states with $Y = 0$ and baryon number $B = 1$ are allowed final states. The allowed reactions are:

2.3. THE LOW ENERGY SCATTERING OF K^+ AND K^- MESONS

$$K^- + p \to K^- + p \quad \text{(a)}$$
$$\to \bar{K}^0 + n \quad \text{(b)}$$
$$\to \Lambda + \pi^0 \quad \text{(c)}$$
$$\to \Sigma^{\mp} + \pi^{\pm} \quad \text{(d)}$$
$$\to \Sigma^0 + \pi^0 \quad \text{(e)}$$
$$\to \Lambda + 2\pi \quad \text{(f)} \qquad (2\text{-}125)$$

The charge exchange reaction (b) is endothermic, because of the mass differences between the K^- and $\bar{K}^0$ mesons and between the neutron and the proton, with a threshold at $T_K = 5.5$ MeV ($p_K = 90$ MeV/c). Hyperon production is exothermic and observed at all energies and the thresholds for the $\Lambda\pi$, $\Sigma\pi$, and $\Lambda 2\pi$ systems lie at 180, 100, and 40 MeV below the K^-N threshold respectively.

The K^-p system is a mixture of $I = 1$ and $I = 0$ eigenstates, while of the possible final systems the $\Lambda\pi^0$ system belongs to $I = 1$ and the $\Sigma\pi$ system is a mixture of $I = 0$ and $I = 1$ states. Ignoring the unimportant $\Lambda + 2\pi$ channel, the transition matrices for low energy scattering divide into a 2×2 matrix for $I = 0$ connecting the K^--N and Σ-π channels, and a 3×3 matrix for $I = 1$, connecting the K^--N, Σ-π and Λ-π channels. As at very low energies only s waves and, to a lesser extent, p waves are important, the analysis can be made by expanding the (2×2) or (3×3) inverse transition matrices in an effective range formulae. The procedure is complicated, as Coulomb corrections and the correction for the K^-, $\bar{K}^0$, and np mass-differences are important.[10]

This analysis can be extended in several directions. At higher momenta near 400 MeV/c, d waves must be added, and the analysis shows

---

[10] Further information can be obtained by using the K^+d amplitudes to deduce the K^+n amplitudes. This requires the use of the impulse approximation and is subject to some uncertainty at low energies.

that the $d_{\frac{3}{2}}$ wave is resonant in the $I = 0$ state, at a total center of mass energy of 1520 MeV. Also a continuation below threshold can be attempted using the analysis outlined in Chapter 5. This is particularly interesting, as the effective range formula predicts a resonance, at a total energy of 1410 MeV, with a width $\Gamma = 37$ MeV in the $I = 0$ state, with negative parity and $J = \frac{1}{2}$. Such a resonance can be identified with the $Y_0^*(1405)$ found in reactions in which the $(\Sigma + \pi)$ system appears in the final state.

The continuation of the p-wave below threshold is also consistent with another known resonance the $Y_1^*(1385)$. This resonance is strongly coupled to the $\Lambda + \pi$ channel, but the coupling to the $\Sigma + \pi$ and the $K^- + N$ channels is small. This indicates that the $Y_1^*(1385)$ is a $(\Lambda + \pi)$ resonance; in contrast the $Y_0^*(1405)$ appears to be a virtual bound state of the $K^- + p$ system. The existence of this complicated structure below the $K^- + N$ threshold has made attempts to determine the $K\Lambda N$ and $K\Sigma N$ coupling constants by dispersion relations difficult, and for this reason these constants are not known with the certainty of the πN coupling constant.

2.4. PHOTOPRODUCTION OF π MESONS FROM NUCLEONS

For photon energies greater than 140 MeV, production of π mesons is possible through the reactions

$$\begin{align}
\gamma + p &\to \pi^0 + p & \text{(a)} \\
&\to \pi^+ + n & \text{(b)} \\
\gamma + n &\to \pi^0 + n & \text{(c)} \\
&\to \pi^- + p & \text{(d)} .
\end{align}$$

(2-126)

2.4. PHOTOPRODUCTION OF π MESONS FROM NUCLEONS

The two neutron reactions (c) and (d) can only be studied indirectly through the interaction of γ rays with deuterons, but the proton reactions (a) and (b) may be observed directly[11] and the low energy total cross sections are shown in Fig. (2-8). The total center of mass energy W is related to the laboratory energy of the γ rays, T_γ by:

$$T_\gamma = \left[\frac{1}{2M_N}(W^2 - M_N^2)\right]^{\frac{1}{2}} . \qquad (2\text{-}127)$$

Using this result, the conspicuous peak in the total cross section measurements at $T_\gamma = 330$ MeV is found to correspond to $W = 1238$ MeV. This is the energy of the $N^*(1236)$ resonance in the $\pi + N$ system; photoproduction at these energies must be largely through this state:

$$\gamma + N \rightarrow N^*(1236) \rightarrow \pi + N .$$

The cross sections can be fitted rather well to a Breit-Wigner formula:

$$\sigma = \frac{2\pi}{q^2} \frac{\Gamma \Gamma_\gamma}{(W-W_R)^2 + \Gamma^2/4} \qquad (2\text{-}128)$$

where Γ_γ is the partial width into the photon channel and where Γ and W_R are the N^* width and resonance energy.

The data at very low energies, near threshold and well away from the resonant state, has been used to provide an independent and interesting check on the πN scattering lengths. Time reversal invariance shows that:

$$|T_{ba}|^2 = |T_{\bar{a}\bar{b}}|^2$$

where $\bar{a}$ and $\bar{b}$ are the same states as a and b, except that the spins of all the particles are reversed. It follows that the cross sections for

---

[11] A bibliography of low energy experiments is given by Donnachie and Shaw (1966).

the scattering of unpolarized beams, summed over the final spin states, for the reactions $a \to b$ and $b \to a$ are related by [see eq. (2-21)]:

$$\mu_a q_a^2 \sigma_{ba} = \mu_b q_b^2 \sigma_{ab} \tag{2-129}$$

where μ_a and μ_b are the spin multiplicites of the states a and b. This relation is known as the principle of detailed balance, for which time reversal invariance is a sufficient but not a necessary condition. Applying eq. (2-129) to the reaction (d), and remembering that the photon has two helicity states (see below), so that the multiplicity of the initial states is 4 and of the final states is 2, we find that:

$$2\sigma(\gamma n \to \pi^- p) = \frac{q^2}{q_\gamma^2} \sigma(\pi^- p \to \gamma n) \tag{2-130}$$

where q and q_γ are the final and initial momenta in the center of mass system.

The two quantities that may be measured at threshold energies are the Panofsky ratio P, that is the ratio of charge exchange scattering to radiative capture:

$$P = \frac{\sigma(\pi^- p \to \pi^0 n_0)}{\sigma(\pi^- p \to \gamma n)} \tag{2-131}$$

and the ratio R of reaction (d) to reaction (b):

$$R = \frac{\sigma(\gamma n \to \pi^- p)}{\sigma(\gamma p \to \pi^+ n)} \tag{2-132}$$

By stopping π^- mesons in hydrogen, the π^0 mesons and γ rays, resulting from capture from the $\pi^- p$ atom formed, may be observed, and the ratio P calculated. The ratio R may be calculated from measurement of photoproduction at deuterons.

From eqs. (2-130), (2-131), and (2-132), the cross section for charge exchange can be expressed in terms of R, P, and the cross section for reaction (b):

2.4. PHOTOPRODUCTION OF π MESONS FROM NUCLEONS

$$\sigma(\pi^- p \to \pi^0 n) = 2PR \left(\frac{q_\gamma^2}{q^2}\right) \sigma(\gamma p \to \pi^+ n) . \qquad (2\text{-}133)$$

The quantities R, P, and $\sigma(\gamma p \to n\pi^+)$ may all be measured at energies corresponding to threshold in the $\pi + N$ system. From the values obtained $(P = 1.52, R = 1.3, q^{-1}\sigma(\gamma p \to \pi^+ n) = 0.010/M_\pi^3)$, the scattering length combination $\left(a_{0+}^{\frac{1}{2}} - a_{0+}^{\frac{3}{2}}\right)$, which determines charge exchange scattering, is calculated to have the value $0.255\ \hbar/m_\pi c$). This value is somewhat lower than those arising from the direct analysis of the low energy πN scattering experiments (0.266 to 0.291), but the difference is probably due to the uncertainties in the corrections required in calculating P and R from the raw data.

Two-Body Scattering in the Helicity Representation

To discuss the angular distribution of the photoproduced pion it is most convenient to work in the helicity representation,[12] introduced by Jacob and Wick (1959). The helicity of a particle is defined as the component of the angular momentum in the direction of motion of the particle. The helicity operator Λ can be written as a scalar product:

$$\Lambda = \frac{\vec{J} \cdot \vec{p}}{|\vec{p}|} \qquad (2\text{-}134)$$

and is an invariant under Lorentz transformation in the direction of $\vec{p}$, the momentum of the particle. The orbital angular momentum of a particle has no component in the direction of motion, so that the eigenvalues λ of Λ, corresponding to single particle eigenstates $|\vec{p}; \lambda>$, can be interpreted as the components of the spin in the direction of $\vec{p}$. For a particle of spin s there are $(2s + 1)$ possible values of λ with $-s \leq \lambda \leq s$, except for massless particles, in which case the only possible values of λ are $\pm s$.

---

[12] An excellent account of the helicity formalism has been given by Martin and Spearman (1970).

Now consider scattering from a two-body state a, consisting of particles 1 and 2, masses M_1, M_2, helicities λ_1, λ_2, to a final two-body state b, consisting of particles 3 and 4, masses M_3, M_4 and helicities λ_3, λ_4. The center of mass momentum in the initial state is $\vec{q}_a(q_a, \theta_a, \phi_a)$ and in the final state $\vec{q}_b(q_b, \theta_b, \phi_b)$. The differential cross section for the collision is again given by eq. (2-23), when T_{ba} the transition matrix is defined as:

$$2iT_{ba} = <\vec{q}_b; \lambda_3, \lambda_4 |S-1| \vec{q}_a; \lambda_1, \lambda_2> . \tag{2-135}$$

A partial wave expansion of T_{ba} can be obtained by introducing a complete set of eigenstates of the total angular momentum j in analogy to the procedure followed in eq. (2-36). The scattering amplitude for scattering into the direction (θ, ϕ), defined as in eq. (2-24):

$$f_{\lambda_3\lambda_4;\lambda_1\lambda_2}(\theta,\phi) = \frac{4\pi}{W} \sqrt{\frac{\omega_b}{\omega_a}} T_{ba}$$

is represented by the series:

$$f_{\lambda_3\lambda_4;\lambda_1\lambda_2}(\theta,\phi) = \frac{1}{2i\sqrt{q_a q_b}} \sqrt{\frac{\omega_b}{\omega_a}} \sum_j (2j+1) <\lambda_3, \lambda_4|(S^j-1)|\lambda_1, \lambda_2> e^{i(\lambda-\mu)\phi} d^j_{\lambda,\mu} \tag{2-136}$$

where $\lambda = \lambda_1 - \lambda_2$, $\mu = \lambda_3 - \lambda_4$ and functions $d^j_{\lambda,\mu}(\theta)$ are defined in Appendix A. One advantage of the helicity formalism is that one summation occurs in eq. (2-136), compared with the multiple sums of eq. (2-50). The functions $d^j_{\lambda,\mu}(\theta)$ are somewhat more complicated than the spherical harmonics appearing in the usual partial wave expansion, but this is a minor difficulty.

A number of symmetry relations are satisfied by the matrix elements of the reduced operator S^j. If the interaction is parity conserving, it can be shown that

$$<-\lambda_3, -\lambda_4 |S^j| -\lambda_1, -\lambda_2> = \frac{\eta_3 \eta_4}{\eta_1 \eta_2} (-1)^{s_3+s_4-s_1-s_2} <\lambda_3, \lambda_4 |S^j| \lambda_1, \lambda_2> \tag{2-137}$$

2.4. PHOTOPRODUCTION OF π MESONS FROM NUCLEONS

where η_i is the intrinsic parity and s_i the spin of the ith particle, and time reversal invariance implies that the S-matrix is symmetrical:

$$\langle \lambda_3, \lambda_4 | S^j | \lambda_1, \lambda_2 \rangle = \langle \lambda_1, \lambda_2 | S^j | \lambda_3, \lambda_4 \rangle . \tag{2-138}$$

We particularize to *elastic scattering* where particles 2 and 4 are of spin 0 and particles 1 and 3 are of spin $\frac{1}{2}$. Then $\lambda_2 = \lambda_4 = 0$ and $\lambda_1 = \lambda_3 = \pm \frac{1}{2}$.

The four possible partial wave amplitudes:

$$f^j_{\lambda_3, \lambda_1}(\theta, \phi) = \frac{1}{2q_a i} \langle \lambda_3 | S^j - 1 | \lambda_1 \rangle \tag{2-139}$$

are reduced to two, by eq. (2-137), if the interaction conserves parity. The two independent elements of the matrix $\langle \lambda_3 | S^j | \lambda_1 \rangle$ will be labeled by subscripts $++$ or $+-$ so that:

$$\langle \tfrac{1}{2} | S^j | \tfrac{1}{2} \rangle = \langle -\tfrac{1}{2} | S^j | -\tfrac{1}{2} \rangle = S^j_{++} = S^j_{--}$$

$$\langle \tfrac{1}{2} | S^j | -\tfrac{1}{2} \rangle = \langle -\tfrac{1}{2} | S^j | +\tfrac{1}{2} \rangle = S^j_{+-} = S^j_{-+} . \tag{2-140}$$

In the same way the independent elements of $f^j_{\lambda_1, \lambda_3}$ are f^j_{++} and f^j_{+-}. The 2×2 S-matrix can be diagonalized by introducing states:

$$|j, m; \pm\rangle = |j, m; \tfrac{1}{2}\rangle \pm |j, m; -\tfrac{1}{2}\rangle . \tag{2-141}$$

The parity of the states $|j, m; \pm\rangle$ is $(-1)^{j+\tfrac{1}{2}}$, where we have not taken into account the intrinsic parities of the colliding particles. In our previous representation where j, ℓ, and m were diagonal, the states with $j = \ell + \tfrac{1}{2}$ had parity $(-1)^{j-\tfrac{1}{2}}$ and those with $j = \ell - \tfrac{1}{2}$ had parity $(-1)^{j+\tfrac{1}{2}}$. This allows us to identify $|j, m; +\rangle$ with a state for which $\ell = j + \tfrac{1}{2}$ and $|j, m; -\rangle$ with a state for which $\ell' = j - \tfrac{1}{2} = \ell + 1$. The diagonal elements of S^j are then $\exp(2i\delta_{\ell+})$ and $\exp(2i\delta_{(\ell+1)-})$, where the phase shifts $\delta_{\ell\pm}$ are those previously introduced; then we have that:

$$S^j_{++} = \frac{1}{2}\left(\exp(2i\delta_{\ell+}) + \exp(2i\delta_{(\ell+1)-})\right), \quad S^j_{+-} = \frac{1}{2}\left(\exp(2i\delta_{\ell+}) - \exp(2i\delta_{(\ell+1)-})\right).$$

(2-142)

Using the explicit forms of $d^j_{\frac{1}{2}\frac{1}{2}}$ (Appendix A), the connection between the scattering amplitudes f_{++}, f_{+-} and the amplitudes f and g can be shown to be:

$$f_{++}(\theta,\phi) = \cos(\theta/2) f(\theta) + g(\theta) \sin(\theta/2)$$

$$f_{+-}(\theta,\phi) = \{\sin(\theta/2) f(\theta) - g(\theta) \cos(\theta/2)\} e^{i\phi} .$$

(2-143)

Photoproduction

As a second example of the helicity formalism, consider the photoproduction of a π^+ meson from a proton:

$$\gamma + p \to \pi^+ + n .$$

The scattering amplitude is given by eq. (2-136). If particle 1 is the proton and 2 is the photon then λ_2 can take the values ± 1, and λ_1 the values $\pm\frac{1}{2}$, so that the initial state can be labeled uniquely by $\lambda = \lambda_1 - \lambda_2$. The final state can be labeled by the helicity of the neutron $\mu = \pm\frac{1}{2}$. The intrinsic parities of the photon and the meson are each -1 and the spins are $s_2 = 1$, $s_1 = \frac{1}{2}$, $s_3 = 0$, $s_4 = \frac{1}{2}$, so that, from eq. (2-137):

$$<-\mu|S^j|-\lambda> = -<\mu|S^j|\lambda> .$$

The matrix elements:

$$f^j_{\mu,\lambda} = \frac{1}{2i\sqrt{q_a q_b}} \sqrt{\frac{\omega_b}{\omega_a}} \{<\mu|S^j|\lambda> - \delta_{\lambda\mu}\}$$

(2-144)

are of the form:

2.4. PHOTOPRODUCTION OF π MESONS FROM NUCLEONS

$$f^j_{\mu,\lambda} = \begin{array}{c|cccc} \mu \backslash \lambda & \frac{3}{2} & \frac{1}{2} & -\frac{1}{2} & -\frac{3}{2} \\ \frac{1}{2} & C^j & A^j & B^j & D^j \\ -\frac{1}{2} & -D^j & -B^j & -A^j & -C^j \end{array} \qquad (2\text{-}145)$$

An alternative description of photoproduction is given by expanding the incoming photon wave into eigenstates of the total angular momentum of the photon, L. These states may be divided into those with parity

TABLE 2.1. The contribution of the multipoles EL and ML to photoproduction.

Photon State	Total Angular Momentum j	Parity	The final π-N state
$E1$	$\frac{1}{2}$	−	$s_{\frac{1}{2}}$
	$\frac{3}{2}$	−	$d_{\frac{3}{2}}$
$M1$	$\frac{1}{2}$	+	$p_{\frac{1}{2}}$
	$\frac{3}{2}$	+	$p_{\frac{3}{2}}$
$E2$	$\frac{3}{2}$	+	$p_{\frac{3}{2}}$
	$\frac{5}{2}$	+	$f_{\frac{5}{2}}$
$M2$	$\frac{3}{2}$	−	$d_{\frac{3}{2}}$
	$\frac{5}{2}$	−	$d_{\frac{5}{2}}$
$E3$	$\frac{5}{2}$	−	$d_{\frac{5}{2}}$
	$\frac{7}{2}$	−	$g_{\frac{7}{2}}$
$M3$	$\frac{5}{2}$	+	$f_{\frac{5}{2}}$
	$\frac{7}{2}$	+	$f_{\frac{7}{2}}$

$(-1)^{L+1}$, which are the magnetic multipole states ML, and those with parity $(-1)^L$, which are the electric multipole states EL; and as the photon has spin 1, the minimum value of L is 1. The possible transitions from the electric and magnetic multipole states conserving both j and the parity are shown in Table 2-1. For the final (meson + nucleon) state, the spectroscopic notation $s, p, d \ldots$ is used for the $\ell = 0, 1, 2$ partial waves, and the subscripts indicate the value of j.

The connection between the helicity states for the photon-nucleon system $|j, m; \lambda\rangle$ and the states for which the total photon angular momentum L and the photon orbital angular momentum ℓ' are diagonal, $|j, m; L, \ell'\rangle$, may be made by combining the photon spin with the photon orbital angular momentum to form L, and then combining L with the nucleon spin to form j. The result is the following:

$$|j,m;\lambda\rangle = \sum_{L,\ell'} \sqrt{\frac{(2\ell'+1)}{(2j+1)}} \, C_{L\frac{1}{2}}(j,\lambda;\lambda_1,\lambda_2) C_{\ell'1}(L,\lambda_1;0,\lambda_1)|j,m;L;\ell'\rangle. \quad (2\text{-}146)$$

Using the explicit values of $C_{\ell'1}(L,\lambda_1;0,\lambda)$ from Table A-2, Appendix A, in eq. (2-146), we find that:

$$|j,m;\lambda\rangle = \sum_L \frac{[L(L+1)(2L+1)]^{\frac{1}{2}}}{\sqrt{2}\,(2j+1)} C_{L\frac{1}{2}}(j,\lambda;\lambda_1,\lambda_2) - \left[\frac{\lambda_1}{\sqrt{L}\sqrt{L+1}}|j,m;L;\ell'=L\rangle\right.$$
$$+ \frac{1}{\sqrt{L+1}\sqrt{2L+1}}|j,m,L;\ell=L+1\rangle + \frac{1}{\sqrt{L}\sqrt{2L+1}}(j,m,L;\ell=L-1\rangle] .$$
$$(2\text{-}147)$$

The intrinsic parity of the photon is -1, so that the parity of the state $|j,m;L;\ell'=L\rangle$ is $(-1)^{L+1}$; thus this state may be identified as a magnetic multipole state which will be denoted by $|j,m;ML\rangle$. The parity of each term in the curly brackets is $(-1)^L$ so these are electric multipole states $|j,m;EL\rangle$. The definition of the states $|j,m;ML\rangle$ and $|j,m;EL\rangle$ with a conventional normalization is:

$$|j,m;ML\rangle \equiv \frac{1}{\sqrt{L}}\frac{1}{\sqrt{L+1}}|j,m;L;\ell'=L\rangle \quad (2\text{-}148)$$

2.4. PHOTOPRODUCTION OF π MESONS FROM NUCLEONS

and:

$$|j,m; EL\rangle = \frac{1}{\sqrt{2L+1}} \left[\frac{1}{\sqrt{L+1}} |j,m; L; \ell = L+1\rangle + \frac{1}{\sqrt{L}} |j,m; L; \ell = L-1\rangle \right].$$

For the final π-N state, the helicity states $|j,m; \lambda\rangle$ can be expressed in terms of states $|j,m; \ell\rangle$ in which the orbital angular momentum ℓ is diagonal:

$$|j,m; \lambda\rangle = \sum_\ell \sqrt{\frac{(2\ell+1)}{(2j+1)}} C_{\ell \frac{1}{2}}(j,\lambda; 0,\lambda) |j,m; \ell\rangle . \tag{2-149}$$

From eqs. (2-144), (2-148) and (2-149), using the expressions for the Clebsch-Gordon coefficients in Table A, Appendix A-1, it is found that:

$$f_{\frac{1}{2},\frac{3}{2}}^{\ell+\frac{1}{2}} = C^{\ell+\frac{1}{2}} = \frac{1}{\sqrt{2}} \sqrt{\ell(\ell+2)} \, [E_{\ell+} - M_{\ell+} - E_{(\ell+1)-} - M_{(\ell+1)-}] \quad (a)$$

$$f_{\frac{1}{2},\frac{1}{2}}^{\ell+\frac{1}{2}} = B^{\ell+\frac{1}{2}} = \frac{1}{\sqrt{2}} [(\ell+2)(M_{(\ell+1)-} - E_{\ell+}) - \ell(M_{\ell+} + E_{(\ell+1)-})] \quad (b)$$

$$f_{\frac{1}{2},-\frac{1}{2}}^{\ell+\frac{1}{2}} = A^{\ell+\frac{1}{2}} = \frac{1}{\sqrt{2}} [\ell(M_{\ell+} - E_{(\ell+1)-}) + (\ell+2)(M_{(\ell+1)-} + E_{\ell+})] \quad (c)$$

$$f_{\frac{1}{2},-\frac{3}{2}}^{\ell-\frac{1}{2}} = D^{\ell+\frac{1}{2}} = \frac{1}{\sqrt{2}} \sqrt{\ell(\ell+2)} \, [M_{\ell+} - E_{\ell+} - E_{(\ell+1)-} - M_{(\ell+1)-}] \quad (d) \tag{2-150}$$

In these equations the magnetic and electric multipole transition amplitudes $M_{\ell\pm}$, $E_{\ell\pm}$ are labeled by the orbital angular momentum of the final state ℓ, the $\pm$ signs having the same significance as in eqs. (2-51) and (2-52), that $j = \ell \pm \frac{1}{2}$.

Angular Distributions

The angular distribution in the center of mass system for an unpolarized beam and target is given by:

$$\frac{d\sigma}{d\Omega} = \frac{q_b \omega_a}{2q_a \omega_b} \sum_\lambda \left| f^j_{\frac{1}{2},\lambda}(\theta,\phi) \right|^2 \quad \text{(a)}$$

$$= \frac{q_b \omega_a}{2q_a \omega_b} \sum_\lambda \left| \sum_j (2j+1) f^j_{\frac{1}{2},\lambda} d^j_{\lambda \frac{1}{2}}(\theta) \right|^2 \quad \text{(b)} \quad (2\text{-}151)$$

where ω_a, ω_b are initial, final reduced energies. The helicity amplitudes $f^j_{\frac{1}{2}\lambda}$ are related to the multipole amplitude by eqs. (2-150). As different amplitudes are required for each of the reactions (a) and (d), fourteen complex numbers are needed to define all the reactions for each value of j. It is expected that a multipole amplitude $E_{\ell\pm}$ or $M_{\ell\pm}$, leading to a final state b with orbital angular momentum ℓ, will be proportional to q_b^ℓ near threshold; and it follows that, for low energies, only multipoles with small values of ℓ will be important. Retaining the terms with $\ell = 0$ and $\ell = 1$, only, and using the expressions for $d^j_{\lambda\mu}$ given in Appendix A, we find that:

$$f_{\frac{1}{2},\frac{3}{2}}(\theta,\phi) = -\frac{3}{\sqrt{2}}(E_{1+} - M_{1+}) \sin(\theta) \cos(\theta/2) \, e^{i\phi}$$

$$f_{\frac{1}{2},\frac{1}{2}}(\theta,\phi) = \frac{1}{\sqrt{2}} [2(M_{1-} + E_{0-}) + (M_{1+} + 3M_{1+})(-1 + 3\cos(\theta))] \cos(\theta/2)$$

$$f_{\frac{1}{2},-\frac{1}{2}}(\theta,\phi) = \frac{1}{\sqrt{2}} [2(M_{1-} - E_{0+}) - (M_{1+} + 3E_{1+})(1 + 3\cos(\theta))] \sin(\theta/2) \, e^{-i\phi}$$

$$f_{\frac{1}{2},-\frac{3}{2}}(\theta,\phi) = -\frac{3}{\sqrt{2}}(E_{1+} - M_{1+}) \sin(\theta) \sin(\theta/2) \, e^{-2i\phi} . \quad (2\text{-}152)$$

Examining terms for large values of ℓ, one can easily see that the partial cross section arising from a single multipole amplitude has the property that the angular distribution depends only on j and L, the total angular momentum and multipole order. For example, both dipole terms E_{0+} and M_{1-} lead to a constant angular distribution, the dipoles with $j = \frac{3}{2}$, E_{2-}, and M_{1+}, to a distribution proportional to $(2 + 3\sin^2\theta)$,

2.4. PHOTOPRODUCTION OF π MESONS FROM NUCLEONS

and the quadrupole terms, E_{1+} and M_{2-}, to a distribution proportional to $(1 + \cos^2\theta)$. At energies close to threshold, the cross section is observed to be isotropic, and as an s-wave final state is expected, the transition must be due to the electric dipole E_{0+}. The static electric dipole for the $\pi^0 p$ system is smaller than that for the $\pi^+ n$ system (in the center of mass system) by a factor (M_π/M_N), and it is expected that π^0 production from the γp reaction will be very small near threshold. This is confirmed by experiments (see Fig. 2-8). Near the resonance energy, the distributions for both π^0 and π^+ production approximated to $(2 + 3\sin^2\theta)$ so that it is concluded that the magnetic dipole M_{1+-}, leading to a p_{33} final state is the dominant term.

The experiments may be analyzed in terms of the multipole amplitudes in much the same way as phase shift analyses for πN scattering are carried out, and the ambiguities that arise may be eliminated in a similar way by measurements of the polarization of the recoil nucleon or by experiments with polarized photons. The number of parameters required at each energy can be reduced by relating the phase of each photoproduction amplitude to the πN phase shifts in the final state. As a preliminary to this we shall consider the reactions in terms of an isotopic spin formalism.

The Electromagnetic Interaction and Isotopic Spin

The electromagnetic interaction does not conserve isotopic spin, but nevertheless behaves in a well-defined manner under an isotopic spin transformation.

The interaction between a charged particle and the electromagnetic field is found by making the replacement (Jauch and Rohrlich, 1955):

$$p^\mu \to p^\mu + Q \cdot A^\mu \qquad (2\text{-}153)$$

in the Hamiltonian for the system,[13] where A^μ is the four-vector potential and p^μ is the four momentum. The operator Q is the charge operator that has the form:

---

[13] This is the assumption of "minimal electromagnetic coupling." Anomalous magnetic moment interactions are supposed to arise indirectly through the strong interactions.

$$Q = \frac{1}{2}(1 + I_3)\, e \qquad \text{(a)}$$

for a nucleon, or:

$$Q = I_3\, e \qquad \text{(b)} \qquad (2\text{-}154)$$

for a π meson, where I_3 is the third component of the isotopic spin. It follows that if the terms quadratic in A^μ are neglected,[14] the transition operator T will be the sum of a term T^S transforming like an isotopic scalar and a term T^{V_3}, that transforms like the third component of an isotopic vector.[15]

The initial γN state which is of isotopic spin $\frac{1}{2}$, can be written $|I_3\rangle$ where $I_3 = +\frac{1}{2}$ for the proton and $-\frac{1}{2}$ for the neutron state. The isotopic spin eigenvectors for the final state, $|I, I_3\rangle$, with $I = \frac{1}{2}$ or $\frac{3}{2}$, are connected with the physical states by eq. (2-84), supplemented by the relations:

$$|\pi^0 + p\rangle = \sqrt{\frac{2}{3}}\,|\tfrac{3}{2},\tfrac{1}{2}\rangle + \frac{1}{\sqrt{3}}\,|\tfrac{1}{2},\tfrac{1}{2}\rangle$$

$$|\pi^+ + n\rangle = \sqrt{\frac{1}{3}}\,|\tfrac{3}{2},\tfrac{1}{2}\rangle - \sqrt{\frac{2}{3}}\,|\tfrac{1}{2},\tfrac{1}{2}\rangle\,. \qquad (2\text{-}155)$$

The scalar operator T^S has matrix elements:

$$\langle I, I_3 | T^S | I_3 \rangle = \delta_{I,\frac{1}{2}} \frac{1}{\sqrt{3}}\, t^0 \qquad (2\text{-}156)$$

where the normalization factor $(1/\sqrt{3})$ is chosen purely for convenience. The vector matrix element can be obtained by using the Wigner-Eckart theorem (see Edmonds, 1957):

---

[14] Transition matrix elements linear in A^μ are proportional to e/c so that those quadratic in A^μ are expected to be smaller by an order of magnitude.

[15] It is difficult to completely eliminate the possibility that terms transforming like a second-rank isotopic spin tensor might also appear, giving rise to transitions in which the isotopic spin changes by two units, $\Delta I = 2$. For a discussion of this possibility see Dombey and Kabir (1966).

2.4. PHOTOPRODUCTION OF π MESONS FROM NUCLEONS

$$<I, I_3 |T^{V_3}| I_3> = C_{\frac{1}{2} 1}(I, I_3; I_3, 0) t^{2I} . \qquad (2\text{-}157)$$

The initial and final values of the third component of the isotopic spin must be identical in order to conserve charge. The physical amplitudes found from eqs. (2-155), (2-156), and (2-157) are as follows:

$$(\gamma + p |T| \pi^0 + p) = \tfrac{2}{3} t^3 - \tfrac{1}{3} t^1 + \tfrac{1}{3} t^0$$

$$(\gamma + p |T| \pi^+ + n) = \sqrt{\tfrac{2}{3}} t^3 + \sqrt{\tfrac{2}{3}} t^1 - \sqrt{\tfrac{2}{3}} t^0$$

$$(\gamma + n |T| \pi^- + p) = \sqrt{\tfrac{2}{3}} t^3 - \sqrt{\tfrac{2}{3}} t^1 - \sqrt{\tfrac{2}{3}} t^0$$

$$(\gamma + n |T| \pi^0 + n) = \tfrac{2}{3} t^3 + \tfrac{1}{3} t^1 + \tfrac{1}{3} t^0 \qquad (2\text{-}158)$$

where T and t^{2I} stand for any of the helicity or multipole amplitudes. For example, each of the multipole amplitudes $E_{\ell\pm}$, $M_{\ell\pm}$ has components $E^{2I}_{\ell\pm}$, $M^{2I}_{\ell\pm}$, and the correct combinations of these to be used in eqs. (2-152), and which determine the angular distributions, are given by eqs. (2-158) for each of the four possible physical processes. The multipoles $E^{0,1}_{\ell\pm}$ $M^{0,1}_{\ell\pm}$ lead to a final state with $I = \tfrac{1}{2}$, and $E^3_{\ell\pm}$, $M^3_{\ell\pm}$ to a state with $I = \tfrac{3}{2}$. The four reactions are determined by only three amplitudes t^i for each helicity state, but if only the γp reactions are considered the data determine just the two combinations t^3 and $(t^1 + 3t^0)$.

As the electromagnetic interaction is weak compared with the strong nuclear intereaction, it may be assumed that the γ-N channel does not contribute to πN elastic scattering, and further that the transition matrix elements for elastic γN scattering are much smaller than those for photoproduction. With these assumptions the reaction matrix, which is real and symmetric, for the coupled γ-N and π-N channels, (for given values of j and the parity) takes the form:

$$K_{\ell\pm} = \begin{pmatrix} & & E & M & | & I=\frac{1}{2} & I=\frac{3}{2} \\ & E & 0 & 0 & | & X^{\frac{1}{2}} & X^{\frac{3}{2}} \\ \gamma\text{-}N & & & & | & & \\ & M & 0 & 0 & | & Y^{\frac{1}{2}} & Y^{\frac{3}{2}} \\ & & -- & -- & | & -- & -- \\ I=\frac{1}{2} & & X^{\frac{1}{2}} & Y^{\frac{1}{2}} & | & \tan\delta^{\frac{1}{2}}_{\ell\pm} & 0 \\ \pi\text{-}N & & & & | & & \\ I=\frac{3}{2} & & X^{\frac{3}{2}} & Y^{\frac{3}{2}} & | & 0 & \tan\delta^{\frac{3}{2}}_{\ell\pm} \end{pmatrix} \quad (2\text{-}159)$$

The submatrix connecting the two π-N channels is diagonal in the isotopic spin and the diagonal elements are the tangents of the (real) phase shifts. The matrix elements X^I and Y^I connect the two isotopic spin states of the πN system, with electric and magnetic multipole photon states respectively. To first order in X^I and Y^I, the transition matrix elements calculated from:

$$T = \frac{2K}{(1-iK)}$$

are found to be:

$$E_{\ell\pm}(I) = X^I \cos\delta^I_{\ell\pm} \, e^{i\delta^{\pm}_{\ell\pm}}$$

$$M_{\ell\pm}(I) = Y^I \cos\delta^I_{\ell\pm} \, e^{i\delta^{\pm}_{\ell\pm}} \quad (2\text{-}160)$$

where $E_{\ell\pm}(I)$ and $M_{\ell\pm}(I)$ are those parts of the multipole amplitudes that lead to a final state of isotopic spin I. These have already been identified, for we found that $E^{0,1}_{\ell\pm}$ and $M^{0,1}_{\ell\pm}$ corresponded to transitions to a final state with $I = \frac{1}{2}$ and $E^3_{\ell\pm}$, $M^3_{\ell\pm}$ to transitions to a final state with $I = \frac{3}{2}$. It follows that $E^{0,1}_{\ell\pm}$, $M^{0,1}_{\ell\pm}$ have the phase $\delta^{\frac{1}{2}}_{\ell\pm}$ so that:

2.4. PHOTOPRODUCTION OF π MESONS FROM NUCLEONS

$$E_{\ell\pm}^{0,1} = |E_{\ell\pm}^{0,1}| \exp\left(i\delta_{\ell\pm}^{\frac{1}{2}}\right), \quad M_{\ell\pm}^{0,1} = |M_{\ell\pm}^{0,1}| \exp\left(i\delta_{\ell\pm}^{\frac{1}{2}}\right) \quad (2\text{-}161)$$

while $E_{\ell\pm}^{3}$, $M_{\ell\pm}^{3}$ have the phase $\delta_{\ell\pm}^{\frac{3}{2}}$:

$$E_{\ell\pm}^{3} = |E_{\ell\pm}^{3}| \exp\left(i\delta_{\ell\pm}^{\frac{3}{2}}\right), \quad M_{\ell\pm}^{3} = |M_{\ell\pm}^{3}| \exp\left(i\delta_{\ell\pm}^{\frac{3}{2}}\right). \quad (2\text{-}162)$$

If the phase shifts $\delta_{\ell\pm}^{I}$ are known from an analysis of π-N scattering data, the number of parameters required to describe photoproduction is reduced by one-half. Conversely, the analysis of photoproduction could be used as an important alternative method of identifying the π-N resonant states *if* the photoproduction data were good enough.

The photoproduction data over the region of the first π-N resonance are not quite plentiful enough for a quantitative and detailed analysis to be performed without the help of some extra theory. The total cross sections at low energies for $\gamma + p \to \pi^0 + p$ and $\gamma + p \to \pi^+ + n$ are illustrated in Fig. 2-7. The most important multipoles have been determined using fixed momentum transfer dispersion relations and are illustrated in Fig. 2-8. With the large multipoles $(M_1 + ...)$ determined, the smaller multipoles can be given a parametric form, the parameters being determined by a least squares fit to the data. Donnachie and Shaw (1967) have in this way determined $(E_{0+}^{1} + 3E_{0+}^{0})$, E_{0+}^{3}, $(M_{1-}^{1} + 3M_{1-}^{0})$, and E_{1+}^{3} from the proton data. Perhaps the most interesting result obtained is that the multipole E_{1+}^{3} vanishes at the energy of the p_{33} resonance in the π-N system. This has been predicted by certain symmetry schemes and in the nonrelativistic quark model. This point will be discussed when these schemes are considered in Chapter 6.

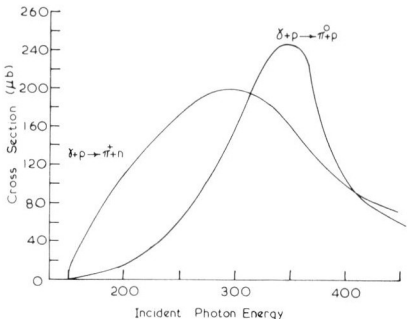

Fig. 2-7. Total cross sections for photoproduction of π^+ and π^0 mesons from protons at energies below 450 MeV.

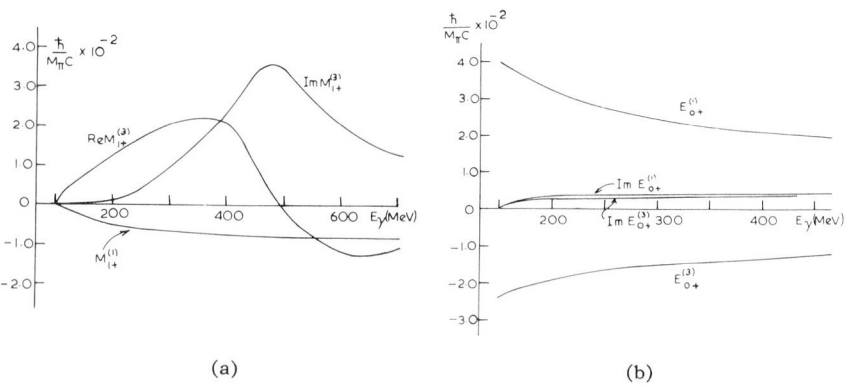

Fig. 2-8. (a) The $M_{1+}^{(1)}$ and $M_{1+}^{(3)}$ transition amplitudes; (b) the $E_{0+}^{(1)}$ and $E_{0+}^{(3)}$ transition amplitudes. (After Donnachie and Shaw, 1966.)

Since 1967, the steady accumulation of data below 500 MeV has led to further multipole analyses, which have confirmed the general conclusions of Donnachie and Shaw and which have allowed the determination of the multipoles with greater precision. Simultaneous fits to data at several energies, using energy-dependent parametrizations of the amplitudes have been made by Walker (1969), by Moorhouse and Rankin (1970), and by Noelle and Pfeil (1971). In contrast, Berends and Weaver (1971) have fitted the data[13] at each energy separately, using the real parts of the multipole amplitudes as parameters and taking the imaginary parts to be consistent with the known πN phase shifts. However in order to have sufficient information to determine a fit, it was found necessary to fix the multipole for the higher partial waves to the values determined from dispersion theory. A similar analysis has also been performed by Noelle et al. (1971).

REFERENCES

H. L. Anderson, W. C. Davidson, M. Gickaman and E. U. Kruse, Phys. Rev. 100, 279 (1955).

R. A. Arndt and M. H. MacGregor, in *Methods of Computational Physics* (Academic Press, New York, 1967), Vol. 6.

F. A. Berends and D. L. Weaver, Nucl. Phys. B30, 575 (1971).

J. M. Blatt and V. F. Weisskopf, *Theoretical Nuclear Physics* (1952).

B. H. Bransden, in *High Energy Physics*, ed. E. H. S. Burhop (Academic Press, New York, 1969), Vol. III.

---

[13] References to the experimental data can be found in the paper by Berends et al. (1971).

D. V. Bugg, P. J. Bussey, D. R. Dance, A. R. Smith, A. A. Carter and J. R. Williams, Nucl. Phys. B26, 588 (1971).

A. A. Carter, J. R. Williams, D. V. Bugg, P. J. Bussey and D. R. Dance, Nucl. Phys. B26, 445 (1971).

S. R. Deans and G. Holliday, Proc. Williamsburg Conference on Intermediate Energy Physics Vol. II, p. 551 (pub. College of William and Mary, Williamsburg, 1966).

A. Donnachie and G. Shaw, Ann. Phys. (N.Y.) 37, 333 (1966).

N. Dombey and P. K. Kabir, Phys. Rev. Letters 17, 730 (1966).

R. A. Donald, W. H. Evans, W. Hart, P. Mason, D. E. Plane and J. C. Reader, Proc. Phys. Soc. 87, 445 (1966).

A. R. Edmonds, *Angular Momentum in Quantum Mechanics* (Princeton University Press, Princeton, 1957).

J. Foote, O. Chamberlain, E. Rodgers and H. Steinger, Phys. Rev. 122, 959 (1961).

M. L. Goldberger and K. M. Watson, *Collision Theory* (John Wiley and Sons, 1964).

S. Goldhaber, W. Shinowsky, G. Goldhaber, W. Lee, T. O'Halloran, T. F. Stubbs, G. M. Pjerrou, D. H. Stork and K. K. Ticho, Phys. Rev. Letters 9, 135 (1962).

J. Hamilton, Phys. Letters 20, 687 (1966).

J. Hamilton and W. S. Woolcock, Phys. Rev. 118, 291 (1960).

L. Van Hove, Phys. Rev. 88, 1358 (1952).

M. M. Hull and F. Lin, Phys. Rev. 139, B630 (1965).

M. Jacob and G. C. Wick, Ann. Phys. (N.Y.) 7, 404 (1959).

REFERENCES

J. M. Jauch and F. Rohrlich, *The Theory of Photons and Electrons* (Addison-Wesley, Reading, Mass., 1955).

W. M. Layson, Nuovo Cimento 27, 724 (1963).

A. D. Martin and R. Perrin, Nuc. Phys. B10, 125 (1969).

A. D. Martin and D. Spearman, *Elementary Particles* (North-Holland, Amsterdam, 1970).

A. Messiah, *Quantum Mechanics* (North Holland, Amsterdam, 1964).

R. G. Moorhouse and W. A. Rankin, Nucl. Phys. B23, 181 (1970).

N. F. Mott and H. S. W. Massey, *Theory of Atomic Collisions* (Oxford University Press, London, 1965). 3rd ed.

R. G. Newton, *Scattering of Waves and Particles* (McGraw-Hill Book Co., New York, 1966).

P. Noelle and W. Pfeil, Nucl. Phys. B31, 1 (1921).

P. Noelle, W. Pfeil and D. Schwela, Nucl. Phys. B26, 461 (1971).

E. S. Pearson and H. O. Hartley, *Biometrica Table for Statisticians* (Cambridge University Press, 1954).

J. A. Poirier and M. Pripstein, Phys. Rev. 130, 1171 (1963).

A. Rittenberg, A. Barbaro-Galtieri, T. Lasinski, A. H. Rosenfeld, T. G. Trippe, M. Roos, C. Bricman, P. Söding, N. Barash-Schmidt and C. G. Wohl, Rev. Mod. Physics 43, S1 (1971).

L. D. Roper and R. M. Wright, Phys. Rev. 138, B190 (1965).

L. D. Roper, R. M. Wright and B. T. Feld, Phys. Rev. 138, B190 (1965).

F. T. Solmitz, Phys. Rev. 94, 1799 (1954).

R. L. Walker, Phys. Rev. 182, 1729 (1969).

CHAPTER 3

FORWARD AND FIXED MOMENTUM TRANSFER DISPERSION RELATIONS

3.1. INTRODUCTION

In this chapter, we shall examine the properties of elastic scattering amplitudes when the energy is allowed to take complex values. It will be shown that the scattering amplitude is an analytic function of the energy, apart from certain cuts and poles, and that this analyticity enables us to write integral relations between the real and imaginary parts of the scattering amplitude, called dispersion relations, that prove to be most powerful aids to the analysis and understanding of experimental data. We shall first discuss briefly some properties of a function $f(\nu)$ of a complex variable ν, that is analytic in the upper half plane, $\text{Im}\,\nu > 0$. By applying Cauchy's theorem we have:

$$f(\nu) = \frac{1}{2\pi i} \oint_C \frac{f(\nu')}{(\nu'-\nu)}\, d\nu' \qquad (3\text{-}1)$$

where the integration is along the closed contour shown in Fig. 3-1, consisting of a segment of the real axis from $\nu' = -L$ to $\nu' = +L$ closed by a semicircle of radius L in the upper half plane, and where ν lies within

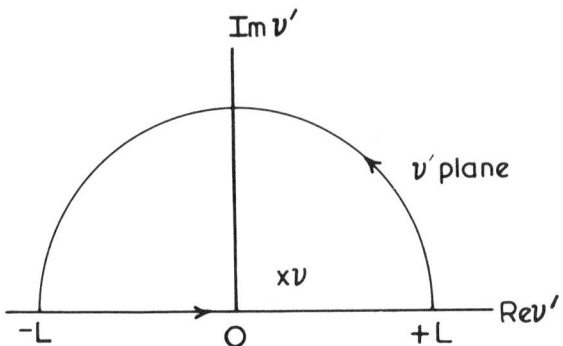

Fig. 3-1. The contour of integration in the complex ν plane for the integral of Eq. (3-1).

the contour C. Provided that $f(\nu)$ vanishes sufficiently rapidly for large $|\nu|$, the semicircle will not contribute to the integral in the limit $L \to \infty$, and:

$$f(\nu) = \frac{1}{2\pi i} \int_{-\infty}^{\infty} \frac{f(\nu')}{(\nu'-\nu)} d\nu' . \qquad (3-2)$$

By writing $\nu \to \nu + i\epsilon$, where ν and ϵ are real, $\epsilon > 0$, and taking the limit $\epsilon \to 0$, we can examine the relation of eq. (3-2) on the real axis. Using the identity:

$$\frac{1}{(\nu'-\nu \mp i\epsilon)} = P \frac{1}{(\nu'-\nu)} \pm i\pi\delta(\nu'-\nu) \qquad (3-3)$$

where P denotes that a principal value integral is to be taken, we obtain:

$$f(\nu) = \frac{1}{i\pi} P \int_{-\infty}^{\infty} \frac{f(\nu')}{(\nu'-\nu)} d\nu' . \qquad (3-4)$$

Taking real and imaginary parts of eq. (3-4), the following dispersion relations[1] known as Hilbert transforms, are found:

---

[1] The mathematical properties of Hilbert transforms are given by Titchmarsh (1948).

3.1. INTRODUCTION

$$\text{Re } f(\nu) = \frac{1}{\pi} P \int_{-\infty}^{\infty} \frac{\text{Im } f(\nu')}{(\nu'-\nu)} d\nu' \quad \text{(a)}$$

$$\text{Im } f(\nu) = -\frac{1}{\pi} P \int_{-\infty}^{\infty} \frac{\text{Re } f(\nu')}{(\nu'-\nu)} d\nu' \quad \text{(b)}$$

(3-5)

These transforms often occur in classical physics in situations where $f(\nu)$ is the Fourier transform of a function $g(t)$, which vanishes for $t<0$, and we shall see that it is in a similar way that dispersion relations enter scattering theory.

If $\nu f(\nu)$ vanishes as $\nu \to \infty$, then a dispersion relation can also be written for this quantity:

$$\nu \text{ Re } f(\nu) = \frac{1}{\pi} P \int_{-\infty}^{\infty} \frac{\nu' \text{ Im } f(\nu')}{(\nu'-\nu)} d\nu' \quad (\nu \text{ real}) \; . \tag{3-6}$$

Combining eqs. (3-5a) and (3-6), it follows that:

$$\int_{-\infty}^{\infty} \text{Im } f(\nu) \, d\nu = 0 \; . \tag{3-7}$$

In general, if $\nu^n f(\nu)$ vanishes as $\nu \to \infty$, the n moment conditions:

$$\int_{-\infty}^{\infty} \nu^i \text{ Im } f(\nu) \, d\nu = 0, \quad i = 0, 1, 2, \ldots (n-1) \tag{3-8}$$

must be satisfied. These conditions are often called "super convergence" relations.

It should be noted that if $f(\nu)$ does not vanish sufficiently rapidly as $\nu \to \infty$, we can write a dispersion relation for $f(\nu)/(\nu-\nu_s)^n$ where n is sufficiently large. In this case, contributions to the Cauchy integral come from the nth order pole at $\nu = \nu_s$. For example if $n = 1$, we find that:

$$\text{Re } f(\nu) = \text{Re } f(\nu_s) + \frac{1}{\pi}(\nu-\nu_s) P \int_{-\infty}^{\infty} \frac{\text{Im } f(\nu')}{(\nu'-\nu)(\nu'-\nu_s)} d\nu' \; . \tag{3-9}$$

The constant term Re $f(\nu_1)$ is known as a subtraction term because eq. (3-9) can be obtained formally by subtracting Re $f(\nu_1)$ from Re $f(\nu)$, with both functions given by eq. (3-5a).

3.2. FORWARD SCATTERING OF NEUTRAL PARTICLES

The transition matrix for elastic two-particle scattering can be shown to possess certain analyticity properties as a function of energy or of momentum transfer, if the general framework of relativistic field theory is correct. In this section we shall obtain a dispersion relation for the elastic scattering amplitude in the forward direction for the simple case in which neutral spinless mesons of mass μ are scattered by spinless target particles of mass M. The energy-momentum vectors of the incident and scattered mesons will be denoted by k_1 and k_2, while those for the initial and final states of the target particle will be p_1 and p_2. Energy and momentum conservation requires that:

$$p_1 + k_1 = p_2 + k_2 \qquad (3\text{-}10)$$

and we must have:

$$p_1^2 = p_2^2 = M^2, \quad k_1^2 = k_2^2 = \mu^2 \ . \qquad (3\text{-}11)$$

Starting from the postulates of field theory,[2] the invariant transition matrix for elastic scattering can be put into the form:

$$<f|T|i> = (2\pi)^{-3} \frac{i}{4} \int d^4x \int d^4y \ e^{i(k_2 \cdot x - k_1 \cdot y)} [\Box_y + \mu^2]$$

$$<d|\theta(x_0 - y_0)\, [j(x),\, \varphi(y)]|b> \ . \qquad (3\text{-}12)$$

---

[2] Some knowledge of the elements of field theory, as discussed for example by Bjorken and Drell (1965), is necessary to follow this section. Some alternative arguments, that lead to the dispersion relations eqs. (3-27) and (3-33), are given in the following chapter.

3.2. FORWARD SCATTERING OF NEUTRAL PARTICLES

In this expression, $|b>$ and $|d>$ are the state vectors of the initial and final states of the target particle, and $\varphi(y)$ is the field operator of the meson which is connected to a source, or current, $j(y)$ through the inhomogeneous Klein-Gordan equation:

$$[\Box_y + \mu^2]\,\varphi(y) = j(y) \ . \tag{3-13}$$

The matrix element $<f|T|i>$ is proportional to the four-dimensional delta function $\delta^4(p_1 + k_1 - p_2 - k_2)$ [see eq. (2-11)]. This factor can be extracted as follows. Using translational invariance, we find that:

$$<d|\theta(x_0-y_0)[j(x),\varphi(y)]|b> = \exp[i(p_2-p_1)\cdot x]<d|\theta(x_0-y_0)[j(0),\varphi(y-x)]|b> \ . \tag{3-14}$$

Changing the variable of integration by the substitution $y' = y-x$, the integration over x can be performed, with the result:

$$<f|T|i> \equiv \delta^4(p_1 + k_2 - p_2 - k_2)\,T_{fi}$$

where:

$$T_{fi} = \frac{\pi}{2} i \int d^4 y'\, \exp(i\,k_1 \cdot y')(\Box_{y'} + \mu^2)<d|\theta(y'_0)[j(0),\varphi(-y')]|b> \ . \tag{3-15}$$

In order to obtain a retarded commutator containing currents only, the operator $(\Box_{y'} + \mu^2)$ can be carried inside the matrix element giving:

$$(\Box_{y'} + \mu^2)<d|\theta(y'_0)[j(0),\varphi(-y')]|b> = <d|\theta(y'_0)[j(0),j(-y')]|b>$$

$$+\ 2<d|\delta(y'_0)[j(0),\dot{\varphi}(-y')]|b>$$

$$+\ <d|\dot{\delta}(y'_0)[j(0),\varphi(-y')]|b>$$

where we have used $[\Box_{y'} + \mu^2]\varphi = j$, and where a dot denotes differentiation with respect to y'_0. We have also used the result $\dot{\theta}(y'_0) = \delta(y'_0)$. An integration by parts then yields:

$$T_{fi} = \frac{\pi}{2} i \int d^4y \, \exp(i\, k_1 \cdot y) \, \{ <d|\theta(y_0)[j(0), j(-y)]|b>$$

$$+ <d|\delta(y_0)[j(0), \dot{\varphi}(-y)]|b> - i\, k_{1_0} <d|\delta(y_0)[j(0), \varphi(-y\,')]|b>\}. \quad (3\text{-}16)$$

The last two terms in curly brackets contain equal time commutators, [because of the factors $\delta(y_0)$]. In general, in a local theory, the current j(0) depends on $\varphi(0)$, but not on the derivatives $\dot{\varphi}(0)$. It follows that the commutator at equal times of j and φ vanishes:

$$\delta(y_0)[j(0), \varphi(y)] = 0 \quad (3\text{-}17)$$

and that this is an expression of causality. The other equal time commutator in eq. (3-16) does not vanish, but it is expected to be proportional to $i\,\delta^3(\vec{y})$ or derivatives of $i\,\delta^3(\vec{y})$.

On using translational invariance once more, the transition matrix T_{fi} can be written in the symmetrical form:

$$T_{fi} = \frac{\pi}{2} i \int d^4y \, \exp\left\{\frac{1}{2} i(k_1+k_2)\cdot y\right\}\left\{<d\left|\theta(y_0)\left[j\left(\frac{y}{2}\right), j\left(-\frac{y}{2}\right)\right]\right|b>\right.$$

$$\left. + <d\left|\delta(y_0)\left[j\left(\frac{y}{2}\right), \dot{\varphi}\left(-\frac{y}{2}\right)\right]\right|b>\right\}. \quad (3\text{-}18)$$

The expression T_{fi} is a function of the two independent scalars that can be formed from $k_1, k_2; p_1, p_2$, with the restrictions of eq. (3-11). The two scalars may be taken to be any two of s, t, and u where:

$$s = (k_1 + p_1)^2 = (k_2 + p_2)^2$$
$$t = (k_1 - k_2)^2 = (p_1 - p_2)^2 \quad (3\text{-}19)$$
$$u = (k_1 - p_2)^2 = (p_1 - k_2)^2.$$

The variables s, t, and u are not independent but are connected by the relation:

$$s + t + u = 2(M^2 + \mu^2).$$

3.2. FORWARD SCATTERING OF NEUTRAL PARTICLES

In the center of mass system, s is the square of the total center of mass energy W, and t is the negative of the square of the momentum transfer. For our present purposes, it is convenient to use in place of s the related variable ν, where:

$$\nu = \frac{1}{4M}(s-u) . \tag{3-20}$$

In the laboratory system, where the target is at rest ($\vec{p}_1 = 0$), ν is related to the energy ω and the momentum $\vec{q}_L$ of incident meson by:

$$\nu = \omega + \frac{1}{4M} t$$

and:

$$\omega = (|\vec{q}_L|^2 + \mu^2)^{\frac{1}{2}} = T_m + \mu \tag{3-21}$$

where T_m is the laboratory kinetic energy of the meson.

We shall start by studying scattering in the forward direction, for which $k_1 = k_2 = (\vec{q}_L, \omega)$ and $t = 0$, so that ν is equal to ω, the meson energy. In this case, T_{fi} given by eq. (3-8) reduces to:

$$T_{fi} = T(\nu) = \frac{\pi}{2} i \int d^4x \exp\{-i(\vec{q}_L \cdot \vec{x} - \nu x_0)\}$$

$$\times \left\{ <M|\theta(x_0)\left[j\left(\tfrac{x}{2}\right), j\left(-\tfrac{x}{2}\right)\right]|M> \right.$$

$$\left. + <M|\left[j\left(\tfrac{x}{2}\right), \dot{\phi}\left(-\tfrac{x}{2}\right)\right]\delta(x_0)|M> \right\} \tag{3-22}$$

where $|M>$ is the state of a single particle of mass M in the rest frame.

Causality and Analyticity

The transition matrix $T(\nu)$ given by eq. (3-22) is now in a form that is convenient for a starting point for the discussion of its analytic properties as a function of ν. The term containing the equal time commutator can be dealt with easily. As we remarked earlier, the equal time commutator

$\delta(x_0)\left[j\left(\frac{x}{2}\right), \dot{\varphi}\left(-\frac{x}{2}\right)\right]$ is proportional to $i\delta^3(\vec{x})$ or to derivatives of $i\delta^3(x)$. It follows that the contribution of this term to $T(\nu)$ is either a constant or a polynomial in ν, and is thus analytic in ν. We shall omit this term in the following discussion.

The first term in $T(\nu)$ contains the retarded commutator $R\left(j\left(\frac{x}{2}\right), j\left(-\frac{x}{2}\right)\right)$ where:

$$R\left(j\left(\tfrac{x}{2}\right), j\left(-\tfrac{x}{2}\right)\right) \equiv i\,\theta(x_0)\left[j\left(\tfrac{x}{2}\right), j\left(-\tfrac{x}{2}\right)\right] \qquad (3\text{-}23)$$

and the integrand vanishes unless $x_0 > 0$. In addition, causality requires that any two field operators $A(x)$, $B(y)$, must commute if the interval $(x-y)$ is spacelike, because measurements made at points separated by spacelike intervals cannot interfere, since signals cannot pass at velocities greater than that of light. It follows that the retarded commutator, and the integrand in eq. (3-22), vanishes unless $|\vec{x}| < x_0$. The dependence of the integral on ν is only through the exponential factor of the form:

$$\exp\{-i\,\vec{n}\cdot\vec{x}\sqrt{\nu^2 - \mu^2} + i\nu x_0\} \qquad (3\text{-}24)$$

where $\vec{n}$ is the unit vector in the direction of $\vec{q}_L$. With the integration region confirmed to $x_0 > 0$ and $|\vec{x}| < x_0$, it is seen that the integral converges and is an analytic function of ν in the upper half plane, with the exception of the region $|\nu| \leq \mu$. These conditions imply that the forward scattering amplitude for scattering of a massless particle with $\mu = 0$ by a particle of mass M will satisfy a dispersion relation of the type in eq. (3-5a), although subtractions may be necessary (Gell-Mann et al., 1955, 1957). For particles of finite mass μ, it is quite difficult to prove the analyticity of the amplitudes in the region $|\nu| \leq \mu$. The expression $T(\nu)$ is clearly analytic in the upper half plane for imaginary values of the mass, $\mu = i\lambda$, and Bogoliubov (1958) has shown that it is possible to establish the analyticity of $T(\nu)$ by analytic continuation in μ from $i\lambda$ to real values (see also Bremmerman et al., 1958; Lehmann, 1959). The rigorous proof is only possible for certain mass ratios (μ/M), a

3.2. FORWARD SCATTERING OF NEUTRAL PARTICLES

condition which includes the case of πN scattering, but not those of NN or KN scattering. As we shall see, empirically, there is every reason to believe that the forward dispersion relations are valid for these processes, although a formal proof has not been discovered.

The Crossing Relation

To make use of a dispersion relation we need to identify the real and imaginary parts of $T(\nu)$ along the whole of the real axis $-\infty < \nu < +\infty$. Physical scattering occurs in the region $\nu > \mu$, and in this region Re T and Im T are directly related to experimental quantities. The amplitude $\overline{T}(\nu)$ is defined, which describes a reaction in which a meson with momentum $-\vec{q}_L$ and energy $-\nu$, in the laboratory system, is scattered in the forward direction from a target of mass M. This amplitude can be obtained by replacing $\vec{q}_L$ by $-\vec{q}_L$ and ν by $-\nu$ in eq. (3-22). As T_{fi} only depends on ν for forward scattering, we see that:

$$\overline{T}(\nu) = T(-\nu) \tag{3-25}$$

and as the retarded commutator is Hermitian, the crossing relation[3] follows:

$$\overline{T}(\nu) = T(-\nu) = T^*(\nu) . \tag{3-26}$$

It should be noted that this relation, by which the physical region of the amplitude $T(\nu)$, $(\nu < -\mu)$ is connected with the physical region of the amplitude $T(\nu)$, $(\nu > \mu)$, is only meaningful because the analyticity of $T(\nu)$ allows a continuation from one region to the other. Using eq. (3-26) the dispersion relation eq. (3-5a) becomes:

$$\text{Re } T(\nu) = \frac{1}{\pi} P \int_0^\infty \text{Im } T(\nu') \left\{ \frac{1}{(\nu'-\nu)} + \frac{1}{(\nu'+\nu)} \right\}$$

$$= \frac{2}{\pi} P \int_0^\mu \frac{\nu' \text{Im } T(\nu')}{(\nu'^2 - \nu^2)} d\nu' + \frac{2}{\pi} P \int_\mu^\infty \frac{\nu' \text{Im } T(\nu')}{(\nu'^2 - \nu^2)} d\nu'. \tag{3-27}$$

---

[3] The crossing relation for the scattering amplitude given in terms of a time ordered product is slightly different from that given here; it is $\overline{T}(\nu) = T(-\nu) = T(+\nu)$.

The Contributions to the Unphysical Region

The next step is to calculate Im $T(\nu')$ in the unphysical region $0<\nu'<\mu$. The amplitude $T(\nu)$ can be split into parts odd and even in ν by:

$$T(\nu) = D(\nu) + i A(\nu) \qquad (3\text{-}28)$$

where:

$$D(\nu) = D(-\nu) \quad \text{and} \quad A(\nu) = -A(-\nu) .$$

To identify D and A, eq. (3-23) is written in the form:

$$R\left(j\left(\tfrac{x}{2}\right), j\left(-\tfrac{x}{2}\right)\right) = \tfrac{i}{2}\left[j\left(\tfrac{x}{2}\right), j\left(-\tfrac{x}{2}\right)\right] + \tfrac{i}{2}\,\epsilon(x_0)\left[j\left(\tfrac{x}{2}\right), j\left(-\tfrac{x}{2}\right)\right] \qquad (3\text{-}29)$$

where $\epsilon(x_0) = 1$ for $x_0 > 0$ and $\epsilon(x_0) = -1$ for $x_0 < 0$. The first term in the right-hand side of eq. (3-29) is odd, and the second term even, in x, and by making the simultaneous interchanges $(\vec{q}_L, \nu) \to (-\vec{q}_L, -\nu)$ and $(\vec{x}, x_0) \to (-\vec{x}, -x_0)$ in eq. (3-22), it is seen that:

$$D(\nu) = \left(\tfrac{\pi}{4}\right)i\int d^4x \,\exp\{-i(\vec{q}_L\cdot\vec{x}-\nu x_0)\} <M\,|\epsilon(x_0)\left[j\left(\tfrac{x}{2}\right), j\left(-\tfrac{x}{2}\right)\right]|M> \quad (a)$$

$$A(\nu) = \left(\tfrac{\pi}{4}\right)\int d^4x \,\exp\{-i(\vec{q}_L\cdot\vec{x}-\nu x_0)\} <M\,|\left[j\left(\tfrac{x}{2}\right), j\left(-\tfrac{x}{2}\right)\right]|M> . \quad (b)$$

$$(3\text{-}30)$$

By taking the complex conjugates of D and A, followed by the change of variables $(\vec{x}, x_0) \to (-\vec{x}, -x_0)$, we have:

$$D^*(\nu) = D(\nu), \; A^*(\nu) = A(\nu)$$

so that D and A are the real and imaginary parts of $T(\nu)$. To explore the imaginary part of $T(\nu)$ further, a complete set of states is inserted in the commutator in eq. (3-30b):

3.2. FORWARD SCATTERING OF NEUTRAL PARTICLES

$$\text{Im } T(\nu) = A(\nu) = \frac{\pi}{4} \sum_n \int d^4x \, \exp\{-i(\vec{q}_L \cdot \vec{x} - \nu x_0)\}$$

$$\times \left\{ <M|j\left(\tfrac{x}{2}\right)|n><n|j\left(-\tfrac{x}{2}\right)|M> - <M|j\left(-\tfrac{x}{2}\right)|n><n|j\left(\tfrac{x}{2}\right)|M> \right\}. \tag{3-31}$$

By using translational invariance, it is possible to perform the integration over x, giving:

$$\text{Im } T(\nu) = 4\pi^5 \sum_n |<M|j(0)|n>|^2 \left\{ \delta\left(\vec{q}_L - \vec{p}_n\right) \delta\left(\nu + M - p_{n_0}\right) \right.$$

$$\left. - \delta\left(\vec{q}_L + \vec{p}_n\right) \delta\left(\nu - M + p_{n_0}\right) \right\} \tag{3-32}$$

where p_n is the total energy-momentum vector of the state n. The delta functions show that both energy and momentum are conserved in the intermediate states n, so that these states are "real" rather than "virtual," and for this reason Im $T(\nu)$ is called the absorptive part of the amplitude. For $\nu \geq \mu$, eq. (3-32) is just an expression of the unitarity of the S-matrix. If the corresponding decomposition is made in the real part of the amplitude, $D(\nu)$, the intermediate states conserve momentum but not energy [because of the factor $\epsilon(x_0)$ in eq. (3-30a)] and so are "virtual" states. In this case $D(\nu)$ is often called the dispersive part of the amplitude.[4]

The value of Im $T(\nu)$ below threshold depends on the nature of the incident and target particles. If both are pseudoscalar mesons, then parity conservation forbids the state n to be a single-particle state. In the equal-mass case, for example $\pi\pi$ scattering, the two-particle states start at $p_{n_0} = 2\mu$ and Im T vanishes unless $|\nu| > \mu$. Accordingly, there is no contribution to the dispersion relation from the unphysical region.

For pseudoscalar meson scattering by a scalar nucleon the lowest state is the single nucleon state of mass M. The next highest state is one nucleon and one meson which contributes only in the physical region $\nu > \mu$.

---

[4] The real and imaginary parts of the amplitude do not coincide with the dispersive and absorptive parts when one or both particles possess spin.

The Single-Particle Contribution

To evaluate the contribution from the single nucleon state, the sum over n must be interpreted according to the closure relation of eq. (2-6):

$$\sum_n \to \left(\frac{1}{p_{n_0}}\right) \int d\vec{p}_n$$

which allows the integration over the delta functions $\delta(\vec{q}_L \pm \vec{p}_n)$ to be performed. In eq. (3-27) we require Im T in the region $\nu > 0$, so that only the second term in eq. (3-32) contributes, and:

$$\text{Im } T(\nu) = -\frac{4\pi^5}{p_0} |<M|j(0)|\vec{q}_L, p_0>|^2 \, \delta(\nu \mp M + p_0) \tag{3-33}$$

where $p_0 = \sqrt{q_L^2 + M^2} = \sqrt{\nu^2 + M^2 - \mu^2}$. The delta function can be re-expressed as a function of ν alone, by using the relation:

$$\frac{1}{p_0} \delta(\nu - M + p_0) = \frac{1}{M} \delta(\nu - \nu_B) \qquad \text{(a)}$$

where:

$$\nu_B = (\mu^2 / 2M) \ . \qquad \text{(b)} \tag{3-34}$$

The matrix element of the current $j(0)$ between single particle states $|p_b>$ and $|p_d>$ must be invariant, and must depend on the only scalar that can be formed from the vectors p_b and p_d, which is $\Delta^2 = (p_b - p_d)^2$. Here $|p_b>$ and $|p_d>$ are single nucleon states with $p_b^2 = p_d^2 = M^2$. We can then define $g(\Delta^2)$ as:

$$g(\Delta^2) \equiv \sqrt{\frac{2}{M}} <p_b|j(0)|p_d> (2\pi)^{\frac{3}{2}} \tag{3-35}$$

and eq. (3-33) becomes:

$$\text{Im } T(\nu) = -\frac{\pi^2}{4} |g(\mu^2)|^2 \, \delta(\nu - \nu_B), \ 0 < \nu < \mu \ . \tag{3-36}$$

3.2. FORWARD SCATTERING OF NEUTRAL PARTICLES

At the point $\Delta^2 = \mu^2$ all three particles described by the vertex function are on the mass shell, and the value of $g(\mu^2)$ at this point is adopted as the definition of the meson-nucleon coupling constant g:

$$g^2 \equiv |g(\mu^2)|^2 \tag{3-37}$$

Inserting eqs. (3-36) and (3-37) into the dispersion relation, eq. (3-27), we find that:

$$\operatorname{Re} T(\nu) = -\frac{\pi g^2 \nu_B}{2(\nu_B^2 - \nu^2)} + \frac{2}{\pi} \int_\mu^\infty \frac{\nu' \operatorname{Im} T(\nu')}{(\nu'^2 - \nu^2)} d\nu' . \tag{3-38}$$

By the optical theorem,[5] in the physical region $\nu > \mu$, $\operatorname{Im} T(\nu)$ is expressible in terms of the total cross section and, with our normalization:

$$\operatorname{Im} T(\nu) = \left(\frac{M q_L}{16\pi^2}\right) \sigma_{\text{total}}(\nu) .$$

The real part of the amplitude can be determined in terms of experimental data through phase shifts, so that only the coupling constant is an unknown parameter, provided always that the integral converges without subtractions.

In deducing the dispersion relation, the analyticity of $T(\nu)$ in the upper half plane was required. Since $\operatorname{Im} T(\nu)$ vanishes for $-\mu < \nu < \mu$ except at the points $\nu = \pm \nu_B$, the representation in eq. (3-38) shows that $T(\nu)$ has a much larger domain of analyticity. Singularities of $T(\nu)$ only occur when ν coincides with ν' in the denominator of the integrand in eq. (3-38). It follows that $T(\nu)$ is analytic in the whole of the ν plane with the exception of cuts from $-\infty$ to $-\mu$ and from $+\mu$ to $+\infty$, together with poles at $\nu = \pm \nu_B$. This is illustrated in Fig. 3-2. The discontinuity across the cuts is $2 \operatorname{Im} T(\nu)$ and $T(\nu)$ is real analytic in the sense that:

$$T^*(\nu^*) = T(\nu) . \tag{3-39}$$

---

[5] See page (29).

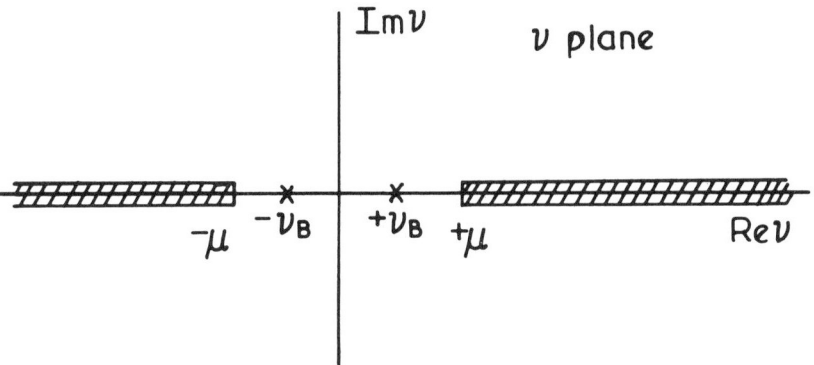

Fig. 3-2. The analytic structure of $T(\nu)$ defined by Eq. (3-38).

From the expression eq. (3-32), we see that Im T changes discontinuously at each production threshold, where a new state enters in the sum over n and each physical threshold is a branch point of the amplitude $T(\nu)$.

3.3. FIXED MOMENTUM TRANSFER DISPERSION RELATIONS

Dispersion relations for fixed nonzero values of the momentum transfer t can be found following similar arguments to those given in the last section. The crossing relation in this case is found from eq. (3-18), by interchanging the incident and the final mesons 1 and 2 and at the same time changing the sign of the energy-momentum vectors:

$$k_1 \to -k_2$$
$$k_2 \to -k_1 \ . \tag{3-40}$$

Using the Hermitian property of the retarded commutator we find that:

$$\langle k_1, p_1 | T | k_2, p_2 \rangle = \langle -k_2, p_1 | T | -k_1, p_2 \rangle^* \ . \tag{3-41}$$

3.3. FIXED MOMENTUM TRANSFER DISPERSION RELATIONS

Under the interchanges [eq. (3-40)], the variables s and u interchange ($s \leftrightarrow u$) while t remains unaltered. From eq. (3-20), it then follows that the crossing-relation in terms of ν and t becomes:

$$T(\nu, t) = T^*(-\nu, t) \ . \tag{3-42}$$

Dispersion relations are then expected to hold[6] that are similar in form to eq. (3-27):

$$\text{Re } T(\nu, t) = \frac{1}{\pi} P \int_0^\infty \text{Im } T(\nu', t) \left\{ \frac{1}{(\nu' - \nu)} + \frac{1}{(\nu' + \nu)} \right\} d\nu' \ . \tag{3-43}$$

For nonforward scattering ν is no longer equal to the total meson energy in the laboratory system, ω, but is given by eq. (3-21). The physical region in eq. (3-43) requires some discussion. For physical scattering ω must be greater than μ, the meson mass, but also we must have $|\cos \theta| \leq 1$ where θ is the angle of scattering. This condition is most easily discussed in terms of the center of mass momentum $\vec{q}$. We use the fact that s is the square of the center of mass energy:

$$s = W^2 = \left[(q^2 + M^2)^{\frac{1}{2}} + (q^2 + \mu^2)^{\frac{1}{2}} \right]^2 \tag{3-44}$$

together with the connection between W and ω[6]:

$$W^2 = M^2 + \mu^2 + 2M\omega$$

to find that:

$$M\nu = q^2 + (q^2 + M^2)^{\frac{1}{2}} (q^2 + \mu^2)^{\frac{1}{2}} - t/4 \ . \tag{3-45}$$

In the center of mass system:

$$t = -2q^2(1 - \cos \theta) \tag{3-46}$$

---

[6] Proofs of fixed t dispersion relations exist for some reactions for certain ranges of t [Hepp (1964); see also the review by Martin (1967)].

and for physical scattering t must be negative. For fixed negative t, with $|\cos\theta| \leq 1$, we must have $q^2 \geq -t/4$ and this corresponds to $\nu > \nu_1$, where:

$$\nu_1 = \frac{1}{M} \{(M^2 - t/4)(\mu^2 - t/4)\}^{\frac{1}{2}} . \tag{3-47}$$

For $\nu < \nu_1$, the contributions from Im $T(\nu, t)$ to the dispersion relation can be investigated by introducing a complete set of states into the retarded commutator as before. The single-particle poles in meson-nucleon scattering occur at $\nu = \nu_B$ where now:

$$\nu_B = \left(\frac{\mu^2}{2M} - \frac{t}{4M}\right) . \tag{3-48}$$

However, this time the continuous contribution due to the (one meson plus one nucleon) state starts at $\nu = \nu_0$, where ν_0 lies below the physical threshold ν_1. At $\nu = \nu_0$ we must have $\omega = \mu$, so that:

$$\nu_0 = \mu + \frac{t}{4M} . \tag{3-49}$$

Between ν_0 and the physical threshold ν_1, Im $T(\nu, t)$ cannot be determined directly from experiment, but it is possible to calculate this quantity from the partial wave series, for a limited range of t. In the center of mass system the scattering amplitude $f(\theta)$ is:

$$f(\theta) = \frac{4\pi}{W} T(\nu, t) = \sum_{\ell=0}^{\infty} (2\ell+1) f_\ell(q) P_\ell(\varkappa) \tag{3-50}$$

where $f_\ell(q)$ is the partial wave amplitude:

$$f_\ell(q) = \frac{1}{q} e^{i\delta_\ell} \sin\delta_\ell \tag{3-51}$$

and $\varkappa = \cos\theta$. It has been shown by Lehmann (1959) that the series for Re $f(\theta)$ converges in the $\varkappa$ plane inside an ellipse with center the origin and with axes $\varkappa_0$ and $\sqrt{(\varkappa_0^2 - 1)}$ where:

3.4. DISPERSION RELATIONS FOR PION-NUCLEON SCATTERING 105

$$x_0^2 = 1 + \frac{2\mu^3(2M+\mu)}{q^2[W^2 - (2\mu-M)^2]} \tag{3-52}$$

while Im $f(\theta)$ converges in a larger ellipse with axes $(2x_0^2 - 1)$ and $2x_0\sqrt{x_0^2 - 1}$. Provided t is such that x is within this ellipse, Im $T(\nu,t)$ can be calculated from the phase shifts using the partial wave series. This method is useful for πN scattering, but for NN or KN scattering the region of analyticity given this way is not large enough to be useful, and other methods must be employed.

3.4. DISPERSION RELATIONS FOR PION-NUCLEON ELASTIC SCATTERING

To find dispersion relations for pion-nucleon scattering, it is necessary to take into account the spin of the nucleon and the different charge states of the particles. Ignoring for the moment the question of charge, we have already seen that two amplitudes are required in the discussion of the scattering of spin 0 by spin $\frac{1}{2}$ particles. In Chapter 2, these amplitudes were taken to be $f(\theta)$ and $g(\theta)$ defined by eqs. (2-54) and (2-52). These amplitudes are not however Lorentz invariants and do not form a natural starting place for the development of dispersion theory.

Invariant Amplitudes

To construct amplitudes that are invariants, the nucleon spin properties are described by four-component Dirac spinors $u(\vec{p}, s)$ where $\vec{p}$ is the momentum and s the spin state of the nucleon. The normalization adopted is that:

$$u^\dagger(\vec{p}; s)\, u(\vec{p}; s) = \frac{E(p)}{M_N}, \quad \bar{u}(\vec{p}, s)\, u(\vec{p}; s) = 1 \ . \tag{3-53}$$

The invariant transition matrix T_{ba} for scattering from an initial state a in which the pion energy-momentum is k_1, the nucleon energy-momentum p_1 and spin s, to a final state b with pion energy-momentum k_2, nucleon energy-momentum p_2 and spin r, can be exhibited in the form:[7]

$$T_{fi} = \frac{M_N}{16\pi^2} \bar{u}(\vec{p}_2, r) \, M(k_1, p_1; k_2, p_2) \, u(\vec{p}_1, s) \qquad (3\text{-}54)$$

where M is a 4×4 matrix. Any matrix such as M can be expanded in terms of the sixteen independent Dirac matrices, $\Gamma = \gamma_\mu, \sigma_{\mu\nu}, \gamma_5 \gamma_\mu, \gamma_5$ and 1:

$$M = A + \sum_\mu \gamma^\mu B_\mu + \sum_{\mu\nu} \sigma^{\mu\nu} C_{\mu\nu} + \sum_\mu D_\mu \gamma^\mu \gamma_5 + E \gamma_5 \, . \qquad (3\text{-}55)$$

Since T_{ba} is to be an invariant, using the transformation properties of the bilinear forms $\bar{u} \, T \, u$ (see Messiah, 1964), the quantities A to E must transform like a scalar, vector, second-rank tensor, pseudovector, and pseudoscalar respectively. The quantities A to E must be constructed from the energy-momentum vectors k_i, p_i. Because of energy-momentum conservation, only three of the k_1, k_2, p_1, p_2 are independent, and suitable combinations may be taken to be:

$$P = \tfrac{1}{2}(p_1 + p_2), \quad Q = \tfrac{1}{2}(k_1 + k_2), \quad K = \tfrac{1}{2}(k_1 - k_2) = \tfrac{1}{2}(p_2 - p_1) \, . \quad (3\text{-}56)$$

No pseudoscalar or pseudovector can be formed from P, Q, and K, so the terms E and D must vanish. The tensor term C must have the general antisymmetrical form:

$$\begin{aligned} C_{\mu\nu} = \{ & (\lambda_1 P + \lambda_2 Q + \lambda_3 K)_\mu \, (\epsilon_1 P + \epsilon_2 Q - \epsilon_3 K)_\nu \\ & - (\epsilon_1 P + \epsilon_2 Q + \epsilon_3 K)_\mu \, (\lambda_1 P + \lambda_2 Q + \lambda_3 K)_\nu \end{aligned} \qquad (3\text{-}57)$$

where λ_i, ϵ_i are constants. On using the Dirac equations satisfied by $\bar{u}(x)$ and $u(p_2)$, all the terms $\left(\sum_{\mu,\nu} \bar{u} \sigma_{\mu\nu} C_{\mu\nu} u \right)$ can be reduced to

---

[7] The normalization factor $(M_N/16\pi^2)$ is introduced to provide a conventional normalization for M.

3.4. DISPERSION RELATIONS FOR PION-NUCLEON SCATTERING

scalar or vector terms like A or B_μ. The term B_μ could be of the form $(B\,Q_\mu + B'\,P_\mu + B''\,K_\mu)$, but terms like $\gamma_\mu\,P_\mu$ and $\gamma_\mu\,K_\mu$ can be reduced to scalar terms like A through the Dirac equation. The only surviving terms are then of the form:

$$M = A + \sum_\mu (\gamma^\mu Q_\mu) B \qquad (3\text{-}58)$$

where A and B are independent invariants, which are functions of two scalars determined by the center of mass energy W and the center of mass scattering angle θ. The fact that the number of independent amplitudes is two already followed from the analysis in Chapter 2, and the next task is to relate A and B to the amplitudes $f(\theta)$ and $g(\theta)$ introduced there.

The elastic scattering amplitude in the center of mass system $f_{fi}(\theta,\phi)$ is:

$$f_{fi}(\theta,\phi) = \frac{4\pi}{W} T_{fi} = \left(\frac{M_N}{4\pi W}\right) [\bar{u}(\vec{q}_f, s_f)\,\{A + \tfrac{1}{2}(q_f+q_i)B\}\,u(\vec{q}_i, s_i)] \qquad (3\text{-}59)$$

where q_f and q_i are the initial and final energy-momentum vectors with:

$$\vec{p}_1 = -\vec{k}_1 = \vec{q}_i;\ \vec{p}_2 = -\vec{k}_2 = \vec{q}_f \qquad (3\text{-}60)$$

and for elastic scattering $|\vec{q}_i| = |\vec{q}_f| = q$. The Dirac spinors for the initial and final nucleon spins s_i and s_f can be reduced to two-dimensional Pauli spinors $X(s_i)$ and $X(s_f)$, using the relation (Sakurai, (1964)):

$$u(\vec{p},s) = \frac{1}{\sqrt{2M_N(p_0+M_N)}} \begin{bmatrix} p_0+M_N & \vec{\sigma}\cdot\vec{p} \\ \vec{\sigma}\cdot\vec{p} & p_0+M_N \end{bmatrix} \begin{pmatrix} X(s) \\ X(s) \end{pmatrix} \qquad (3\text{-}61)$$

It easily follows that:

$$f_{fi}(\theta,\phi) = X^\dagger(s_f)\left[f_1 + f_2\,\frac{1}{q^2}\,(\vec{\sigma}\cdot\vec{q}_i)(\vec{\sigma}\cdot\vec{q}_f)\right] X(s_i) \qquad (3\text{-}62)$$

where f_1 and f_2 are defined by:

$$f_1 = \left(\frac{E+M_N}{8\pi W}\right)[A + (W-M_N)B]$$

$$f_2 = -\left(\frac{E-M_N}{8\pi W}\right)[A - (W+M_N)B] \tag{3-63}$$

where E is the center of mass energy of the nucleon:

$$E = (M_N^2 + q^2)^{\frac{1}{2}}. \tag{3-64}$$

On using the relation:

$$(\vec{\sigma}\cdot\vec{q}_i)(\vec{\sigma}\cdot\vec{q}_f) = \vec{q}_i\cdot\vec{q}_f + i\vec{\sigma}\cdot(\vec{q}_i\times\vec{q}_f) \tag{3-65}$$

and comparing it with eq. (2-54), f_1 and f_2 can be related to the amplitudes f and g introduced in Chapter 2:

$$f = f_1 + f_2 \cos\theta$$

$$g = f_2 \sin\theta. \tag{3-66}$$

Using this result, f_1 and f_2, and hence A and B, can be expressed in terms of the partial wave amplitudes $f_{\ell\pm}$ by:

$$f_1 = \sum_{\ell=0}^{\infty} f_{\ell+} P'_{\ell+1}(\cos\theta) - \sum_{\ell=2}^{\infty} f_{\ell-} P'_{\ell-1}(\cos\theta)$$

$$f_2 = \sum_{\ell=1}^{\infty} (f_{\ell-} - f_{\ell+}) P'_{\ell}(\cos\theta). \tag{3-67}$$

The inverse relations to eq. (3-63), giving A and B in terms of f_1 and f_2, from which a partial wave decomposition of A and B can be made, are:

$$A = 4\pi\left[\left(\frac{W+M_N}{E+M_N}\right)f_1 - \left(\frac{W-M_N}{E-M_N}\right)f_2\right] \quad \text{(a)}$$

$$B = 4\pi\left[\left(\frac{1}{E+M_N}\right)f_1 + \left(\frac{1}{E-M_N}\right)f_2\right] \quad \text{(b)}. \tag{3-68}$$

3.4. DISPERSION RELATIONS FOR PION-NUCLEON SCATTERING

Corresponding expansions into invariant amplitudes can be carried out for other two-particle scattering processes. Important cases are nucleon-nucleon scattering, for which the invariant amplitudes were obtained by Goldberger et al., (1957), and photoproduction of mesons from nucleons, discussed by Chew et al., (1957b). The general problem of determining invariant amplitudes for particles of arbitrary spin is complicated but a solution has been given by Barut et al., (1963), and more recently the problem has been considered by Scadron and Jones (1968).

Crossing Symmetry

To this point we have ignored the charge properties of the pions and nucleons. In fact separate amplitudes must be introduced for $\pi^0 p$, $\pi^+ p$ and for $\pi^- p$ scattering. It can then be shown, either from perturbation theory, or from the reduction formula, that under the interchange $s \leftrightarrow u$, or $\nu \leftrightarrow -\nu$ the $\pi^0 + P$ amplitude crosses into itself, while the $\pi^+ + p$ amplitude crosses to the $\pi^- + p$ amplitude (see Fig. 3-3), and vice versa. The crossing relation for the amplitude is then:

$$T^{(\mu)}(-\nu, t) = T^{(-\mu)*}(\nu, t) \qquad (3\text{-}69)$$

where $\mu = \pm 1$ for $\pi^{\pm}$ scattering and $\mu = 0$ for π^0 scattering. The corresponding relations for the odd and even parts of the amplitude $T^{(\pm)}$ defined by:

$$T^{(\pm)} = \tfrac{1}{2}(T^{(\pm 1)} \pm T^{(\mp 1)}) \qquad (3\text{-}70)$$

become:

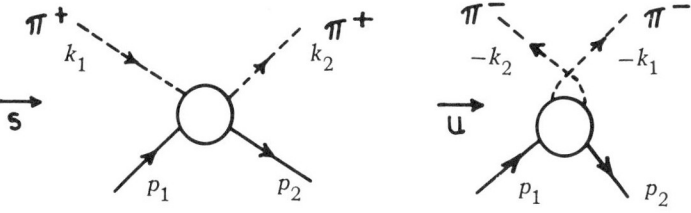

Fig. 3-3. πN amplitudes connected by $s \leftrightarrow u$ crossing.

$$T^{(\pm)}(-\nu, t) = \pm T^{(\pm)*}(\nu, t) \ .$$

Each amplitude $T^{(\pm)}(\nu, t)$ can be expressed in terms of the invariant amplitudes $A^{(\pm)}$ and $B^{(\pm)}$, and these have the crossing relations:

$$A^{(\pm)}(-\nu, t) = \pm A^{(\pm)*}(\nu, t)$$

$$B^{(\pm)}(-\nu, t) = \mp B^{(\pm)*}(\nu, t) \ . \qquad (3\text{-}71)$$

The Dispersion Relations

A dispersion relation can be written for each of the four independent amplitudes $A^{(\pm)}$, $B^{(\pm)}$. If charge independence is assumed, the amplitudes for π^0-p scattering can be obtained from $T^{(\pm)}$, but if only charge symmetry is imposed, the amplitude T^0 is independent and can be expressed in terms of the two further invariant amplitudes $A^{(0)}$ and $B^{(0)}$. If the crossing conditions are incorporated, the dispersion relations for fixed momentum transfer are:

$$\operatorname{Re} A^{(\pm)}(\nu, t) = \text{(single particle contributions)}$$

$$+ \frac{1}{\pi} P \int_{\nu_1}^{\infty} d\nu' \operatorname{Im} A^{(\pm)}(\nu, t) \left(\frac{1}{(\nu'-\nu)} \pm \frac{1}{(\nu'+\nu)} \right)$$

$$\operatorname{Re} B^{(\pm)}(\nu, t) = \text{(single particle contributions)}$$

$$+ \frac{1}{\pi} P \int_{\nu_1}^{\infty} d\nu' \operatorname{Im} B^{(\pm)}(\nu, t) \left(\frac{1}{(\nu'-\nu)} \mp \frac{1}{(\nu'+\nu)} \right) \quad (3\text{-}72)$$

where the possibility that subtractions may be required to make the integrals converge has been ignored, and where ν_1 is defined by eq. (3-47).

The only single-particle contributions arise from the nucleon intermediate state at $\nu = \pm \nu_B$ where, from eq. (3-48) with $\mu = M_\pi$ and $M = M_N$:

$$\nu_B = \left(\frac{M_\pi^2}{2M_N} - \frac{t}{4M_N} \right) \ . \qquad (3\text{-}73)$$

3.4. DISPERSION RELATIONS FOR PION-NUCLEON SCATTERING

By charge conservation in $\pi^- + p$ scattering the intermediate state represents a neutron and no corresponding contribution arises in the $\pi^+ + p$ system.

The single particle term can be evaluated by similar arguments to those leading from eq. (3-33) to eq. (3-38). In the present case, as the π meson is pseudoscalar, the most general form for the $\pi^- $-$p$ vertex must be:

$$(2\pi)^3 <p_2|j(0)|p_1> = i\sqrt{2}\, g(\Delta^2)(\bar{u}(\vec{p}_2)\,\gamma_5\,u(\vec{p}_1))\,M_N \qquad (3\text{-}74)$$

where p_2 is the four momentum of the neutron and p_1 of the proton, and $\Delta^2 = (p_2 - p_1)^2$. The π-N coupling constant[8] is then defined as:

$$g_{\pi NN} = g(M_\pi^2) \; . \qquad (3\text{-}75)$$

By charge conservation no corresponding contribution arises from the $\pi^+ + p$ system as noted above.

In making the detailed evaluation of single neutron contribution using eq. (3-74), it is seen that the sum over the spin state of the neutron is of the form:

$$\sum_{\text{spin}} \bar{u}(\vec{p}_2)\,\gamma_5\,u(\vec{p}_n)\,\bar{u}(\vec{p}_n)\,\gamma_5\,u(\vec{p}_1)$$

which can be reduced to the expression:

$$-\bar{u}(\vec{p}_2)\left(\frac{\gamma \cdot p_n}{2M_N}\right) u(\vec{p}_1) \; .$$

This shows that the single nucleon pole in the dispersion relation contains a factor of γ_μ. It must on this account only contribute to the amplitude $B^{(\pm)}$ and not to $A^{(\pm)}$. The final form of the dispersion relation is then:

---

[8] The subscripts on $g_{\pi NN}^2$ are dropped in what follows, except when we need to distinguish two different couplings.

$$\text{Re } A^{(\pm)}(\nu, t) = \frac{1}{\pi} P \int_{\nu_1}^{\infty} d\nu' \text{ Im } A^{(\pm)}(\nu', t) \left(\frac{1}{(\nu'-\nu)} \pm \frac{1}{(\nu'+\nu)}\right)$$

$$\text{Re } B^{(\pm)}(\nu, t) = \frac{g^2}{2M_N} \left(\frac{1}{(\nu_B-\nu)} \mp \frac{1}{(\nu_B+\nu)}\right)$$

$$+ \frac{1}{\pi} P \int_{\nu_1}^{\infty} d\nu' \text{ Im } B^{(\pm)}(\nu', t) \left(\frac{1}{(\nu'-\nu)} \mp \frac{1}{(\nu'+\nu)}\right). \quad (3\text{-}76)$$

π-N Forward Dispersion Relations

The forward scattering amplitude ($t = 0$) in the center of mass system is:

$$f^{(\pm)}(\nu, 0) = \frac{4\pi}{W} T(\nu, 0) = \frac{M_N}{4\pi W} \{A^{(\pm)}(\nu, 0) + \nu B^{(\pm)}(\nu, 0)\} . \quad (3\text{-}77)$$

Because of the extra pole at $W = 0$ that occurs in $f^{(\pm)}$ it is more convenient to employ the forward scattering amplitude in the *laboratory system* $F^{(\pm)}(\nu)$, which is connected with $f^{(\pm)}(\nu)$ by:

$$F^{(\pm)}(\nu) = \frac{q_L}{q} f^{(\pm)}(\nu, 0) . \quad (3\text{-}78)$$

On using the relation of eq. (2-95):

$$\frac{q_L}{q} = \frac{W}{M_N}$$

we find that:

$$F^{(\pm)}(\nu) = \frac{1}{4\pi} \{A^{(\pm)}(\nu, 0) + \nu B^{(\pm)}(\nu, 0)\} . \quad (3\text{-}79)$$

The dispersion relations for $F^{(\pm)}$ are now easily found from eq. (3-76):

3.4. DISPERSION RELATIONS FOR PION-NUCLEON SCATTERING

$$\text{Re } F^{(+)}(\nu) = \frac{g^2 \nu^2}{4\pi M_N(\nu_B^2 - \nu^2)} - \frac{1}{2\pi^2} P \int_{M_\pi}^{\infty} \text{Im } B^{(+)}(\nu', 0) \, d\nu'$$

$$+ \frac{2}{\pi} P \int_{M_\pi}^{\infty} d\nu' \, \frac{\nu' \text{Im } F^{(+)}(\nu')}{(\nu'^2 - \nu^2)} \quad (3\text{-}80a)$$

$$\text{Re } F^{(-)}(\nu) = \frac{g^2 \nu \nu_B}{4\pi M_N(\nu_B^2 - \nu^2)} + \frac{2\nu}{\pi} P \int_{M_\pi}^{\infty} d\nu' \, \frac{\text{Im } F^{(-)}(\nu')}{(\nu'^2 - \nu^2)}. \quad (3\text{-}80b)$$

Using the optical theorem, together with eq. (3-68), Im $F^{\pm}$ can be expressed in terms of $\sigma_\pm$, the total cross sections for π^+- and π^--proton scattering:

$$\text{Im } F^{(\pm)}(\nu) = \frac{q_L}{8\pi} (\sigma_-(\nu) \pm \sigma_+(\nu)). \quad (3\text{-}81)$$

Empirically σ_+ and σ_- appear to become constant and to approach the same value at high energies. This is an example of the Pomeranchuk theorem, to be discussed in Chapter 8. The cross sections above 10 GeV/c have been fitted to the following forms (Foley et al., 1967):

$$\frac{1}{2}(\sigma_-(\nu) - \sigma_+(\nu)) = \frac{A}{(q_2)^n}, \quad \begin{array}{l} A = 1.925 \pm 0.28 \\ n = 0.31 \end{array}$$

$$\frac{1}{2}(\sigma_-(\nu) + \sigma_+(\nu)) = B + \frac{A}{(q_2)^n}, \quad \begin{array}{l} A = 18.5 \pm 4.25 \\ B = 22.12 \pm 0.94, \end{array} \quad n = 0.69 \pm 0.21 \quad (3\text{-}82)$$

where q_L is measured in units of GeV/c and the cross sections are in millibarns (Foley et al., 1967). In view of this, we expect that the integral over Im $F^{(+)}$ in eq. (3-80) will certainly diverge. To cure this trouble the dispersion relation is written for the combination $\{F^{(+)}(\nu) - F^{(+)}(\nu_s)\}$ where the subtraction is at an arbitrary point ν_s. We find in place of eq. (3-80a):

$$\text{Re } F^{(+)}(\nu) - \text{Re } F^{(+)}(\nu_1) = \frac{g^2 \, \nu_B^2 (\nu^2 - \nu_s^2)}{4\pi \, M_N (\nu_B^2 - \nu^2)(\nu_B^2 - \nu_s^2)}$$

$$+ \frac{2}{\pi} (\nu^2 - \nu_s^2) P \int_{M_\pi}^{\infty} d\nu' \, \frac{\nu' \, \text{Im } F^{(+)}(\nu')}{(\nu'^2 - \nu_s^2)(\nu'^2 - \nu^2)}$$

which is now convergent. (3-83)

If a corresponding subtraction is made in eq. (3-80b) and the subtraction point is chosen to be the threshold for elastic scattering, $\nu_s = M_\pi$, the two dispersion relations eqs. (3-80b) and (3-83) can be combined to yield the following standard form, for the π^+ and π^- proton scattering amplitudes $F_\pm(\nu)$, where (Goldberger et al., 1955) $F_\pm(\nu) = F^{(+)} \mp F^{(-)}$:

$$\text{Re } F_\pm(\nu) - \frac{1}{2}\left(1 \pm \frac{\nu}{M_\pi}\right) \text{Re } F_\pm(M_\pi) - \frac{1}{2}\left(1 \mp \frac{\nu}{M_\pi}\right) \text{Re } F_\mp(M_\pi)$$

$$= \pm \frac{2f^2}{M_\pi^2} \frac{q_L^2}{(\nu \mp \nu_B)} + \frac{q_L^2}{2\pi^2} \int_{M_\pi}^{\infty} d\nu' \, \frac{1}{q_L} \left(\frac{\sigma_\pm(\nu')}{(\nu' - \nu)} + \frac{\sigma_\mp(\nu')}{(\nu' + \nu)} \right) \quad (3\text{-}84)$$

where q_L is the laboratory momentum of the pion, and we have used the optical theorem, eq. (3-81). The constant f is the effective (pseudovector) coupling constant defined as (see Section 7.1):

$$f^2 = \frac{g^2}{4\pi} \left(\frac{M_\pi}{2M_N}\right)^2. \quad (3\text{-}85)$$

Applications of π-N Dispersion Relations

The amplitudes $F_\pm(\nu)$ appearing on the left-hand side of the dispersion relation, eq. (3-84), can easily be expressed in terms of the phase shifts using eqs. (3-67), (3-68), and (3-79):

3.4. DISPERSION RELATIONS FOR PION-NUCLEON SCATTERING

$$\text{Re } F_+(\nu) = \frac{q_L}{2q^2} \left\{ \sin\left(2\delta_0^{\frac{3}{2}}\right) + 2 \sin\left(2\delta_{1+}^{\frac{3}{2}}\right) + \sin\left(2\delta_{1-}^{\frac{3}{2}}\right) + \ldots \right\}$$

$$\text{Re } F_-(\nu) = \frac{q_L}{6q^2} \left\{ \sin\left(2\delta_0^{\frac{3}{2}}\right) + 2 \sin\left(2\delta_0^{\frac{1}{2}}\right) + 2 \sin\left(2\delta_{1+}^{\frac{3}{2}}\right) \right.$$

$$\left. + 4 \sin\left(2\delta_{1+}^{\frac{1}{2}}\right) + \ldots \right\}. \tag{3-86}$$

The subtraction constants $\text{Re } F_\pm(M_\pi)$ are, in the same way, related to the s-wave scattering lengths:

$$\text{Re } F_+(M_\pi) = \left(1 + \frac{M_\pi}{M_N}\right) a_{0+}^{\frac{3}{2}}$$

$$\text{Re } F_-(M_\pi) = \frac{1}{3} \left(1 + \frac{M_\pi}{M_N}\right) \left(a_{0+}^{\frac{3}{2}} + 2a_{0+}^{\frac{1}{2}}\right). \tag{3-87}$$

The integrals over the experimental cross sections converge quite rapidly, and, at least for moderate values of ν (< 1 GeV), the uncertainties which arise from the fact that the cross sections have only been measured up to momenta of ~ 26 GeV/c are small. In the region for which phase shift analyses are available, everything is known in the dispersion relation except for the single parameter f^2. A fit can only be obtained if the low energy phase shifts of the Fermi type, in which the p_{33} wave is resonant, are used. Historically this served to eliminate the alternative phase shift solutions (Anderson et al., 1955), before polarization measurements were available. The value of $f^2 \approx 0.08$.

The correct set of phase shifts having in this way been established, the dispersion relations may be used to find f^2 accurately. Probably the best method for the determination of f^2 is that of Woolcock (1961) and Hamilton and Woolcock (1963), who used the dispersion relation of eq. (3-76) for the amplitude B in the forward direction. If B_+ and B_- are the B amplitudes for π^+- and π^--proton scattering, we find by subtracting the relations for $B^{(+)}$ and $B^{(-)}$ that:

$$\text{Re } B_+(\nu, 0) = -\frac{8\pi M_N f^2}{\nu_B(\nu+\nu_B)} + \frac{1}{\pi} P \int_{M_N}^{\infty} d\nu' \left[\frac{\text{Im } B_+(\nu', 0)}{(\nu'-\nu)} - \frac{\text{Im } B_-(\nu', 0)}{(\nu'+\nu)} \right].$$
(3-88)

The difference between Re B_+ and the integral term was computed over a π meson laboratory energy range from 15 to 200 MeV and fitted to the pole term to find f^2. The reason for the accuracy of this method can be seen by examining the partial wave expansion of the amplitude $B_+(\nu, 0)$:

$$B_+(\nu, 0) = \frac{4\pi f_0^{\frac{3}{2}}}{(E+M_N)} + \frac{4\pi f_{1-}^{\frac{3}{2}}}{(E-M_N)} - \frac{8\pi}{q^2} M_N \left(2 - \frac{E}{M_N}\right) f_{1+}^{\frac{3}{2}} + \dots \quad (3\text{-}89)$$

Because of the denominator $(E+M_N)$, the s-wave makes a negligible contribution to $B(\nu, 0)$, and over the low energy region $f_{1-}^{\frac{3}{2}}$ is small for both isotopic spin states, so that B_+ is dominated by the resonant partial wave $f_{1+}^{\frac{3}{2}}$, which is known very accurately. Under the integral Im B_+ is also dominated by $f_{1+}^{\frac{3}{2}}$. To calculate Im B_-, the $I = \frac{1}{2}$ amplitudes are required, which are not known so accurately, but the contribution from this term is an order of magnitude smaller than that from Im B_+. Estimates of Im B_+ must be made above the region in which the phase shifts are known accurately, but the contributions from higher energies are quite small. The best value for f^2 is:

$$f^2 = 0.081 \pm 0.003 \; . \tag{3-90}$$

The value of f^2 having been determined, the fixed momentum dispersion relation may be used in a variety of ways to determine the s- and p-wave π-N scattering lengths with greater precision. For example, the forward dispersion relation of eq. (3-80), evaluated for small y where $y = (\nu-M_\pi)$, provides effective range formulae for the s-wave phase shifts of the type:

3.4. DISPERSION RELATIONS FOR PION-NUCLEON SCATTERING

$$\left(\sin\left(2\delta_0^{\frac{1}{2}}\right) + 2\sin\left(\delta_0^{\frac{3}{2}}\right)\right)\frac{W}{2q(M_N + M_\pi)} = \left(a_{0+}^{\frac{1}{2}} + 2a_{0+}^{\frac{3}{2}}\right) + C^+(y)\, q_L^{\ 2}$$

$$\left(\sin\left(2\delta_0^{\frac{1}{2}}\right) - \sin\left(2\delta_0^{\frac{3}{2}}\right)\right)\frac{W}{2q(M_N + M_\pi)} = \left(a_{0+}^{\frac{1}{2}} - a_{0+}^{\frac{3}{2}}\right) + C^-(y)\, q_L^{\ 2} \quad (3\text{-}91)$$

where $C^{\pm}(\omega)$ are empirically found to be constant below $\sim$ 45 MeV. As the usual power series expansion of $\delta_0^{\ I}$ converges slowly, much more accurate values of $a_{0+}^{\frac{1}{2}}$ and $a_{0+}^{\frac{3}{2}}$ can be found by fitting the data with the form eq. (3-91). The results of a fit of recent data below 45 MeV, using this formula (Hamilton, 1966), were given in Chapter 2 (page 64). Relations involving both s- and p-wave scattering lengths have been found from the dispersion relations for $B(\nu, 0)$ and $\frac{\partial B(\nu, t)}{\partial t}\big|_{t=0}$, when $\nu \to M_\pi$ and a consistent set of scattering lengths has been computed by Hamilton and Woolcock (1963). The partial wave amplitude at energies away from the threshold can also be obtained either by examining the dispersion relations for the derivatives of the scattering amplitude with respect to t, in the forward direction, or else by using the fixed momentum transfer dispersion relations over a finite interval of t. In this way, Chew et al., (1957a) and later Höhler et al., (1960, 1962) were able to give a relativistic description of the p_{33} resonance. Because of the poor convergence of the partial wave series away from the physical scattering region, neither of these methods appears capable of very accurate results, and it seems better to proceed from dispersion relations for the partial wave amplitudes themselves, which will be discussed in the next chapter.

Validity of Dispersion Relations at Higher Energies

The forward dispersion relations are extremely well satisfied over the whole of the low energy region containing the p_{33} resonance. It is important to test these relations at higher energies. If the relations were found to fail at high energies, this might imply that interactions at small distances were not casual, but would more likely imply that our ideas about the extrapolation of the cross sections required for computing the dispersion integral were in error. A sensitive test of the forward dispersion relations is given by the unsubtracted relation for the amplitude $F^{(-)}(\nu)$ given by eq. (3-80b). From eq. (2-85) it is seen that $F^{(-)}(\nu)$ is directly related to the amplitude for charge exchange scattering in the forward direction:

$$F^{(-)}(\nu) = \frac{1}{\sqrt{2}} <\pi^- + p|T|\pi^0 + n> . \qquad (3\text{-}92)$$

The differential cross section for charge exchange scattering in the forward direction has been measured up to ~ 20 GeV/c and combined with the measurements of the total cross sections σ_- and σ_+ this allows the calculation of the modulus of the left-hand side of the dispersion relation, since:

$$\frac{d\sigma^{C.Ex.}(\theta=0)}{d\Omega} = |\sqrt{2}\, F^{(-)}(\nu)|^2 = 2|\mathrm{Re}\, F^{(-)}(\nu)|^2 + \frac{q_L^2}{8\pi^2} |\sigma_-(\nu) - \sigma_+(\nu)|^2 . \qquad (3\text{-}93)$$

The integral on the right hand side of the dispersion relation, eq. (3-80b), can also be calculated from the measured total cross section together with the extrapolation, eq. (3-82). The function $|\mathrm{Re}\, F^{(-)}|^2$ given by the dispersion relation agrees with that given by eq. (3-93) and is within the experimental uncertainties over the whole energy range up to 20 GeV/c (Amblard et al., 1964; Höhler et al., 1966).

The same dispersion relation evaluated at $\nu = M_\pi$ provides a sum rule for the scattering length combination $\left(a_{0+}^{\frac{1}{2}} - a_{0+}^{\frac{3}{2}}\right)$:

$$\frac{1}{3M_\pi}\left(1 + \frac{M_\pi}{M_N}\right)\left(a_{0+}^{\frac{1}{2}} - a_{0+}^{\frac{3}{2}}\right) = \frac{2f^2}{M_\pi^2\left(\frac{M_\pi^2}{4M_N^2} - 1\right)} + \frac{1}{4\pi^2}\int_{M_\pi}^{\infty}\frac{d\nu'}{q_L'}\left[\sigma_-(\nu') - \sigma_+(\nu)\right]$$

$$(3\text{-}94)$$

3.4. DISPERSION RELATIONS FOR PION-NUCLEON SCATTERING

If the integral is computed from the experimental cross sections and the extrapolation of eq. (3-92), it is found that (Samaranayake and Woolcock, 1965) $(a_{0+}^{\frac{1}{2}} - a_{0+}^{\frac{3}{2}}) = 0.292 \pm 0.020 \frac{\hbar}{M_\pi c}$ which is significantly larger than the value $0.27 \pm 0.007 \frac{\hbar}{M_\pi c}$ found directly from the low energy data (see page 65). This disagreement may be associated with the method of extracting electromagnetic effects to obtain the charge independent amplitudes at low energies or, more likely, may be due to the inaccuracy of the extrapolation of the cross sections to energies beyond those measured.

A further test both of the dispersion relations and of the high energy extrapolation is possible using measurements of the ratio of the real to the imaginary parts of the elastic scattering amplitudes in the forward direction:

$$a_\pm = \text{Re } F_\pm / \text{Im } F_\pm . \tag{3-95}$$

The $a_\pm$ are measured by observing the interference at very small angles between Coulomb and nuclear scattering. The results of Foley et al., (1965, 1967), which extend from 8 to 26 GeV/c for $\pi^- p$ and from 8 to 20 GeV/c for $\pi^+ p$ scattering, can all be represented by the linear fit:

$$a_+(\nu) = -0.292 + 0.008\nu$$

$$a_-(\nu) = -0.158 + 0.002\nu \tag{3-96}$$

where ν is measured in GeV. As $\text{Im } F_\pm$ can be calculated from the total cross sections by the optical theorem, a comparison between $\text{Re } F^\pm$ given by the dispersion relation and that calculated from the measured values is possible. In the comparison, the high energy total cross sections were fitted, in order to extrapolate the integral in the dispersion integral, to the forms:

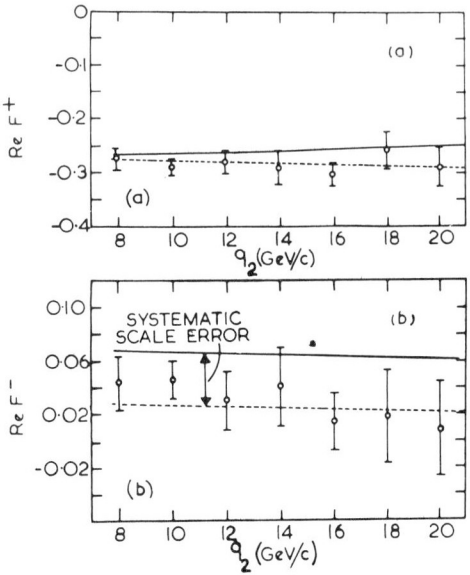

Fig. 3-4. The real parts of the amplitudes $F^{\pm}$. (1) Solid lines are calculated from the dispersion relations. (2) The individual points arise from experimental measurements of the $F^{\pm}$. (3) Dashed curve in (b) is an estimate of possible systematic error. (4) Dotted curve in (a) is the result of a subtraction in the dispersion relation at 20 GeV/c (from Foley et al., 1967).

$$\sigma_{\pm}(\nu) = (a + b_{\pm} / q_L^{n_{\pm}}) \text{ mb} \qquad (3\text{-}97)$$

with

$$a = 22.60 \pm 0.4, b_{+} = 25.9 \pm 9.6, b_{-} = 19.6 \pm 1.8,$$

$$n_{+} = 1.06 \pm 0.24, \text{ and } n_{-} = 0.67 \pm 0.08.$$

The data from 8 to 20 GeV/c are shown in Fig. 3-4 together with the fitted curves. The real part of the amplitude F^{+} was computed from the dispersion relation, eq. (3-83), and that of F^{-} from the unsubtracted relation, eq. (3-80b). In each case the value of the coupling constant was given by eq. (3-85) with $f^2 = 0.081$, and from the low energy data Re $F^{+}(M_{\pi})$ was taken to be -0.002 (in units with $M_{\pi} = \hbar = c = 1$). In Fig. 3-4, the solid curves show the dispersion relation predictions and the

3.5. DISPERSION RELATIONS FOR $K^{\pm}N$ SCATTERING

individually plotted points arise from the direct measurements of Re $F^{\pm}(\nu)$. The agreement is very satisfactory for F^+ but less so for F^-; but if allowance is made for the possible systematic errors (the dotted curve in the figures), agreement can be reached for F^- as well. If a further subtraction is made in the dispersion relation for F^+ at 26 GeV/c, the agreement becomes even better, and F^+, calculated from the dispersion integral becomes independent of the high energy behavior of the cross sections (the dotted line in Fig. 3-4a). This result confirms the fact that the dispersion relations are valid up to energies of at least 20 GeV and substantiates the underlying idea of causality on which the theory is based. Only charge symmetry is used in this work, but the charge exchange data referred to earlier are in good agreement with the predictions from the measurements of $F_{\pm}$, and, within the experimental errors, charge independence is verified up to 20 GeV.

As we have seen the agreement between Re $F^{\pm}$, as predicted by the dispersion relation, and the experimental values is better for the subtracted than for the unsubtracted case. This is due to the fact that the dispersion relation is more dependent on the form of high energy extrapolation such as eq. (3-97). This implies that some light can be cast on the asymptotic forms of the cross sections by requiring that the predicted and measured Re $F^{\pm}$ agree in the energy interval below 20 GeV. Working along these lines, Gajdicar and Moffat (1967) have obtained evidence for a reversal of sign of the cross section difference $(\sigma_- - \sigma_+)$ at some energy greater than 20 GeV. Such a result, if substantiated, will have important implications for the theories of high energy scattering discussed in Chapter 8.

3.5. FORWARD DISPERSION RELATIONS FOR $K^{\pm}N$ SCATTERING

Dispersion relations for KN scattering can be written following the procedure of the last section. If $F^{\pm}(\nu)$ is the forward scattering amplitude in the laboratory system for K^+ or K^- scattering on protons, and we make use of analyticity and crossing, we find the dispersion relations:

$$\text{Re } F^{\pm}(\nu) = \frac{1}{\pi} P \int_0^{\infty} d\nu' \left[\frac{\text{Im } F^{\pm}(\nu')}{(\nu'-\nu)} + \frac{\text{Im } F^{\mp}(\nu')}{(\nu'+\nu)} \right]. \quad (3\text{-}98)$$

Physical scattering occurs for $\nu > M_K$ where M_K is the mass of the K meson, and in this region $\text{Im } F^{\pm}(\nu)$ is related to the observed total K^+ and K^--p cross sections by the optical theorem. The K^+-p system has hypercharge $Y = 2$, and since there are no stable particles with this value of the hypercharge, there are no single-particle contributions to $\text{Im } F^+(\nu)$, and $\text{Im } F^+(\nu)$ vanishes for $\nu < M_K$. In the K^--p system on the other hand, single-particle contributions arise from both Σ and Λ intermediate single-particle states, giving pole contributions to the dispersion relations that are proportional to the $K\Lambda N$ and $K\Sigma N$ coupling constants.

From the discussion in Chapter 2, we know that $\text{Im } F^-$ does not vanish at the K^--p threshold, because of the exothermic hyperon production reactions. The lowest threshold is the Λ-π threshold at $\nu = \nu_0$, where:

$$\nu_0 = \frac{1}{2M_N} \left[(M_{\pi} + M_{\Lambda})^2 - (M_N^2 + M_K^2) \right] \quad (3\text{-}99)$$

and contributions arise to the dispersion relation from the region $\nu_0 < \nu < M_K$. Putting these contributions together and subtracting the dispersion relation for K^+p from that for K^-p scattering to ensure convergence, we find finally:

$$\text{Re } [F^-(\nu) - F^+(\nu)] = \frac{2\nu\Gamma_n}{\nu_n^2 - \nu^2} + \frac{2\nu\Gamma_{\Sigma}}{\nu_{\Sigma}^2 - \nu^2} + \frac{2}{\pi} P \int_{\nu_0}^{M_K} \frac{\nu \text{ Im } F^-(\nu')}{(\nu'^2 - \nu^2)} d\nu'$$

$$+ \frac{1}{2\pi^2} P \int_{M_K}^{\infty} d\nu' \frac{q_L \nu}{(\nu'^2 - \nu^2)} (\sigma^-(\nu') - \sigma^+(\nu')). \quad (3\text{-}100)$$

In this equation ν_Λ and ν_Σ are the Λ and Σ pole positions, and Γ_Λ, Γ_Σ are proportional to the coupling constants $g^2_{K\Lambda N}$ and $g^2_{K\Sigma N}$. The determination of these constants has been the subject of much work, which has been summarized by Bransden (1969), by Martin et al., (1969a, b), and by Queen et al., (1969). Grave difficulties arise in evaluation of the dispersion relations because it is necessary to estimate $\text{Im } F^-(\nu)$ in the unphysical region below the K^--p threshold. The basis for this estimate is usually an extrapolation of the amplitude through the physical threshold using the M matrix discussed in Chapter 3. The results indicate that $g^2_{K\Lambda N}$ lies somewhere in the range $13 > g^2_{K\Lambda N} > 5$ while $g^2_{K\Sigma N}$ is small. Values near the upper end of this range would be consistent with those predicted by $SU3$ invariance using a D/F ratio of 1.5, but further data are required before a definite conclusion can be reached.

REFERENCES

B. Amblard, P. Borgeaud, Y. Ducros, P. Falk-Vairant, O. Guisan, W. Laskar, P. Sonderegger, A. Stirling, M. Yoert, A. Tran Ha and S. D. Warshaw, Phys. Letters 10, 138 (1964).

H. L. Anderson, W. C. Davidson and U. E. Kruse, Phys. Rev. 100, 339 (1955).

A. O. Barut, I. Muzenich and D. N. Williams, Phys. Rev. 130, 442 (1963).

J. D. Bjorken and S. D. Drell, *Relativistic Quantum Fields* (McGraw-Hill Book Co., New York, 1965).

N. N. Bogoliubov, Unpublished, (1958).

B. H. Bransden, in *High Energy Physics*, ed. E. H. S. Burhop (Academic Press, New York, 1969), Vol. III.

H. J. Bremmerman, R. Oehme and J. G. Taylor, Phys. Rev. 109, 2178 (1958).

G. F. Chew, M. L. Goldberger, F. E. Low and Y. Nambu, Phys. Rev. 106, 1337 (1957a). Phys. Rev. 106, 1345 (1957b).

K. J. Foley, R. S. Gilmore, R. S. Jones, S. J. Lindenbaum, W. A. Love, S. Ozaki, E. H. Willen, R. Yamaka and L. C. L. Yuan, Phys. Rev. Letters 14, 862 (1965).

K. J. Foley, R. S. Jones, S. J. Lindenbaum, W. A. Love, S. Ozaki, E. D. Platner, C. A. Quarles and E. H. Willen, Phys. Rev. Letters 19, 193, 330 (1967).

T. J. Gajdicar and J. W. Moffat, Phys. Rev. Letters 18, 1154 (1967).

M. Gell-Mann, M. L. Goldberger and W. Thirring, Phys. Rev. 95, 1612 (1951).

M. L. Goldberger, H. Mirjazaura and R. Oehme, Phys. Rev. 99, 986 (1955).

M. L. Goldberger, Y. Nambu and R. Oehme, Ann. Phys. (N.Y.) 2, 226 (1957).

J. Hamilton and W. S. Woolcock, Rev. Mod. Phys. 35, 737 (1963).

J. Hamilton, Phys. Letters 20, 687 (1966).

K. Hepp, Helv. Phys. Acta 37, 639 (1964).

G. Höhler, J. Baacke, J. Giesecke and N. Zovko, Proc. Roy. Soc. A289, 500 (1966).

G. Höhler and J. Baacke, Phys. Letters 18, 181 (1965).

G. Höhler and K. Dietz, Z. f. Physk. 160, 453 (1960).

G. Höhler and K. Dietz, Proc. Int. Conf. on High Energy Phys., CERN (1962).

H. Lehmann, Nuovo Cimento Suppl. 1, 14, 153 (1959).

REFERENCES

A. Martin, *Proc. 1967 Int. Conf. on Particles and Fields* (Interscience, New York, 1967, ed. R. Hagen, G. Guralnik and V. S. Mathur.

A. D. Martin, N. M. Queen and G. Violini, Nuc. Phys. B13, 481 (1969a).

A. D. Martin, N. M. Queen and G. Violini, Phys. Letters 29B, 311 (1969b).

A. Messiah, *Quantum Mechanics* (North Holland, Amsterdam, 1964).

N. M. Queen, M. Restignoli and G. Violini, Fort. der Physik 17, 467 (1969).

J. J. Sakurai, *Invariance Principles and Elementary Particles* (Princeton University Press, 1964).

U. K. Samaranayake and W. S. Woolcock, Phys. Rev. Letters 15, 936 (1965).

M. D. Scadron and H. F. Jones, Phys. Rev. 173, 1734 (1968).

E. C. Titchmarsh, *The Theory of Fourier Integrals* (Clarendon Press, Oxford, 1948).

W. S. Woolcock, Thesis, London University (1961).

CHAPTER 4

ANALYTIC PROPERTIES OF SCATTERING AMPLITUDES

In the last chapter, it was shown that the scattering amplitudes for two-body collisions at fixed momentum transfer were analytic functions of energy except for cuts and poles on the real axis. This analyticity was sufficient for the derivation of certain dispersion relations, which were of great importance in the analysis of scattering data. A further advance in understanding can be achieved by studying the continuation of the momentum transfer to complex values, and indeed it is possible that the analytic properties of the scattering amplitude as a function of all its variables may be sufficient to determine the dynamics of the strong interactions of elementary particles, when used in conjunction with the crossing and unitarity properties of the S-matrix.

To extend the analytic properties already discussed, the methods of rigorous field theory can be used. This procedure has proved extremely difficult and, although significant progress has been made (Martin, 1967; Bros et al., 1965, 1967), more extensive results have come by examining the problem from other points of view. If it is conjectured that the perturbation series expansions of the scattering amplitude display the correct analytic structure of the theory (despite the lack of convergence of these series), then a term-by-term examination of the series will reveal this structure. Much progress has been made in this direction and many results have been shown to be true to all orders in the perturbation series. A different approach has become known as axiomatic S-matrix theory. In this theory certain fundamental postulates about the S-matrix, including unitarity, crossing, and relativistic invariance are made from the start. It is then supposed that the S-matrix is an analytic function of all its variables, except for such singularities as are required by the initial postulates. The extensive results found in this way are equivalent to those obtained from the perturbation series, but have the advantage of not depending on the concept of the local field, which may turn out to be an inappropriate vehicle for the description of elementary particle interactions. It is not our object to discuss any of these approaches in detail in this work [good accounts have been given recently in the books by Eden et al., (1966) and Chew (1966)], but some of the main results that are important for the analysis of the pion-nucleon system will be summarized.

4.1. POTENTIAL SCATTERING

Before examining the relativistic S-matrix, it is very instructive to consider the simple, but useful, model of nonrelativistic elastic scattering of a particle by a central potential.[1] In general, the analytic properties of the scattering amplitude depend on the particular potential assumed. For our purposes, the most relevant potential is that suggested by Yukawa, who showed that the exchange of a meson of mass M between two nucleons gave rise, in second order perturbation theory, to a potential:

$$V(r) = A\, r^{-1} \exp(-Mr) . \qquad (4\text{-}1)$$

The results to be reviewed will hold for all potentials that can be represented as a superposition of Yukawa potentials:

---

[1] Proof of the analytic properties about to be described may be consulted in the work by d'Alfaro and Regge (1965) or in the review of Martin (1965).

4.1. POTENTIAL SCATTERING

$$V(r) = r^{-1} \int_M^\infty \sigma(\lambda) \exp(-\lambda r) \, dr \qquad (4\text{-}2)$$

where $\sigma(\lambda)$ is a distribution, limited by the requirement that the integral I exists where:

$$I = \int_0^\infty r \, |V(r)| \, dr \, .$$

The Schrödinger equation for the wave function, $\psi(\vec{r})$, of a particle of mass M moving in a potential $V(r)$ is:

$$[\nabla^2 + s - V(r)] \, \psi(\vec{r}) = 0 \qquad (4\text{-}3)$$

where s is the energy.[2] The solution of eq. (4-3), which for large r is asymptotic to an incident plane wave and an outgoing spherical wave, is:[3]

$$\psi(\vec{r}) = \exp(i \vec{q} \cdot \vec{r}) + \int G_0(\vec{r}, \vec{r}\,') \, V(r') \, \psi(\vec{r}\,') \, d\vec{r}\,' \qquad (4\text{-}4)$$

where $\vec{q}$ is the incident momentum with $q^2 = s$, and $G_0(\vec{r}, \vec{r}\,')$ is the Green's function for a free particle:

$$G_0(\vec{r}, \vec{r}\,') = -\frac{1}{4\pi} \frac{e^{iq|\vec{r}-\vec{r}\,'|}}{|\vec{r}-\vec{r}\,'|} \, . \qquad (4\text{-}5)$$

For large r, the asymptotic form of $\psi(\vec{r})$ is:

$$\psi(\vec{r}) \underset{r \to \infty}{\sim} \exp(i \vec{q} \cdot \vec{r}) + f(s, \theta) \, r^{-1} \exp(i q r) \qquad (4\text{-}6)$$

where $f(s, \theta)$ is the scattering amplitude, which is a function of both s and the scattering angle θ. From eqs. (4-4), (4-5), and (4-6) it follows that:

---

[2] We shall use units for which $2M = \hbar = 1$.

[3] It is assumed that the reader is familiar with elementary nonrelativistic scattering theory as presented in, for example, Messiah (1964).

$$f(s, \theta) = -\frac{1}{4\pi} \int d\vec{r} \, \exp(-i\,\vec{q}\,'\cdot \vec{r}) \, V(r) \, \psi(\vec{r}) \tag{4-7}$$

where $\vec{q}\,'$ is the momentum of the scattered particle and $q\,'^2 = q^2$.

Fixed Momentum Transfer

It is convenient to consider f as a function of s and t, rather than of s and θ, where t is the negative of the square of the momentum transfer:

$$t = -(\vec{q}\,' - \vec{q})^2 = -2s(1 - \cos\theta) \ . \tag{4-8}$$

The variables s and t are of the nonrelativistic analogues of the s and t introduced in Chapter 3. The amplitude $f(s, t)$ can be exhibited as a series (the Born series) by iterating eq. (4-4) for $\psi(\vec{r})$ and using eq. (4-7), and the analytic properties of each term can be examined (Klein and Zeemach, 1958). Alternatively the Fredholm solution of eq. (4-4) can be used to provide the same information (Khuri, 1957). It is found for fixed negative real values of t in the range:

$$-4M^2 \leq t \leq 0$$

that $f(s, t)$ is analytic in the complex s plane, except for a cut extending along the real axis in the interval $0 \leq s \leq \infty$ and for poles at the energies of bound states of the system on the negative real axis. The sheet on which $f(s, t)$ is analytic is called the physical sheet, and if f is considered as a function of q rather than of s, it corresponds to the upper half q plane, $\mathrm{Im}\, q > 0$. The amplitude $f(q, t)$ in the lower half q plane corresponds to the second Riemann sheet of s reached by continuing $f(s, t)$ through the cut in the s plane. It will be recognized that this structure is similar to that of the relativistic amplitude discussed in Chapter 3, with the difference that in that case, there is a second cut along the negative s axis which is required by crossing symmetry.

It is easily verified that $f(s, t)$ satisfies the relation:

$$f^*(s^*, t) = f(s, t)$$

4.1. POTENTIAL SCATTERING

from which it follows that $f(s, t)$ is real along the real axis below the cut, and the discontinuity across the cut is $2\mathrm{Im}\, f(s, t)$. It can also be shown that for large $|s|$, $f(s, t)$ approaches the first Born approximation $f_B(t)$, obtained by substituting the incident plane wave in the right-hand side of eq. (4-7):

$$f(s, t) \underset{|s|\to\infty}{\sim} f_B(t)$$

$$f_B(t) = -\frac{1}{4\pi} \int dr \, \exp\left[i(\vec{q}-\vec{q}\,')\cdot \vec{r}\right] V(r) = \int_{M^2}^{\infty} d\lambda \, \frac{\sigma(\lambda)}{\lambda^2 - t} \, . \qquad (4\text{-}9)$$

The Born approximation $f_B(t)$ is just the Fourier transform of the potential, and is an analytic function in the t plane cut along the real axis in the interval $M^2 \leq t \leq \infty$.

These properties are sufficient for us to write the dispersion relation:

$$f(s, t) - f_B(t) = \sum_i \frac{R_i(t)}{(s+s_i)} + \frac{1}{\pi} \int_0^\infty \frac{[f(s', t)]_s}{(s'-s-i\epsilon)} ds' \, . \qquad (4\text{-}10)$$

In this equation we have written the discontinuity across the cut as $2i[f(s, t)]_s$ and for $-4M^2 \leq t \leq 0$, this is equal to $2i \,\mathrm{Im}\, f(s, t)$. Later when we continue $[f(s, t)]_s$ to complex values of t, it will no longer be possible to identify the discontinuity with Im f. To see that the pole positions at $s = -s_i$ correspond to the energies of bound states, we remember that the physical sheet maps onto the upper half q plane, so that the poles in the q plane are on the positive imaginary axis at positions $q = i|\sqrt{s_i}|$. From eq. (4-6) it is seen that $\psi(r)$ is normalizable at such points, so that the energies $-s_i$ must be the eigenvalues of the Schrödinger equation (4-3). As the bound states occur for definite values of the angular momentum, the residues $R_i(t)$ must be proportional to $P_\ell(\cos \theta) = P_\ell(1 + t/2s)$.

Analyticity in the Momentum Transfer

With the use of the partial wave expansion, the amplitude can be shown to be analytic in a limited region in the complex t plane, corresponding to the interior of the Lehmann ellipse (see Chapter 3). This is not sufficient for a dispersion relation to be written, but from the Born series or from the Fredholm solution, it has been shown that the discontinuity $[f(s,t)]_s$ is in fact analytic in the entire complex t plane (for real s), cut along the real axis in the interval $4M^2 \le t \le \infty$. A dispersion relation for $f(s,t)$ can now be written, provided that $f(s,t)$ behaves suitably for large $|t|$. Using the method of complex angular momenta we shall show later that $[f(s,t)]_s$ is bounded by a polynomial in t for large $|t|$. If the polynomial is of order N, we can write a dispersion relation for $[f(s,t)]_s$ of the form:

$$[f(s,t)]_s = \sum_{n=0}^{N-1} \left[\frac{1}{n!} C_n(s) t^n\right] + \frac{t^N}{\pi} P \int_{4m^2}^{\infty} \frac{\rho(s,t')}{t'^N(t'-t-i\epsilon)} dt' . \quad (4\text{-}11)$$

The subtraction constants $C_n(s)$ are functions of s, and are known as single spectral functions, while the discontinuity across the cut in the t plane $\rho(s,t)$ is known as a double spectral function. The functions $C(s)$ and $\rho(s,t)$ are real. Inserting eq. (4-11) into eq. (4-10), a representation of $f(s,t)$ as a function of both complex s and complex t is found, which is known as the Mandelstam representation:

$$f(s,t) - f_B(t) = \sum_i \frac{R_i(t)}{(s+s_i)} + \sum_{n=0}^{N-1} \left[\frac{t^N}{\pi n!} \int_0^\infty \frac{C_n(s') ds'}{(s'-s-i\epsilon)}\right]$$

$$+ \frac{t^N}{\pi^2} \int_0^\infty ds' \int_{4m^2}^\infty dt' \frac{\rho(s',t')}{t'^N(s'-s-i\epsilon)(t'-t-i\epsilon)} . \quad (4\text{-}12)$$

The order of integration is (4-12) can be reversed, so that for fixed s we have:

$$f(s,t) = \frac{1}{\pi} \int_{m^2}^\infty dt' \frac{[f(s,t')]_t}{(t'-t-i\epsilon)} . \quad (4\text{-}13)$$

For simplicity we have supposed there are no bound states and that $N = 0$ and $[f(s,t)]_t$ denotes the discontinuity across the t cut. It should be noticed that the t cut starts at $t = M^2$, rather than $4M^2$, because of the cut in $f_B(t)$. We have:

4.1. POTENTIAL SCATTERING

$$[f(s,t)]_t = [f_B(t)]_t + \frac{1}{\pi}\int_0^\infty ds' \frac{\rho(s',t)}{(s'-s)} \qquad (4\text{-}14)$$

and as $\rho = 0$ for $t < 4M^2$, it will be noticed that $[f(s,t)]_t = [f_B(t)]_t$ in the interval $M^2 \leq t \leq 4M^2$.

The most remarkable property of the Mandelstam representation is that if $f_B(t)$ is given (which is equivalent to being given the potential), the double spectral function can be constructed[4] by inserting eq. (4-14) into the unitarity condition:

$$\text{Im } f(s,t) = \frac{\sqrt{s}}{4\pi} \int d\Omega''(\theta'',\phi'') \, f^*(s,t_1) \, f(s,t_2) \qquad (4\text{-}15)$$

where:

$$t_1 = -(\vec{q}'-\vec{q}'')^2, \; t_2 = -(\vec{q}-\vec{q}'')^2, \; q''^2 = q^2$$

and (θ'',ϕ'') are the polar angles of $\vec{q}''$.

Once the double spectral function is known, the scattering amplitude can be determined from eq. (4-12), provided that no subtraction terms in t are required. Even if subtractions are required, the scattering amplitude can be determined in principle by one of the several methods. For example, for sufficiently large s, $f(s,t)$ approaches the Born term $f_B(t)$ so that no subtractions are required for large s, and the amplitude can be determined. The scattering amplitude for all s can now be found uniquely by analytic continuation in s. Other methods can also be devised based on the partial wave dispersion relations discussed below. We see that for potential scattering, the Mandelstam representation, together with the unitarity equation, constitute dynamical equations, which take the place of Schrödinger's equation. It is the hope of many theorists, but by no means proved, that a knowledge of the analytic structure of relativistic scattering amplitudes together with unitarity and the crossing property (which has no nonrelativistic analogue) may be in a similar way sufficient to determine elementary particle dynamics.

---

[4] Blankenbecler et al., (1960).

Partial Wave Amplitudes and Complex Angular Momentum

In obtaining the dispersion relation for $[f(s,t)]_s$, a polynomial bound in t was assumed. To examine this further, and in addition to obtain some results that will be important for the discussion of high energy scattering in Chapter 8, it is necessary to discuss the partial wave expansion of $f(s,t)$. The wave function can be expanded in the form:

$$\psi(r) = \sum_{\ell=0}^{\infty} r^{-1} \chi_\ell(r) P_\ell(\cos\theta) \tag{4-16}$$

and χ_ℓ obeys the radial Schrödinger equation:[5]

$$\frac{d^2 \chi_\ell(r)}{dr^2} + \left[s - V(r) - \frac{\ell(\ell+1)}{r^2} \right] \chi_\ell(r) = 0 . \tag{4-17}$$

The corresponding expansion of the scattering amplitude $f(s,\theta)$ is expanded as:

$$f(s,\theta) = \sum_{\ell=0}^{\infty} (2\ell + 1) f_\ell(s) P_\ell(\cos\theta) \tag{4-18}$$

and the boundary conditions satisfied by the $\chi_\ell(r)$ are:

$$\chi_\ell(0) = 0, \quad \chi_\ell(r) \sim q^{-1} i^\ell \sin(qr - \tfrac{1}{2}\ell\pi) + f_\ell(s) e^{iqr}, \tag{4-19}$$

for $r=0$ and $r \to \infty$ respectively. From Chapter 2 we know that the unitarity condition shows that $f_\ell(s)$ can be expressed in terms of a real phase shift δ_ℓ by:

$$f_\ell(s) = \frac{1}{2iq} \left(e^{2i\delta_\ell} - 1 \right) . \tag{4-20}$$

From eq. (4-18):

---

[5] There are two independent solutions of (4-17). The physical solution is the one for which $r^{-1} \chi_\ell(r)$ is finite for small r and satisfies the boundary condition $\lim_{r \to 0} r^{-\ell-1} \chi_\ell(r) = 0$.

4.1. POTENTIAL SCATTERING

$$f_\ell(s) = \frac{1}{2} \int_{-1}^{+1} d\cos\theta \, P_\ell(\cos\theta) \, f(s,\theta)$$

$$= -\frac{1}{4s} \int_0^{-4s} dt \, P_\ell\left(1 + \frac{t}{2s}\right) f(s,t) \qquad (4\text{-}21)$$

where in the second line the integration variable has been changed from $\cos\theta$ to t. Referring to eq. (4-10), it is seen that $f_\ell(s)$ will possess the "right-hand cut" along the real axis extending over the interval $0 \leq s < \infty$ together with those bound state poles at $s = s_i$ that occur in the angular momentum state ℓ; but in addition, from eq. (4-13), $f_\ell(s)$ is seen to possess a further cut, the "left-hand cut," along the real axis extending over the interval $-\infty < s < -\frac{M^2}{4}$ (Fig. 4-1). As $f_\ell(s)$ is real analytic, the discontinuity across the cuts is $2i \, \text{Im} \, f_\ell(s)$. On the right-hand cut, the discontinuity can be related to the partial cross section through the unitarity conditions:

$$\text{Im} \, f_\ell(s) = \sqrt{s} \, f_\ell^*(s) \, f_\ell(s) \qquad (4\text{-}22)$$

while on the left-hand cut the discontinuity is determined by the potential. In place of eq. (4-18), the asymptotic form of $X_\ell(r)$ for large r can be written as:

$$X_\ell(r) \underset{r \to \infty}{\sim} [\phi^-(\ell, s) \, e^{iqr} + \phi^+(\ell, s) \, e^{-iqr}] \qquad (4\text{-}23)$$

This differs from $X_\ell(r)$ defined in eq. (4-19) by a complex constant.

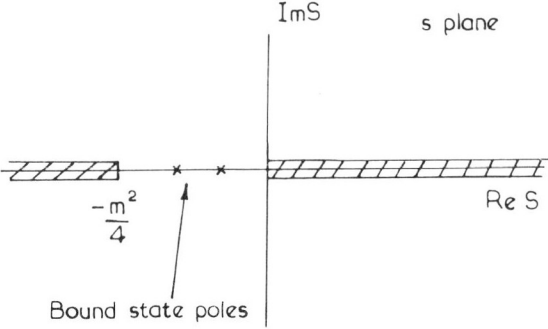

Fig. 4-1. The analytic structure of the partial wave amplitude $f_\ell(s)$.

ϕ^- and ϕ^+ are called Jost functions, and it is evident from eqs. (4-19) and (4-20) that they are connected with the phase shifts by:

$$S(\ell, s) = e^{2i\delta_\ell} = (-1)^{\ell+1} \frac{\phi^-(\ell, s)}{\phi^+(\ell, s)} \qquad (4\text{-}24)$$

where $S(\ell, s)$ is the S-matrix. Now solutions like $X_\ell(s, r)$ of a differential equation are generally analytic functions of the parameters, in this case s and ℓ.[6] This circumstance allows the analytic properties of the Jost functions and of the partial wave amplitudes to be determined directly. For example by constructing integral equations for the Jost functions in terms of the potential and X_ℓ, it can be shown that these functions are analytic in s and ℓ except for certain singularities. As a function of q (for fixed integral ℓ), $\phi^+(\ell, q)$ is analytic in the q plane except for a cut along the imaginary axis in the interval $-\infty < \text{Im} q < -M/2$. From eq. (4-20) and the Schrödinger equation we have:

$$\phi^+(\ell, q) = [\phi^-(\ell, q^*)]^* \qquad (4\text{-}25)$$

so that $\phi^-(\ell, q)$ is analytic except for a cut along the positive imaginary q axis in the interval $\frac{M}{2} \le \text{Im} q < \infty$. As a function of s, $\phi^\pm(\ell, s)$ must have a square root branch point at $s = 0$ arising from the relation $q = \sqrt{s}$. Taking the branch cut along the positive real axis, the physical sheet of $\phi^+(\ell, s)$ is defined as that for which $\text{Im} q > 0$; for the unphysical sheet $\text{Im} q < 0$. The cuts in the q plane then map onto the interval $-\infty < s < M^2/4$, on the physical sheet for ϕ^- and on the unphysical sheet for ϕ^+.

From eqs. (4-20) and (4-24), we see that:

$$f_\ell(s) = \frac{i}{2q} \left[\frac{\phi^+(\ell, s) + (-1)^\ell \phi^-(\ell, s)}{\phi^+(\ell, s)} \right] \qquad (4\text{-}26)$$

---

[6] This holds by a theorem of Poincaré if the boundary conditions are independent of the parameters. The boundary condition at the origin is independent of s, so that $X_\ell(s, r)$ is an (entire) analytic function of s. Slightly different boundary conditions are required for the case of complex ℓ.

4.1. POTENTIAL SCATTERING

and from the properties of ϕ^+ and ϕ^-, $f_\ell(s)$ is seen to possess the right- and left-hand cuts that we obtained earlier from eqs. (4-10) and (4-13).

The amplitude will have a pole if $\phi^+(\ell, s)$ vanishes: in that case $\chi_\ell(s, r) \underset{r \to \infty}{\approx} \phi^-(\ell, s) \, e^{iqr}$. If this pole is on the physical sheet, $\mathrm{Im}\, q > 0$, $\chi_\ell(s, r)$ will be a damped exponential and a normalizable function, which must correspond to a physical boundstate, with s real and negative.[7] These are the only poles on the physical sheet, but on the unphysical sheet there may be poles anywhere. If near the real positive s axis, they correspond to resonances, which were mentioned in Chapter 2 and which will be discussed fully in Chapter 5.

Complex Angular Momentum

Consider for the moment a square well potential, rather than a superposition of Yukawa potentials: then for each integer value of ℓ, there may be bound states as shown in Fig. 4-2. We are free to draw interpolating curves between the bound states at different ℓ, in various ways. However the bound states are given by the solution of:

$$\phi^+(\ell, s) = 0 \qquad (4\text{-}27)$$

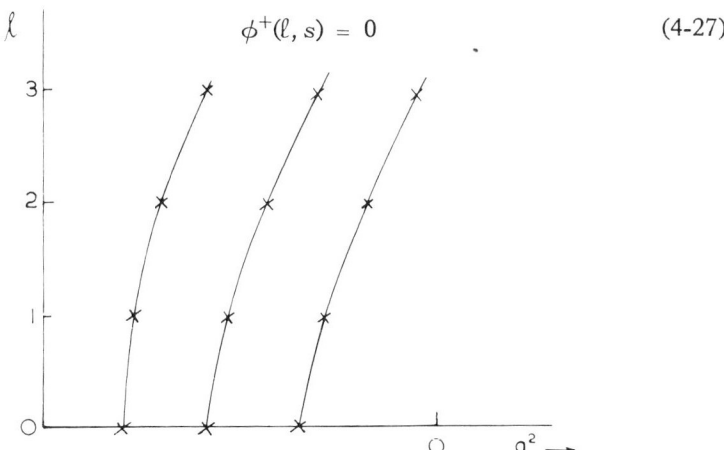

Fig. 4-2. The bound states of a square well potential given by the zero of $\phi^+(\ell, q^2)$, connected by interpolating curves.

---

[7] This is essentially the reason for the terminology "physical" and "unphysical" sheets.

for integer ℓ, and a unique interpolation[8] is obtained by solving this equation for noninteger values of ℓ, with solutions:

$$\ell = a(s) . \qquad (4\text{-}28)$$

For negative real s, $a(s)$ is real (associated with the reality of ϕ^+).[9] For $s > 0$, $\phi^+(\ell, s)$ becomes complex and the solution, eq. (4-28), becomes a complex number.

These poles of the amplitude $f(\ell, s)$ as a function of complex ℓ (real s) are called Regge poles and the functions $a(s)$ as functions of s are called Regge trajectories. For a Yukawa potential, it can be shown that $f(\ell, s)$ is meromorphic for $\operatorname{Re} \ell > -\frac{1}{2}$ and that the Regge poles are confined to the region $\operatorname{Im} \ell > 0$. For large $|\ell|$, $f(\ell, s)$ decreases like an exponential in the real direction and behaves like a power of ℓ in the imaginary direction. Because of the analyticity of $f(\ell, s)$, and because $P_\ell(\mu)$

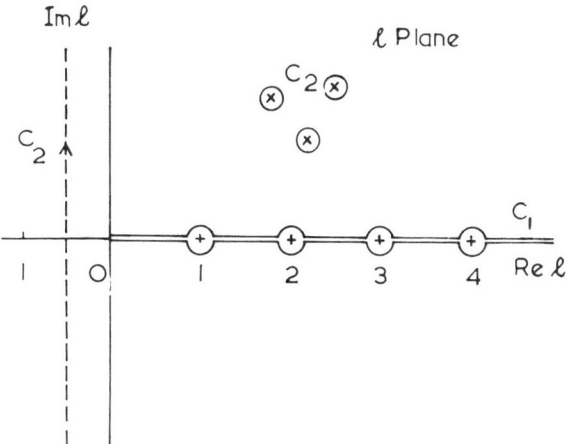

Fig. 4-3. The contour C_1 for the Sommerfeld-Watson integral, Eq. (4-29). If the contour is opened into a line integral C_2, each of the Regge poles marked by crosses must be encircled as shown.

---

[8] See d'Alfaro and Regge (1965).

[9] For $s < 0$, $\chi_\ell(s, r)$ is real and the exponentials $e^{\pm iqr}$ are real; so the coefficients $\phi^\pm(\ell, s)$ in eq. (4-23) are also real. The zero of $\phi^+(\ell, s)$ is a single zero at ℓ integer; from analyticity in ℓ the single zero cannot turn into a double zero for ℓ noninteger, which would be required for a to be complex.

4.1. POTENTIAL SCATTERING

is analytic in the μ plane cut from $-\infty < \mu < -1$ (Morse and Feshbach, 1953), the partial wave series of eq. (4-18) can be expressed as a contour integral (Fig. 4-3), known as the Sommerfeld-Watson transformation:

$$f(s,\theta) = \frac{1}{2} i \int_{C_1} \frac{(2\ell+1) f(\ell,s) P_\ell(-\cos\theta)}{\sin \pi \ell} d\ell . \qquad (4\text{-}29)$$

The residues of the poles of $(1/\sin \pi \ell)$ at integer ℓ are $(-1)^\ell/\pi$, and the factor $(-1)^\ell$ is taken care of by writing $P_\ell(-\cos\theta)$, which is equal to $(-1)^\ell P_\ell(\cos\theta)$ for integer ℓ. The contour C_1 can now be opened into a line integral along Re $\ell = -\frac{1}{2}$ and closed by an infinite semicircle. Because $f(\ell,s)$ vanishes as $|\ell| \to \infty$ and because:

$$|P_\ell(\cos\theta)| < \frac{C}{\sqrt{|\ell| \sin\theta}} e^{\mathrm{Im}\, \ell(\pi-\theta)}, \; |\ell| \to \infty, \; \theta \text{ real}$$

the contribution of the semicircle to the integral is zero. In opening the contour, we must allow for the poles in $f(\ell,s)$ by encircling each of them as shown in Fig. 4-3. We find that:

$$f(s,\theta) = \sum_i \frac{\beta_i(s) P_{a_i(s)}(-\cos\theta)}{\sin \pi a_i(s)} + \frac{i}{2} \int_{-\frac{1}{2}-i\infty}^{-\frac{1}{2}+i\infty} d\ell \frac{(2\ell+1) f(\ell,s) P_\ell(-\cos\theta)}{\sin \pi \ell} \qquad (4\text{-}30)$$

where $\beta_i(s)$ is $-(2\ell+1)\pi$ times the residue of $f(\ell,s)$ at the pole $\ell = a_i(s)$.

This representation can be used to continue $f(s,\theta)$ to complex values of θ or of t. In particular, the behavior of $f(s,t)$ or of $[f(s,t)]_s$ at large $|t|$ can be studied.

Using the result:

$$|P_{i\lambda-\frac{1}{2}}(-\cos\theta)| < \frac{C}{\sqrt{|\sin\theta|}} \frac{e^{\lambda(\pi-\mathrm{Re}\,\theta)}}{\sqrt{\lambda}}, \; \lambda \text{ real}$$

the integral term in eq. (4-30) can be seen to behave like $|t|^{-\frac{1}{2}}$ for large t ($\cos\theta = 1 + t/2s$); also for large $|t|$, $P_{a_i(s)}(\cos\theta)$ behaves like

$t^{a_i(s)}$, so that the asymptotic form of $f(s,t)$ depends on the pole with the largest value of Re $a_i(s)$. As it can be shown that there are a finite number of poles, this establishes the polynomial bound which was necessary for the establishment of the dispersion relation, eq. (4-11), and the Mandelstam representation, eq. (4-12).

We now consider one of the Regge pole terms of eq. (4-30) and see how it may give rise to a bound state. A Regge pole term is:

$$\frac{\beta(s) P_{a(s)}(-\cos\theta)}{\sin \pi a(s)} . \tag{4-31}$$

This gives a pole in $f(s,\theta)$ (that is a bound state), when ℓ is an integer. We have previously noted this property (page 130) and that it occurs for a negative value of s, say $-s_i$:

$$a(-s_i) = \ell, \text{ where } \ell \text{ is integer }.$$

We can expand the eq. (4-31) about this value and get:

$$f(s,\theta) = \frac{\beta(-s_i) P_\ell(-\cos\theta)}{(s+s_i)\pi \left(\frac{\partial a}{\partial s}\right)_{s=s_i} \cos(\pi\ell)} + \ldots . \tag{4-32}$$

On comparing eq. (4-32) with eq. (4-18), we see that eq. (4-32) represents a pole of $f_\ell(s)$ in the s plane at $s = -s_i$ with residue:

$$\beta(-s_i)\left[(2\ell+1)\pi\left(\frac{da}{ds}\right)_{s=s_i}\right]^{-1} .$$

A resonance arises when for some values s_R of s:

$$a(s_R) = \ell_1 + i\eta_1$$

where ℓ_1 is an integer and η_1 is small and:

$$\left.\frac{da}{ds}\right|_{s=s_R} = \gamma, \gamma \text{ real }.$$

4.1. POTENTIAL SCATTERING

Then for $s \approx s_R$, $\alpha(s) = (s-s_R)\gamma + \ell_1 + i\eta_1$, and eq. (4-32) becomes:

$$\frac{\beta(s_R) P_{\ell_1}(\cos\theta)}{\pi[\gamma(s-s_R) + i\eta_1]} \tag{4-33}$$

and:

$$f_{\ell_1}(s) \approx \frac{\beta(s_R)}{(2\ell_1+1)\pi[\gamma(s-s_R) + i\eta_1]} \tag{4-34}$$

This is in the form of a Breit-Wigner resonance with half width (η_1/γ). If the background is negligible, unitarity shows that:

$$\frac{(2\ell_1+1)\pi\eta_1}{q_R \beta} = -1 \quad . \tag{4-35}$$

Both the bound and resonance situation can occur on the one Regge trajectory as illustrated in Fig. 4-4.

The importance of the idea of expressing scattering amplitudes in terms of Regge poles is not only that in this way the Mandelstam representation is given a meaning, but, as we shall see, particles can be

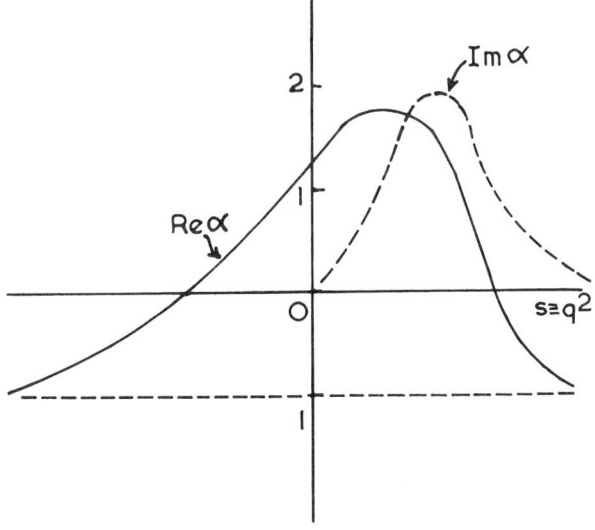

Fig. 4·4

Fig. 4-4. Typical Regge trajectory for scattering by a Yukawa well.

classified according to the trajectories on which they lie, and also, when combined with crossing, these ideas give rise to a theory of high energy scattering, which is outlined in Chapter 8.

4.2. THE RELATIVISTIC S-MATRIX

Elementary particle scattering is necessarily much more complicated than potential scattering; in particular production processes of arbitrary complexity are possible if sufficient energy is available, and this complexity will be reflected in the analytic structure. Nevertheless, there are reasons to believe that the scattering amplitude for a two-body scattering process $A + B \to C + D$ satisfies the Mandelstam representation, in a crossing symmetric form, for certain values of the masses, and that for other mass values, the Mandelstam representation may be a good approximation. Later in this chapter we will show that the assumption of the Mandelstam representation allows the derivation of partial wave dispersion relations, which have been used, for example, in the prediction of π-N phase shifts. Before this, we shall sketch briefly some of the analytic properties of the relativistic amplitude, referring the interested reader for a full discussion to the books by Chew (1966) and Eden et al., (1966).

Crossing

Consider for simplicity the interaction of particles without spin or charge. If the particles are labeled 1, 2, 3, and 4, then the elastic scattering process:

$$1 + 2 \to 3 + 4 \tag{4-35}$$

is represented by Fig. 4-5a, where the four momentum, p_i of each particle is shown. We have the usual constraints that:

$$p_1 + p_2 = p_3 + p_4 \quad \text{and} \quad p_i^2 = M_i^2$$

4.2. THE RELATIVISTIC S-MATRIX

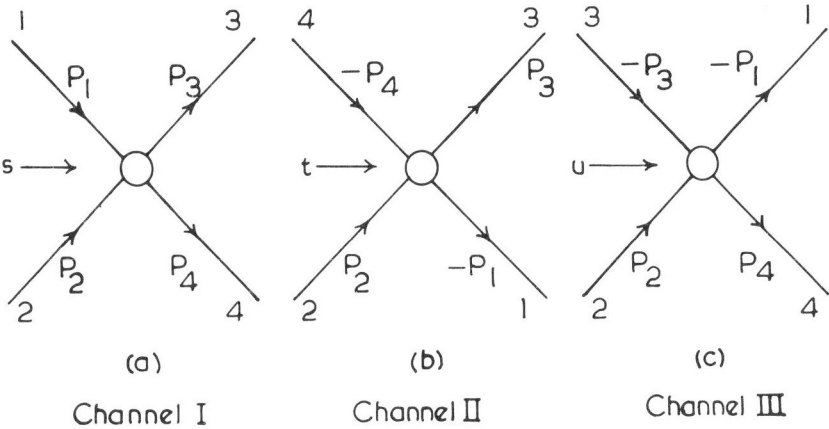

Fig. 4-5. Amplitudes connected by crossing.

where M_i is the mass of particle i. In place of the p_i, we shall usually employ the invariants s, t, and u introduced in Chapter 3 page (94). The invariant s is the square of the center of mass energy for the reaction of Fig. 4-5a, which we shall refer to as Channel I.

If the reduction formulae are used, as we saw in Chapter 3, the same invariant scattering amplitude[10] $T(s,t)$ describes the reaction of Fig. 4-5a and that shown in Fig. 4-5b, which is Channel II:

$$4 + 2 \to 1 + 3 \ . \tag{4-36}$$

Equally well we can "cross" the lines representing particles 1 and 3 and obtain the reaction (Channel III):

$$3 + 2 \to 1 + 4 \tag{4-37}$$

and the same scattering amplitude will describe this reaction (see Fig. 4-5c). The regions of the independent variables s and t, for which physical scattering is possible in each channel, do not overlap, and, as

---

[10] We deal here with amplitudes defined by the time ordered product rather than by the retarded product, and the crossing relation does not involve complex conjugation.

we have remarked before, for this crossing property to be meaningful an analytic continuation from one region to the other must be possible. The existence of sufficient analyticity for this continuation has been proved from field theory by Bros et al., (1965).

In the physical region of Channel II, the variable t represents the square of the center of mass energy, while in the physical region of Channel III, u is this quantity.

The crossing property can be generalized to a production amplitude. If such an amplitude is represented by a diagram such as that in Fig. 4-5a but with n lines entering or leaving the figure, then one invariant amplitude, a function of the $(3n-10)$ independent scalars that can be formed from the four-momenta $p_1, p_2 \ldots p_n$, describes all the processes that can be described by crossing the n lines in any way.

Normal Thresholds and Poles

It will be now assumed that the scattering amplitude is an analytic function of all its variables except for certain singularities, which are required by the existence of particles and by the unitarity condition. We have already seen, in the last chapter, that the scattering amplitude must possess a pole if a stable particle exists having the same quantum numbers as those of one of the reaction channels. The pole is on the real energy axis for the appropriate channel, for example if Fig. 4-5a represents π-N scattering, then the nucleon will give rise to a pole at $s = +M_N^2$ in Channel I and a pole at $u = M_N^2$ in Channel III. The contribution to the amplitude $T(s,t)$ from these poles will be of the form:

$$\frac{g^2}{s-M_N^2} + \frac{g^2}{u-M_N^2} \ . \tag{4-38}$$

As unstable particles (resonances) are observed, these must give rise to further poles, which, as will be demonstrated in Chapter 5, are on an unphysical sheet of the amplitude at points away from the real energy axis in any channel.

4.2. THE RELATIVISTIC S-MATRIX

The next singularities of the amplitude that we have seen to be present are the branch points that arise at the thresholds for each new physical process. We restate the argument here. Below all thresholds, by examining the absorptive part of the amplitude, as we did in Chapter 3, we can show that the scattering amplitude is real, so that below the Channel I thresholds:[11]

$$T_I(s, t) - T_I^*(s, t) = 0 \ . \tag{4-39}$$

At the elastic scattering threshold, which in the case of particles of equal mass is at $s = 4M^2$, the unitarity condition states that:

$$T_I(s, t) - T_I^*(s, t) = \sum_j T_{I \to j}(s, t) \, T_{j \to I}^*(s, t) \tag{4-40}$$

where the sum runs over all energetically possible states; and in the case of elastic scattering, this reduces to an integration over the angular variable of the intermediate state, as in eq. (4-15). The discontinuous change from the right-hand side of eq. (4-39) to the right-hand side of eq. (4-40) implies a singularity, in this case a branch point in $T_I(s, t)$. When the energy reaches $s = (3M)^2$, the production of a further particle becomes possible and the sum over j in the unitarity equation changes discontinuously to include the corresponding terms, giving rise to a branch point at $s = (3M)^2$. Continuing this process, we find in the equal-mass case, a series of branch points, called normal thresholds, at $s = (2M)^2, (3M)^2, (4M)^2 \ldots$. Similar series of branch points arise in $T(s, t)$ from the physical thresholds in Channels II and III at thresholds in (for the equal-mass case) $t = (2M)^2, (3M)^2, (4M)^2 \ldots$ and $u = (2M)^2, (3M)^2, (4M)^2$.

If the cuts associated with the branch points in Channel I are drawn along the real s axis, the physical amplitude is defined as the boundary value of $T(s, t)$ obtained by approaching the cut from above:

---

[11] When we consider $T(s, t)$ in any particular channel, we attach a subscript I, II, or III.

$$T_I(s,t) = \underset{\epsilon \to 0^+}{\ell t}\, T(s+i\epsilon, t) \;.$$

Keeping in mind that $T(s,t)$ is real analytic in the sense that $T^*(s^*, t) = T(s, t)$, it is seen that $T_I^*(s, t)$, in eq. (4-40), is the boundary value of $T(s, t)$ obtained by approaching the cut from below:

$$T_I^*(s,t) = \underset{\epsilon \to 0^+}{\ell t}\, T(s-i\epsilon, t)$$

thus the left-hand side of the unitarity equation represents the discontinuity across the cut, and the equation shows how this discontinuity can be calculated.

Each term in the right-hand side of eq. (4-40) will itself satisfy a unitarity equation and possess poles and thresholds both in the direct and crossed channels, and these singularities give rise to further singularities in T on the left-hand side. In fact the unitarity equations can be iterated to build up a set of singularities, which can be shown to be the same as those given by the terms of the perturbation expansion of field theory.[12] The assumption that the only singularities of the scattering amplitude are those given by the particle poles, and by those implied by the unitarity equations, has been termed the hypothesis of maximum analyticity of the first kind (Chew, 1966).

The Box Diagram

The lowest (second) order terms in perturbation theory are of the form of eq. (4-38) and account for the stable single particle poles. For equal-mass scalar particles there is an identical term in Channel II of the form $g^2/(t-M^2)$ which must be added to the poles at $s = M^2$ and $u = M^2$, while for equal-mass pseudoscalar particles no corresponding poles are allowed in any of the channels by parity conservation. The next higher terms are of fourth order, and we examine the singularities associated with

---

[12] Except that the unstable particle poles have to be included in the perturbation expression as well as the stable particles.

4.2. THE RELATIVISTIC S-MATRIX

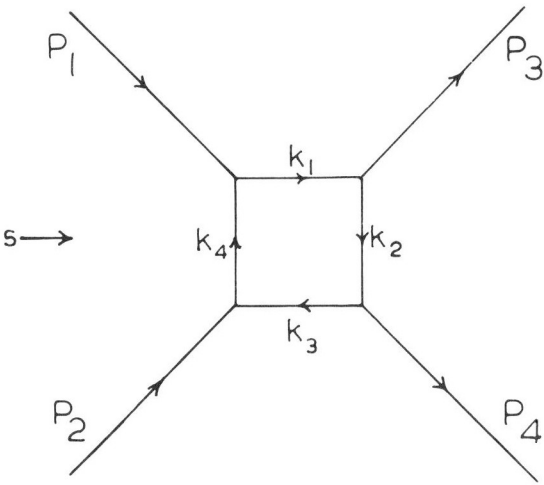

Fig. 4-6. A fourth order box diagram.

the Feynman diagram shown in Fig. 4-6 in the case of scalar particles of equal mass. The contribution of this box diagram to the scattering amplitude is (see for example Bjorken and Drell, 1965):

$$T(s,t) = g^4 \left(\frac{1}{2\pi}\right)^6 \int d^4 k_1 \int d^4 k_2 \int d^4 k_3 \int d^4 k_4 \, \delta(p_1+k_4-k_1)$$

$$\times \delta(k_1-k_2-p_3) \, \delta(k_2-k_3-p_4) \, \delta(p_2+k_3-k_4) \left[\prod_{i=1}^{4} (+k_i^2-M^2-i\epsilon)\right]^{-1}. \tag{4-41}$$

Integration over the delta functions that conserve energy and momenta at each vertex gives (omitting the overall energy-momentum conservation factor $\delta(p_1+p_2-p_3-p_4)$):

$$T(s,t) = \frac{g^4}{(2\pi)^6} \int d^4 k_1 \left[\prod_{i=1}^{4} (k_i^2-M^2-i\epsilon)\right]^{-1} \tag{4-42}$$

where k_2^2, k_3^2 and k_4^2 are now expressed in terms of the p_i and k_i. This integral is most easily reduced using the Feynman formula:

$$\prod_{i=1}^{4}\left(\frac{1}{A_i}\right) = (n-1)! \int_0^1 \left(\prod_{i=1}^{4} da_i\right) \delta\left(\sum_{j=1}^{n} a_j - 1\right)\left(\sum_{j=1}^{n} a_j A_j\right)^{-n} \quad (4\text{-}43)$$

giving:

$$T(s,t) = \frac{g^4}{(2\pi)^6}(3!) \int_0^1 da_1 \int_0^1 da_2 \int_0^1 da_3 \int_0^1 da_4 \int dk_1^4 \frac{\delta(a_1+a_2+a_3+a_4-1)}{\left[\sum_{j=1}^{4} a_j(k_j^2 - M^2 - i\epsilon)\right]^4}$$

$$(4\text{-}44)$$

In general, the singularities of the integral in a multiple integral can be avoided by deforming the contour of integration suitably. There are two cases for which this cannot be done and for which the singularity will be transmitted to the integral. The first case occurs when the singularity in the integrand is at an end point of the contour of integration and so cannot be avoided, and the second when two singularities coincide, "pinching" the contour of integration between them.

Looking at eq. (4-42), one can see that the integral converges when $k_1^\mu \to \pm\infty$, and no singularities of $T(s,t)$ arise from the boundary of the integration region. The other way in which $T(s,t)$ can become singular is for two poles in the integral to coincide, that is, for:

$$k_i^2 = M^2 \quad \text{and} \quad k_j^2 = M^2 \quad i \neq j$$

at some value of s and t.[13] It is impossible to realize this condition for real k_1, when k_i and k_j meet at a vertex, and the singularities can only arise either when:

$$k_1^2 = M^2, \; k_3^2 = M^2 \quad \text{(a)}$$

or when:

$$k_4^2 = M^2, \; k_2^2 = M^2 \quad \text{(b)} \quad . \quad (4\text{-}45)$$

---

[13] The $(i\epsilon)$ terms will be included in M^2 from now on.

4.2. THE RELATIVISTIC S-MATRIX

Expressing k_3 in terms of k_1, p_2, and p_1, it is found that $(k_1+k_3) = (p_1+p_2)$, from which $s = (k_1+k_3)^2$. The condition in eq. (4-45a) then can only be satisfied for $(k_1+k_3)^2 > 4M^2$, or $s > 4M^2$, so that k_1 and k_3 are physical momenta, and the intermediate two-particle state in Fig. 4-5 is a real state in Channel I, provided that the condition of eq. (4-45a) is satisfied. In a similar way, eq. (4-45b) is satisfied when k_2 and k_4 represent physical particles in Channel II and in this case $t > 4M^2$. To find the boundary curve of the region of analyticity, we examine eq. (4-44). For two singularities to coincide, we require the denominator

$$D = \sum_{j=1}^{4} a_j (k_j^2 - M^2)$$

to vanish, and D to be stationary with respect to variations in a_i and k_1^μ:

$$\frac{\partial D}{\partial a_i} = 0, \; i = 1, 2, 3, 4 \qquad \frac{\partial D}{\partial k_1^\mu} = 0, \; \mu = 0, 1, 2, 3 \; . \qquad (4\text{-}46)$$

The first of these conditions repeats eqs. (4-45a and b), and the second becomes:

$$\sum_{i=1}^{4} a_i k_{i\mu} \frac{\delta k_i^\nu}{\partial k_1^\mu} = 0, \; \mu = 0, 1, 2, 3 \; . \qquad (4\text{-}47)$$

The eqs. (4-46) and (4-47) are known as Landau equations, and take a similar form for diagrams other than Fig. 4-5. In the present case, eq. (4-47) becomes:

$$\sum_{i=1}^{4} a_i k_i^\mu = 0, \; \mu = 0, 1, 2, 3 \; .$$

The scalar products of these equations with each of the k_i can be taken in turn, and by defining $y_{ij} = (k_i \cdot k_j) / M^2$, the condition for a nontrivial solution of these equations is given by the vanishing of the determinant of the coefficients of the a_i:

$$\begin{vmatrix} 1 & y_{12} & y_{13} & y_{14} \\ y_{12} & 1 & y_{23} & y_{24} \\ y_{13} & y_{23} & 1 & y_{34} \\ y_{14} & y_{24} & y_{34} & 1 \end{vmatrix} = 0 \ . \tag{4-48}$$

In terms of s and t, it is found that:

$$y_{24} = \frac{s}{2M^2} - 1, \quad y_{13} = \frac{t}{2M^2} - 1$$

and $y_{12} = y_{14} = y_{23} = y_{34} = \frac{1}{2}$. The boundary curve of the region analytic in s and t is then, from eq. (4-48):

$$(s-4M^2)(t-4M^2) = 4M^4 \ . \tag{4-49}$$

For $t < 4M^2$, $T(s,t)$ will be analytic except for a cut along the real s axis, in the interval $4M^2 \leq s < \infty$. The branch point at $s = 4M^2$ is of course a normal threshold of the type we have already considered. Provided that $T(s,t)$ vanishes sufficiently rapidly for large s, a single variable dispersion relation can be written in s:

$$T(s,t) = \frac{1}{\pi} \int_{4M^2}^{\infty} \frac{[T(s',t)]_s}{(s'-s-i\epsilon)} ds' \tag{4-50}$$

where $2i[T(s',t)]_s$ is the discontinuity across the cut. For real $s' > 4M^2$, $[T(s',t)]_s$ will be analytic in t, with the exception of a cut along the real t axis, in the interval $t_1 \leq t < \infty$, where t_1 is the root of eq. (4-49) for $s = s'$. This discontinuity $[T(s',t)]_s$, then satisfies the equation:

$$[T(s',t)]_s = \frac{1}{\pi} \int_{4M^2}^{\infty} \frac{\rho_{12}(s',t')}{(t'-t-i\epsilon)} dt' \tag{4-51}$$

where $2i\rho_{12}(s',t')$ is the discontinuity across the cut, and the question of subtractions has again been ignored. Combining eq. (4-50) and eq. (4-51), we find a double dispersion relation for the amplitude given by the box diagram of Fig. 4-7:

4.2. THE RELATIVISTIC S-MATRIX

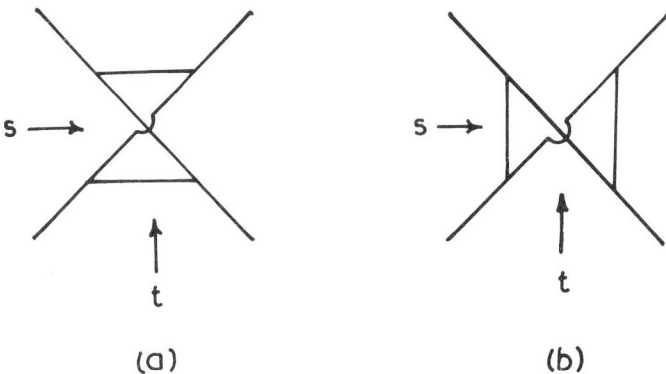

Fig. 4-7. Box diagram representing fourth order terms obtained from that of Fig. 4-6 by interchanging (a) t and u (b) s and u.

$$T(s, t) = \frac{1}{\pi^2} \int_{4M^2}^{\infty} ds' \int_{4M^2}^{\infty} dt' \frac{\rho_{12}(s', t')}{(s'-s-i\epsilon)(t'-t-i\epsilon)} \quad (4\text{-}52)$$

where the double spectral function ρ_{12} vanishes outside the boundary curve, eq. (4-49). Because of crossing symmetry, two further box diagrams are associated with the fourth order amplitude $T(s, t)$. These are shown in Fig. 4-7. Corresponding to these diagrams are contributions to the amplitude identical in form to eq. (4-52) but, with the variable s, t, u interchanged, in terms of spectral functions $\rho_{32}(u, t)$ and $\rho_{13}(s, u)$. The complete representation of the three fourth order diagrams taken together is as follows:

$$T(s, t) = \frac{1}{\pi^2} \int_{4M^2}^{\infty} ds' \int_{4M^2}^{\infty} dt' \frac{\rho_{12}(s', t')}{(s'-s)(t'-t)} + \frac{1}{\pi^2} \int_{4M^2}^{\infty} du' \int_{4M^2}^{\infty} dt' \frac{\rho_{32}(u', t')}{(u'-u)(t'-t)}$$

$$+ \frac{1}{\pi^2} \int_{4M^2}^{\infty} du' \int_{4M^2}^{\infty} ds' \frac{\rho_{13}(s', u')}{(s'-s)(u'-u)} \quad (4\text{-}53)$$

where the factors $i\epsilon$ have been suppressed. This is the relativistic version of the Mandelstam representation discussed in section 4.1. For the scattering of identical particles (for example in $\pi\pi$ scattering) the three

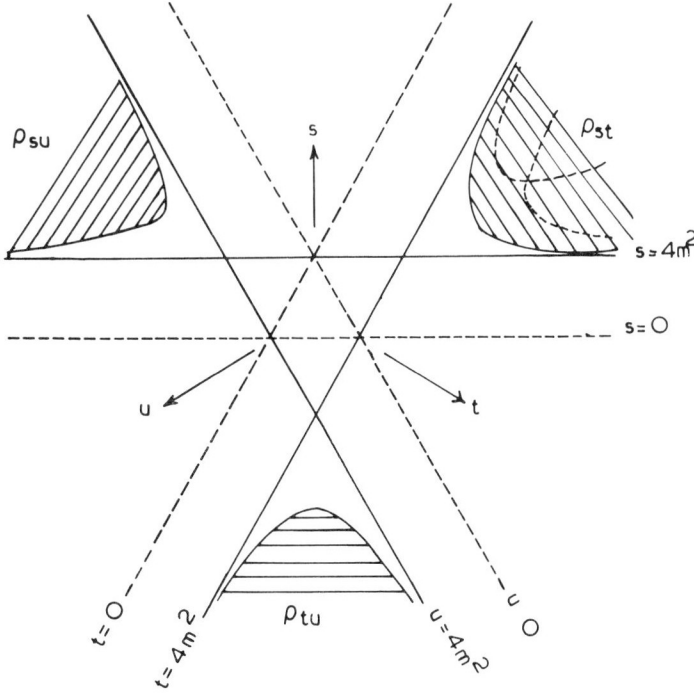

Fig. 4-8. The Mandelstam plot. The shaded regions indicate approximately where the three double spectral functions for the interaction of equal mass particles fail to vanish. The form of the boundary of $\rho_{12}(S, t)$ for $\pi\pi$ scattering is shown by the broken line.

functions ρ_{12}, ρ_{32}, and ρ_{13} are equal. The regions of physical scattering and the regions where ρ_{12}, ρ_{32}, and ρ_{13} fail to vanish are shown in Fig. 4-8. In general, subtractions may be required, and these will be discussed as needed in particular examples. The Mandelstam representation has been shown to hold for a large class of Feynman graphs provided that the masses of the particles concerned fall into certain ranges. The boundaries of the double spectral function will of course depend on the masses of the particles in each case. For some mass values even the fourth order diagrams fail to obey the Mandelstam representation, because singularities develop at points below the normal thresholds.

4.3. PARTIAL WAVE DISPERSION RELATIONS

The Mandelstam representation almost certainly holds for $\pi\pi$ scattering and probably also holds for πN scattering. A large analyticity domain for $\pi\pi$ scattering has been shown to hold using rigorous field theoretical methods by Martin (1966a, b), and for πN scattering by Somers (1967), but these domains are still much less extensive than that given by the Mandelstam representation.

Up to this point, we have discussed spinless particles. It is assumed that for particles with spin the Mandelstam representation will hold independently for the invariant amplitudes into which the full amplitude may be decomposed. For example, in πN scattering each of the functions $A(s, t)$ and $B(s, t)$ introduced in Chapter 3 will be represented in terms of different double spectral functions. The helicity formalism can also be used to construct amplitudes for which the Mandelstam representation holds (Hara, 1964). A difficulty, which arises in some cases, occurs because decompositions of the complete scattering amplitude are such that, in general, each invariant function is multiplied by some function of the momentum. At the points where these external factors have zeros, the invariant amplitudes must possess compensating singularities if the total amplitude is to remain finite. These singularities, known as "kinematic" singularities have nothing to do with the dynamics of the system, and it is desirable to choose amplitudes for which they do not occur.

4.3. PARTIAL WAVE DISPERSION RELATIONS FOR πN SCATTERING

The Mandelstam Representation

To describe the elastic scattering of π mesons by nucleons four invariant amplitudes $A^{\pm}(s, t)$, $B^{\pm}(s, t)$ were introduced in Chapter 3. Here we consider these amplitudes as functions of the variables s and t, rather than of ν and t, the variables used earlier. The three channels connected by crossing, and illustrated in Fig. 4-5, are:

I	$\pi + N \to \pi + N$	Physical region $s > (M_\pi + M_N)^2$, $t < 0$
II	$\pi + \pi \to N + \bar{N}$	Physical region $t > 4M_\pi^2$, $s < 0$
III	$\pi + N \to \pi + N$	Physical region $u > (M_\pi + M_N)^2$, $t < 0$.

It should be remembered that u is not independent, but is given in terms of s and t by:

$$u = 2M_\pi^2 + 2M_N^2 - s - t . \tag{4-54}$$

The crossing conditions expressed in terms of s, t, and u have been given in eq. (3-71).

The interchange $s \leftrightarrow u$ corresponds to the interchange of the two II mesons, so that in Channel II the crossing relations reduce to a statement of the Pauli principle. Assuming no subtractions, the Mandelstam representations of the amplitudes $B_\pm$ are:

$$B^\pm(s,t) = \frac{g^2}{M_N - s} \mp \frac{g^2}{M_N^2 - u} + \frac{1}{\pi^2} \int_{s_0}^\infty ds' \int_{t_0}^\infty dt' \frac{b^\pm_{12}(s',t')}{(s'-s)(t'-t)}$$

$$+ \frac{1}{\pi^2} \int_{u_0}^\infty du' \int_{t_0}^\infty dt' \frac{b^\pm_{32}(u',t')}{(u'-u)(t'-t)} + \frac{1}{\pi^2} \int_{u_0}^\infty du' \int_{s_0}^\infty ds' \frac{b^\pm_{13}(s',u')}{(s'-s)(u'-u)} .$$

$$\tag{4-55}$$

The lower limit of integration over s' is the lowest normal threshold in Channel I at $s_0 = (M_N + M_\pi)^2$, and in the same way $u_0 = (M_N + M_\pi)^2$ is the lowest threshold in Channel III. In Channel II the lowest threshold is that for a state of two π mesons, so that $t_0 = 4M_\pi^2$. The amplitudes $A^\pm(s,t)$ have similar representations in terms of double spectral functions $a^\pm_{12}$, $a^\pm_{23}$, and $a^\pm_{13}$; but in this case the pole terms are absent. Corresponding to the crossing relations of eq. (3-71), the following relations exist between the double spectral functions:

$$a^\pm_{12}(s,t) = \pm a^\pm_{32}(s,t), \quad b^\pm_{12}(s,t) = \mp b^\pm_{32}(s,t) \tag{4-56}$$

$$a^\pm_{13}(s,u) = \pm a_{13}(u,s), \quad b^\pm_{13}(s,u) = \mp b^\pm_{13}(u,s) . \tag{4-57}$$

4.3. PARTIAL WAVE DISPERSION RELATIONS

The fixed momentum transfer dispersion relations, discussed in Chapter 3, can easily be recovered from the Mandelstam representation. For example, if we define b_1 and b_2 so that:

$$b_1^{\pm}(s,t) = \frac{1}{\pi} \int_{t_0}^{\infty} dt' \frac{b_{12}^{\pm}(s,t')}{(t'-t)} + \frac{1}{2\pi} \int_{u_0}^{\infty} du' \frac{b_{13}^{\pm}(s,u')}{(u'-u)}$$

$$b_2^{\pm}(s,t) = \frac{1}{\pi} \int_{t_0}^{\infty} dt' \frac{b_{32}^{\pm}(u,t')}{(t'-t)} + \frac{1}{2\pi} \int_{s_0}^{\infty} ds' \frac{b_{13}^{\pm}(s',u)}{(s'-s)} \quad (4\text{-}58)$$

then eq. (4-56) can be written as:

$$B^{\pm}(s,t) = \frac{g^2}{M_N - s} \mp \frac{g^2}{M_N - u} + \frac{1}{\pi} \int_{s_0}^{\infty} ds' \frac{b_1^{\pm}(s',t)}{(s'-s)} + \frac{1}{\pi} \int_{u_0}^{\infty} du' \frac{b_2^{\pm}(u',t)}{(u'-u)}. \quad (4\text{-}59)$$

For real negative t, we may set $b_1^{\pm}(s,t) = \text{Im } B^{\pm}(s,t)$ and, by crossing, $b_2^{\pm}(s,t) = \mp \text{Im } B^{\pm}(s,t)$. Then on changing the variables from s, s' to ν, ν', the fixed momentum dispersion relations is obtained in the form of eq. (3-76). The amplitudes $A^{\pm}$ can be treated in the same way.

Partial Waves

The partial wave amplitudes $f_{\ell\pm}^I$ can be expressed in terms of f_1^I, f^I by the inverse of eq. (3-67), which is:

$$f_{\ell\pm}^I(s) = \frac{1}{2} \int_{-1}^{+1} d\mu \, [P_\ell(\mu) f_1^I(s,\mu) + P_{\ell+1}(\mu) f_2^I(s,\mu)] \quad (4\text{-}60)$$

where $\mu = \cos\theta$. The amplitudes f_1^I, $f_2^{\pm}$ are given directly in terms of A^I and B^I by eq. (3-63), where I denotes the isotopic spin channel, and finally, the amplitudes A^I, B^I are given in terms of $A^{\pm}$ by:

$$A^{\frac{3}{2}} = A^+ - A^-, \quad A^{\frac{1}{2}} = 2A^- + A^+ \quad (4\text{-}61)$$

and the same equations relate B^I to $B^{\pm}$.

The center of mass momentum q and the angle of scattering θ are given in terms of s and t by:

$$q^2 = \frac{1}{4s}[s - (M_\pi + M_N)^2][s - (M_\pi - M_N)^2] \qquad \text{(a)}$$

$$\cos\theta = 1 + \frac{t}{2q^2}. \qquad \text{(b)} \qquad (4\text{-}62)$$

We are now in a position to examine the singularities of $f^I_{\ell\pm}(s)$ in the complex s plane. These arise from three sources: $f^I_{\ell\pm}(s)$ will be singular at the points where A^I or B^I are singular in s, at the points where the integration over $\cos\theta$ from $+1$ to -1 meets a cut in A^I or B^I, and at points where the kinematical factors in eq. (3-63) are singular. The singularities in $f^I_{\ell\pm}(s)$, which have been shown in Fig. 4-9, are found as follows (MacDowell, 1959).

(a) **The Pole Terms.** The nucleon pole at $s = M_N^2$ occurs in the amplitude $B^{\frac{1}{2}}$ only [using eqs. (4-56) and (4-61)] and contributes only to the helicity amplitude $f_2^{\frac{1}{2}}$, because of the factor $(W - M_N)$ in eq. (3-63). As this term is independent of $\cos\theta$, it follows from eq. (4-60) that it gives rise to a pole in the partial wave amplitude $f_2^{\frac{1}{2}}$ only, that is, in the P_{11} amplitude. A contribution also arises from the pole in $B^{\frac{1}{2}}$ to the kinematic singularities described under (e) below, in the amplitudes $f_{0+}^{\frac{1}{2}}$ and $f_1^{\frac{1}{2}}$.

The second pole at $u = M_N^2$ produces singularities at values of s and $\cos\theta$ related by eqs. (4-54) and (4-62b):

$$s + 2q^2(\cos\theta - 1) = M_N^2 + 2M_\pi^2. \qquad (4\text{-}63)$$

As there are two values of s for each value of q^2, some care must be taken in determining the cuts in s. The solutions of eq. (4-62a) are:

4.3. PARTIAL WAVE DISPERSION RELATIONS

$$s = W^2 = \left[\sqrt{M_N^2 + q^2} \pm \sqrt{M_\pi^2 + q^2}\right]^2 \qquad (4\text{-}64)$$

and for all q (complex or real), s lies outside the circle $s = |M_N^2 - M_\pi^2| e^{i\phi}$ if the upper sign is taken, and inside the circle if the lower sign is taken. The equation showing the position of the singularities, eq. (4-63), becomes, in terms of q^2:

$$\pm \sqrt{M_N^2 + q^2}\sqrt{M_\pi^2 + q^2} = \tfrac{1}{2} M_\pi^2 - q^2 \cos\theta \; .$$

There will be two solutions of this equation for each sign, and, in each of the total of four cases, the corresponding values of s as $\cos\theta$ varies between $+1$ and -1 will give the positions of the cuts in $f_{\ell\pm}^I(s)$. The detailed solution shows that these cuts are along the real s axis in the intervals:

(a) $\quad -\infty < s < \infty$

(b) $\quad \dfrac{1}{M_N^2}(M_N^2 - M_\pi^2)^2 < s < M_N^2 + 2M_\pi^2 \; . \qquad (4\text{-}65)$

All the amplitudes $f_{\ell\pm}^I$ will contain these cuts. The discontinuities across the cuts can be directly evaluated and are proportional to the coupling constant g^2. In perturbation theory, the first of these cuts corresponds to intermediate states containing a nucleon-antinucleon pair and the second, often called the "short cut," corresponds to an exchange of a nucleon.

(b) The Physical Cut. The cuts in $A_\pm$, $B_\pm$ from $s = (M_\pi + M_N)^2$ to infinity are reproduced in the partial wave amplitudes $f_{\ell\pm}^I$; the discontinuity across this cut is $2i\,\text{Im}\,f_{\ell\pm}^I$, which is known directly in terms of phase shifts and inelasticity parameters:

$$\text{Im}\,f_{\ell\pm}^I = \frac{1}{2q}\left(1 - \eta_{\ell\pm}^I \cos 2\delta_{\ell\pm}^I\right) \; . \qquad (4\text{-}66)$$

This is an expression of the unitarity equation, which can also be written in the form:

$$\text{Im } f_{\ell\pm}^I = q \, |f_{\ell\pm}^I|^2 \, R(s) \tag{4-67}$$

where below the first inelastic threshold $R(s) = 1$, and above this threshold:

$$R = \frac{\sigma_{\ell\pm}^I(T)}{\sigma_{\ell\pm}^I(\text{el})} = 2 \left(\frac{1 - \eta_{\ell\pm}^I \cos 2\delta_{\ell\pm}^I}{1 + \eta_{\ell\pm}^I - 2\eta_{\ell\pm}^I \cos 2\delta_{\ell\pm}^I} \right) \tag{4-68}$$

where $\sigma_{\ell\pm}^I(T)$, $\sigma_{\ell\pm}^I(\text{el})$ are the total and elastic cross sections for each partial wave.

(c) **The Cut from Channel III.** The cut from $u = (M_N + M)^2$ to infinity in $A_\pm$ and $B_\pm$ can be explored in the same way as for the pole $u = M_N^2$. It gives rise to a single cut in s in the interval.

(d) **The Cut from Channel II.** In the variable t, the cut in $A_\pm$ and $B_\pm$ runs from the threshold at $t_0 = 4M^2$ to infinity. The singularities in $f_{\ell\pm}^I(s)$ will occur at values of s for which $4M_\pi^2 < t < \infty$ and at the same time $-1 \leq \cos\theta \leq +1$.

Rearranging eq. (4-62b) as:

$$q^2 = \frac{t}{2(\cos\theta - 1)}$$

we see that the cut in $f_{\ell\pm}^I$ as a function of q^2 is along the real axis in the interval $-\infty < q^2 < M_\Pi^2$. The corresponding values of s come from eq. (4-64). There are two cases: for $q^2 < -M_N^2$, s ranges over the interval $-\infty < s < 0$, while for $-M_N^2 < q^2 < M_\pi^2$, s lies on the circle:

$$s = (M_N^2 - M_\pi^2) \, e^{i\phi} \tag{a}$$

where:

$$\tan\phi = \pm \sqrt{\frac{|q|^2 - M_\pi^2}{M_N^2 - |q^2|}} \, . \tag{b} \tag{4-69}$$

4.3. PARTIAL WAVE DISPERSION RELATIONS

(e) **The Kinematical Singularities.** The kinematical factors in eq. (3-63) are functions of $W = \sqrt{s}$, giving rise to a cut in the interval $-\infty < s < 0$. In addition, the factors $\left(\dfrac{E \pm M_N}{W}\right)$ when written as functions of W contain $(1/W^2)$ as a factor, which may give rise to a singularity in the s plane at $s = 0$. Some recent work (Jakob and Steine, 1969; Peterson, 1970) has shown that $f_{\ell\pm}^I(s)$ near $s = 0$ is determined by high energy backward $\pi N \to \pi N$ scattering. This scattering can be calculated in terms of the exchange of the Δ Regge trajectory, from which it is found that $f_0(s) \propto s^{-0.69}$ near $s = 0$. If desired, these kinematical singularities can be removed by using the function $s^{\frac{3}{2}} f_{\ell\pm}^I(s)$ in place of $f_{\ell\pm}^I(s)$, but at the expense of a less convergent behavior at large s.

This completes the discussion of the singularities of $f_{\ell\pm}^I$ as a function of s, shown in Fig. 4-9, if, of course, it is possible to consider $f_{\ell\pm}^I$ as a function of q^2 or of W. In the former case, the cuts are entirely on the real axis. The physical cut runs from $q^2 = 0$ to infinity and the left-hand cuts are from $q^2 = -\infty$ to $q^2 = -M_\pi^2$. This is a simple structure, but the discontinuities across the left-hand cuts are not so readily evaluated as in the s plane. The cut structure in the W plane

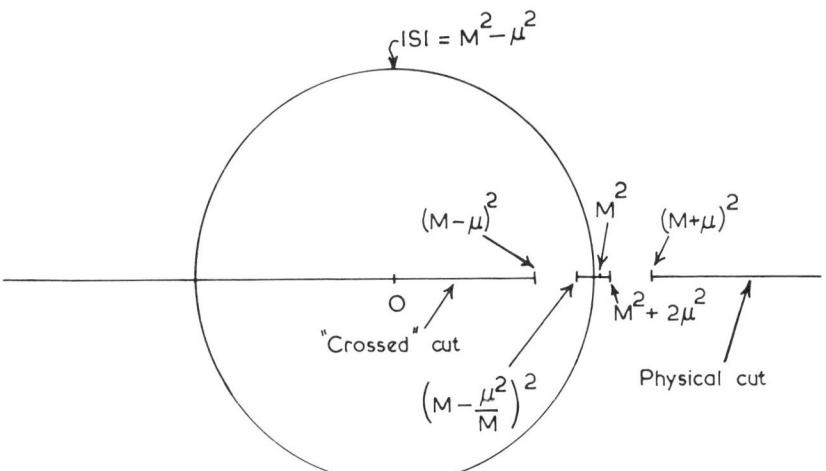

Fig. 4-9. The singularities in the partial wave amplitudes for πN scattering $f_\ell(s)$.

is somewhat more complicated than in the s plane [MacDowell (1959)] but in this case the partial wave amplitudes possess the interesting property known as MacDowell symmetry:

$$f^I_{\ell\pm}(W) = -f^I_{(\ell+1)_\mp}(-W) \tag{4-70}$$

which is easily proved from eqs. (4-60) and (3-63).

MacDowell symmetry is of particular importance in the Regge approach to backward scattering, where it implies that the baryon trajectories appear not singly but in complex conjugate pairs, each with opposite parity.

The Dispersion Relation

Starting from the Mandelstam representation for A and B and using eqs. (3-76) and (4-60), it can be shown that, as in the case of potential scattering, the partial wave amplitudes are real analytic:

$$f^{I*}_{\ell\pm}(s) = f^I_{\ell\pm}(s^*) \tag{4-71}$$

and by applying Cauchy's theorem to contours bordering the cuts, a dispersion relation follows, in the usual way:

$$\text{Re } f^I_{\ell\pm}(s) = \frac{R_1}{s-M_N^2} + \frac{R_2}{s} + \frac{1}{\pi} P \int_{s_0}^{\infty} ds' \, \frac{\text{Im } f^I_{\ell\pm}(s')}{(s'-s)}$$

$$+ \frac{1}{\pi} P \int_{\left(\frac{|M_N^2 - M_\pi^2|}{M_N^2}\right)}^{M_N^2 + 2M_\pi^2} ds' \, \frac{\text{Im } f^I_{\ell\pm}(s')}{(s'-s)}$$

$$+ \frac{1}{\pi} P \int_{-\infty}^{(M_N - M_\pi)^2} ds' \, \frac{\text{Im } f^I_{\ell\pm}(s')}{(s'-s)} + \frac{1}{2\pi i} \int_C ds' \, \frac{\rho_{\ell\pm}(s')}{(s'-s)}. \tag{4-72}$$

4.3. PARTIAL WAVE DISPERSION RELATIONS

The last integral in eq. (4-72) represents the contribution from the circle cut, and $\rho_{\ell\pm}(s')$ the discontinuity, unlike the other discontinuities, is complex. The residues of the poles at $s = M^2$ and $s = 0$ are R_1 and R_2 respectively, and, as we have seen, R_1 vanishes except in the P_{11} state.

In writing eq. (4-72), it has been assumed that no subtractions are necessary. The general question of subtractions in partial wave dispersion relation has been discussed by Kinoshita (1967), who shows that under some rather weak conditions, no more that one subtraction is required for any value of ℓ. The conditions are that:

$$|f_\ell(s)| \leq \exp\left[c(\log|s|)^{2-\epsilon}\right], \quad \epsilon \to 0$$

and that the number of times the sign of the discontinuity Im $f_\ell(s)$ along the left-hand cut changes should not increase more rapidly than:

$$c_1 (\log|s|)^{1-\epsilon}$$

as $s \to -\infty$. In the applications considered here we shall assume unless it is otherwise stated, that no subtractions are required, although they may be introduced to improve convergence.

The N/D Method

For potential scattering we noted that the discontinuity across the left-hand cut was determined by the potential. In an analogous way, the left-hand discontinuities in elementary particle scattering can be considered as representing the forces in the problem. If these can be determined, either exactly or approximately, then it is possible to construct the amplitude in the physical region from the dispersion relation, together with the unitarity equation (4-6). A method often employed is known as the N/D method and is a version of the Wiener-Hopf method for the solution of integral equations (Noble, 1958).

To illustrate the technique, consider a partial wave amplitude $f_\ell(s)$ with a physical (right-hand) cut from s_0 to infinity and a left-hand cut along the real axis in the interval $-\infty < s < s_1$. We write:

$$f_\ell(s) = N(s) / D(s) \qquad (4\text{-}73)$$

where $N(s)$ is a function analytic in the s plane with the left-hand cut only and $D(s)$ is analytic in s with the right-hand cut only. Then the following relations hold:

$$\text{Im } N(s) = \text{Re } D(s) \text{ Im } f_\ell(s), \quad -\infty < s \leq s_1 \qquad (a)$$

$$\text{Im } N(s) = 0, \; s > s_1 \qquad (b)$$

$$\text{Im } D(s) = \text{Re } N(s) \text{ Im } f_\ell^{-1}(s), \; s_0 \leq s < \infty \qquad (c)$$

$$\text{Im } D(s) = 0, \; s < s_0 \qquad (d) \qquad (4\text{-}74)$$

Both $N(s)$ and $D(s)$ satisfy dispersion relations; for $N(s)$ we have:

$$\text{Re } N(s) = \frac{1}{\pi} P \int_{-\infty}^{s_1} ds' \frac{\text{Im } N(s')}{(s'-s)} = \frac{1}{\pi} P \int_{-\infty}^{s_1} ds' \frac{\text{Re } D(s') \text{ Im } f_\ell(s')}{(s'-s)} \qquad (4\text{-}75)$$

while $D(s)$ satisfies:

$$\text{Re } D(s) = \text{Re } D(s_2) + \frac{1}{\pi} (s-s_2) P \int_{s_0}^{\infty} ds' \frac{\text{Im } D(s')}{(s'-s)(s'-s_2)} \qquad (4\text{-}76)$$

where a subtraction has been performed at $s = s_2$. To normalize the equations, $\text{Re } D(s_2)$ can be put equal to unity. The unitarity condition for $\text{Im } f_\ell^{-1}(s)$ takes a particularly simple form; we have [see eq. (4-67)]:

$$\text{Im } f_\ell^{-1}(s) = -q(s) R(s) \qquad (4\text{-}77)$$

where $R(s)$ is given by eq. (4-68). Combining eqs. (4-74c), (4-75), and (4-76), one finds an integral equation for $\text{Re } D(s)$:

$$\text{Re } D(s) = 1 + \frac{1}{\pi} (s-s_2) P \int_{-\infty}^{s_1} ds' K(s, s') \text{ Re } D(s') \qquad (4\text{-}78)$$

4.3. PARTIAL WAVE DISPERSION RELATIONS

where the kernel $K(s, s')$ is:

$$K(s, s') = -\frac{1}{\pi} P \int_{s_0}^{\infty} ds'' \frac{q(s'') R(s'')}{(s''-s)(s''-s_2)} \frac{\operatorname{Im} f_\ell(s')}{(s'-s'')}. \quad (4\text{-}79)$$

The kernel depends on the discontinuity along the left-hand cut $\operatorname{Im} f_\ell(s')$ and the inelasticity $R(s)$, both of which must be known. Whether eq. (4-78) has a solution depends on the behavior of $\operatorname{Im} f_\ell(s')$. For example, if $\operatorname{Im} f(s')$ vanishes sufficiently rapidly as $s' \to -\infty$, and if $R(s) \to$ constant as $s \to +\infty$, a solution exists, and it is easy to solve eq. (4-78) numerically, although due care must be taken in evaluating the principal value integrals.

A similar integral equation can be obtained for $N(s)$; by eliminating $D(s)$ between eqs. (4-75) and (4-76), we have:

$$\operatorname{Re} N(s) = \frac{1}{\pi} P \int_{-\infty}^{s_1} ds' \frac{\operatorname{Im} f_\ell(s')}{(s'-s)} + \frac{1}{\pi} \int_{s_0}^{\infty} ds' \, \overline{K}(s, s') \operatorname{Re} N(s') \quad (4\text{-}80)$$

with:

$$K(s, s') = -\frac{1}{\pi} P \int_{-\infty}^{s_1} ds'' \left[\frac{(s''-s_2)\operatorname{Im} f_\ell(s'')}{(s''-s)(s'-s'')} \right] \frac{q(s') R(s')}{(s'-s_2)}. \quad (4\text{-}81)$$

The equation has the advantage that the kernel is nonsingular provided that the subtraction point s_2 is chosen outside the range $s_0 < s < \infty$.

When either $D(s)$ or $N(s)$ is known, the amplitude can be found by quadratures. The solution is not unique, as it is possible to add poles at $D(s)$ without violating any of the conditions placed upon the amplitude $f_\ell(s)$. These poles generate poles in the amplitude itself on a nonphysical sheet and are known as C.D.D. poles (Castillejo, Dalitz, and Dyson, 1956).

This ambiguity can only be eliminated if further information, such as the high energy behavior of $f_\ell(s)$, is known. This is the case for potential scattering, where N and D are uniquely determined in terms of the Jost functions.

If it turns out that the calculated function $D(s)$ has a zero, a pole will also appear in the amplitude itself. If the zero is below the right-hand cut, the pole is on the real axis (because Im $D = 0$) and must correspond to a bound state. If the zero is in the region $s > s_0$, at the point $\bar{s}$, then for s near $\bar{s}$ we have Re $D = a(s-\bar{s})$ and:

$$f_\ell(s) = \frac{N(\bar{s})}{a(s-\bar{s}) - iq(\bar{s})\, R(\bar{s})\, N(\bar{s})} \tag{4-82}$$

which has the form of a Breit-Wigner resonance at the point s with width $\Gamma = -q(\bar{s})\, R(\bar{s})\, N(\bar{s})$. These poles representing bound states and resonances appear as part of the solution and do not have to be inserted at the beginning.

REFERENCES

A. D'Alfaro and T. Regge, *Potential Scattering* (North Holland, Amsterdam, 1965).

R. Blankenbecler, M. L. Goldberger, N. N. Khuri and S. B. Treiman, Ann. Phys. (N.Y.) 10, 1 (1960).

J. D. Bjorken and S. D. Drell, *Relativistic Quantum Fields* (McGraw-Hill, 1965a).

J. D. Bjorken and S. D. Drell, *Relativistic Quantum Mechanics* (McGraw-Hill, 1965b).

J. Bros, H. Epstein and V. Glasser, Comm. Math. Phys. 1, 240 (1965).

J. Bros, H. Epstein and V. Glasser, Comm. Math. Phys. 6, 77 (1967).

L. Castillejo, R. H. Dalitz and F. J. Dyson, Phys. Rev. 101, 453 (1956).

G. F. Chew, *The Analytic S-matrix* (Benjamin, New York, 1966).

R. J. Eden, P. V. Landshoff, D. I. Olive and J. C. Polkinghorne, *The Analytic S-matrix* (Cambridge University Press, Cambridge (1966)).

REFERENCES

M. L. Goldberger and K. M. Watson, *Collision Theory* (John Wiley, New York, 1964).

Y. Hara, Phys. Rev. 136, B507 (1964).

H. P. Jakob and F. Steine, Zeit. F. Physik 228, 353 (1969).

N. N. Khuri, Phys. Rev. 107, 1148 (1957).

T. Kinoshita, Phys. Rev. 154, 1438 (1967).

A. Klein and C. Zeemach, Nuovo Cimento 10, 1078 (1958).

A. Martin, Progress in Elementary Particles and Cosmic Ray Physics Vol. 8, p. 1 (1965).

A. Martin, Nuovo Cimento A 42, 930 (1966a).

A. Martin, Nuovo Cimento A 44, 1219 (1966b).

A. Martin, *Proc. 1967 Int. Conf. on Particles and Fields* (Interscience, New York, 1967) ed. C. R. Hagen, G. Guralrick and V. A. Mather, p. 252.

S. W. MacDowell, Phys. Rev. 116, 776 (1959).

A. Messiah, *Quantum Mechanics* (North-Holland, Amsterdam, 1964).

P. M. Morse and H. Feshbach, *Methods of Theoretical Physics* (McGraw-Hill, New York, 1953).

B. Noble, *Methods based on the Wiener-Hopf Technique for the Solution of Partial Differential Equations* (Pergamon Press, London, 1958).

J. L. Peterson, Nucl. Phys. B 15, 549 (1970).

G. Somers, Nuovo Cimento A 48, 92 (1967).

CHAPTER 5

FORMATION OF NUCLEON RESONANCES

5.1. BARYON RESONANCES, FORMATION AND PRODUCTION

Characteristic of hadron experiments is the occurence of particularly strong interactions in some channels at some energies. These resonances may first be seen for example as peaks in total cross sections in πN or $K^- N$ scattering experiments, or as departures from expected distributions in phase space (Dalitz) plots in three (or more) body final states (see Chapter 10). Further investigation shows that the resonances occur in states for which all the strong interaction quantum numbers $(J, P, I,$ etc.) are good. Consequently they may be regarded as particles with masses high enough to be formed from, and to decay into, other hadrons. As the decay is hadronic the lifetimes are very short, being $\sim 10^{-22}$ to $\sim 10^{-23}$ sec with corresponding resonance widths of ~ 10 to ~ 100 MeV. Among baryon resonances the archetypical first is the Δ or $N^*(1236)$, otherwise known as the p_{33}, a $p_{\frac{3}{2}}$ $I = \frac{3}{2}$ resonance in the πN system at a center of mass energy of approximately 1236 MeV (de Hoffmann et al., 1954). We consider it as an unstable $J^P = \frac{3}{2}^+$ particle of mass 1236 MeV/c^2 which decays into a pion and a nucleon. The first meson resonances discovered were the ρ (750), a p-wave $I = 1$ $\pi\pi$ resonance (Erwin et al., 1961) and the $K^*(890)$ (Alston et al., 1961), likewise a $J^P = 1^-$ particle, both in bubble chamber experiments at Berkeley.

The existence of the ρ meson was foreshadowed by theoretical nucleon form factor investigations (Frazer and Fulco, 1959, 1960), suggesting a strong $J = 1$ $\pi\pi$ interaction. (For an even earlier suggestion of the ρ from πN scattering, see Takeda, 1955.)

It is useful when analyzing experiments concerning the occurence of resonances to distinguish between production and formation experiments. Production of a resonance occurs when the resonance appears in a final state accompanied by one or more other particles; formation occurs when it is the only particle in a final state. For example the $N^*_{\frac{3}{2}}(1236)$ may be either produced or formed in pion-nucleon scattering experiments, depending on the energy of the interaction. In Fig. 5-1, we show schematically formation (5-1a) and production (5-1b) of the N^*.

In this chapter we are concerned with the *formation* of baryon, particularly nucleon, resonances. The formation of N^* resonances is experimentally investigated by pion-nucleon scattering and the resulting differential and polarization cross sections can (with some difficulties

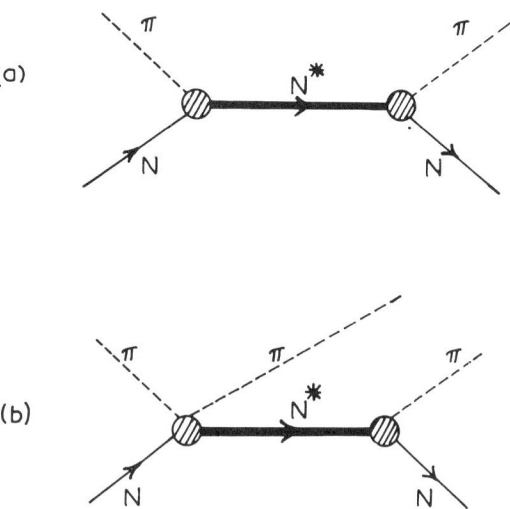

Fig. 5-1. The formation (a) and production (b) of an N^* resonance in a pion-nucleon experiment.

5.2. RESONANCE THEORY

and uncertainties to be described below) be analyzed into partial wave amplitudes of good J, P, and I. We have already seen how this can be done, in the case of the lowest energy member of the N^* family, in Chapter 2. The behavior of these partial wave amplitudes as functions of energy can be used to judge whether there are resonances in that amplitude, that is whether or not there exist unstable baryons of a certain energy and the given J, P, and I. Before describing the methods and results of partial wave amplitude, or phase shift, analysis, we discuss in the next section (5.2) some aspects of the theory of resonances that are particularly relevant to deciding whether a given partial wave amplitude is resonant or not.

5.2. RESONANCE THEORY

Elastic Resonances

We suppose a two-particle scattering process with only one open channel, a. This could be because the energy is too low for any other channel, or because the couplings to other channels vanish. An example of such an elastic state (to a good approximation) is πN scattering in the p_{33} ($J^P = \frac{3}{2}^+$, $I = \frac{3}{2}$) state up to 1500 MeV, center of mass energy, which contains the elastic resonance $N^*_{\frac{3}{2}}(1236)$. Another is the p-wave ($J^P = 1^-$, $I = 1$) $\pi\pi$ interaction up to center of mass energy 1 GeV, which contains the elastic resonance $\rho(750)$.

The T-matrix for the channel a as defined in eq. (2-39) is:

$$T_{aa} = \sin\delta_a \, e^{i\delta_a} \tag{5-1}$$

where δ_a is the phase shift. In this one-channel, elastic case we shall drop the suffix a.

A simple type of resonance (the term coming from obvious analogies to other physical situations) occurs when $\delta_a = \delta$ passes through $\pi/2$ rapidly enough as a function of energy to lead to a peak in the cross section. This is not a general definition. We shall see below that the rate of variation of δ with energy is inversely proportional to the time delay in the scattering process. A large time delay, associated with the formation of a short-lived state, is a necessary condition for resonance. We also wish to define a resonance so as to maintain the correspondence with particles. We do this by imposing the condition that a resonance, like a particle, must be associated with a pole; in the case of a resonance the pole is on an unphysical sheet of the analytic scattering amplitude. If both the conditions, that of time delay and that of correspondence with a pole, are satisfied, then a resonance exists. In this general definition the concept of the phase shift passing through $\pi/2$ at the resonance energy does not hold.

However, many elementary particle resonances occur when δ is at or near $\pi/2$ and this makes a convenient entry point into the theory. Let W_π be the center of mass energy at which $\delta = \frac{\pi}{2}$ and expand $\tan \delta$ as a power series in $W_\pi - W$:

$$\tan \delta = \frac{\gamma_\pi/2}{W_\pi - W} + A + \ldots \quad . \tag{5-2}$$

We may write eq. (5-2) as $\tan \delta = \tan(\phi + \delta')$ where:

$$\tan \phi = A, \quad \tan \delta' = \frac{\gamma/2}{W_R - W}$$

and:

$$W_R = W_\pi + \frac{1}{2} \gamma_\pi \sin\phi \cos\phi, \quad \gamma = \gamma_\pi \cos^2\phi \tag{5-3}$$

$$T = e^{2i\phi} \frac{\gamma/2}{W_R - W - i\gamma/2} + e^{i\phi} \sin\phi \quad . \tag{5-4}$$

5.2. RESONANCE THEORY

Equation (5-4) is the general Breit-Wigner resonance form for elastic scattering. Expression T can be regarded as a function of (complex) energy W and the form eq. (5-4), has a pole in $T(W)$ at the complex energy $W = W_R - i\gamma/2$.

For elastic scattering in channel a with orbital angular momentum ℓ and total angular momentum J the asymptotic wave function for center of mass momentum q is:

$$\psi(r) = \frac{\sin(qr - \ell\pi/2)}{qr} + (-i)^\ell T_{aa}^J \frac{e^{iqr}}{qr} . \tag{5-5}$$

Equation (5-5) is the defining equation of the transition element T_{aa}^J from which the expression eq. (5-1) in terms of the phase shift δ_{aa} is derived. When the center of mass energy W is near the pole at $W_R - i\gamma/2$ it follows from eqs. (5-4) and (5-5) that:

$$\psi(\mathbf{r}, t) \propto \frac{1}{W_R - W - i\gamma/2} \frac{e^{iqr}}{r} e^{-iWt} \tag{5-6}$$

where the proper time-variation of an energy eigenstate has been displayed. In practice all wave trains must be of finite duration, and a finite wave packet $\phi(\mathbf{r}, t)$ can be constructed by combining solutions of different energies so that:

$$\phi(\mathbf{r}, t) = \int_{-\infty}^{\infty} dW\, B(W) \frac{1}{W_R - W - i\gamma/2} \frac{e^{iqr}}{r} e^{-iWt} \tag{5-7}$$

where $B(W)$ is a slowly varying function of energy, zero for $W < 0$. The whole contribution to the integral in eq. (5-7) comes from the pole (on completion of the contour by an infinite semicircle in the lower or upper half W plane) and we find that:

$$\left.\begin{aligned} \phi(\mathbf{r}, t) &= B(W_R)\, e^{-iW_R t}\, e^{-\gamma t/2} \frac{e^{iqr}}{r}, & t > 0 \\ \phi(\mathbf{r}, t) &= 0 & , t < 0 \end{aligned}\right\} \tag{5-8}$$

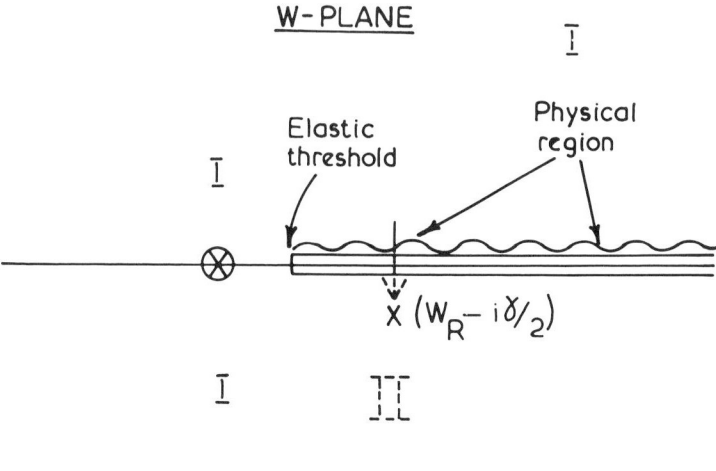

Fig. 5-2. The W plane for the function $T(W)$, showing the cut between the physical sheet I, and the unphysical sheet II. The wavy line shows the region of physical measurement of $T(W)$; ⊗ is a bound state pole on sheet I, and X is a resonance pole on sheet II.

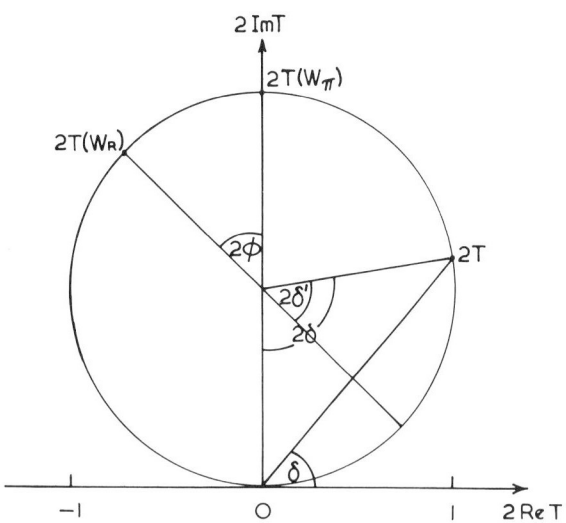

Fig. 5-3. Argand diagram for the resonance scattering amplitude with only one channel. Plotted is $2T$, denoted by $2T$. The locus of T is a circle, the unitarity circle, center $(0, i)$ radius 1. W_R is the resonance energy where $\delta' = \delta - \phi$ is $\pi/2$.

5.2. RESONANCE THEORY

where $\phi(\mathbf{r}, t)$ represents a metastable state, formed in a collision at $t = 0$, decaying with mean lifetime $\tau = 1/\gamma$, and $\gamma > 0$ (otherwise there is an acausal accretion of probability).

This leads to the conclusion that the pole in T, eq. (5-4), is below the real axis on the second or unphysical sheet determined by the branch cut in $T(W)$ starting at the elastic threshold and extending along the positive real axis. This is illustrated in Fig. 5-2. Contrastingly a bound state, $\gamma = 0$, is on the real axis on sheet I.

In Fig. 5-3 we plot the Argand diagram for the scattering amplitude associated with an elastic resonance state, where ϕ is approximately constant and $\delta = \phi + \delta'$. In the diagram δ, δ' pass through $\pi/2$ at W_π, W_R respectively, and since

$$\frac{d\delta'}{dW} = \frac{\gamma/2}{(W_R - W)^2 + (\gamma/2)^2} ,$$

W_R is the point of fastest variation in δ' or δ. Since $\gamma > 0$, the circle is described counterclockwise. We *define* the resonance energy W_R as the energy of fastest variation of δ; W_R is also, by eq. (5-4), the real part of the position of the pole in the T-matrix. However, this latter point depends on the assumed constancy of γ and ϕ, that is, on the neglect of higher terms in eq. (5-2). The term γ is the resonance width and specifies the rapidity of the maximum variation of γ:

$$\left.\frac{d\delta}{dW}\right|_{max} = \left.\frac{d\delta'}{dW}\right|_{max} = \frac{2}{\gamma} = 2\tau ,$$

τ being the resonance lifetime.

The P_{33} resonance has $\phi \simeq 0$, and the resonance energy is at $\delta = \pi/2$.

Inelastic Resonances

Generally in elementary particle resonances there are many channels effectively open from which the resonance may be formed or into which it may decay. For definiteness we consider a multi-channel situation with labels $a = 1, 2 \ldots n$ for the channels open in the energy range of interest, each channel involving two particles, their center of mass momentum being q_a for channel a.

Confining our attention to the case of short range forces, the asymptotic wave function in channel β corresponding to an incident wave in channel a is (for total angular momentum J and orbital angular momentum ℓ_β in channel β):

$$\psi_{\beta a}(r) \sim \delta_{\beta a} \frac{\sin(q_a r - \ell_a \pi/2)}{q_a r} + \sqrt{\frac{\omega_\beta \, q_\beta}{\omega_a \, q_a}} \, T^J_{\beta a}(W) \frac{\exp i(q_\beta r - \ell_\beta \pi/2)}{q_\beta r} \quad (5\text{-}9)$$

where, if the two particles in channel a have masses m'_a and m''_a, $\omega_a \equiv \sqrt{m'^2_a + q_a^2} \, \sqrt{m''^2_a + q_a^2} \, / W$ is the reduced energy in channel a. Equation (5-9) is the defining equation of the transition matrix or T-matrix, $T^J_{\beta a}(W)$ and the corresponding defining equation of the S-matrix is:

$$\psi_{\beta a}(r) \sim \delta_{\beta a} \frac{\exp(-i q_a r)}{q_a r} + (-1)^{\ell_\beta - 1} \sqrt{\frac{\omega_\beta \, q_\beta}{\omega_a \, q_a}} \, S^J_{\beta a}(W) \frac{\exp(i q_\beta r)}{q_\beta r} \quad (5\text{-}10)$$

so that:

$$S^J_{\beta a}(W) = \delta_{\beta a} + 2i \, T^J_{\beta a}(W) \, . \quad (5\text{-}11)$$

We derived the T-matrix formula for elastic resonances from a rather elementary point of view. We shall now derive the corresponding results for inelastic resonances by starting from a more general standpoint. We associate resonance phenomena with the existence of eigenstates of the complete Hamiltonian, for which there are asymptotically only outgoing waves. The condition for outgoing waves in all channels may be stated by means of the boundary conditions:

$$\left(\frac{\partial \psi_\beta}{\partial r} - i q_\beta \psi_\beta \right) \bigg|_{r=\infty} = 0 \quad \beta = 1, 2 \ldots n \, . \quad (5\text{-}12)$$

5.2. RESONANCE THEORY

This boundary condition is not satisfied by the first term of eq. (5-9) or eq. (5-10), and so the eigenvalue W of the total Hamiltonian must correspond to the situation where $T(W)$ and $S(W)$ have an isolated singularity, the simplest possibility being a pole at $W = \overline{W}$. The outflow of probability in channel β is given by:

$$\frac{-i}{2\omega_\beta} (\psi_\beta^* \nabla \psi_\beta - (\nabla \psi_\beta^*) \psi_\beta) = r_\beta \frac{q_\beta}{\omega_\beta} |\psi_\beta|^2 \qquad (5\text{-}13)$$

where r_β is the unit radial vector. For real $\overline{W}$ above the channel thresholds, the outflow of probability would be positive in all channels. This is inconsistent with the conservation of probability so that the eigenvalue $\overline{W}$ cannot be real:

$$\overline{W} = W_R - i\gamma/2 \qquad (5\text{-}14)$$

where γ is positive, corresponding to a decay of probability for the eigenstate (see the section "Elastic Resonances," above). In the neighborhood of such an isolated resonance pole, the leading term of the T-matrix element is:

$$T_{\beta a} = \frac{C_\beta C_a}{W_R - W - i\gamma/2} \qquad (5\text{-}15)$$

where the amplitudes C_a are in general complex. The pole is below the real axis on an unphysical sheet of $T_{a\beta}$. The factorization property of the residue of the pole corresponds to the nondegeneracy of the eigenstate; in fact in channel space the eigenstate is given by the column matrix (C_a) and the form eq. (5-15) leads to the property that $\det(T)$ or $\det(S)$ have only a simple pole at $W = W_R$. This form also corresponds to symmetry for the T-matrix or S-matrix.

$$T_{a\beta} = T_{\beta a} \quad \text{and} \quad S_{a\beta} = S_{\beta a} \;. \qquad (5\text{-}16)$$

This symmetry corresponds to the assumption of time reversal invariance, which we have tacitly assumed to hold.

Generally with the inclusion of nonresonant background scattering the T-matrix element is:

$$T_{\beta a} = D_{\beta a} + \frac{C_\beta C_a}{W_R - W - i\frac{\gamma}{2}} \tag{5-17}$$

where D is also a symmetric matrix and is assumed, for the purposes of the present model situation, to be constant or slowly varying. In that case the unitarity condition:

$$T^*_{\gamma a} T_{\gamma \beta} = \frac{i}{2}(T^*_{\beta a} - T_{a\beta}) \tag{5-18}$$

requires that:

$$D^*_{\gamma a} D_{\gamma \beta} = \frac{i}{2}(D^*_{\beta a} - D_{\beta a}) \quad \text{(a)}$$

$$C^*_\gamma C_\gamma = \frac{\gamma}{2} \quad \text{(b)}$$

$$D^*_{\gamma a} C_\gamma = \frac{1}{2i}(C_a - C^*_a) \quad \text{(c)} \tag{5-19}$$

If we write $C_a = R_a e^{i\phi_a} \sqrt{\frac{\gamma}{2}}$ where R_a is real, then:

$$T_{\beta a} = D_{\beta a} + \frac{R_\beta R_a e^{i(\phi_a + \phi_\beta)} \gamma/2}{W_R - W - i\gamma/2} \tag{5-20}$$

$$\sum R_a^2 = 1 \tag{5-21}$$

and:

$$\gamma_a = R_a^2 \gamma \tag{5-22}$$

is the partial width for channel a.

From the formation experiments which we consider in this chapter, we often know (as for example in πN scattering) an elastic scattering amplitude T_{aa}, and it is consequently of interest to plot the Argand diagram of T_{aa} of eq. (5-20) for the idealized case when R_a, ϕ_a, γ, D_{aa} are

5.2. RESONANCE THEORY

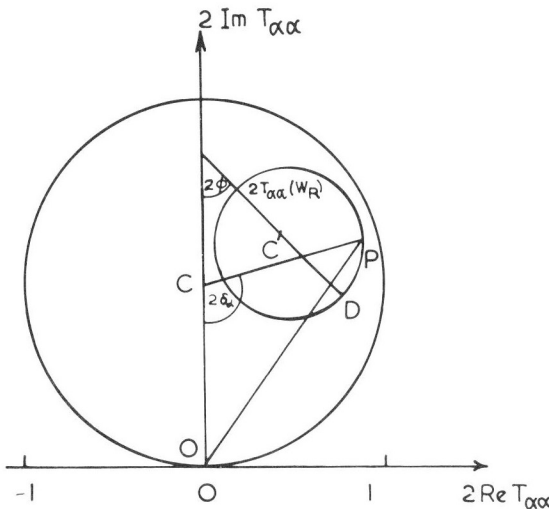

Fig. 5-4. Argand diagram of the elastic scattering amplitude $T_{\alpha\alpha}(W)$; $P = 2 T_{\alpha\alpha}(W)$. For an idealized resonance (see text) the locus of $P = 2 T_{\alpha\alpha}(W)$ is a counterclockwise circle, center C' and radius γ_α/γ. Also, W_R is the resonance energy, γ the total width, γ_α the partial width into channel α, and ϕ_α the phase associated with the α-channel (see text). At the point D, $T_{\alpha\alpha} = D_{\alpha\alpha}$. Further, $T_{\alpha\alpha} = \frac{1}{2i}(\eta_\alpha \exp 2i\delta_\alpha - 1)$, where $\angle OCP = 2\delta_\alpha$, $CP = \eta_\alpha$.

all constant. From the previous elastic resonance diagram it is obvious that $T_{\alpha\alpha}$ describes a counterclockwise circle of radius R_α^2 wholly [by eq. (5-18)] within or on the unitarity circle, as shown in Fig. 5-4. If only elastic channel information is available, then a mark of resonance is the fairly rapid description of a counter-clockwise circle by the complex amplitude of that channel. In practice the circle might be distorted due to movement of background.

K-Matrix, Eigenphases, and Threshold Behavior

For r outside the range of the (short-range) forces, the wave function in channel β corresponding to an incident wave in channel α is (for total angular momentum J and orbital angular momentum ℓ_α in channel a):

$$\psi_{\beta a}(r) = \delta_{\beta a} j_{\ell_a}(q_a r) + (-i)^{\ell_\beta} \sqrt{\frac{\omega_\beta \, q_\beta}{\omega_a \, q_a}} \, T^J_{\beta a}(W) \, \frac{\exp i \, q_\beta r}{q_\beta r} \qquad (5\text{-}23)$$

The asymptotic form of eq. (5-23) is eq. (5-9), which is obtained from eq. (5-23) by noting that the spherical Bessel function is given asymptotically by:

$$j_\ell(x) \sim x^{-1} \sin(x - \ell \pi/2) \qquad (5\text{-}24)$$

Equation (5-23) is the scattering wave function and as such is the solution of the wave equation (in the exterior region) with outgoing waves in all channels $\beta \ne a$. There are solutions of the wave equation with other boundary conditions. The stationary wave solution:

$$\psi^{stat}_{\beta a}(r) = \delta_{\beta a} j_\ell(q_a r) + \sqrt{\frac{\omega_\beta \, q_\beta}{\omega_a \, q_a}} \, K^J_{\beta a}(W) \, \eta_\ell(q_\beta r) \qquad (5\text{-}25)$$

defines the K-matrix. The asymptotic form of the Bessel function $\eta_\ell(x)$ is:

$$\eta_\ell(x) \sim x^{-1} \cos(x - \ell \pi/2) \ .$$

The T-matrix is given in terms of the K-matrix by:

$$\mathbf{T}^{-1} = \mathbf{K}^{-1} - i \qquad (5\text{-}26)$$

and the S-matrix by:

$$\mathbf{S} = (1 + i\mathbf{K})(1 - i\mathbf{K})^{-1} \ . \qquad (5\text{-}27)$$

The K-matrix is real and symmetric. Since $\mathbf{K}$ is real and symmetric it can be diagonalized by an orthogonal matrix $\mathbf{R}$; let the resulting elements of the diagonal matrix be $\tan \Delta_A$, $A = 1, 2, \ldots n$. Then:

$$K_{\beta a} = \sum_{A=1}^{n} R_{A\beta} (\tan \Delta_A) \, R_{Aa} \qquad (5\text{-}28)$$

The Δ_A are known as the eigenphases belonging to the eigenchannels A. Equations (5-26), (5-27) show that $\mathbf{T}$ and $\mathbf{S}$ are also diagonalized by $\mathbf{R}$ and that:

$$T_{\beta\alpha} = \sum_{A=1}^{n} R_{A\beta} R_{A\alpha} \frac{1}{2i} (\exp(2i\Delta_A) - 1) \qquad (5\text{-}29)$$

$$S_{\beta\alpha} = \sum_{A=1}^{n} R_{A\beta} R_{A\alpha} \exp(2i\Delta_A) \qquad (5\text{-}30)$$

For the one-channel case, $n = 1$, the K, T, and S matrices are necessarily diagonal, and the eigenphase is the phase shift.

If one of the eigenphases, say $A = R$, passes through $\pi/2$, and all other quantities vary slowly as a function of energy, then, using similar arguments as for the elastic phase δ in the section "Elastic Resonances" above, eq. (5-29) can be rewritten:

$$T_{\beta\alpha} = R_{R\beta} R_{R\alpha} \frac{e^{2i\phi} \gamma/2}{W_R - W - i\gamma/2} + e^{i\phi} \sin\phi$$

$$+ \sum_{A}' R_{A\beta} R_{A\alpha} \exp(i\Delta_A) \sin\Delta_A \qquad (5\text{-}31)$$

where $\sum_{A}'$ denotes that the eigenchannel R is omitted from the summation. Equation (5-31) is a particular form of eq. (5-20) in which all the C_α have the same phase $\phi_\alpha = \phi$. The essential assumption made in deriving eq. (5-31) is that only one eigenphase varies rapidly in the neighborhood of a resonance; this is, in principle, an exceptional situation. The resonant eigenphase would then increase by an amount π across the resonance, and would then have to cross all the nonresonant eigenphases (mod π) in the energy range of this resonance. This situation would violate Wigner's "no crossing" theorem for eigenphases. What happens in the normal situation is that eigenphases vary rapidly in the vicinity of a resonance so as to avoid crossing, and that various eigenphases each in turn take the role of "resonant" eigenphase (Goebel and McVoy, 1967; Dalitz and Moorhouse, 1970).

The properties of the Bessel functions in eq. (5-25) are such that $q^{-\ell} j_\ell(qr)$ and $q^{\ell+1} \eta_\ell(qr)$ are functions of q^2 analytic and without singularities and finite and nonzero at $q = 0$. The analytic properties of ψ as a function of W (which follow from the analytic theory of differential equations) show that:

$$R(W) = q^{-\ell-\frac{1}{2}} K(W) q^{-\ell-\frac{1}{2}} \qquad (5\text{-}32a)$$

is an analytic function of W^2 with no singularities[1] for $\operatorname{Re} q_\alpha^2 > 0$. In eq. (5-32) $q^{-\ell-\frac{1}{2}}$ is the diagonal matrix with matrix elements $q_\alpha^{-\ell_\alpha-\frac{1}{2}}$. It follows from eq. (5-32a) that:

$$K(W) = q^{\ell+\frac{1}{2}} R(W) q^{\ell+\frac{1}{2}} \qquad (5\text{-}32b)$$

has branch cuts at the thresholds $q_\alpha^2 = 0$ and threshold behavior given by eq. (5-32b).

Using the analyticity of $R(W)$, the K-matrix may be continued below one or more thresholds of the n open channels. Suppose that we continue below the channel β, then in eq. (5-32) q_β becomes $i|q_\beta|$, and though the continued K-matrix is symmetric, $K_{\alpha\beta}$ is no longer real. In an energy region in which p of the original n open channels are closed, a new reaction matrix of dimensions $m = n-p$ may be defined by eqs. (5-32b) and (5-26) and denoted by $\bar{K}$. The required connection between $\bar{K}$ and the continued K is obtained by the condition that the T-matrix for the m open channels must have the same value whether computed from $\bar{K}$ or K. We denote the closed channels by $a, b, \ldots$ and find that:

$$\bar{K}_{\beta a} = K_{\beta a} + i K_{\beta a} (1 - i K)^{-1}_{ab} K_{ba} \ . \qquad (5\text{-}33)$$

This relation is obvious for two channels, and a general proof may be given by expanding the T-matrix in powers of K (Dalitz, 1963).

---

[1] Sufficiently far below the thresholds $q_\alpha^2 = 0$, $R(W)$ has dynamical branch cuts arising from the physical forces which give rise to the reactions (Dalitz and Tuan, 1960).

5.2. RESONANCE THEORY

Höhler Diagrams

The forward elastic scattering amplitude in channel a is given by:

$$f_{aa}(0) = \frac{1}{q} \sum_{\ell, J} \left(J + \frac{1}{2}\right) T_{aa}(\ell, J) \qquad (5\text{-}34)$$

where $T_{aa}(\ell, J)$ is the T-matrix element of this section where the dependence on the orbital angular momentum ℓ and the total angular momentum J has been written explicitly. It has been pointed out by Adair (1959), Cronin (1960), and Hohler et al., (1961) that the circular path traced by the partial wave amplitude $T_{aa}(\ell, J)$ in the neighborhood of a resonance can, in some cases, be clearly seen in the Argand diagram of $f_{aa}(0)$, plotted as a function of energy. The higher J resonances in a particular energy region will be more evident. For $\pi^{\pm}$-p or K-p elastic scattering $f_{aa}(0)$ can be plotted by using the optical theorem and forward dispersion relations.

K-Matrix Poles and T-Matrix Poles

In the sections on elastic and inelastic resonances above, the correspondence of resonances to poles on unphysical sheets of the T-matrix elements has been demonstrated. In the elastic case the phase shift *may* pass through $\pi/2$ at or near the resonance energy as illustrated in Fig. 5-3. But it is evident that if $|\phi|$ is large enough the phase shift δ need not pass through $\pi/2$ anywhere in the vicinity of a resonance; for example if ϕ is large and negative the resonance can occur at small δ, and δ need never reach $\pi/2$ at all. In the inelastic case, the analogue of δ, at a particular value of energy W, is that eigenphase which happens at W to be playing the role of resonant eigenphase (see the section "K-matrix, Eigenphases, and Threshold Behavior"). There *may* be a W at which the analogue eigenphase passes through $\pi/2$. If, in the elastic case δ or in the inelastic case, an eigenphase Δ passes through $\pi/2$ at or near the resonance energy, then eq. (5-28) shows that the K-matrix has a pole at the energy where δ (or Δ) $= \pi/2$. This pole in the

K-matrix is of course on the real axis. However the existence of a resonance does not necessarily imply the existence of a real axis K-matrix pole.

We now consider some types of poles which may occur, and it will be sufficient for our purpose to consider two two-particle channels: 1, containing particle masses m_1, m_1', and 2, containing particle masses m_2, m_2'. We take the threshold of channel 1 as the first threshold; so we can consider resonances near the elastic threshold, 1, and near an inelastic threshold, 2.

In order to display analyticity properties it is convenient to define matrices T' and K' by:

$$T' = q^{-\frac{1}{2}} T q^{-\frac{1}{2}}, \quad K' = q^{-\frac{1}{2}} K q^{-\frac{1}{2}} \qquad (5\text{-}35a)$$

so that the connections between T' and K', and between K' and the matrix R (regular and nonzero at thresholds), are, from eqs. (5-26) and (5-32):

$$T'^{-1} = K'^{-1} - iq$$
$$K' = q^\ell R q^\ell \qquad (5\text{-}35b)$$

The elements of the T'-matrix are analytic functions of W^2 with branch cuts at the channel thresholds, $q_\alpha^2 = 0$. The center of mass momentum in channel α is given by:

$$q_\alpha = [(W^2 - (m_\alpha + m_\alpha')^2)(W^2 - (m_\alpha - m_\alpha')^2) / 4W^2]^{\frac{1}{2}}. \qquad (5\text{-}36)$$

Consequently from eq. (5-35) there is a square root branch point at each threshold in each element of the T' matrix. For the present two threshold model these branch points define three unphysical sheets, II, III, and IV, besides the physical sheet I, which may be defined as follows:

5.2. RESONANCE THEORY

Sheet	Sign Im (q_1, q_2)
I	$(+, +)$
II	$(-, +)$
III	$(-, -)$
IV	$(+, -)$

The connections of sheets II, III, and IV with sheet I are illustrated in Fig. 5-5. (See also Frazer and Hendry, 1964.)

First of all, as remarked previously, a bound state is represented by a pole on sheet I, on the real axis below the first physical threshold. We refer to eq. (5-23), from which it is seen that the condition for there to be an energy for which $\psi_{aa}(r)$ consists entirely of damped exponential waves, $\exp(-|q_a|r)$, (and thus corresponds to a bound state at that energy) is that T_{aa} has a pole at that energy with $q_a = +i|q_a|$ — that is, a pole on the real axis of sheet I. If the bound state is sufficiently close to threshold, there can be a simple connection with the scattering length and the K-matrix through eq. (5-35). For scattering in channel 1, $T' = T_{11}'$ and $K' = K_{11}'$ [where K_{11}' is the reduced K-matrix of eq. (5-33)] are appropriate with:

$$T'^{-1} = K'^{-1} - iq_1 \ .$$

S-PLANE

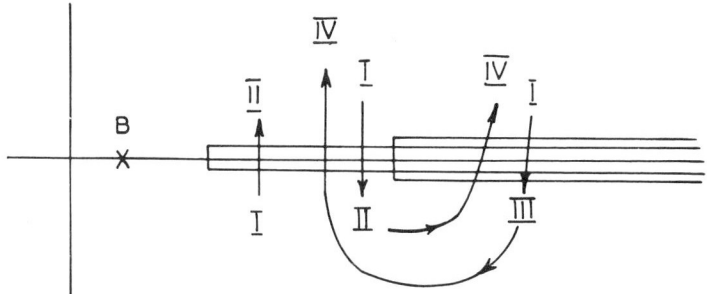

Fig. 5-5. Some analyticity properties of $T_{\alpha\beta}'$ in the $s = W^2$ plane. There are square root branch points at 1 and 2 giving rise to 4 sheets. Sheet I is the physical sheet, and sheet III is reached by traversing both branch cuts as shown. The connections of sheets II and IV with I are also shown. B is the position of a possible bound state pole.

If δ is the phase shift and a the scattering length, then:

$$K'^{-1} = q_1 \cot\delta = \frac{1}{a} + rq_1^2 + \ldots \tag{5-37}$$

$$T'^{-1} = \frac{1}{a} - iq_1 \ . \tag{5-38}$$

Thus for large enough negative values of a there will be a pole on the real axis of sheet I, a bound state pole, near threshold. For large enough positive values of a there will be a pole on the real axis on sheet II. Such a pole is sometimes called an anti-bound state.

Near inelastic thresholds we may have states arising in an analogous way to the type of bound state just considered. As we have seen resonances often correspond to poles in the K-matrix. They may be present in some original K-matrix, or they may only arise when one reduces the K-matrix to the number of open channels by the vanishing of the denominator in eq. (5-33). For example, let us reduce the K-matrix from two channels to one:

$$\bar{K}_{11}' = K_{11}' + iK_{12}' \frac{q_2}{1 - iq_2 K_{22}'} K_{21}' \tag{5-39}$$

where, in eq. (5-39), q_2 is to be continued below threshold by $q_2 = +i|q_2|$. We see that for a large negative K_{22} at threshold 2, there will be a resonance a short distance below threshold 2. As the vanishing of this denominator is precisely the defining condition for a bound state in the absence of channel 1, we call it a virtual bound state (Dalitz, 1963). For a narrow resonance the corresponding pole in the T-matrix is near this real axis K-matrix pole and given by $T_{11}'^{-1} = \bar{K}_{11}'^{-1} - iq_2 = 0$. Since sign $(\text{Im } q_2) = +$, this pole must be on sheet I or II, of which I is impossible [since it would lead to a positive exponential time dependence; see eq. (5-8) above]. The virtual bound state pole in T is consequently below threshold 2, on sheet II as one would expect.

5.2. RESONANCE THEORY

Generally poles near the real axis on sheet II below threshold 2, or on sheet III above threshold 2, give rise to normal Breit-Wigner resonances with, if near threshold, the modification due to the threshold behavior [eq. (5-32)]. The effect of poles on II above 2 or on III below 2 or on IV is generally muted and not of Breit-Wigner form because of distance from the physical sheet. A pole on sheet IV *near threshold* is nearer the physical region and may show as a threshold rise in a cross section. However, in general, the occurence of a pole on one unphysical sheet implies the occurence of accompanying poles on other unphysical sheets. These associated poles are all attached to the same resonance phenomenon (Eden and Taylor; 1963; Dalitz and Rajasekaran, 1963; Nauenberg and Nearing, 1964; Davies and Moorhouse, 1967).

Isolated Reggeon State

The characterization of a reggeon state in the s-channel, defined as the state associated with a single Regge trajectory as discussed in Chapter 4, is closely related to that adopted above for a resonance state. First, the total angular momentum of the state is considered explicitly as a variable, so that we consider $S(J, W)$ and the asymptotic wavefunction [eq. (5-10)]. Again we ask for the eigenstate which is picked out by the outgoing wave condition [eq. (5-12)], but for a real physical value for W. This is not possible unless we allow J to become complex, and the eigenstate for the reggeon (in the s-channel) corresponds to a pole in $S(J, W)$ at the complex value:

$$J = a(W) = a_R(W) + i\beta(W) \ . \tag{5-40}$$

In this neighborhood, the S-matrix is dominated by the term:

$$S(J,W) \approx N \frac{CC^+}{a(W)-J} = N \frac{CC^+}{a_R(W)+i\beta(W)-J} \tag{5-41}$$

where N is a numerical factor including the signature factor (cf. Frautschi, 1963; Squires, 1963). Again, the residue of the pole factorizes for the multichannel case, this being the condition that the eigenstate be nondegenerate. The pole may approximate to a *real physical value* J for the angular momentum for energy W_0, given by:

$$\alpha_R(W_0) - J = 0 \ . \tag{5-42}$$

When the reggeon state α is narrow in W, this corresponds to the case of $\beta(W)$ small. In this case, the approximation:

$$\alpha_R(W) - J = \alpha_R(W) - \alpha_R(W_0) = (W - W_0)\, \alpha'_R(W_0) \tag{5-43}$$

may be made and we retrieve the Breit-Wigner expression:

$$S(J, W) \sim -\frac{N\, CC^+}{\alpha'_R(W_0)\,\{W_0 - W - i\beta(W_0)/\alpha'_R(W_0)\}} \ . \tag{5-44}$$

For states of narrow width, there is thus a close association between the resonances and the reggeons. This need no longer be the case when the W- and J-poles lie far from the real axis.

Equation (5-42) gives a means to define a family of reggeon states, corresponding to its solutions for the positive integral (or half-integral) values of J. The characterization of the family in this way is unique when the analytic properties of the scattering amplitude are specified in the right-hand J plane (analyticity, and the absence of essential singularities at infinity). Most generally, eq. (5-42) defines a sequence of reggeons with spin values increasing in steps of 2 (known as a Regge trajectory), because of the signature factor (Chapter 8) which goes with the Regge amplitude and makes the residue zero for the J values in between; however, the two trajectories with opposite signature can overlap completely in suitable circumstances. These trajectories generally start at some lowest J-value J_0, then correspond to states $J_0 + 2, J_0 + 4, \ldots$. The trajectory may include only a finite number of reggeon states (which is the normal situation for a Regge trajectory corresponding to a static potential interaction, for example). As the spin increases, the normal expectation is that the width of the states should increase, so that the J-poles move farther and farther from the real J-axis, until their individual influence has little importance for $S(J, W)$ (although their mean properties will still be important). However, on the leading meson trajectory, the present indications are (Focacci et al., 1966; Baud et al., 1969 and 1970) that the reggeon (or resonance) states may remain narrow up to rather high J values.

Such trajectories are well known in molecular and nuclear physics, where they correspond to the rotational excitations of low-lying levels. Particularly long series of such rotational levels are known for diatomic molecules (cf. Herzberg, 1950); for symmetrical systems such as H-H, these separate into even and odd trajectories, but for nonsymmetrical systems such as H-D, both even and odd spins belong to the same trajectory. The rotational series known for nuclei are much shorter; perhaps the best example is provided by Be^8, where Bose statistics limits the rotational levels to even J, and the level sequence $J = 0, 2, 4\ldots$ is known up to $J = 8$ (Darriulat et al., 1965).

In general, we may conveniently turn this relationship about, and *define* the reggeon states on a Regge trajectory to be the *rotational excitations* of the lowest state on the trajectory. In the case of elementary particles, such a viewpoint accords rather well with the L-excitation quark model, discussed in Chapter 6.

The Criterion of Time-Delay

Another physical notion which bears directly on the property of resonance is that of time-delay in a collision. In the literature, this quantity has been discussed in two ways, which have been proved equivalent by Smith (1960).

The most direct procedure, that of Goldberger and Watson (1964), is to obtain an expression for the time-delay between the arrival of the incident wave packet and the departure of the wave packet from the collision region. This procedure uses only the asymptotic wave-function and so leads to a direct relation between the time-delay Q and the S-matrix:

$$Q = -i \frac{dS}{dW} S^{-1} . \qquad (5\text{-}45)$$

For the one-channel problem, with elastic scattering, we have $S = \exp(2i\delta)$ and so:

$$Q = 2\frac{d\delta}{dW} . \qquad (5\text{-}46)$$

As we shall see later, the expression eq. (5-45) is actually quite general, holding valid for the multichannel case, in which case Q becomes a lifetime matrix.

From this notion, Wigner (1955a) derived an inequality, which must be satisfied for the scattering from any interaction region where the interaction vanishes outside a definite radius a. The time advance of the outgoing particle relative to the ingoing particle is then limited by the size of the region and the velocity of the particle; classically, this could not exceed $2a/v$. The more detailed quantum treatment led Wigner to replace this by the limit $(2a + 1/q)/v$. With eq. (5-46) for the time-delay, this led to Wigner's inequality:

$$\frac{d\delta}{dW} \geq -\frac{1}{v}\left(a + \frac{1}{2q}\right) . \qquad (5\text{-}47)$$

This is generally referred to as Wigner's causality condition. A subsequent discussion by Wigner (1955b) shows that its derivation depends ultimately on the fact that probabilities must be positive and must sum to unity; the expression of causality in Wigner's derivation which leads to the characteristic property of his R-functions [from which he had first derived the inequality eq. (5-47)] depends on the fact that if the outgoing wave packet emerges too soon it will interfere with the arriving wave packet, and this would generally lead to probabilities exceeding unity. It is still not completely clear what is the relationship of this expression of the causality principle with the analyticity properties that are usually discussed as the expression of the causality principle and which are concerned only with the asymptotic form of the scattering wave-function.

The second approach is that of Smith (1960), who measures the delay time by considering the excess of particles in the interaction region over the number that would be there if there were no interaction. If there is a

5.2. RESONANCE THEORY

delay time, the particles must spend a correspondingly longer time within this region, so that the intensity of particles in this region must be correspondingly high relative to the intensity in the incident beam. The analogy between this notion and the Q-factor which characterizes an electromagnetic resonator in classical electromagnetism is the reason for Smith's choice of Q to denote this delay time. Smith's expression for Q (for the case of one channel) is given by:

$$Q = \lim_{R \to \infty} \left\{ \text{Av} \left[\int_0^R \left(\psi^\dagger \psi - \psi^\dagger_{as} \psi_{as} \right) dV \right] \right\} \tag{5-48}$$

where $\psi \to \psi_{as}$ as $r \to \infty$, and ψ_{as} is normalized to unit incident flux. The limit R in the inner integration could be taken to infinity, but this integration leads to some terms dependent on R which oscillate finitely. The notation "Av" refers to the process of eliminating these oscillatory terms by replacing them by their average value as function of R in the asymptotic region; after this step, the upper limit R can be taken to infinity to give a uniquely defined quantity. It was demonstrated by Smith that this expression, eq. (5-48), for Q was equivalent to the expression eq. (5-45).

For an isolated resonance, we have:

$$\tan \delta = \frac{\Gamma}{2(W_R - W)} \tag{5-49}$$

for which case we find at once, using eq. (5-46):

$$Q = \frac{\Gamma}{(W_R - W)^2 + \Gamma^2/4} \tag{5-50}$$

On resonance, this takes the value:

$$Q = 4/\Gamma \tag{5-51}$$

whereas its average over the energy spectrum of the resonance gives the value $2/\Gamma$. For a narrow Breit-Wigner resonance, the time-delay in the scattering is thus twice its lifetime, $1/\Gamma$, and can therefore be very long. The Wigner condition, eq. (5-47), then indicates that Γ must necessarily be positive for a narrow resonance, and this requires that the resonance pole at $(W_0 - i\Gamma/2)$ should lie on the unphysical sheet of the W-plane, below the physical W-axis, a result well known to stem from the causality principle as discussed on pages 173 and 174.

There is one feature of this characterization of a resonance, or metastable state, which appears rather unsatisfactory. The time-delay Q is a localized property in the sense that it refers to the behavior of δ only in an infinitesimal energy range about the energy W. If the phase shift δ varies with a rapid but small (perhaps only a few degrees) increase across the energy W_0, the time-delay can become very long at W_0, even though none of the other features of resonance appear. An example of this rapid behavior may be provided by the following delay-time expression:

$$Q = a \exp\{-(W-W_0)^2/\gamma^2\} . \qquad (5-52)$$

Integration of eq. (5-45) for the S-matrix with this expression for Q then gives:

$$S = \exp(2i\phi) \exp\left(ia\gamma \int_{-\infty}^{(W-W_0)/\gamma} \exp(-X^2) dX\right) \qquad (5-53)$$

where the phase shift δ has the value ϕ below this increase at W_0, and the value $\left(\phi + \frac{a\gamma}{2\sqrt{\pi}}\right)$ above. This expression, eq. (5-53), for the S-matrix has no singularities on the finite W-plane, but does have an essential singularity at infinity.[2] An example of quite a different kind is provided by the rapid increase of δ with W at an S-wave threshold W_t,

---

[2] Another example would be provided by the delay-time $Q(W)$ appropriate to relativistic S-matrix models of the type discussed by Calucci, Fonda, and Ghirardi (1968). These models are designed to give a phase shift increase by π across the energy W_0, but they could equally well be adjusted to give a rapid increase in δ by amount $a\pi$. With these models, the only singularities on the physical sheet for the S-matrix are the usual left-hand branch cuts appropriate to the interaction processes effective.

5.2. RESONANCE THEORY

like $a(W-W_t)^{\frac{1}{2}}$, for which the delay-time increases indefinitely as the energy W falls to W_t. However, here there is an obvious physical interpretation for the effect; the colliding particles stay together for a long time simply because of the increasingly low velocity available for their separation as W falls to the threshold energy for that channel. The essential point is that, although narrow resonances lead to long delay-times, there may be other circumstances which can lead to long delay-times.

This discussion of the delay-time has been generalized by Smith (1960) to the multichannel case. This leads to the introduction of a "lifetime matrix" Q, whose element $Q_{\beta a}$ for the transition $a \to \beta$ is given by the following generalization of the expression eq. (5-48):

$$Q_{\beta a} = \lim_{R \to \infty} \left\{ \text{Av} \left[\int^R \psi_\beta^+ \psi_a \, dV - R \left(v_a^{-1} \delta_{a\beta} + \sum_\gamma S_{a\gamma} v_\gamma^{-1} S_{\gamma\beta}^\dagger \right) \right] \right\} \quad (5\text{-}54)$$

In this expression, ψ_a denotes the (multichannel) wave-function describing an incident wave with unit flux in channel a, and outgoing waves in all channels, and v_β denotes the relative velocity of the two particles in channel β at energy W, when they are far separated. For all channels, the integration is to be taken out to radial distance R, and the terms subtracted in eq. (5-54) represent the behavior of the integral for large R, after the operation "Av" which eliminates (by averaging) the terms resulting from the integration, which oscillate finitely as R increases indefinitely. After this operation "Av," the integrals are convergent in the limit $R \to \infty$. Smith shows by direct calculation that this lifetime matrix Q is related with the S-matrix by an expression of the same form as eq. (5-45):

$$Q = -i \frac{dS}{dW} S^\dagger \quad \text{(a)}$$

$$= -i \frac{dS}{dW} S^{-1} \quad \text{(b)} \quad (5\text{-}55)$$

the latter expression making use of the unitarity relation:

$$S^\dagger S = SS^\dagger = 1 \, . \quad (5\text{-}56)$$

The fact that Q is a hermitian matrix follows directly from the expression eq. (5-55a) taken together with eq. (5-56).[3] Hence Q has eigenstates in channel space, corresponding to real eigenvalues q_A. Smith points out that these eigenstates of Q correspond to metastable states for the system, and that the eigenvalues q_A give their characteristic lifetimes.

The relation of delay-time to this lifetime matrix needs a little discussion. For inelastic processes (as well as for elastic scattering) Eisenbud (1948) had obtained the following expression for $(\Delta t)_{\beta a}$, which represents the time interval between the outgoing wave in channel β and the ingoing wave in channel a:

$$(\Delta t)_{\beta a} = \text{Re}\left(-i\left(\frac{dS}{dE}\right)_{a\beta}(S_{a\beta})^{-1}\right). \tag{5-57}$$

Smith pointed out that the sum:

$$\tau_a = \sum_\beta P_{a\beta}(\Delta t)_{\beta a} \tag{5-58}$$

represents the weighted average of the delay-times for all the outgoing channels (including channel a) when the incident wave is in channel a, where the weight $P_{a\beta}$ denotes the relative probability for the excitation of channel β in this case. Since this probability is given in terms of the S-matrix by $P_{a\beta} = |S_{a\beta}|^2$, this average delay-time is given by:

$$\begin{aligned}
\tau_a &= \sum_\beta S^*_{a\beta} S_{a\beta} \text{Re}\left\{-i\left(\frac{dS}{dW}\right)_{a\beta}(S_{a\beta})^{-1}\right\} & \text{(a)} \\
&= \sum_\beta \text{Re} -\left\{i\left(\frac{dS}{dW}\right)_{a\beta} S^\dagger_{\beta a}\right\} & \text{(b)} \\
&= Q_{aa} & \text{(c)}
\end{aligned} \tag{5-59}$$

---

[3] Generally dS/dE and $dS^\dagger/dE$ do not commute with either $S^\dagger$ or S. However, because of the unitarity relation eq. (5-56), the following equalities do relate these quantities:

$$\frac{dS^\dagger}{dW}S + S^\dagger\frac{dS}{dW} = 0 = \frac{dS}{dW}S^\dagger + S\frac{dS^\dagger}{dW}$$

and the latter of these relations must be invoked to show that $Q^\dagger = Q$, when Q is taken in the form of eq. (5-55a).

5.2. RESONANCE THEORY

the last step being taken in consequence of eq. (5-55a), the operation Re on the right-hand side of eq. (5-59b) being unnecessary in view of the reality of Q_{aa}, which follows from the hermitian property of the expression eq. (5-55a). Smith has also discussed the form taken by the causality condition of Wigner for the multichannel case, giving lower bounds for the eigenvalues q_A of the lifetime matrix Q.

We shall now derive a simple expression (Dalitz and Moorhouse, 1970) for the average delay-time defined as:

$$\bar{\tau} = \frac{1}{n} \sum_{\alpha=1}^{n} \tau_\alpha = \frac{1}{n} \operatorname{Tr} Q . \tag{5-60}$$

We shall show that:

$$\bar{\tau} = -\frac{i}{n} \frac{d}{dW} \ell n \, (\det S) .$$

Let L be the unitary matrix that diagonalizes S, so that:

$$S = L^\dagger S_d L \tag{5-61}$$

where S_d is the diagonal matrix with matrix elements S_A. As we have seen in the section on "K-matrix Poles and T-matrix Poles" above the energy dependence of L may be exceedingly rapid and complicated in the neighborhood of a resonance energy:

$$\frac{dS}{dW} = \frac{dL^\dagger}{dW} S_d L + L^\dagger \frac{dS_d}{dW} L + L^\dagger S_d \frac{dL}{dW} \tag{5-62}$$

$$\operatorname{Tr} Q = -L i \operatorname{Tr} \left(\frac{dS}{dW} S^\dagger \right)$$

$$= -i \operatorname{Tr} \left\{ \left(\frac{dL^\dagger}{dW} S_d L + L^\dagger \frac{dS_d}{dW} L + L^\dagger S_d \frac{dL}{dW} \right) (L^\dagger S_d^* L) \right\} \ldots \tag{5-63}$$

Making repeated use of the cyclic relation $\operatorname{Tr}(AB \ldots XY) = \operatorname{Tr}(YAB \ldots X)$ and of the unitary properties:

$$L^\dagger L = L L^\dagger = 1$$

the first and third terms of the Trace in eq. (5-63) reduce to the form:

$$\text{Tr}\left(L\frac{dL^{\dagger}}{dW} + \frac{dL}{dW}L^{\dagger}\right) = \text{Tr}\left(\frac{d}{dW}(LL^{\dagger})\right) = 0$$

since $S_d S_d^* = S_d^* S_d = 1$. We are then left with the central term of eq. (5-63), which reduces to:

$$\text{Tr } Q = -i\,\text{Tr}\left(\frac{dS_d}{dW}S_d^*\right) = -i\sum_{A=1}^{n}\frac{dS_A}{dW}S_A^* \,. \qquad (5\text{-}64)$$

Now $S_A^* = \frac{1}{S_A}$, as can be seen directly from the fact that eq. (5-61) preserves the unitarity of S_d or from eq. (5-30), so that:

$$\text{Tr } Q = -i\sum_{A=1}^{n}\frac{d \ln S_A}{dW} = -i\frac{d}{dW}\ln\left(\prod_{i=1}^{n}S_A\right) = -i\frac{d}{dW}\ln(\det S_d)\,. \quad (5\text{-}65)$$

$$\bar{\tau} = -\frac{i}{n}\frac{d}{dW}\ln\det S\,. \qquad (5\text{-}66)$$

What is a Resonance?

If we had definite underlying theories of the primary interactions, most of the discussion about the interpretation of scattering interactions in terms of resonances would be irrelevant. In this situation, we would be able to make predictions from the theory, which could be compared directly with experiment, a procedure that would require only the S- or T-matrix elements for the physical (W) axis. The nature of the S-matrix on the complex W plane, and on its unphysical sheets, would then be only of secondary importance.

The question on the nature of resonance arises because we are in the position that we can only do phenomenology. This means that we have to simplify our representation of the data by focusing attention on outstanding features of it and by endeavoring to give some definite characterization of them. One of these features is the production or formation of particles, this concept including very unstable as well as quite stable objects. It is the very unstable objects that are the elementary particle resonances, and their formation is signaled by extended time-delay in emergence of the reaction products. It is because of this very direct physical significance of time-delay that we have chosen the energy of maximum time-delay as the resonance energy as discussed in the section "Elastic Resonances." There may be other reasons for time-delay besides particle formation, as mentioned in the section "The Criterion of Time-Delay." So before identifying a time-delay as a particle or resonance one should ask whether it be associated with a pole on an unphysical sheet. However, the information that gives an indication of a large time-delay — typically the fast traversal by the amplitude of a resonance circle as described in the sections "Elastic Resonances" and "Inelastic Resonances" — is the same information whose extrapolation on to the unphysical sheet informs us as to the existence of a pole. Consequently, in practice the position is that a time-delay, or resonancelike circle in an amplitude of good quantum numbers (J, P, isospin etc.), is taken, in the absence of any other explanation such as an S-wave threshold, to be evidence of resonance.

5.3. PARTIAL WAVE ANALYSIS OF ELASTIC SCATTERING EXPERIMENTS

The analysis of pion-nucleon elastic scattering experiments has been in recent years one of the most fruitful means of discovery of new resonances. We shall now discuss the methods of analysis with particular reference to the pion-nucleon case. First we shall review a few basic facts and formulae, and for these we make general reference to Chapter 2; also there is in Chapter 3 a discussion related to the whole of this section 5.3.

In pion-nucleon scattering there are three independent elements of the density matrix, a knowledge of which at a given energy is necessary (in the absence of further assumptions on theory) and (nearly) sufficient to determine the partial wave amplitudes. We illustrate by writing the center of mass amplitude for the scattering of pions by nucleons with initial and final components (in the same direction) of spin m, m' in terms of the spin-flip amplitude $g(\theta)$ and non-spin-flip amplitude $f(\theta)$ as:

$$f_{m'm}(\theta,\phi) = f(\theta) \delta_{m'm} + i g(\theta) <m'|\boldsymbol{\sigma}|m> \cdot \mathbf{n} \qquad (5\text{-}67)$$

$$\vec{n} = \vec{q}_i \times \vec{q}_f / |\vec{q}_i \times \vec{q}_f| \qquad (5\text{-}68)$$

where $\vec{q}_i$, $\vec{q}_f$ are the initial and final pion momentum, $q = |\vec{q}_i| = |\vec{q}_f|$; $f(\theta)$, $g(\theta)$ are expressed in terms of the partial wave scattering amplitudes $f_{\ell\pm}$ for $j = \ell \pm \frac{1}{2}$ by:

$$f(\theta) = \sum_\ell \{(\ell+1) f_{\ell+} + \ell f_{\ell-}\} P_\ell(\cos\theta), \quad g(\theta) = \sum_\ell \{f_{\ell+} - f_{\ell-}\} \sin\theta \, P'_\ell(\cos\theta) \qquad (5\text{-}69)$$

$$f_{\ell\pm} = \frac{1}{2iq} \left(\eta_{\ell\pm} e^{2i\delta_{\ell\pm}} - 1\right). \qquad (5\text{-}70)$$

We can summarize the information to be obtained from experiment by considering scattering by polarized protons where the polarization of the protons is $\vec{P}_i$. Setting $\vec{P}_i = 0$ gives the corresponding quantities for scattering from an unpolarized target. The differential cross section and the polarization of the recoil nucleon is given (see eq. (2-109)) by:

$$\frac{d\sigma}{d\Omega} = |f(\theta)|^2 + |g(\theta)|^2 + 2 \, \mathrm{Im} \, [f(\theta) g^*(\theta)] (\vec{n} \cdot \vec{P}_i) \qquad (5\text{-}71a)$$

$$\frac{d\sigma}{d\Omega} \vec{P}_f = 2 \, \mathrm{Im} \, [f(\theta) g^*(\theta)] \vec{n} - 2 \, \mathrm{Re} \, [f(\theta) g^*(\theta)] (\vec{n} \times \vec{P}_i)$$
$$+ (|f(\theta)|^2 + |g(\theta)|^2)(\vec{n} \cdot \vec{P}_i)\vec{n} - (|f(\theta)|^2 - |g(\theta)|^2)(\vec{n} \times (\vec{n} \times \vec{P}_i)) .$$
$$(5\text{-}71b)$$

From single scattering and double scattering (recoil nucleon polarization detection) experiments one can thus determine $|f(\theta)|$, $|g(\theta)|$ and their relative phase.[4]

---

[4] At a given angle θ there are the differential cross-section $|f|^2 + |g|^2$ and three polarization quantities: $\mathrm{Im}\,[fg^*]$, $\mathrm{Re}\,[fg^*]$, and $|f|^2 - |g|^2$. Measurement of all these will determine f and g apart from a common phase, indeterminable

5.3. ANALYSIS OF ELASTIC SCATTERING EXPERIMENTS

The information to be obtained from total cross sections can be explicitly expressed in terms of partial waves by:

$$\sigma_{total} = \frac{4\pi}{q} \sum_{\ell=0}^{\infty} \{(\ell+1) \text{ Im } f_{\ell+} + \ell \text{ Im } f_{\ell-}\} . \tag{5-72}$$

Resonances often show as peaks or shoulders in graphs of the total cross section as a function of energy (Fig. 2-3), and it should be noted that because of unitarity (expressed in eq. (5-70) by $|\eta_{\ell\pm}| \le 1$) the maximum contribution from partial wave of angular momentum j to the total cross section is $\frac{4\pi}{q^2}(j+\frac{1}{2})$.

In the forward direction, $\theta = 0$, the absolute phase can be determined by use of the optical theorem (see eq. (2-29)) — $\text{Im } f(0) = \frac{q}{4\pi}\sigma_{total}$ and forward dispersion relations to determine the $f(0)$. One may also, in principle, obtain the absolute phase near the forward direction from the Coulomb scattering which is important at small enough angles. At this moment in fact, extremely little data of the required accuracy is available in that angular region. The Coulomb scattering corrections were discussed in Chapter 2.

Bowcock, Cottingham and Ng (1970) have drawn attention to the continuous multiplicity of solutions arising from lack of knowledge of the absolute phase. Under the transformation $f(\theta) \to f(\theta) \exp(i\phi(\cos\theta))$, $g(\theta) \to g(\theta) \exp(i\phi(\cos\theta))$ with $\phi(1) = 0$, the right hand sides of eq. (5-71) are invariant as well as $\text{Im } f(0)$ and $\text{Re } f(0)$. There is a large class of functions $\phi(\cos\theta)$ satisfying reasonable analyticity conditions in $(\cos\theta)$. Now this problem of the continuous multiplicity has nearly always been resolved in practice by taking only a finite number of partial waves moti-

---

from eq. (5-71). Measurement of $|f|^2 + |g|^2$, $\text{Im } [fg^*]$, and $[\text{Re } fg^*]$ will determine f, g (always apart from the common phase) but without distinction as to which is f and which is g. Measurement of $|f|^2 + |g|^2$, $\text{Im } [fg^*]$, and $|f|^2 - |g|^2$ will leave the relative phase of f, g undetermined by π.

vated by the finite range of the forces. (We note that multiplication by $\exp(i\phi(\cos\theta))$ will transform a solution with a finite number of partial waves into one with an infinite number.) One can see on inspection of eq. (5-69) that knowledge of $|f(\theta)|$, $|g(\theta)|$, their relative phase and the assumption of a finite number of partial waves will determine $f_{\ell\pm}$ and their relative phases[5] — the absolute phase being found from the optical theorem.

Though the assumption of a finite number of partial waves is qualitatively correct and Bowcock, Cottingham and Ng (1970) have shown that their transformation does not annihilate or create N^* resonances found using the assumption, it seems desirable to use theory to find the high partial waves (Burckhardt 1972) — especially as theory, with phenomenological input, is likely to be effective in these very peripheral interactions. Alcock and Cottingham (1970) have used the Mandelstam double spectral representation to generate higher partial waves in an attempt to definitively resolve the continuous multiplicity problem. Another method is to expand in other than a partial wave series. (See Cutkowsky and Deo (1968); R. C. Miller et al., (1972) and references therein.) All discussion in the remainder of this chapter premises that the higher partial waves are 'known' (to be zero or some determinable value) and that consequently the problem of continuous multiplicity of solutions does not arise.

From a resonance point of view the most important quantities are partial wave amplitudes in a pure isospin state. We note that the partial wave amplitudes $f_{\ell\pm}$ for $\pi^{\pm}p$ elastic scattering or charge exchange scattering, $\pi^- + p \to \pi^0 + n$, are given in terms of pure isospin scattering amplitudes for $I = \frac{3}{2}$ and $I = \frac{1}{2}$ by eq. (2-87). From the foregoing discussion we see that it is only necessary to perform the complete range of scattering experiments for two of these processes, for example the $\pi^+ p$

---

[5] The principle is readily seen by taking $\max(\ell) = 1$ in eq. (5-69).

5.3. ANALYSIS OF ELASTIC SCATTERING EXPERIMENTS

and $\pi^- p$ elastic processes, to determine the pure isospin partial wave amplitudes. However, the inevitable experimental errors alone would make the measurement of all three quantities desirable.

In practice, with the present state of experimental technique, there are, at or near a given energy, many gaps in the data. There may be missing points in the angular range, and information on recoil nucleon polarization from polarized targets is currently (1972) not available. This leaves only an incomplete knowledge of $|f|^2 + |g|^2$ and $\text{Im}\,[fg^*]$, which is not sufficient to determine the partial wave amplitudes without further assumptions. It now seems that remarkable successes can be achieved by using the heuristic principle of smooth behavior of the partial wave amplitudes as a function of energy. How this principle is used and the resulting limitations and uncertainties will be described in this section.

Polynomial Fitting

If one has an angular distribution of data, such as a differential cross section or polarization at a given energy, then a useful first step is to analyze into powers of $\cos\theta$ or orthogonal polynomials such as $P_n(\cos\theta)$. For example:

$$\frac{d\sigma}{d\Omega}(\theta) = \sum_{n=0}^{N} a_n (\cos\theta)^n ; \quad \frac{d\sigma}{d\Omega}(\theta) = \sum_{m=0}^{M} c_m P_m(\cos\theta) \, . \quad (5\text{-}73)$$

For a given experimental angular distribution $E_i = \frac{d\sigma}{d\Omega}(\theta_i)$ at K angles, θ_i with experimental errors Δ_i, the a_n can be determined by the usual process of curve-fitting. We form:

$$F = \sum_{i=1}^{K} \left[\frac{E_i - \sum_{n=0}^{N} a_n (\cos\theta_i)^n}{\Delta_i} \right]^2 \quad (5\text{-}74)$$

and find the required a_n as those values for which F is a minimum. To find the required degree of the series one fits repeatedly for increasing values of N until there is no significant decrease in $\chi^2 = F/(K-N)$; $K-N$ is the number of degrees of freedom of the fit. Series in orthogonal polynomials such as $P_m(\cos\theta)$ have the advantage that, for K values of $\cos\theta_i$ uniformly distributed between -1 and 1 with uniform errors Δ_i, there are small (vanishing in the limit $K \to \infty$) correlations between the c_m. That is, if M is increased to $M+1$, then c_m for $m \leq M$ are the same in the new fit as in the old. This advantage is decreased if the $\cos\theta_i$ are markedly nonuniform, for example because of gaps in the angular range of measurement.

Curve-fitting of this type can show up bad data values as points which polynomials of no reasonable degree can fit. These are often obvious by eye. More valuably, it is a first step towards a complete analysis into partial waves:

$$\frac{d\sigma}{d\Omega} = |f(\theta)|^2 + |g(\theta)|^2 = \sum_{m=0} c_m P_m(\cos\theta); \quad |\vec{P}| \cdot \frac{d\sigma}{d\Omega} = 2\,\text{Im}\,[fg^*]$$

$$= \sin\theta \sum_{m=0} d_m P_m(\cos\theta)$$

$$|f(\theta)|^2 - |g(\theta)|^2 = \sum_{m=0} e_m P_m(\cos\theta) \tag{5-75}$$

$$c_m = \sum_{\ell^\pm,\ell'^\pm} \gamma_m(\ell^\pm, \ell'^\pm)\,\text{Re}(f_{\ell\pm}\cdot f^*_{\ell'\pm}),\; d_m = \sum_{\ell^\pm,\ell'^\pm} \delta_m(\ell^\pm, \ell'^\pm)\,\text{Im}(f_{\ell\pm}, f^*_{\ell'\pm})$$

$$e_m = \sum_{\ell^\pm,\ell'^\pm} \epsilon_m(\ell^\pm, \ell'^\pm)\,\text{Re}(f_{\ell\pm}, f^*_{\ell'\pm}) \;. \tag{5-76}$$

The $\gamma_m(\ell^\pm, \ell'^\pm)$, $\delta_P(\ell^\pm, \ell'^\pm)$ have been evaluated by Ollson and Trower (1969) up to $m = 11$ and $P = 8$. Knowledge of $c_m\,d_m\,(e_m)$ obviously gives the upper limit of the angular momentum of nonnegligible amplitudes. And it may give a subset of partial wave amplitudes containing the subset of important partial wave amplitudes. Inspection of the $c_m, d_m (e_m)$ over a range of energies may give a first glimpse of what partial wave amplitudes are resonant.

5.3. ANALYSIS OF ELASTIC SCATTERING EXPERIMENTS

Ambiguities in Phase Shift Analysis

We have seen that because data on polarization of recoil nucleons from scattering of pions by polarized targets is not available, it is not possible to determine the partial wave amplitudes at a single energy without some other principle. Some ambiguities due to lack of data are named. Suppose that at some energy we only have the scattering differential cross section from an unpolarized target and that $f_{\ell\pm}$ are a set of partial wave amplitudes that fit the data. Then the *Minami ambiguity* is the existence of another set of partial wave amplitudes, $f^M_{\ell\pm}$ such that $f^M_{(\ell+1)-} = f_{\ell+}$, $f^M_{(\ell-1)+} = f_{\ell-}$, which give exactly the same differential cross section, as can be seen by explicit substitution in eqs. (5-69) and (5-71a) with $P_i = 0$. The two solutions in the Minami ambiguity are obtained one from the other by parity interchange. For scattering experiments from a *polarized* target (eq. (5-71a) with $P_i \neq 0$) then, there is the *generalized Minami ambiguity* between the solutions $f_{\ell\pm}$ and $f^M_{\ell\pm}$ such that $f^M_{(\ell+1)-} = -f^*_{\ell+}$ and $f^M_{(\ell-1)+} = -f^*_{\ell-}$. A measurement of one of the rotation parameters, A or R, resolves this ambiguity.[6]

One can eliminate these ambiguities by resorting to the hypothesis of smooth variation with energy of partial wave amplitudes in a crude form. One looks at partial wave amplitudes nearby (generally lower) in energy, and the demand of smoothness or similarity in the solutions at neighboring energies is sufficient to resolve the ambiguities. It is plain that this is an iterative process, which descends to threshold, where ambiguities may be resolved by observation of threshold behavior (power of the center of mass momentum) as discussed in Chapter 2. Interference with Coulomb scattering, particularly at low energies where it is easily observed, also resolves ambiguities (Chapter 2).

---

[6] One may remark that if one took the generalized Minami ambiguous solution in the neighborhood of the 33 resonance one would obtain a partial wave amplitude with a rapidly descending phase shift. This would be very suspicious because of the Wigner theorem, eq. (5-47), limiting the descents of phase shifts for potentials of a given range.

Discrete Energy Phase Shift Analysis

The procedure in discrete energy phase shift analysis is as follows. The range of interaction between pion and nucleon is approximately $1/2M_\pi$; so if q is the center of mass momentum, partial waves with $\ell > L_q = q/2M_\pi$ are expected to be small, and this seems to be confirmed by polynomial analysis, as described above.

One first limits the number of unknown partial waves by taking some ansatz (for example, zero) for the higher ones. One then finds all the solutions (that is satisfactory fits of the partial wave amplitudes to the data) at each energy by minimizing:

$$F = \sum_i \left(\frac{E_i - T_i}{\Delta_i}\right)^2 \tag{5-77}$$

where E_i, Δ_i are the experimentally measured quantity (differential cross section, etc.) and its error, and T_i is the corresponding calculated quantity containing the partial wave amplitudes to be found. Large errors Δ_i lead to large numbers of solutions. Taking one solution at the lowest energy, one selects the solution(s) "continuous" with it at the next energy, and so on until the highest energy. One thus generates a number of possible paths from lowest to highest energy, each path leading through one of the solutions at each energy. Paths may be found by inspection (Bareyre et al., 1965, 1967; or Johnson and Steiner, 1967) by a minimum path length computer program.

More complicated methods may be used, as in the work of Auvil et al., (1964), Donnachie (1967), Lovelace (1967) and Donnachie et al., (1968). A division of partial waves into three types is made: (1) lower ℓ partial waves; (2) intermediate ℓ partial waves, taken from partial wave dispersion relations with assumed knowledge of the longer-range forces; (3) high partial waves, set equal to zero. Solutions at each energy are found for the lower partial waves, and paths are found from the lowest to the highest energy, each path close to one of the solutions at each energy. Each path is found from partial wave dispersion relations with left-hand

5.4. RESONANCES: RESULTS OF PARTIAL WAVE ANALYSIS

cut discontinuity (or forces) parametrized, the parameters being determined by a best fit to the solutions. The smooth paths define new partial wave amplitudes at each energy, and these values are used as input to the fitting (to data) program and new solutions are found. The procedure is then iterated. At some stage the previously theoretically determined amplitudes of type (b) are allowed to vary freely.

We delay comment on these methods until after the next section.

Energy-Dependent Phase Shift Analysis

We have seen the complications that arise in the single-energy type of phase shift analysis, in imposing smooth variation from energy to energy. In the energy-dependent type of phase shift analysis one seeks to overcome this difficulty by parametrizing each partial wave amplitude as a function of energy and fitting the data at many energies simultaneously. The technical price that has to be paid for this built in smoothness with energy is a large increase in the number of parameters that have to be simultaneously fitted to the data. The energy range that can be covered by this method is limited by the speed and capacity of existent computers, since the larger the energy range the greater the number of parameters necessary for the description of the energy variation of each partial wave amplitude.

A simple but effective method is to develop the elastic phase shifts, $\delta_{\ell\pm}$, and absorption parameters, $\eta_{\ell\pm}$, as power series in the center of mass momentum, q. (A brief discussion of this method as applied to the low energy region was given in Chapter 2.) This was used by Roper (1964), Roper and Wright (1965), and Roper, Wright, and Feld (1965), with some refinement for partial waves known to be resonant. Another method is to use dispersion relations (Chapter 4) for the *inverse* partial wave amplitudes $f_{\ell\pm}^{-1}$, parametrizing the left-hand cut and the inelasticity. This was done by Bransden, Moorhouse, and O'Donnell (1964, 1965 a, b), with parameters used for the left-hand cut variable, thus not assuming knowledge of the forces driving the π-N system. Both these analyses assumed the higher partial waves to be zero.

Obviously there are a great many reasonable parametrizations. For example, one may parametrize each partial wave amplitude as possible Breit-Wigner resonances and variable background (Davies, 1970; Ayed et al., 1972). Or one could use dispersion relations in $\cos\theta$. (Kane and Spearman 1963). Roychoudhury, Perrin, and Bransden (1970) [see also Bransden and Ogden (1971)] have developed a special method for the energy region between 2 and 5 GeV/c incident pion momentum. Below this region a description of the amplitude in terms of resonance-dominated partial wave amplitudes seems appropriate, and it is in this region below ~ 2 GeV/c that the methods described above have been applied. Above ~ 5 GeV/c a description of the amplitudes in terms of crossed-channel processes, such as Regge pole exchange (Chapter 8) is economical. Roychoudhury et al., parametrize the $A(s, t)$, $B(s, t)$ amplitudes as a sum of two terms: (1) crossed (t and u) channel Regge poles, and (2) a remainder term being the (remainder of) partial wave amplitudes suitably parametrized.

5.4. RESONANCES: ASSESSMENT OF RESULTS OF PARTIAL WAVE ANALYSIS

Interpretation of Strong Interaction

According to the analyses described in the last section most partial waves (of angular momentum and energy satisfying $\ell < q/2M_\pi$) exhibit strong interactions insofar as there is at least one energy, in the neighborhood of which the amplitude varies relatively quickly as a function of energy. Characteristically at these energies the partial wave amplitude describes a counter-clockwise circle in the Argand diagram. This is illustrated in Fig. 5-6. If we limit the term resonance to phenomena associated with T or K-matrix poles, then we shall *presume* that where major portions of such circles exist, there is a resonance. In just a comparatively few cases where the energy variation is both well known and characteristic (see section 5.2 above), or where there is good evidence from other channels (see section 5.2) and section 5.5 below) can we be *confident* that there is a resonance.

We wish to distinguish clearly between the interpretation of a strong interaction in a partial wave as a resonance, and the question of the existence of that strong interaction. We shall discuss the latter point in the following section.

Reliability of Partial Wave Analyses

We have seen how, at a given energy, lack of some experimental measurements together with errors on existing ones gives rise to a multiplicity of possible partial wave solutions, and how the principle of energy continuity reduces this number. In the whole energy range from an (N^*) mass of ~ 1 GeV to ~ 2 GeV, the phase shift analyses are in broad agreement on the strong interactions on resonances to be found therein. This is a remarkable fact, and one might inquire whether some types of solution may have been missed because of insufficient search in parameter space and/or associated latent theoretical or other prejudice. In this connection, the analyzers are working in a climate of opinion which favors "established resonances." Secondly, there is theoretical prejudice from partial wave dispersion relations, used with some assumptions on the long-range forces, overt in the case of the analysis discussed in the last paragraph of the section on "Single Energy Phase Shift Analysis," but which may possibly exist in some others. This is because the solution at higher energy depends on the solution at lower energy, in particular at the energy T_π (pion kinetic energy) ~ 300 MeV corresponding to total mass ~ 1300 MeV. There is some uncertainty in the solution at about this energy and the tendency is to favor the solution indicated by partial wave dispersion relations. [For discussions of this point see Bransden et al., (1965a), and Roper, Wright and Feld (1965).] An additional criticism is that nearly all the existing analyses set higher partial waves equal to zero, and there might well be a nonnegligible combined effect from these waves. This type of situation can be illustrated at lower energies, for example at $T_\pi \sim 300$ MeV, where f-waves cannot be determined from the available data alone; but if f-waves are put in from theory (hopefully not too unreliable since involving long-range forces only for f-waves at 300 MeV) then the d-waves change considerably (Kane and Spearman, 1963).

Despite these latent reservations we consider that the balance of probabilities is in favor of the correctness of the strong interactions or resonances (though not necessarily all the less important waves) found by present phase shift analyses. This consideration is partly from the great amount of searching for solutions performed, and partly from the coincidence of the results with observations of strong interaction on resonance phenomena in other channels.

Evidence from Other Channels

This is available for just a few of our resonances or strong interactions. We consider first the γN channel in formation of N^* in photon-nucleon interactions. Resonancelike photoproduction has been observed at the mass of the $N^*_{\frac{3}{2}}(1236)$, $N^*_{\frac{1}{2}}(1520)$, $N^*_{\frac{1}{2}}(1690)$, $N^*_{\frac{3}{2}}(1920)$, and higher resonances. There is no question on the existence of the $N^*_{\frac{3}{2}}(1236)$. However one must be careful of adducing the other photoproduction phenomena as evidence for the correctness of the *partial wave analysis*; strong interactions are evident at these energies from π-N differential and total cross sections without analysis, and from final state interactions one would expect enhancement of photoproduction at these energies. To support the partial wave analysis assignments of the resonances or strong interactions we require an independent analysis of pion photoproduction, for which not enough data yet exists, though there are strong indications of the $N^*_{\frac{1}{2}}(1520)$, $J^P = \frac{3}{2}^-$.

It is from other channels that we have two special cases which support the partial wave analysis, and particularly strongly because the states concerned are of low angular momentum, which are more difficult to resolve.

5.4. RESONANCES: RESULTS OF PARTIAL WAVE ANALYSIS

First there is the $I = \frac{1}{2}$, $J^P = \frac{1}{2}^-$, s_{11} πN system, with a resonance mass from partial wave analysis in the range ~1500 ~1560 MeV; the πN amplitude shows a particularly strong energy dependence in the absorption parameter. Now experiments also show (nearly) isotropic η production in the reaction:

$$\pi^- + p \to \eta + n$$

particularly strong just above threshold (1458 MeV), and which can be shown to be s-wave from the threshold excitation. This is sufficient to confirm the correctness of the s_{11} wave given by the partial wave analysis in this energy region. Moreover, investigations of the η-production support a resonance hypothesis for this wave (Hendry and Moorhouse, 1965; and Davies and Moorhouse, 1967).

Secondly, there is the $I = \frac{1}{2}$, $J^P = \frac{1}{2}^+$, p_{11} πN system, with a strong interaction centered in the energy range ~1400 to ~1500 MeV; the partial wave analysis reveals also in this case a rather strong energy dependence in the absorption parameter. But there is evidence (Foley et al., 1967; Anderson et al., 1966; Chadwick et al., 1962) from missing mass experiments in p-p and π-p collisions of the production of a system, say (R), with invariant energy (mass) centered about 1420 MeV:

$$p + p \to p + (R); \quad \pi + p \to \pi + (R) .$$

The system (R) is produced much more preferentially forward than any of the resonances such as $N_1^*(1520)$ which are also observed, thus strongly suggesting the mechanism of diffraction dissociation of the nucleon so that (R) has the quantum numbers $I = \frac{1}{2}$, $J^P = \frac{1}{2}^+$. Though there is evidence for a system (R) with the same quantum numbers as the p_{11} strong interaction of the partial wave analysis, we cannot

Table 5-1. Resonance Results of Different Partial Wave Analyses

Partial Wave	1[a]			2[b]			3[c]		
$L_{2I,2J} J^{P,I}$	Mass MeV/c^2	Γ_{tot} MeV	Γ_{el}/Γ_{tot}	Mass MeV/c^2	Γ_{tot} MeV	Γ_{el}/Γ_{tot}	Mass MeV/c^2	Γ_{tot} MeV	Γ_{el}/Γ_{tot}
$p_{33}(1)\ \frac{3^+}{2},\frac{3}{2}$	1235	129	1.00				1230	110	1.0
$s_{31}(1)\ \frac{1^-}{2},\frac{3}{2}$	1620	140	0.35	1617	141	0.28	1614	160	0.32
$d_{33}(1)\ \frac{3^-}{2},\frac{3}{2}$	1700	260	0.16	1650	188	0.12	1723	207	0.16
$p_{33}(2)\ \frac{3^+}{2},\frac{3}{2}$	1680	220	0.10	—	—	—	1890	130	0.17
$f_{35}(1)\ \frac{5^+}{2},\frac{3}{2}$	1875	250	0.18	1840	136	0.20	1879	259	0.53
$p_{31}(1)\ \frac{1^+}{2},\frac{3}{2}$	1900	200	0.33	1914	290	0.18	1797	230	0.18
$f_{37}(1)\ \frac{3^+}{2},\frac{3}{2}$	1925	200	0.40	1935	196	0.50	1921	232	0.40
$d_{35}(1)\ \frac{5^-}{2},\frac{3}{2}$	?2200	?600	?0.25	—	—	—	1870	160	0.10
$p_{11}(1)\ \frac{1^+}{2},\frac{1}{2}$	1470	220	0.65	1460	390	0.50	1427	236	0.52
$d_{13}(1)\ \frac{3^-}{2},\frac{1}{2}$	1520	120	0.58	1512	106	0.45	1528	119	0.57
$s_{11}(1)\ \frac{1^-}{2},\frac{1}{2}$	1500	50	0.25	1502	36(?)	0.36	1523	78	0.33
$d_{15}(1)\ \frac{5^-}{2},\frac{1}{2}$	1683	150	0.45	1670	115	0.50	1655	141	0.40
$f_{15}(1)\ \frac{5^+}{2},\frac{1}{2}$	1688	140	0.65	1685	104	0.54	1679	133	0.60

5.4. RESONANCES: RESULTS OF PARTIAL WAVE ANALYSIS

Table 5-1 (Continued)

Partial Wave $L_{2I,2J}J^{P},I$	1[a] Mass MeV/c^2	Γ_{tot} MeV	Γ_{el}/Γ_{tot}	2[b] Mass MeV/c^2	Γ_{tot} MeV	Γ_{el}/Γ_{tot}	3[c] Mass MeV/c^2	Γ_{tot} MeV	Γ_{el}/Γ_{tot}
$s_{11}(2)$ $\frac{1^-}{2},\frac{1}{2}$	1670	120	0.50	1766	404	0.56	1681	148	0.50
$d_{13}(2)$ $\frac{3^-}{2},\frac{1}{2}$	?	?	?	—	—	—	1730	130	0.10
$p_{13}(1)$ $\frac{3^+}{2},\frac{1}{2}$	1850	300	0.25	1844	450	0.40	1691	101	0.17
$p_{11}(2)$ $\frac{1^+}{2},\frac{1}{2}$	1720	160	0.20	1770	445	0.43	1755	183	0.15
$d_{13}(3)$ $\frac{3^-}{2},\frac{1}{2}$	2075	150	0.30	—	—	—	—	—	—
$d_{15}(2)$ $\frac{5^-}{2},\frac{1}{2}$	2100	150	0.20	—	—	—	2055	170	0.09
f_{17} $\frac{7^+}{2},\frac{1}{2}$	2000	200	0.15	—	—	—	2048	183	0.06
g_{17} $\frac{7^-}{2},\frac{1}{2}$	2225	150	0.35	1906?	319?	.06?	2150	322	0.19
$f_{15}(2)$ $\frac{5^+}{2},\frac{1}{2}$	?2175	?150	?0.25	—	—	—	2001	56	0.09

a. Almehed and Lovelace, 1972.
b. Davies, 1970. (The date of this analysis is 1968, and it does not include a large amount of later data, used in a and c.)
c. Ayed, Bareyre and Lemoigne, 1973.

Notation: We use the notation $L_{2I,2J}(1)$, $L_{2I,2J}(2)$, ... for the lowest resonance, second lowest resonance... in the $L_{2I,2J}$ state, for example $p_{11}(1)$, $p_{11}(2)$.... The motivation for this notation is the present liability to variation of quoted masses, though obviously the notation is only useful when the existence and quantum numbers of resonances up to a common mass are rather certain.

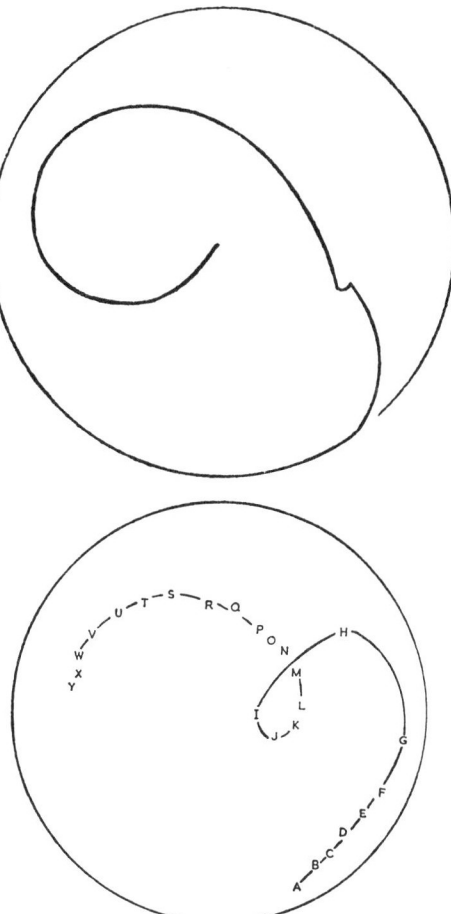

Fig. 5-6. Argand diagram of the S_{11} wave from (a) Davies (1969, 1970); (b) Donnachie, Kirsopp, and Lovelace (1968).

adduce independent evidence that the strong interaction is a resonance.[7] There is also a body of more tentative supporting evidence such as Morgan's (1968) isobar model analysis of the $I = \frac{1}{2}$ $\pi + N \to 2\pi + N$ processes for a second p_{11} resonance (Table 5-1).

---

[7] Gellert et al., (1966) from an analysis of the momenta in the events $p + p \to p + p + \pi^+ + \pi^-$, present a case for the phenomenon (R) in p-p collisions being merely a reflection of strong $N^*(1236)$ production. However one is left with the problem of finding a mechanism in the π-p collisions.

5.4. RESONANCES: RESULTS OF PARTIAL WAVE ANALYSIS

Resonance Results

We show the Argand diagram of the s_{11} wave in Fig. 5-6a from the analysis of Davies and in Fig. 5-6b from the analysis of Donnachie, Kirsopp, and Lovelace (1968). These two figures illustrate a not untypical difference in the results of partial wave analysis and also a rather difficult case of separating the resonance loops (see section on "Inelastic Resonances") from the background. Figure 5-6a contains two clear loops which are sufficiently rapidly traversed with increasing energy to be regarded as arcs of two resonance circles. The fact that these are only arcs and not complete circles is attributed to the influence of nonresonant background and the interference of one resonance with the other. Though Fig. 5-6b is compatible with the existence of two resonances, one would not necessarily deduce them on this diagram alone. It should be stated that evidence from other partial wave analyses (not illustrated here) and from inelastic channels, as discussed in the section "Evidence from other Channels" above, make the existence of two s_{11} resonances rather certain.

In Table 5-1 we list the resonance parameters reported by Almehed and Lovelace (1972) where, presumably, following earlier practice by Lovelace (1968) the problem of background is rather arbitrarily resolved by taking the energy of minimum η, corresponding to maximum absorption in the partial wave, as the resonance energy for the inelastic resonance.[8] Also listed in Table 5-1 are some resonance parameters of Davies (1970). Here the partial wave amplitudes are parametrized by eqs. (5-17), (5-20), with possibly more than one resonance and a slowly varying background term. Consequently the resonance position on the Argand diagram, with corresponding energy W_R, is according to the Wigner condition (see section on "Elastic Resonances") namely the point of fastest variation

---

[8] Dalitz and Moorhouse (1970) show that this is equivalent to an assumption that the phase of the resonance in the πN channel is equal to the phase of the background in the πN channel.

with energy on the loop, and W_R would be the energy of maximum delay-time if the scattering were elastic (see section on "The Criterion of Time-Delay"). The third set of resonance parameters in Table 5-1 is that of Ayed, Bareyre and Lemoigne (1973) with the resonance position and width again estimated from the rate of variation with energy. In Table 5-1 a dash denotes that though the energy region has been analyzed, the resonance indicated has not been observed very distinctly. A blank denotes that the energy region has not been analyzed.

REFERENCES

R. K. Adair, Phys. Rev. 113, 338 (1959).

J. W. Alcock and W. N. Cottingham, Nucl. Phys. B31, 443 (1971).

S. Almehed and C. Lovelace, Nucl. Phys. B40, 157 (1972).

M. Alston, L. W. Alvarez, P. Eberhard, M. L. Good, W. Graziano, H. Ticho and S. G. Wojcicki, Phys. Rev. Letters 6, 300 (1961).

R. Ayed, P. Bareyre and Y. Lemoigne, Proc. of XVI International Conf. on High Energy Physics (Ed. A. Roberts, National Accelerator Laboratory, 1973).

E. W. Anderson, et al., Phys. Rev. Letters 16, 885 (1966).

P. Auvil, A. Donnachie, A. T. Lea and C. Lovelace, Phys. Letters 12, 76 (1964).

P. Bareyre, C. Bricman, A. V. Stirling and G. Villet, Phys. Letters 18, 342 (1965).

P. Bareyre, C. Bricman and G. Villet, Phys. Rev. 165, 1730 (1967).

REFERENCES

R. Baud, et al., Phys. Letters 31B, 549 (1970).

B. H. Bransden and P. J. Ogden, Nucl. Phys. B26, 511 (1971).

B. H. Bransden, R. G. Moorhouse and P. J. O'Donnell, Phys. Letters, 11, 339 (1964).

B. H. Bransden, R. G. Moorhouse and P. J. O'Donnell, Phys. Rev. 139, B1566 (1965a).

B. H. Bransden, R. G. Moorhouse and P. J. O'Donnell, Phys. Letters 19, 420 (1965b).

J. Bowcock, W. N. Cottingham and P. Ng, (unpublished, 1970).

H. Burckhardt, Nuovo Cimento 10A, 379 (1972).

G. Calucci, L. Fonda and G. C. Ghirardi, Phys. Rev. 166, 1219 (1968).

G. B. Chadwick, et al., Phys. Rev. 128, 1823 (1962).

J. W. Cronin, Phys. Rev. 118, 824 (1960).

R. Cutkowsky and A. Deo, Phys. Rev. 174, 1859 (1968).

R. H. Dalitz and S. F. Tuan, Ann. Physics (New York), 10, 307 (1960).

R. H. Dalitz, Ann. Rev. Nucl. Science 13, 339 (1963).

R. H. Dalitz and G. Rajasekaran, Phys. Letters 7, 273 (1963).

R. H. Dalitz and R. G. Moorhouse, Phys. Letters 14, 159 (1965).

R. H. Dalitz and R. G. Moorhouse, Proc. Roy. Soc. (1970).

P. Darriulat, G. Igo, H. G. Pugh and H. D. Holngren, Phys. Rev. 137, B315 (1965).

A. T. Davies and R. G. Moorhouse, Nuovo Cimento 52A, 1112 (1967).

A. T. Davies, Nuclear Phys. B21, 359 (1970).

F. de Hoffman, N. Metropolis, E. F. Alei and H. A. Bethe, Phys. Rev. 95, 1586 (1954).

A. Donnachie, Proc. Scottish Univ. Summer Sch. (ed. T. W. Preist and L. L. J. Vick, Oliver and Boyd, Edinburgh, 1966).

A. Donnachie, R. G. Kirsopp and C. Lovelace, Phys. Letters 26B, 161 (1968).

R. J. Eden and J. R. Taylor, Phys. Rev. Letters 11, 516 (1963).

L. Eisenbud, Phys. Rev. 73, 1407 (1946).

A. R. Erwin, R. March, W. Walzer and E. West, Phys. Rev. Letters 6, 628 (1961).

M. N. Focacci, et al., Phys. Rev. Letters 17, 890 (1966).

K. T. Foley, et al., Phys. Rev. Letters 19, 397 (1967).

S. C. Frautschi, *Regge Poles and S-Matrix Theory* (Benjamin Inc., New York, 1963).

W. R. Frazer and G. Fulco, Phys. Letters 2, 364 (1959).

W. R. Frazer and G. Fulco, Phys. Rev. 117, 1609 (1960).

W. R. Frazer and A. W. Hendry, Phys. Rev. 134, B 1307 (1964).

E. Gellert, et al., Phys. Rev. Letters 17, 884 (1966).

C. J. Goebel and K. W. McVoy, Phys. Rev. 164, 1932 (1967).

M. L. Goldberger and K. M. Watson, *Collision Theory* (Benjamin Inc., New York, 1963).

A. W. Hendry and R. G. Moorhouse, Phys. Letters 18, 171 (1965).

G. Herzberg, *Molecular Spectra and Molecular Structure I: Spectra of Diatomic Molecules* (Van Nostrand, New York, 1950).

REFERENCES

G. Hohler, G. Ebel and J. Zwingenberger, Proc. Aix-en-Provence Intern. Conf. Elementary Particles, Saclay, France (1961).

C. H. Johnson and H. M. Steiner, Univ. Calif. Radiation Lab., Reports U.C.R.L. 18001 (1967).

G. L. Kane and T. D. Spearman, Phys. Rev. Letters 11, 45 (1963).

C. Lovelace, *Proc. 1967 Heidelberg Conference on High Energy Interactions* (North Holland, Amsterdam, 1968).

D. Morgan, Phys. Rev. 166, 1731 (1968).

R. C. Miller, et al., Nucl. Phys. B37, 401 (1972).

M. Nauenberg and J. C. Nearing, Phys. Rev. Letters 12, 63 (1964).

L. E. Olsson and W. P. Trower, Univ. Illinois, Urbana, Physics Dept. Tech. Reports 146 (1969).

L. D. Roper, Phys. Rev. Letters 12, 342 (1964).

L. D. Roper and R. M. Wright, Phys. Rev. 135, B921 (1965).

L. D. Roper, R. M. Wright and B. T. Feld, Phys. Rev. 138, B190 (1965).

R. K. Roychoudhury, R. Perrin and B. H. Bransden, Nucl. Phys. B22, 532 (1970).

F. T. Smith, Phys. Rev. 118, 349 (1960).

E. J. Squires, *Complex Angular Momentum and Particle Physics* (Benjamin Inc., New York 1950) (1963).

G. Takeda, Phys. Rev. 100, 440 (1955).

E. P. Wigner, Phys. Rev. 98, 145 (1955a).

E. P. Wigner, Am. J. Phys. 23, 371 (1955b).

CHAPTER 6

SYMMETRIES AND CLASSIFICATION OF PARTICLES AND RESONANCES

6.1. ISOSPIN, AN SU2 SYMMETRY

As outlined in Chapter 1, the interaction of pions and nucleons is charge-independent. The symmetry appears to be perturbed only by comparatively weak forces of the order of the electromagnetic interactions. It is a reasonable hypothesis that the specifically hadronic interactions are exactly charge-independent, and that the observed deviations are due to the electromagnetic interactions. The charge independence can be best described as a symmetry in *isospin*, which is mathematically identical to spin in ordinary space. The pion is assigned total isospin 1, giving three independent isospin states of the pion. These are linear combinations of the three physical states of the pion – π^+, π^-, and π^0. The nucleon is assigned total spin $\frac{1}{2}$ with two independent isospin states that are linear combinations of p and n. Charge independence is then the statement that hadronic interactions are invariant under rotations in isospin space.

Slightly more generally all properties of pions and nucleons are invariant under rotations in isospin space, except for perturbation due to weak and electromagnetic interactions. This means that all pion masses should be approximately equal, since for example the π^+ state can be rotated (in isospin space) into the π^0 state. Similarly the proton mass should be approximately equal to the neutron mass. The figures are $\{m(\pi^\pm) - m(\pi^0)\}/m(\pi^0) = 3.4\%$,[1] $\{m(n) - m(p)\}/m(n) = 0.14\%$. It seems that the isospin invariance, otherwise isospin symmetry, is good for all hadronic interactions and hadrons. The Σ particles have isospin 1 and $\{m(\Sigma^+) - m(\Sigma^-)\}/m(\Sigma^-) = 6.6\%$. The pions are said to form an *isospin multiplet* of isospin 1, the nucleons an isospin multiplet of isospin $\frac{1}{2}$. There are elementary particles and resonances belonging to isospin multiplets with $I = 0$ ($\eta, n, \ldots$), $I = \frac{1}{2}$ ($K, \bar{K}, N, \ldots$) $I = 1$ ($\pi, \Sigma \ldots$) and $I = \frac{3}{2}$ ($\Delta, \Omega^-, \ldots$) but none, on present evidence, with higher isospin. This remarkable limitation is connected with multiplet limitation in SU3, which will be discussed subsequently.

SU2 Transformations

We now consider the mathematical formulation of rotations in isospin space. This is identical to the transformation of spin states under rotations in ordinary space. However it is useful to write the results explicitly, both for completeness of presentation and to lay a foundation for the discussion of SU3 subsequently.

Let $\psi_\alpha (\alpha = 1, 2)$ be the two-component nucleon isospinor (conventionally, as discussed below $\psi = \begin{pmatrix} 1 \\ 0 \end{pmatrix}$ and $\begin{pmatrix} 0 \\ 1 \end{pmatrix}$ represent proton and neutron states respectively). The pion is an isospin 1 state, that is a vector, in the three-dimensional isospin space, so it is represented by $\phi_i = (\phi_1, \phi_2, \phi_3)$ being the components along the 1, 2, 3 axes of isospin space. Then any transition operator for strong interactions, such as a Lagrangian or Hamiltonian, must be invariant under a rotation of these isospinors and isovectors in isospin space. Such a rotation, like a rotation of spinors and vectors in ordinary spin space, is given, for a particular rotation R, by:

---

[1] π^+ and π^- are charge conjugate particles, which is also a symmetry of electromagnetic interactions so that $m(\pi^+) = m(\pi^-)$.

6.1. ISOSPIN, AN SU2 SYMMETRY

$$\psi_\alpha' = V_{\alpha\beta}(R)\,\psi_\beta \qquad\qquad \phi_i' = W_{ij}(R)\,\phi_j \;. \qquad (6\text{-}1)$$

In eq. (6-1) the set of 2×2 matrices V and the set of 3×3 matrices W are each an *irreducible representation* of the unitary unimodular group of 2×2 matrices U. [We remind the reader[2] that for V to be a *representation*: to each U there corresponds a matrix V and the matrices V also have the group property of eq. (6-3). For the representation to be *reducible* there is a fixed matrix H so that for all $V: H^{-1} V H = \begin{pmatrix} V_1 & V_3 \\ 0 & V_2 \end{pmatrix}$ or $\begin{pmatrix} V_1 & 0 \\ V_3 & V_2 \end{pmatrix}$ where V_1, V_2, V_3 are matrices and 0 denotes a matrix of zeros. It is readily verified that the set of matrices V_1, and also the set V_2, form representation of the group. The representation is said to be *irreducible* if it cannot be reduced in this way to smaller representations.] The matrices U are unitary, that is the Hermitian conjugate is the inverse, and unimodular:

$$U^+ U = 1 = \begin{pmatrix} 1 & 0 \\ 0 & 1 \end{pmatrix} ;\; \det U = 1 \;. \qquad (6\text{-}2)$$

The set of all such matrices forms a group, since U^{-1} exists, and if U_1, U_2 are unitary and unimodular:

$$U_1 U_2 = U_3 \qquad (6\text{-}3)$$

where U_3 is also unitary and unimodular. The group is known as the group SU2, where U is a mnemonic for unitary and 2 for two-dimensional, with S denoting unimodular. The set of *all* unitary 2×2 matrices including, but not restricted to the unimodular is also a group, known as U(2).

---

[2] A knowledge of group theory is not necessary for this chapter but we refer any reader wishing to know more to Hammermesh (1962) or, on a more advanced level, to Weyl (1946).

CLASSIFICATION OF PARTICLES AND RESONANCES

Since V is a 2×2 matrix, it can obviously be taken to be identical with U. (It can be shown that V must be unitary and unimodular.) It is advantageous to consider infinitesimal unitary transformations. The general form is:

$$V = 1 + i\tau_j \Delta a_j$$

where Δa_j are real infinitesimal quantities, τ_j are 2×2 matrices (with summation over j understood), and 1 is the unit 2×2 matrix. We find the conditions on τ_j so that eq. (6-2) be satisfied to first order in arbitrary Δa_j. Equation (6-2) gives:

$$(1 - i\tau_j^+ \Delta a_j)(1 + i\tau_j \Delta a_j) = 1 \qquad \det(1 + i\tau_j \Delta a_j) = 1$$

or:

$$(\tau_j^+ - \tau_j)\Delta a_j = 0 \qquad \operatorname{Tr}(\tau_j \Delta a_j) = 0$$

or:

$$\tau_j^+ = \tau_j \qquad \operatorname{Tr}(\tau_j) = 0 \ .$$

There are only three independent traceless hermitian 2×2 matrices, and without loss of generality we can take these as:

$$\tau_1 = \begin{pmatrix} 0 & 1 \\ 1 & 0 \end{pmatrix} \tau_2 = \begin{pmatrix} 0 & -i \\ i & 0 \end{pmatrix} \tau_3 = \begin{pmatrix} 1 & 0 \\ 0 & -1 \end{pmatrix} . \qquad (6\text{-}4a)$$

These are the Pauli (iso)spin matrices. The proton and neutron states must be orthogonal isospinors; the convention is that a proton state be a multiple of the isospinor $\begin{pmatrix} 1 \\ 0 \end{pmatrix}$ and a neutron a multiple of $\begin{pmatrix} 0 \\ 1 \end{pmatrix}$. Another convention is that states where charge is a good quantum number, such as proton or neutron states, have also 3-*axis component* of isospin as a good quantum number. Consequently rotations of isospin space about the 3-axis do not change these states, so τ_3 must correspond to such rotations (since $V = 1 + i\tau_3 \Delta a_3$ preserves a proton (neutron) state as a

6.1. ISOSPIN, AN SU2 SYMMETRY

proton (neutron) state, whereas for example $V = 1 + i\tau_1 \Delta a_1$ does not). The expressions $V = 1 + i\tau_1 \Delta a_1$, $1 + i\tau_2 \Delta a_2$, and $1 + i\tau_3 \Delta a_3$ are isospinor rotations about the 1, 2, and 3 axes respectively in isospin space. The charge operator Q is given by:

$$Q/e = I_3 + \frac{1}{2}$$

where $I_1 = \frac{1}{2}\tau_1$, $I_2 = \frac{1}{2}\tau_2$, $I_3 = \frac{1}{2}\tau_3$ are, for a single nucleon state, the three isospin operator components.

The matrices (6-4a) can be verified to obey the commutation relations:

$$[\tau_i, \tau_j] = 2 i \epsilon_{ijk} \tau_k \tag{6-5a}$$

where $\epsilon_{ijk} = 0$ if any two of i, j, k are equal, and $\epsilon_{ijk} = \pm 1$ accordingly as i, j, k form an even or odd permutation of 1, 2, 3. These are just the usual angular momentum commutation relations and could have been expressed as $[I_1, I_2] = i I_3$ and cyclic permutations thereof, or:

$$[I_i, I_j] = i \epsilon_{ijk} I_k . \tag{6-5b}$$

Equation (6-5a) is the defining equation of a *Lie algebra*, associated with the Lie group SU2, with structure constants ϵ_{ijk} and *generators* τ_i. [A set of linear operators, such as quantum mechanical operators, is said to form a Lie algebra if (a) any linear combination with complex coefficients of members of the set belongs to the set, and (b) the commutator of any two members of the set belongs to the set. Linear combinations of the generators form the particular Lie algebra, which is thus characterized by the structure constants.] The expressions of eq. (6-4a) are a 2×2 representation of the generators; 3×3 matrices T_i obeying eq. (6-5) and thus forming a 3×3 representation of the generators are:

$$T_1 = 2\begin{pmatrix} 0 & 0 & 0 \\ 0 & 0 & -i \\ 0 & i & 0 \end{pmatrix}, \quad T_2 = 2\begin{pmatrix} 0 & 0 & +i \\ 0 & 0 & 0 \\ -i & 0 & 0 \end{pmatrix}, \quad T_3 = 2\begin{pmatrix} 0 & -i & 0 \\ +i & 0 & 0 \\ 0 & 0 & 0 \end{pmatrix}. \tag{6-4b}$$

The expressions $W = 1 + i T_1 \Delta a_1$, $1 + i T_2 \Delta a_2$, and $1 + i T_3 \Delta a_3$ are isovector rotations about the 1, 2, and 3 axes respectively, and $I_1 = \frac{1}{2} T_1$, $I_2 = \frac{1}{2} T_2$, and $I_3 = \frac{1}{2} T_3$ are, for a single pion state, the three isospin operator components.

The terms ϕ_1, ϕ_2, ϕ_3 are the components of the isovector pion field along the 1, 2, and 3 axes, transforming under eq. (6-1). It is immediate that:

$$I_3 \phi_3 = 0, \quad I_3(\phi_1 + i \phi_2) = +(\phi_1 + i \phi_2), \quad I_3(\phi_1 - i \phi_2) = -(\phi_1 - i \phi_2) \qquad (6\text{-}6)$$

so that $Q/e = I_3$, and ϕ_3 corresponds to the π^0 field and $\phi_1 \pm i \phi_2$ to the $\pi^{\pm}$ field, possible phase factors apart.

We have found the form of V and W of eq. (6-1), for the case when the unitary transformations are infinitesimal:

$$V = 1 + i \tau_j \Delta a_j \qquad (6\text{-}7a)$$

and these form a 2×2 irreducible representation (and are indeed identical with) the (infinitesimal) unimodular unitary group:

$$W = 1 + i T_j \Delta a_j \qquad (6\text{-}7b)$$

and these form a 3×3 irreducible representation of the (infinitesimal) unimodular unitary group. [The fact that eq. (6-7b) is a representation is obvious from the fact that there is a one-to-one correspondence between eqs. (6-7a) and (6-7b). We leave the proof of irreducibility of eqs. (6-7a) and (6-7b) to the interested reader.] From eqs. (6-7b) and (6-4b) it is evident that eqs. (6-7) are rotations through infinitesimal angles $2\Delta a_1$, $2\Delta a_2$, and $2\Delta a_3$ about the 1, 2, and 3 axes respectively in isospin space.

The general finite transformations, eq. (6-1), and the complete representations V and W of the group $SU2$ are generated by iteration of eqs. (6-7). (It is a property of Lie groups that any member of the group can be reached by a product of members infinitesimally different from unity.) In

6.1. ISOSPIN, AN SU2 SYMMETRY

eqs. (6-7) let $\Delta a_i = a_i/n$ where a_i is finite and n large; Δa_i are infinitesimal for $n \to \infty$. Then n repeated transformations, eqs. (6-7), give, as $n \to \infty$:

$$V = \lim_{n \to \infty} \prod_n (1+i\tau_j a_j/n); \quad W = \lim_{n \to \infty} \prod_n (1+i T_j a_j/n)$$

$$V = \exp(i\tau_j a_j) \qquad W = \exp(i T_j a_j) \qquad (6\text{-}8)$$

It is easy to show that the expression:

$$\psi^\dagger \tau_i \psi \, \phi_i \qquad (6\text{-}9a)$$

is an isospin invariant. To do this we use the infinitesimal transformations and the fact that the j, k element of T_i is given by:

$$(T_i)_{jk} = -2i\, \epsilon_{ijk} . \qquad (6\text{-}10)$$

Equation (6-10) follows immediately from eq. (6-4b). Consequently:

$$\phi_i' = \phi_i + i(T_j)_{ik} \Delta a_j \phi_k = \phi_i + 2\epsilon_{jik} \Delta a_j \phi_k$$

$$\psi^{\dagger'}(\tau_i)\psi' = \psi^\dagger(\tau_i)\psi + \psi^\dagger\, i[\tau_i, \tau_j]\, \Delta a_j \psi$$

$$= \psi^\dagger(\tau_i)\psi - 2\psi^\dagger \tau_k \psi\, \Delta a_j\, \epsilon_{ijk}$$

It follows that to first order Δa_j:

$$\psi^{\dagger'} \tau_i \psi' \phi_i' = \psi^\dagger \tau_k \psi\, \phi_k .$$

This proves the invariance for infinitesimal transformations, and thus for a general transformation, since this can be built from a series of infinitesimal transformations. Since ϕ_i transforms like an isovector, the isospin invariance of eq. (6-9a) proves that $\psi^\dagger \tau_i \psi$ transforms like a vector and the expression is often written:

$$\psi^\dagger \vec{\tau}\psi \cdot \vec{\phi} . \qquad (6\text{-}9b)$$

6.2. YOUNG TABLEAUX AND THE PERMUTATION SYMMETRY OF WAVE FUNCTIONS

The properties of many particle wave functions under permutations of the particles play a vital role in physics, distinguishing between particles obeying Fermi statistics and particles obeying Bose statistics. But permutation symmetries also play an important part in classifying the irreducible representations of SU2, SU3, ... These two roles of the permutation group are closely connected, in a rather simple and elementary way, which we will illustrate here using SU2 wave functions.

The representation of SU2 by the unitary, unimodular 2×2 matrices V of eqs. (6-1) and (6-7) is known as the *basic representation*;[3] *basis vectors* in that representation are the spinors $\begin{pmatrix} 1 \\ 0 \end{pmatrix}$ and $\begin{pmatrix} 0 \\ 1 \end{pmatrix}$. If we translate into isospin terms:

$$p = \begin{pmatrix} 1 \\ 0 \end{pmatrix} \quad \text{and} \quad n = \begin{pmatrix} 0 \\ 1 \end{pmatrix}. \qquad (6\text{-}11)$$

Let us label the particles in one or more particle states by numbers 1, 2, 3... . Then we write down one and more particle states as follows:

$$\left. \begin{array}{ll} p(1), \quad n(1) & (1\text{a}) \\ p(1)\, p(1), \quad \frac{1}{\sqrt{2}}\{p(1)\, n(2) + p(2)\, n(1)\}, \quad n(1)\, n(2) & (2\text{a}) \\ \frac{1}{\sqrt{2}}\{p(1)\, n(2) - p(2)\, n(1)\} & (2\text{b}) \end{array} \right\} \qquad (6\text{-}12)$$

The states (1a) transform by V under SU2 transformations so that:

$$p(1) \to V\, p(1) = \begin{pmatrix} V_{11} \\ V_{21} \end{pmatrix}; \quad n(1) \to V\, n(1) = \begin{pmatrix} V_{12} \\ V_{22} \end{pmatrix}$$

so that:

---

[3] More generally the basic representation of SUn is that of unitary, unimodular $n \times n$ matrices.

6.2. YOUNG TABLEAUX

$$\chi_1 p(1) + \chi_2 n(1) \to \chi_1 V p(1) + \chi_2 V n(1) = \chi_1' p(1) + \chi_2' n(1) \quad (6\text{-}13)$$

where:

$$\chi_\alpha' = V_{\alpha\beta} \chi_\beta . \quad (6\text{-}14)$$

However, the two-particle wave functions (2a) have been constructed to be an isovector. That is they transform amongst themselves under SU2 as follows:

$$\chi_1 \{p(1) p(2)\} + \chi_2 \frac{1}{\sqrt{2}} \{p(1) n(2) + p(2) n(1)\} + \chi_3 \{n(1) n(2)\}$$

$$\to \chi_1 \{[V p(1)] [V p(2)]\} + \chi_2 \frac{1}{\sqrt{2}} \{[V p(1)] [V n(2)]$$

$$+ [V p(2)] [V n(1)]\} + \chi_3 \{[V n(1)] [V n(2)]\} = \chi_1' \{p(1) p(2)\}$$

$$+ \chi_2' \frac{1}{\sqrt{2}} \{p(1) n(2) + p(2) n(1)\} + \chi_3' \{n(1) n(2)\} \quad (6\text{-}15)$$

where:

$$\chi_i' = \tilde{W}_{ij} \chi_j \quad (6\text{-}16)$$

and $\tilde{W}$ transforms by a unitary transformation to W of eqs. (6-1) and (6-7).[4]

So both (1a) and (2a) of eq. (6-12) are bases of irreducible representation of SU2. So also of course is the isoscalar (2b); it transforms into a unimodular multiple of itself giving rise to a 1×1 matrix representation of SU2. We note that (2b) is *antisymmetric* under the interchange of particles $1 \leftrightarrow 2$ and (2a) is symmetric under $1 \leftrightarrow 2$. This connection of behavior under permutation of particles and bases of irreducible representations extends to multiparticle wave functions and is most simply described by the use of Young tableaux.

---

[4] It is easy to see this by taking V to be the generator transformations of eq. (6-7a), when an explicit evaluation of eq. (6-15) shows $\tilde{W} = X^\dagger W X$ where X transforms from a basis of vectors along the x, y, z axes to the corresponding spherical harmonic basis.

Young Tableaux

Associated with every irreducible representation of SU2 (generally SUm); which can be formed from an n-particle wave function is a Young tableau of n squares arranged in two (generally m) rows starting from the left. A Young tableau [see Hammermesh (1962, Chapter 7); Weyl (1946)] specifies the dimensions of the irreducible representation and also the transformations of the wave functions under permutation of the particles. The general Young tableau for SUm has λ_k squares in row k, with m rows ($\leq n$) so that $\lambda_1 \geq \lambda_2 \geq ... \geq \lambda_m$, and an example for SU6 with $n = 9$ is:

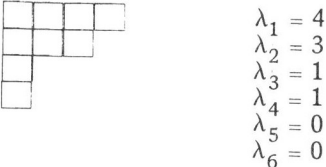

$\lambda_1 = 4$
$\lambda_2 = 3$
$\lambda_3 = 1$
$\lambda_4 = 1$
$\lambda_5 = 0$
$\lambda_6 = 0$

The permutation properties of wave functions, given by the Young tableaux, are in general complicated, being bases of matrix representations of the symmetric group S_n (that is the group of permutations of n objects), but a Young tableau of one row only ($\lambda_2 = \lambda_3 = ... = \lambda_p = 0$) is associated with a totally *symmetric* wave function, and a Young tableau of one column only is associated with a totally *antisymmetric* wave function.

We now illustrate some of the properties of Young tableaux in the case of SU2. Some Young tableaux and their associated wave functions are:

$$\lambda_1 = 1, \ \lambda_2 = 0 : p(1), n(1) \qquad (6\text{-}16\text{a})$$

$$\lambda_1 = 2, \ \lambda_2 = 0 : p(1)\,p(2), \frac{1}{\sqrt{2}}\{p(1)\,n(2) + p(2)\,n(1)\}, \ n(1)\,n(2) \qquad (6\text{-}16\text{b})$$

$$\lambda_1 = \lambda_2 = 1 : \frac{1}{\sqrt{2}}\{p(1)\,n(2) - p(2)\,n(1)\} \ . \qquad (6\text{-}16\text{c})$$

6.2. YOUNG TABLEAUX

It is a theorem that the dimensions of the irreducible representation associated with a Young tableau is given by:

$$\text{SU2}: d = \lambda_1 - \lambda_2 + 1 \ . \tag{6-17}$$

The dimension of a representation with isospin quantum number I is $2I + 1$. So from eq. (6-17):

$$I = (\lambda_1 - \lambda_2)/2 \ . \tag{6-18}$$

We now see how we could have built wave functions of eq. (6-16b) entirely from a knowledge of the Young tableau ☐☐ . The fact that it is a single row tableau tells us that the wave functions must be symmetric, and eq. (6-17) indicates that there are three wave functions (basis functions of the $d = 3$ dimensional representation). Consequently the wave functions must be eq. (6-16b), or three independent linear combinations of eq. (6-13). [Of course, the functions of eq. (6-13) are chosen to be eigenfunctions of I_3.] In the case of the Young tableaux ☐ , since it is a single column we know that the wave functions must be totally antisymmetric, and eq. (6-17) (redundantly) shows that $d = 1$; the wave function is thus fixed as eq. (6-16c).

We will now construct three-particle wave functions using Young tableaux. These are:

☐☐☐   ☐☐
            ☐   (6-19)

(a)   (b)

By eq. (6-17) there are four wave functions associated with (6-19a), and each is totally symmetric. If as usual we construct them to be eigenfunctions of I_3 or charge (and this is easily achieved by having the number of protons as good quantum number for each wave function), we must get (with the sign convention that isospin raising and lowering operators are positive):

$$p(1)\,p(2)\,p(3),\,\frac{1}{\sqrt{3}}\{p(1)\,p(2)\,n(3)+p(1)\,n(2)\,p(3)+n(1)\,p(2)\,p(3)\},\,\frac{1}{\sqrt{3}}\{n(1)\,n(2)\,p(3)$$

$$+n(1)\,p(2)\,n(3)+p(1)\,n(2)\,n(3)\},\,n(1)\,n(2)\,n(3)\;.$$

(6-20)

Equation (6-18) gives $I = \frac{3}{2}$ for the multiplet (6-20).

The Young tableau (6-19b) gives slightly more complicated behavior of wave functions under permutation of the particles. Since we shall need these properties (see below, section 9.3), we now expound them. Each Young tableau of (6-19) specifies an irreducible representation of the symmetric group S_3 the group of permutations of three objects (which in our case can be thought of as the particles). The dimensions of the associated representations are given by the number of ways of labeling the squares by 1, 2, and 3, so that the numbers are in ascending order in each row starting from the left and in each column starting from the top. In the case of (6-19a) there is just one such arrangement: ⬚1⬚2⬚3⬚. Consequently this matrix representation of S_3 is of dimension 1×1. We already know this since each wave function of eq. (6-20) transforms into itself under any interchange of particle numbers, so that each permutation is represented by the identity transformation, equivalent to the number 1.

But corresponding to (6-19b) we have two possible arrangements:

$$\begin{array}{|c|c|}\hline 1 & 2 \\\hline 3 \\\hline\end{array} \qquad\qquad \begin{array}{|c|c|}\hline 1 & 3 \\\hline 2 \\\hline\end{array} \tag{6-21}$$

leading to a 2×2 irreducible representation of S_3 and thus two SU2 multiplets [each of dimension $d = 2$ by eq. (6-14)] to form a basis for this S_3 representation. These wave functions are:

$$P_1 = \frac{1}{\sqrt{6}}(n(1)\,p(2)+n(2)\,p(1))\,p(3)-\sqrt{\frac{2}{3}}\,p(1)\,p(2)\,n(3),\; N_1 = -\frac{1}{\sqrt{6}}(p(1)\,n(2)$$

$$+p(2)\,n(1))\,n(3) + \sqrt{\frac{2}{3}}\,n(1)\,n(2)\,p(3) \tag{6-22}$$

$$P_2 = \frac{1}{\sqrt{2}}(n(1)\,p(2)-n(2)\,p(1))\,p(3),\; N_2 = -\frac{1}{\sqrt{2}}(p(1)\,n(2)-p(2)\,n(1))\,n(3)\;.$$

6.2. YOUNG TABLEAUX

P_1, N_1 form a basis for a 2×2 irreducible representation of SU2, so that when $p(1)$, $n(1)$ and $p(2)$, $n(2)$ and $p(3)$ each transform by eq. (6-13), then any linear combination of P_1 and N_1 transforms into a linear combination of P_1 and N_1 by a 2×2 matrix representation of SU2 equivalent to V.[5] Similarly for P_2, N_2 — in other words, P_1, N_1 form an isospin doublet and so do P_2, N_2. (P_1, N_1 can be formed by combining particles 1 and 2 into an isospin triplet and then adding particle 3 to form a doublet; P_2, N_2 by first combining 1 and 2 to form a singlet.)

P_1, P_2 (N_1, N_2) form a basis for a 2×2 irreducible matrix representation of S_3. If P is a permutation of 1 2 3, and that permutation is applied to the particle labels in the wave functions P_1, P_2 (N_1, N_2), then:

$$\begin{pmatrix} P_1 \\ P_2 \end{pmatrix} \to Q_P \begin{pmatrix} P_1 \\ P_2 \end{pmatrix}, \begin{pmatrix} N_1 \\ N_2 \end{pmatrix} \to Q_P \begin{pmatrix} N_1 \\ N_2 \end{pmatrix}. \qquad (6\text{-}23)$$

If (ij) denotes that permutation which is the interchange of i and j, then we find by explicit calculation:

$$Q_{(1\ 2)} = \begin{pmatrix} 1 & 0 \\ 0 & -1 \end{pmatrix} Q_{(2\ 3)} = \begin{pmatrix} -\frac{1}{2} & \frac{\sqrt{3}}{2} \\ \frac{\sqrt{3}}{2} & \frac{1}{2} \end{pmatrix} Q_{(1\ 3)} = \begin{pmatrix} -\frac{1}{2} & -\frac{\sqrt{3}}{2} \\ -\frac{\sqrt{3}}{2} & \frac{1}{2} \end{pmatrix}. \quad (6\text{-}24)$$

All six permutations, P, can be obtained by repeated application of (1 2) and (2 3) and thus all Q_P are expressible as repeated products of $Q_{(1\ 2)}$ and $Q_{(2\ 3)}$.

These properties will be used when we come to consider the symmetry under particle interchange of total wave functions which are products of isospin (or unitary spin) wave functions, spin wave functions, and space wave functions.

---

[5] It is easily verified explicitly, by taking V to be the infinitesimal generator transformations, that P_1, N_1 transform by V, and P_2, N_2. Generally matrix representations V', V are said to be *equivalent* if they are connected by $V' = H^{-1} V H$ where H is the same matrix for all V', V.

6.3. THE EIGHTFOLD WAY, AN SU3 SYMMETRY

The Lie algebra associated with SUn has n^2-1 generators. Thus SU3 has 8 generators $F_1 \ldots F_8$, obeying the commutation relations:

$$[F_i, F_j] = i f_{ijk} F_k \tag{6-25}$$

where the f_{ijk} are the totally antisymmetric (that is antisymmetric under interchange of any two indices) structure constants the nonzero ones being given by:

$$f_{123} = 1, f_{147} = f_{246} = f_{257} = f_{345} = \frac{1}{2}, f_{156} = f_{367} = -\frac{1}{2}, f_{458} = f_{678} = \frac{\sqrt{3}}{2}$$
$$\tag{6-26}$$

and by permutation of the indices using the antisymmetry condition.

The basic representation of SU3 is of course the group of 3×3 unitary, unimodular matrices itself. Gell-Mann's original 1961 representation (Gell-Mann and Ne'eman, 1964) for the generators $F_i = \frac{1}{2}\lambda_i$ in this representation is very convenient, as it is closely related to the representation, eq. (6-4a) of the isospin generators:

$$\lambda_1 = \begin{pmatrix} 0 & 1 & 0 \\ 1 & 0 & 0 \\ 0 & 0 & 0 \end{pmatrix} \quad \lambda_2 = \begin{pmatrix} 0 & -i & 0 \\ i & 0 & 0 \\ 0 & 0 & 0 \end{pmatrix} \quad \lambda_3 = \begin{pmatrix} 1 & 0 & 0 \\ 0 & -1 & 0 \\ 0 & 0 & 0 \end{pmatrix}$$

$$\lambda_4 = \begin{pmatrix} 0 & 0 & 1 \\ 0 & 0 & 0 \\ 1 & 0 & 0 \end{pmatrix} \quad \lambda_5 = \begin{pmatrix} 0 & 0 & -i \\ 0 & 0 & 0 \\ i & 0 & 0 \end{pmatrix} \quad \lambda_6 = \begin{pmatrix} 0 & 0 & 0 \\ 0 & 0 & 1 \\ 0 & 1 & 0 \end{pmatrix}$$

$$\lambda_7 = \begin{pmatrix} 0 & 0 & 0 \\ 0 & 0 & -i \\ 0 & i & 0 \end{pmatrix} \quad \lambda_8 = \begin{pmatrix} \frac{1}{\sqrt{3}} & 0 & 0 \\ 0 & \frac{1}{\sqrt{3}} & 0 \\ 0 & 0 & \frac{-2}{\sqrt{3}} \end{pmatrix} \tag{6-27}$$

The matrices $\lambda_1 \lambda_2 \lambda_3$ are similar to the representation of the isospin generators $\tau_1 \tau_2 \tau_3$ and have the same commutation relations.

6.3. THE EIGHTFOLD WAY, AN SU3 SYMMETRY

Indeed, since $f_{123} = 1$ and f_{ijk} are totally antisymmetric under permutation of the indices ijk, F_1, F_2, and F_3 have, from eq. (6-25), the same commutation relations as I_1, I_2, and I_3 given by eq. (6-5b). One is thus enabled to make the identification:

$$F_1 = I_1, F_2 = I_2, F_3 = I_3 \tag{6-28}$$

and to incorporate isospin within the larger SU3 symmetry of the eightfold way. Thus if all interactions were invariant under SU3 transformations, they would be invariant under isospin transformations. The physical situation is that both symmetries are inexact, but that SU2, isospin, is a better symmetry than SU3, the eightfold way. SU3 is broken by hadronic interactions which produce mass splittings between members of the same SU3 multiplet of the order of 200 MeV, while isospin is broken by electromagnetic strength interactions with mass splittings of the order of 2 MeV.

As implied by its name, the eightfold way was invented largely as a classification scheme for the eight (semi)stable baryons N, Λ, Σ, Ξ and for the approximately eight pseudoscalar and also vector mesons. Since that time more decuplets, octets, and singlets (with the last two regarded as a nonet in the case of mesons) have been found [see for example Harari (1969)], but no established examples of other multiplets. So it is desirable to consider first the dimensions of the irreducible representations of SU3, and thus the numbers of particles that can be accomodated in SU3 multiplets. We use Young tableaux and the dimension formula, which we state without proof:

$$d = \frac{1}{2}(\lambda_1 - \lambda_2 + 1)(\lambda_1 - \lambda_3 + 2)(\lambda_2 - \lambda_3 + 1) \tag{6-29}$$

where $\lambda_1, \lambda_2, \lambda_3$ are the number of squares[6] in rows 1, 2, 3 respectively of the Young tableau. The first few Young tableaux, together with the dimension of the corresponding irreducible representation of SU3, are:

---

[6] This is a standard notation for the Young tableau numbers. No confusion should arise with the Gell-Mann matrices λ_i, $i = 1...8$ of eq. (6-27).

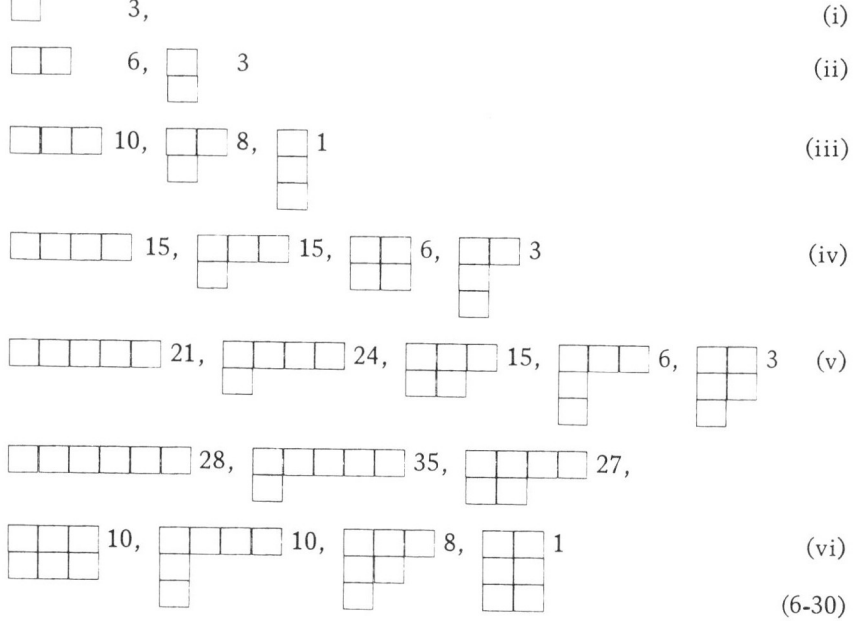

(6-30)

It is among those Young tableaux with 3, 6, 9, ... squares that we find decuplets, octets, and singlets, and in the absence of strong evidence for twenty-seven-fold, thirty-five-fold, etc., representations, physicists generally look to the simplest possibility — that there are only three square Young tableaux for the eightfold way.

SU3 Classification of Baryons Using Mathematical Quarks

The last observation leads[7] to a mathematical scheme of classification using *three* basic triplets of SU3 corresponding to the *three* squares of the Young tableaux. In SU2 there is a basic doublet which we denoted by p, n where $p = \begin{pmatrix} 1 \\ 0 \end{pmatrix}$ and $n = \begin{pmatrix} 0 \\ 1 \end{pmatrix}$. In SU3 there is a basic triplet denoted by p, n, λ where:

---

[7] See M. Gell-Mann, "The Eightfold Way: A Theory of Strong Interaction Symmetry," reprinted in Gell-Mann and Ne'eman (1964); also see Gell-Mann (1962). Earlier work on a different triplet model was done by Sakata (1956).

6.3. THE EIGHTFOLD WAY, AN SU3 SYMMETRY

$$p = \begin{pmatrix} 1 \\ 0 \\ 0 \end{pmatrix} \quad n = \begin{pmatrix} 0 \\ 1 \\ 0 \end{pmatrix} \quad \lambda = \begin{pmatrix} 0 \\ 0 \\ 1 \end{pmatrix}. \tag{6-31}$$

As discussed (eqs. (6-27) ff.) the isospin in the 3×3 representation is given by $I_1 = \frac{1}{2} \lambda_1$, $I_2 = \frac{1}{2} \lambda_2$, $I_3 = \frac{1}{2} \lambda_3$. This gives immediately that p, n are an isospin doublet, and λ is an isospin singlet.

Just as in section 6.2 we constructed SU2 multiplets using products of p, n doublets, so we can construct SU3 multiplets by using products of the triplets. A general triplet member — that is, either p or n or λ — is denoted by q and called a *quark*. To construct the multiplets corresponding to eq. (6-30, iii) we need products of three quarks such as $p(1)\, p(2)\, p(3)$. It must be noted that *in this section* the triplets, or quarks, $\{p(1), n(1), \lambda(1)\}$, $\{p(2), n(2), \lambda(2)\}$, $\{p(3), n(3), \lambda(3)\}$ out of which we shall construct the multiplets are not physically tangible, but are mathematical, entities. [In (slightly) mathematical language, we are constructing a twenty-seven-dimensional vector space from products of three three-dimensional vector spaces, and finding the bases of irreducible representations of SU3 in that twenty-seven-dimensional space.] In particular, the names p, n, λ are merely mnemonics indicating that these abstract vectors have the same isospin (and strangeness) as the physical proton, neutron, and lambda. With this underlying knowledge, the less mathematical reader will lose nothing by thinking of multiplet components like $p(1)\, p(2)\, p(3)$ as three-particle wave functions.

☐☐☐ As it is simplest, we first construct the decuplet from 3 quarks. The decuplet, ☐☐☐, is totally symmetric under permutations of the quark label; there are only ten totally symmetric functions, and knowing that p, n have isospin $\frac{1}{2}$, and λ has isospin 0, we easily arrange the functions in isospin multiplets corresponding to the particles of the physical decuplet:

$I = \frac{3}{2}$, $\Delta(1236)$:

$$n(1)\,n(2)\,n(3),\ \frac{1}{\sqrt{3}}\{p(1)\,n(2)\,n(3)+n(1)\,p(2)\,n(3)+n(1)\,n(2)\,p(3)\},$$

$$\frac{1}{\sqrt{3}}\{p(1)\,p(2)\,n(3)+p(1)\,n(2)\,p(3)+n(1)\,p(2)\,p(3)\},\ p(1)\,p(2)\,p(3)$$

$I = 1$, $Y_1^*(1385)$:

$$\frac{1}{\sqrt{3}}\{n(1)\,n(2)\,\lambda(3)+n(1)\,\lambda(2)\,n(3)+\lambda(1)\,n(2)\,n(3)\},\ \frac{1}{\sqrt{6}}\{(p(1)\,n(2)+p(2)\,n(1))\,\lambda(3)$$

$$+(p(1)\,n(3)+p(3)\,n(1))\,\lambda(2)+(p(2)\,n(3)+p(3)\,n(2))\,\lambda(1)\},\ \frac{1}{\sqrt{3}}\{p(1)\,p(2)\,\lambda(3)$$

$$+p(1)\,\lambda(2)\,p(3)+\lambda(1)\,p(2)\,p(3)\}$$

$I = \frac{1}{2}$, $\Xi_{\frac{1}{2}}^*(1530)$:

$$\frac{1}{\sqrt{3}}\{n(1)\,\lambda(2)\,\lambda(3)+\lambda(1)\,n(2)\,\lambda(3)+\lambda(1)\,\lambda(2)\,n(3)\},\ \frac{1}{\sqrt{3}}\{p(1)\,\lambda(2)\,\lambda(3)$$

$$+\lambda(1)\,p(2)\,\lambda(3)+\lambda(1)\,\lambda(2)\,p(3)\}$$

$I = 0$, $\Omega^-(1674)$:

$$\lambda(1)\,\lambda(2)\,\lambda(3)\ .\tag{6-32}$$

This isospin structure, eq. (6-32), of the decuplet came as a necessary consequence of the permutation symmetry of the quark functions. The known properties of the physical decuplet particles help us to make some necessary assignments of quantum numbers to the basic quarks. We already know that I_3 is an additive quantum number in the sense that the eigenvalue of I_3 for a three (or many) quark function is the sum of the eigenvalues of I_3 for the individual quarks. On any simple quark theory

6.3. THE EIGHTFOLD WAY, AN SU3 SYMMETRY

we would expect the charge, Q, and strangeness S or hypercharge, Y, to be additive in the same sense. Examination of eq. (6-32) gives:

$$\text{Charge }\{p, n, \lambda\} = e\left\{\frac{2}{3}, -\frac{1}{3}, -\frac{1}{3}\right\}$$
$$\text{Strangeness }\{p, n, \lambda\} = \{0, 0, -1\}. \tag{6-33}$$

Thus in the basic 3×3 representation:

$$Q/e = \frac{1}{3}\begin{pmatrix} 2 & 0 & 0 \\ 0 & -1 & 0 \\ 0 & 0 & -1 \end{pmatrix} = \frac{1}{2}\left(\lambda_3 + \frac{1}{\sqrt{3}}\lambda_8\right) \tag{6-34a}$$

and more generally Q is expressed in terms of the SU3 generators, for any represention, by:

$$Q/e = F_3 + \frac{1}{\sqrt{3}} F_8. \tag{6-34b}$$

From the relation (Chapter 1, p. 17) $Q/e = I_3 + \frac{1}{2}Y$ we see that, in any representation:

$$Y = \frac{2}{\sqrt{3}} F_8 \tag{6-35a}$$

and in the 3×3 representation:

$$Y = \frac{1}{\sqrt{3}} \lambda_8 = \frac{1}{3}\begin{pmatrix} 1 & 0 & 0 \\ 0 & 1 & 0 \\ 0 & 0 & -2 \end{pmatrix} \tag{6-35b}$$

and assigning baryon number $B = \frac{1}{3}$ for each quark and using $Y = B + S$:

$$S = \begin{pmatrix} 0 & 0 & 0 \\ 0 & 0 & 0 \\ 0 & 0 & -1 \end{pmatrix} \quad B = \begin{pmatrix} \frac{1}{3} & 0 & 0 \\ 0 & \frac{1}{3} & 0 \\ 0 & 0 & \frac{1}{3} \end{pmatrix} \tag{6-36}$$

agreeing with eq. (6-33).

The members of SU3 multiplets can be labeled by the eigenvalues of (I_3, Y) and are often displayed graphically (weight diagrams), as in Fig. 6-1(a) for the decuplet.

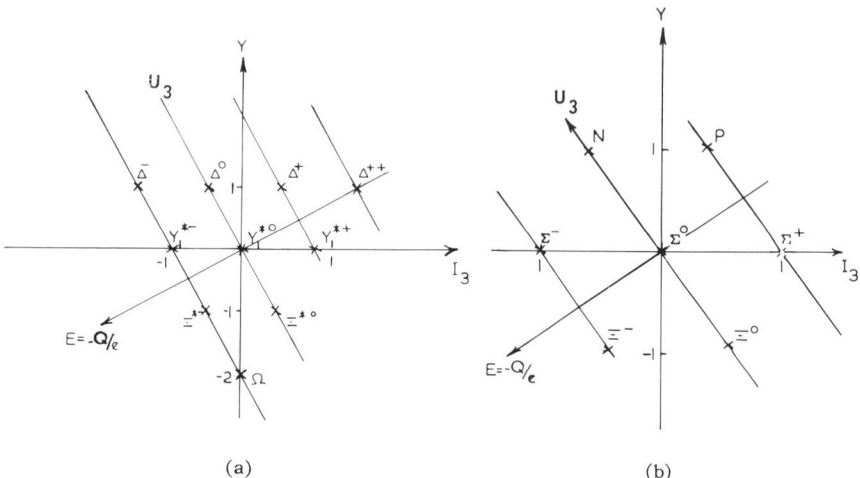

Fig. 6-1. (a) SU3 decuplet, (b) SU3 octet, in the (I_3, Y) plane; the heavy star denotes 2 octet members. Also marked are the U_3 and $E = Q/e$ axes (section 6.7). The particle labels are those of the lowest energy baryon octet and decuplet. Particles on a horizontal line belong to an isospin multiplet; particles of the same charge, joined in the diagram by sloping lines, belong to a U-spin multiplet.

Another simple quark function to construct is the SU3 singlet corresponding to the Young tableau , giving rise to totally antisymmetric quark functions. Such a function cannot contain two quarks of the same kind, and must be:

$$\frac{1}{\sqrt{6}} \{ p(1) n(2) \lambda(3) + p(3) n(1) \lambda(2) + p(2) n(3) \lambda(1) - p(2) n(1) \lambda(3) - p(3) n(2) \lambda(1)$$
$$- p(1) n(3) \lambda(2) \} \ . \tag{6-37}$$

The quark functions for an octet are more complicated. The Young tableau is and, as explained above in section 6.2, this means that the quark functions form a basis for a 2×2 matrix representation of the symmetric group S_3. Thus there are two independent octets (which we will call O_1 and O_2) of quark functions making $2 \times 8 = 16$ functions. This completes the count of 27 independent 3-quark functions since 10 (decuplet) + 16 (octet) + 1 (singlet) = 27.

6.3. THE EIGHTFOLD WAY, AN SU3 SYMMETRY

We can reinterpret the SU2 spinors p, n of eq. (6-22) to be SU3 triplets of eq. (6-31), that is, quarks. In that case, since P_1, P_2 (and also N_1, N_2) form the 2×2 irreducible representation of S_3 under permutation of the quark labels 1, 2, 3, they are *necessarily* octet functions. This establishes P_i, N_i ($i = 1, 2$) of eq. (6-22) as $I = \frac{1}{2}$, $Y = 1$ members of the octets O_i ($i = 1, 2$). Thus the P, N functions must correspond to the physical proton and neutron. We will see later, when we incorporate spin, how the apparent choice between P_1, N_1 and P_2, N_2 (or generally O_1, O_2) functions disappears.

We may now further resolve the I, Y octet structure. The functions P_i, N_i were constructed out of ppn, nnp respectively, and we may form quark functions with just the same properties under permutation of 1, 2, 3 out of $\lambda\lambda p$, $\lambda\lambda n$. Addition of quark charges shows total charge (in units of e) of 0, -1 respectively; addition of quark hypercharge eq. (6-35b) shows that $Y = -1$ for both; since λ has isospin zero, the functions necessarily form an isospin doublet. We call the functions Ξ_i^0, Ξ_i^- ($i = 1, 2$), and they are given explicitly below.

So far we have discovered $I = \frac{1}{2}$, $Y = 1$ and $I = \frac{1}{2}$, $Y = -1$ submultiplets, making four octet members. The I, Y structure of the other four is given by noting that we can form quark functions with $pp\lambda$ having, analogously to eq. (6-22), the correct permutation properties. Such functions, since λ has isospin zero, must necessarily be members of an $I = 1$, $Y = 0$ submultiplet. These quark functions, given below, are called Σ_i^+, Σ_i^0, Σ_i^- ($i = 1, 2$). This leaves one member of the octet, which then must have isospin $I = 0$ and so can only be formed from a $pn\lambda$ combination giving $Y = 0$. These functions are called Λ_i ($i = 1, 2$). This completes the determination of the I, Y structure of an SU3 octet, shown in Fig. 6-1(b).

We now write explicitly the quark octet wave functions O_1 and O_2, forming a basis for the 2×2 representation eq. (6-24) of S_3. The only functions which are not given by direct analogy with eq. (6-22) are Σ_i^0, Λ_i, which can be constructed from knowledge of the isospin, and use of eq. (6-24):

$$N_1 = -\frac{1}{\sqrt{6}}\{(p(1)\,n(2) + p(2)\,n(1))\,n(3) - 2n(1)\,n(2)\,p(3)\}$$

$$P_1 = \frac{1}{\sqrt{6}}\{(n(1)\,p(2) + n(2)\,p(1))\,p(3) - 2p(1)\,p(2)\,n(3)\}$$

$$\Sigma_1^- = \frac{1}{\sqrt{6}}\{(n(1)\,\lambda(2) + n(2)\,\lambda(1))\,n(3) - 2n(1)\,n(2)\,\lambda(3)\}$$

$$\Sigma_1^+ = \frac{1}{\sqrt{6}}\{(p(1)\,\lambda(2) + p(2)\,\lambda(1))\,p(3) - 2p(1)\,p(2)\,\lambda(3)\}$$

$$\Sigma_1^0 = -\frac{1}{\sqrt{12}}\{2(p(1)\,n(2) + p(2)\,n(1))\,\lambda(3) - (p(2)\,n(3) + p(3)\,n(2))\,\lambda(1) - (p(1)\,n(3)$$
$$+ p(3)\,n(1))\,\lambda(2)\}$$

$$\Lambda_1 = \frac{1}{2}\{(p(3)\,n(2) - p(2)\,n(3))\,\lambda(1) + (p(3)\,n(1) - p(1)\,n(3))\,\lambda(2)\}$$

$$\Xi_1^- = \frac{1}{\sqrt{6}}\{(n(1)\,\lambda(2) + n(2)\,\lambda(1))\,\lambda(3) - 2\lambda(1)\,\lambda(2)\,n(3)\}$$

$$\Xi_1^0 = \frac{1}{\sqrt{6}}\{(p(1)\,\lambda(2) + p(2)\,\lambda(1))\,\lambda(3) - 2\lambda(1)\,\lambda(2)\,p(3)\} \qquad (6\text{-}38\text{a})$$

$$N_2 = -\frac{1}{\sqrt{2}}(p(1)\,n(2) - p(2)\,n(1))\,n(3) \qquad P_2 = \frac{1}{\sqrt{2}}(n(1)\,p(2) - n(2)\,p(1))\,p(3)$$

$$\Sigma_2^- = -\frac{1}{\sqrt{2}}(n(1)\,\lambda(2) - n(2)\,\lambda(1))\,n(3) \qquad \Sigma_2^+ = -\frac{1}{\sqrt{2}}(p(1)\,\lambda(2) - p(2)\,\lambda(1))\,p(3)$$

$$\Sigma_2^0 = -\frac{1}{2}\{(p(1)\,n(3) + p(3)\,n(1))\,\lambda(2) - (p(2)\,n(3) + p(3)\,n(2))\,\lambda(1)\}$$

$$\Lambda_2 = \frac{1}{\sqrt{12}}\{2(n(1)\,p(2) - n(2)\,p(1))\,\lambda(3) - (n(2)\,p(3) - n(3)\,p(2))\,\lambda(1) + (n(1)\,p(3)$$
$$- n(3)\,p(1))\,\lambda(2)\}$$

$$\Xi_2^- = \frac{1}{\sqrt{2}}(n(1)\,\lambda(2) - n(2)\,\lambda(1))\,\lambda(3) \qquad \Xi_2^0 = \frac{1}{\sqrt{2}}(p(1)\,\lambda(2) - p(2)\,\lambda(1))\,\lambda(3)$$

$$(6\text{-}38\text{b})$$

6.3. THE EIGHTFOLD WAY, AN SU3 SYMMETRY

Anti-Quarks (Charge-Conjugate Quarks)

Since, in eq. (6-37), baryon number $\frac{1}{3}$ is attributed to every quark, mesons cannot be built out of the quark functions considered so far. (Though they could, of course, be built out of 3-quark wave functions with *different* quarks with a different baryon number and strangeness assignment.) The simplest method of building mesons is to introduce anti-quarks with the usual properties associated with that prefix in elementary particle physics, namely, opposite charge, baryon number, strangeness, and hypercharge. The anti-quarks are denoted generally by $\bar{q}$ and specifically by $\bar{p}, \bar{n}, \bar{\lambda}$, and like the quarks are mathematical objects, so far as this section is concerned.

The general quark triplet $q = q_1, q_2, q_3$ where $q_1 = p, q_2 = n, q_3 = \lambda$ transforms under SU3 by:

$$q_i \to U_{ij} q_j \qquad (6-39)$$

where U is a 3×3 unitary unimodular matrix. But the matrices U^*, where $*$ denotes the operation of taking the complex conjugate of such matrix elements, also form a group representing SU3. This is the complex conjugate representation coinciding with the adjoint representation[8] and also known as the contragredient representation. We define the anti-quark by its transformation properties, so that when the quark triplet transforms by eq. (6-39), the anti-quark triplet $\bar{q} = \bar{q}_1, \bar{q}_2, \bar{q}_3$ where $\bar{q}_1 = \bar{p}, \bar{q}_2 = \bar{n}, \bar{q}_3 = \bar{\lambda}$ transforms under SU3 by:

$$\bar{q}_i \to U^*_{ij} \bar{q}_j \quad . \qquad (6-40)$$

The matrices U can be generated by the infinitesimal transformations given in terms of the generators λ_ν [eq. (6-27)] and infinitesimal real displacements Δa_ν:

$$U = 1 + i \lambda_\nu \Delta a_\nu \quad . \qquad (6-41)$$

---

[8] The complex conjugate representation coincides with the adjoint representation for unitary matrices. The adjoint representation is defined as U^{-1T}, where T denotes transpose.

Then the matrices U^* are given by:

$$U^* = 1 + i\bar{\lambda}_\nu \Delta a_\nu$$
$$\bar{\lambda}_\nu = -\lambda_\nu^*.$$
(6-42)

(The representation $\frac{1}{2}\lambda_i$ of the generators obeys the commutation relations of eq. (6-25); by applying the charge conjugation operation $*$ to eq. (6-25) it is readily seen that $\frac{1}{2}\bar{\lambda}_\nu$ also satisfy it.) The $\bar{\lambda}_\nu$ are the generators in the U^* representation. In the U^* or anti-quark representation from eqs. (6-34) and (6-35):

$$Q/e = F_3 + \frac{1}{\sqrt{3}} F_8 = \frac{1}{2}\left(\bar{\lambda}_3 + \frac{1}{\sqrt{3}}\bar{\lambda}_8\right) = \frac{1}{3}\begin{pmatrix} -2 & 0 & 0 \\ 0 & 1 & 0 \\ 0 & 0 & 1 \end{pmatrix}$$ (6-43)

$$Y = \frac{1}{\sqrt{3}} F_8 = \frac{1}{\sqrt{3}}\bar{\lambda}_8 = \frac{1}{3}\begin{pmatrix} -1 & 0 & 0 \\ 0 & -1 & 0 \\ 0 & 0 & 2 \end{pmatrix}.$$ (6-44)

Applying these operators to $\bar{p} = (1\ 0\ 0)$, $\bar{n} = (0\ 1\ 0)$ or $\bar{\lambda} = (0\ 0\ 1)$, we see that charge and hypercharge reverse sign from quark to anti-quark. The relation $Y = B + S$ is preserved by changing the sign of B and S in going from quark to anti-quark. The *charge conjugation* operation effects the interchange $\bar{p} \leftrightarrow p$, $\bar{n} \leftrightarrow n$, $\bar{\lambda} \leftrightarrow \lambda$.

Though irreducible representations of SU3 are only completely specified by the partition numbers $\lambda = \lambda_1 - \lambda_2$, $\mu = \lambda_2 - \lambda_3$ pertaining to the Young tableaux, where confusion between two irreducible representations does not arise, they are sometimes labeled just by their dimensionality. Thus one may write $\{1\}, \{8\}, \{10\}$ for ⬚, ⬚⬚, ⬚⬚⬚ . Similarly for the basic, quark, representation ⬚ one writes $\{3\}$. The anti-quark representation is labeled $\{\bar{3}\}$ and it can also be obtained as follows:

Consider the irreducible representation of SU3 specified by the Young tableau ⬚ . By eq. (6-29) it has dimension $d = 3$. Since it is a two-square

6.3. THE EIGHTFOLD WAY, AN SU3 SYMMETRY

column the SU3 multiplet functions are antisymmetric functions of two quarks. These are:

$$\frac{1}{\sqrt{2}}(n(1)\lambda(2) - n(2)\lambda(1)), \quad \frac{1}{\sqrt{2}}(\lambda(1)p(2) - \lambda(2)p(1)), \quad \frac{1}{\sqrt{2}}(p(1)n(2) - p(2)n(1)) \quad (6\text{-}45)$$

and application of the infinitesimal transformations of eq. (6-41) to the individual $p(1)$, $n(1)$, $\lambda(1)$, $p(2)$, $n(2)$, $\lambda(2)$ of eq. (6-45) shows that the triplet, eq. (6-45), transforms according to eq. (6-42), that is, like the triplet $\bar{p}, \bar{n}, \bar{\lambda}$. So the Young tableau ☐ specifies a $\{\bar{3}\}$ representation of SU3.

It can be helpful to think in the following way. A complete column in a Young tableaux for SU3 is ☐ ; the $\{\bar{3}\}$ representation has one missing square or "hole" ☐ indicated by dotted lines, and this "hole" is equivalent to an anti-quark. More generally the irreducible representation specified by a Young tableau, with squares representing quarks, is equivalent to the Young tableaux of holes, with the holes representing anti-quarks. Symbolically, for example:

☐ = ☐ = $\{\bar{3}\}$; ☐☐☐ = ☐☐☐ = $\{\overline{10}\}$;

$\{8\}$ = ☐☐ = ☐☐ = $\{8\}$.

We may combine a quark function with an anti-quark function to form a meson, since the resulting $\bar{q}q$ function has baryon number zero. There are nine such $\bar{q}q$ functions; what SU3 multiplets do they form? Multiplication of SU3 (SUn) representations and multiplets is achieved by combining Young tableaux using certain simple rules (Hammermesh, 1962), giving in this case:

$$\Box \otimes \Box = \Box \oplus \Box\Box \qquad (6\text{-}46)$$
$$\bar{q} \quad q \qquad \bar{q}q \quad \bar{q}q$$

Thus the $\bar{q}q$ functions can be formed into an SU3 singlet and an SU3 octet.

Nonets of Mesons using Mathematical Quarks

The isospin generators ("Pauli matrices") for anti-quarks are:

$$\bar{\lambda}_1, \bar{\lambda}_2, \bar{\lambda}_3 = \begin{pmatrix} 0 & -i & 0 \\ -i & 0 & 0 \\ 0 & 0 & 0 \end{pmatrix}, \begin{pmatrix} 0 & -i & 0 \\ i & 0 & 0 \\ 0 & 0 & 0 \end{pmatrix}, \begin{pmatrix} -1 & 0 & 0 \\ 0 & 1 & 0 \\ 0 & 0 & 0 \end{pmatrix}$$

showing that $-\bar{n}, \bar{p}$ is an isotopic doublet transforming like p, n and that of course $\bar{\lambda}$ is an isoscalar. Any SU3 singlet must be constructed out of the two isospin singlets available, $\frac{1}{\sqrt{2}}(\bar{p}p + \bar{n}n)$ and $\bar{\lambda}\lambda$. We show that the required combination is (where $q_1, q_2, q_3 = p, n, \lambda$):

$$\frac{1}{\sqrt{3}}(\bar{p}(1)p(2) + \bar{n}(1)n(2) + \bar{\lambda}(1)\lambda(2)) = \frac{1}{\sqrt{3}} \bar{q}_i(1) q_i(2) . \qquad (6\text{-}47)$$

Applying the SU3 transformations, eqs. (6-39), (6-40):

$$\frac{1}{\sqrt{3}} \bar{q}_i(1) q_i(2) \to \frac{1}{\sqrt{3}} (U^*_{ij} \bar{q}_j(1))(U_{ik} q_k(2)) = \frac{1}{\sqrt{3}} \bar{q}_j(1) q_k(2) (U^+)_{ji} U_{ik}$$

$$= \frac{1}{\sqrt{3}} \bar{q}_j(1) q_j(2) . \qquad (6\text{-}48)$$

The last step follows because U is unitary. It is simple to construct I, Y multiplets of the SU3 octet:

$I = \frac{1}{2}, Y = 1$ $\bar{\lambda}(1) n(2), \bar{\lambda}(1) p(2)$

$I = \frac{1}{2}, Y = -1$ $\bar{p}(1)\lambda(2), -\bar{n}(1)\lambda(2)$

$I = 1, Y = 0$ $\bar{p}(1) n(2), \frac{1}{\sqrt{2}}(\bar{p}(1) p(2) - \bar{n}(1) n(2)), -\bar{n}(1) p(2)$

$I = 0, Y = 0$ $\frac{1}{\sqrt{6}}(\bar{p}(1) p(2) + \bar{n}(1) n(2) - 2\bar{\lambda}(1)\lambda(2)) . \qquad (6\text{-}49)$

The $I = Y = 0$ quark-anti-quark function follows from the observations that it must be formed from $\frac{1}{\sqrt{2}}(\bar{p}p + \bar{n}n)$ and $\bar{\lambda}\lambda$, and must be orthogonal[9] to the SU3 singlet wave function, eq. (6-47).

We refer to the nine mesons together as a *nonet*. Physical examples [given in the order $(I, Y) = (1, 0), (\frac{1}{2}, 1), (\frac{1}{2}, -1) (0, 0) (0, 0)$] are the nonet of pseudoscalar mesons π, K, $\bar{K}$, η, (X^0), of vector mesons $\rho(760)$, $K^*(890)$, $\bar{K}^*(890)$, $\omega(788)$, $\phi(1020)$ (see Table 6-2) and of $J^P = 2^+$ mesons $A_2(1300)$, $K(1420)$, $K^*(1420)$, $f(1250)$, $f'(1500)$. We cannot necessarily assign physical $I = Y = 0$ particles to SU3 octets or SU3 singlets. Looked at from the SU3 point of view, there may be significant though not dominant hadronic forces which break the SU3 symmetry, and lead to the physical isospin zero particles being mixtures of SU3 singlet and SU3 octet wave functions. When one works out the mixing for ω, ϕ, one arrives at the simple, and possibly suggestive, answer:

$$\omega \simeq \frac{1}{\sqrt{2}}(\bar{p}p + \bar{n}n), \quad \phi \simeq \bar{\lambda}\lambda \ .$$

It is remarkable that the mathematical quark classification (in its simplest form) for mesons leads us to nonets and no other multiplets. Experimentally, only SU3 singlets and octets are observed (in the above examples with $J^P = 1^-$ and $J^P = 2^+$, also closely associated in mass so that the nonet description seems apt). If we had been guided by a more general SU3, we might reasonably have expected decuplets, in analogy with the baryons.

---

[9] This orthogonality is in a sense a defining property of multiplets. Each member of a multiplet is an eigenfunction of Casimir operators with the same set of eigenvalues within a given multiplet and a different set for each different multiplet. The orthogonality follows immediately. The (one only) Casimir operator for isospin multiplets is I^2 (see section 6.1 above).

6.4. SPINNING QUARKS AND SU6

In the particle classification we have so far not included particle spin. This can be done by attributing spin $\frac{1}{2}$ to each quark. Let the spin functions corresponding to spin $\pm\frac{1}{2}$ along some chosen (fixed) axis of quantization be α, β. We now have six-component *mathematical* quarks: $p(1)\alpha(1), n(1)\alpha(1), \lambda(1)\alpha(1), p(1)\beta(1), n(1)\beta(1), \lambda(1)\beta(1)$. In order to have the correct SU3 properties we must build baryons out of three-quark functions, and mesons out of quark-anti-quark functions.

Baryons

There will be $6^3 = 216$ independent three-quark functions. We form our functions as multiplets of SU6 and, equally, multiplets (of dimension 1 and 2) of the symmetric group S_3 — the permutation of the quark labels 1, 2, 3. For three quarks there are three Young tableaux: ▢▢▢ , ▢▢/▢ , ▢/▢/▢ . For SUn a Young tableau has up to n rows of squares with λ_i squares in the ith row. The dimension of the corresponding irreducible representation of SUn is:

$$d = \prod_{i=1}^{n-1} \frac{1}{i^{n-i}} \prod_{j=i+1}^{n} (\lambda_i - \lambda_j + j - i) \ . \qquad (6\text{-}50)$$

Putting $n = 6$ in eq. (6-50) we find the following dimensions:

▢▢▢ $d = 56$, ▢▢/▢ $d = 70$, ▢/▢/▢ $d = 20$. (6-51)

The {56} and {20} give rise to one-dimensional representations of S_3, the quark functions being symmetrical and anti-symmetrical respectively. The {70} is seen by the structure of its Young tableau to correspond to the same two-dimensional irreducible representation of S_3 [equivalent to that specified by eq. (6-24)] that we have had before in the case of the same Young tableau for SU2 and for SU3. We shall use these permutation symmetry properties to find the SU6 (quark function) multiplets as sums of products of SU3 quark functions and spin quark functions.

6.4. SPINNING QUARKS AND SU6

To do this we need the permutation properties of the quark spin functions. We have already discussed this for the SU2 isospin functions, eqs. (6-20) and (6-22). The S (total quark spin) $= \frac{3}{2}$ quartet functions are symmetric:

$$Q^{-\frac{3}{2}} = \beta(1)\beta(2)\beta(3), \quad Q^{-\frac{1}{2}} = \frac{1}{\sqrt{3}}(\beta(1)\beta(2)a(3) + \beta(2)\beta(3)a(1) + \beta(3)\beta(1)a(2)),$$

$$Q^{\frac{1}{2}} = \frac{1}{\sqrt{3}}(\beta(1)a(2)a(3) + \beta(2)a(3)a(1) + \beta(3)a(1)a(2)), \quad Q^{\frac{3}{2}} = a(1)a(2)a(3)$$

(6-52)

and there are two $S = \frac{1}{2}$ doublets:

$$D_1^{-\frac{1}{2}} = -\frac{1}{\sqrt{6}}(a(1)\beta(2) + a(2)\beta(1))\beta(3) + \sqrt{\frac{2}{3}}\beta(1)\beta(2)a(3)$$

$$D_1^{\frac{1}{2}} = \frac{1}{\sqrt{6}}(\beta(1)a(2) + \beta(2)a(1))a(3) - \sqrt{\frac{2}{3}}a(1)a(2)\beta(3)$$

$$D_2^{-\frac{1}{2}} = -\frac{1}{\sqrt{2}}(a(1)\beta(2) - a(2)\beta(1))\beta(3) \quad D_2^{\frac{1}{2}} = \frac{1}{\sqrt{2}}(\beta(1)a(2) - \beta(2)a(1))a(3).$$

(6-53)

The $D_1^{-\frac{1}{2}}, D_2^{-\frac{1}{2}}$ (and also $D_1^{\frac{1}{2}}, D_2^{\frac{1}{2}}$) transform under permutation of 1, 2, 3 by eqs. (6-23)-(6-24).

We can now find the three-quark functions of six-component quarks in the $\{56\}$ of SU6. Denote a quark decuplet function from eq. (6-32) by Δ, an octet function from eq. (6-38a) by O_1 and from eq. (6-38b) by O_2, and the singlet function, eq. (6-37), by S. We know that each $\{56\}$ quark function must be symmetric under permutation of quark number. The SU3 anti-symmetric singlet S has no possible spin partner with which it can form a symmetric combination. The product of a decuplet and a quartet function, ΔQ, is obviously totally symmetric and there are $10 \times 4 = 40$ such quark functions.

From the octet and doublet we can form four independent types of combination, which we choose as $\frac{1}{\sqrt{2}}(O_1 D_1 + O_2 D_2)$; $\frac{1}{\sqrt{2}}(O_1 D_2 - O_2 D_1)$; $\frac{1}{\sqrt{2}}(O_2 D_2 - O_1 D_1)$; $\frac{1}{\sqrt{2}}(O_1 D_2 + O_2 D_1)$. Under permutation O and D transform by the orthogonal 2×2 matrices Q of eq. (6-24):

$$\begin{pmatrix} O_1 \\ O_2 \end{pmatrix} \rightarrow Q \begin{pmatrix} O_1 \\ O_2 \end{pmatrix} \quad \begin{pmatrix} D_1 \\ D_2 \end{pmatrix} \rightarrow Q \begin{pmatrix} D_1 \\ D_2 \end{pmatrix} \tag{6-54}$$

$$O_1 D_1 + O_2 D_2 = (O_1 \ O_2) \begin{pmatrix} D_1 \\ D_2 \end{pmatrix} \rightarrow (O_1 \ O_2) Q^T Q \begin{pmatrix} D_1 \\ D_2 \end{pmatrix} = (O_1 \ O_2) \begin{pmatrix} D_1 \\ D_2 \end{pmatrix}$$

Consequently $\frac{1}{\sqrt{2}}(O_1 D_1 + O_2 D_2)$ is a symmetric function. We can show by explicit calculations using eqs. (6-54) and (6-24) that $\frac{1}{2}(O_1 D_2 - O_2 D_1)$ is anti-symmetric and that the pair $\frac{1}{\sqrt{2}}(O_2 D_2 - O_1 D_1)$, $\frac{1}{\sqrt{2}}(O_1 D_2 + O_2 D_1)$ transform like O_1, O_2 or D_1, D_2 by eqs. (6-54), (6-24).

This leaves ΔQ and $\frac{1}{\sqrt{2}}(O_1 D_1 + O_2 D_2)$ as the only totally symmetric 3-quark functions, with $10 \times 4 + 8 \times 2 = 56$ totally symmetric functions. [We have in fact by our enumeration derived the dimensionality of the ☐☐☐ = {56} multiplet of SU6, without needing to use the general formula, eq. (6-50).] The resulting quark functions are one decuplet of spin $\frac{3}{2}$ and one octet of spin $\frac{1}{2}$:

$$\{56\} = (10, 4) + (8, 2) . \tag{6-55}$$

This accounts remarkably for the lowest-mass decuplet and the lowest-mass octet, the physical particles of which are set forth in Table 6-1. As an example of the quark functions of the {56} we write down the quark function corresponding to a proton of spin $+\frac{1}{2}$ along the axis of quantization:

6.4. SPINNING QUARKS AND SU6

Y \ (SU3), (SU2)	(8,2)	(10,4)
1	$N_{\frac{1}{2}}(939)$	$N^*_{\frac{3}{2}}(1236)$
0	$\Sigma_1(1193), \Lambda_0(1116)$	$Y^*_1(1385)$
-1	$\Xi_{\frac{1}{2}}(1318)$	$\Xi^*_{\frac{1}{2}}(1530)$
-2		$\Omega^-_0(1672)$

Table 6-1. The physical particles corresponding to the ▢▢▢, fifty-six-dimensional multiplet of SU6. The top heading is the dimensions of the irreducible representations of SU3 and SU2 (particle spin). The suffix attached to each particle name gives the isospin I. The numbers in brackets give the approximate median mass in MeV/c^2 of the isospin multiplet.

$$P = \frac{1}{\sqrt{2}}\left(P_1 D_1^{\frac{1}{2}} + P_2 D_2^{\frac{1}{2}}\right)$$

$$= \frac{1}{6\sqrt{2}}\{(n(1)p(2) + n(2)p(1))p(3) - 2p(1)p(2)n(3)\}\{(\beta(1)a(2) + \beta(2)a(1))a(3)$$

$$- 2a(1)a(2)\beta(3)\} + \frac{1}{2\sqrt{2}}(n(1)p(2) - n(2)p(1))p(3)(\beta(1)a(2) - \beta(2)a(1))a(3) \quad (6\text{-}56)$$

To find the quark functions in the ▢▢ representation of SU6, we list the S_3 doublets that can be formed [see eqs. (6-51) ff.]:

$$\{70\}_1 = \Delta D_1 \quad O_1 Q \quad S D_2 \quad \frac{1}{\sqrt{2}}(O_2 D_2 - O_1 D_1)$$

$$\{70\}_2 = \Delta D_2 \quad O_2 Q \quad S D_1 \quad \frac{1}{\sqrt{2}}(O_1 D_2 + O_2 D_1) . \quad (6\text{-}57)$$

Each row of eq. (6-57) contains $10 \times 2 + 8 \times 4 + 1 \times 2 + 8 \times 2 = 70$ functions forming a multiplet of SU6:

$$\{70\} = (10, 2) + (8, 4) + (8, 2) + (1, 2) . \qquad (6\text{-}58)$$

The quark functions in the ⊟ representation of SU6 are given by the totally anti-symmetric combinations:

$$\tfrac{1}{2}(O_1 D_2 - O_2 D_1), \, SQ \qquad (6\text{-}59)$$

$$\{20\} = (8, 2) + (1, 4) . \qquad (6\text{-}60)$$

The total number of quark functions enumerated is $6^3 = 216 = 56 + 2 \times 70 + 20$.

Mesons

With spin $\tfrac{1}{2}$ quarks one can form thirty-six mesons out of a quark-anti-quark pair. The anti-quarks are $\bar{p}a, \bar{n}a, \bar{\lambda}a, \bar{p}\beta, \bar{n}\beta, \bar{\lambda}\beta$. It is obvious that the thirty-six quark functions can be formed into a spin singlet nonet and a spin triplet nonet. For a reason that will become evident, we choose to make the quark functions *antisymmetric* under the interchange of the (anti-)quark labels 1, 2; this corresponds to anti-commutation of the operators creating (annihilating) the mathematical quark and those creating (annihilating) anti-quarks. To satisfy this let the nonet functions of eqs. (6-47) and (6-49) be N_i ($i = 0$, singlet; $i = 1, \ldots,$ 8 octet), and form N_i^S and N_i^A, which are symmetrized and antisymmetrized respectively with respect to the interchange $1 \leftrightarrow 2$. The quark functions are:

$$N_i^S \tfrac{1}{2}(a(1)\beta(2) - a(2)\beta(1)) \qquad (6\text{-}61)$$

$$N_i^A \beta(1)\beta(2), \; N_i^A \tfrac{1}{2}(\beta(1)a(2) + \beta(2)a(1)), \; N_i^A a(1)a(2) . \qquad (6\text{-}62)$$

More explicitly the functions corresponding for example to the π_0 and ω are:

6.4. SPINNING QUARKS AND SU6

$$\frac{1}{2\sqrt{2}}\{(\bar{p}(1)\,p(2)-\bar{n}(1)\,n(2))+(\bar{p}(2)\,p(1)-\bar{n}(2)\,n(1))\}\,(\alpha(1)\,\beta(2)-\alpha(2)\,\beta(1)) \tag{6-63a}$$

$$\frac{1}{2\sqrt{2}}\{(\bar{p}(1)\,p(2)+\bar{n}(1)\,n(2))-(\bar{p}(2)\,p(1)+\bar{n}(2)\,n(1))\}\,(\alpha(1)\,\beta(2)+\alpha(2)\,\beta(1))\,. \tag{6-63b}$$

If we apply the operation of charge conjugation $(\bar{p} \leftrightarrow p,\; \bar{n} \leftrightarrow n,\; \bar{\lambda} \leftrightarrow \lambda)$ to the neutral, $Y = 0$ quark functions, we obtain:

$$\text{spin } 0 : C = +1, \quad \text{spin } 1 : C = -1\,. \tag{6-64}$$

This is correct for π^0, η^0 and X, which we wish to assign to the spin 0 nonet and for ρ^0, ω and ϕ, which we wish to assign to the spin 1 nonet. Also for $Y = 0$ multiplets $G = C(-1)^I$ is also verified. We assume opposite parity for quark and anti-quark resulting in $P = -1$ for all the quark functions. We may then assign the lowest-mass mesons to eq. (6-61) (pseudoscalar mesons) and eq. (6-62) (vector mesons) as shown in Table 6-2.

(SU3), (SU2) Y	(8, 1) (1, 1)	(8, 3) (1, 3)
1	$K_{\frac{1}{2}}(496)$	$K^*_{\frac{1}{2}}(890)$
0	$\pi_1(137),\ \eta_0(549),\ X_0(958)$	$\rho_1(760),\ \omega_0(783)\ \phi_0(1019)$
−1	$\bar{K}_{\frac{1}{2}}(496)$	$K^*_{\frac{1}{2}}(890)$

Table 6-2. The physical particles corresponding to the $\bar{q}q$ model for mesons. The top heading is the dimensions of the irreducible representations of SU3 and SU2 (particle spin). The suffix attached to each particle name gives the isospin I. The numbers in brackets give the approximate median mass in MeV/c^2 of the isospin multiplet.

6.5. THE L-EXCITATION QUARK MODEL

The use of mathematical quark functions for classifying hadrons, with three quarks for baryons and quark-anti-quark for mesons, imposes limitations stronger than those of SU3 by restricting baryons to decuplets, octets, and singlets and mesons to nonets. Also the symmetric quark functions with spin $\frac{1}{2}$ quarks or baryons correspond to the lowest-mass octet and decuplet, and the (antisymmetric) quark-anti-quark functions to the lowest-mass meson nonets. These successes suggest that one should try to extend the quark functions so as to classify higher-mass hadrons. Since these high mass states also include higher spin states, there are two obvious methods. One is the addition of more quarks so that, for example, a baryon might be composed as a $qqqq\bar{q}$ object. However this method involves higher SU3 multiplets such as the {27} [see eq. (6-30)], and very large SU6 multiplets indeed. These complications seem unattractive just as long as such higher multiplets remain unobserved. Lacking these higher SU3 multiplets is the L-excitation model of quarks (Greenberg, 1964; Dalitz, 1965) in which an extra angular momentum, having the characteristics of an orbital angular momentum of physical quarks, is added.

Several points of view are possible on the L-excitation quark model. One may regard the quarks composing an elementary particle as physical objects, having orbital angular momentum and obeying a dynamical equation, the eigenvalue of energy for which is the mass of the elementary particle. Among this class is the *nonrelativistic quark model* (Dalitz, 1965, 1966, 1967, 1968; Mitra and Ross, 1967; Faiman and Hendry, 1968; Greenberg and Resnikoff, 1968) in which the quarks are heavy enough to have nonrelativistic internal motion. If M_Q is the quark mass and a_Q is the interaction range of the quarks, the condition for nonrelativistic internal motion is:

$$\hbar/a_Q \ll M_Q c \ .$$

6.5. THE L-EXCITATION QUARK MODEL

For $M_Q c^2 > 5$ GeV, which we know to be necessary from failure to produce quarks in acceleration experiments (Dorfan et al., 1965), this is satisfied for reasonable potential ranges of order $a_Q \sim 10^{-14}$ cm. Such heavy quarks would be strongly bound in order to give the observed elementary particle masses, and the interaction must be such that three quarks and a quark-anti-quark pair are strongly bound while two quarks are not bound, or are bound much more weakly. In the model there are multiplets of particles corresponding to different SU3, quark spin, and quark orbital angular momentum quantum numbers. Forces can be postulated that give correct mass differences for particles in different multiplets (Mitra and Ross, 1967; Faiman and Hendry, 1968; Mitra, 1968).

A *relativistic* quark model in which one has light quarks, say $M_Q c^2 \sim 300$ MeV, can be conceived. Explicit calculations with such models are more difficult.

Another view point is to reject the idea of physical quarks so that the "orbital" angular momentum is just another internal quantum number associated with the *mathematical* quarks. Such internal quantum numbers like mathematical quark spin and mathematical quark "orbital" angular momentum, may of course appear in mass formulae. From some operational standpoints there is no difference between the mathematical and physical quark picture; for example one might use successful mass formulae suggested by a physical quark model in a mathematical quark context. It should be understood that wherever we use the language and results of the nonrelativistic quark model, it is for conceptual purposes and does not imply the necessary existence of physically realizable single quarks.

Symmetrical Quark Model for Baryons

We may write a quark model wave function for baryons symbolically as:

$$f(1,2,3) = \sum f_{SU_3}(1,2,3) \times f_S(1,2,3) \times f_L(1,2,3) \times f_R(1,2,3) \quad (6\text{-}65)$$

where 1, 2, 3 are the quark numbers, $f_{SU_3}(1, 2, 3)$ are SU3 quark functions [see eq. (6-38)], $f_S(1, 2, 3)$ are quark spin functions [see eqs. (6-52), and (6-53)], $f_L(1, 2, 3)$ are quark orbital functions, and $f_R(1, 2, 3)$ are quark radial functions.

The baryon octet and decuplet (Table 6-1) have quark functions [such as eq. (6-56)] $\sum$ SU3(1, 2, 3) × S(1, 2, 3), totally symmetric with zero orbital angular momentum. A possible generalization is to consider only those quark functions which have $f_{SU3} \, f_S \, f_L$ totally symmetric; this is the L-excitation *symmetric* quark model.[10]

If, as a guide, one assumes a nonrelativistic quark model with harmonic quark wave functions (1 s), (1 p), (2 s), (1 d) ... (in order of excitation energy), then the ground state configuration of three quarks is $(1\,s)^3$ and the next excited state $(1\,s)^2 (1\,p)$. The first multiplies the {56} SU6 quark functions to give the particles of Table 6-1. For the second with total orbital angular momentum $L = 1$ there are two independent wave functions (apart from the center of mass motion). These can be taken as:

$$\phi_1 = (\vec{r}_1 + \vec{r}_2 - 2\vec{r}_3) \, \phi(\vec{r}_1 \vec{r}_2 \vec{r}_3), \quad \phi_2 = \sqrt{3} \, (\vec{r}_1 - \vec{r}_2) \, \phi(\vec{r}_1 \vec{r}_2 \vec{r}_3) \qquad (6\text{-}66)$$

where $\vec{r}_1, \vec{r}_2, \vec{r}_3$ are the quark position vectors and where $\phi(\vec{r}_1 \vec{r}_2 \vec{r}_3)$ is symmetric under interchange of 1, 2, 3. The ϕ_1, ϕ_2 transform by eq. (6-23) under permutation of 1, 2, 3; so to form a totally symmetrical function, we must combine with the quark functions of the SU6 {70} [eq. (6-57)] that also transform by eq. (6-23):

$$\frac{1}{\sqrt{2}} \, (\{70\}_1 \, \vec{\phi}_1 + \{70\}_2 \, \vec{\phi}_2) \ . \qquad (6\text{-}67)$$

---

[10] For physical quarks obeying Fermi statistics this implies an antisymmetric radial function (for example, for the baryon octet and decuplet). Besides the fact that rather unusual forces are required to give the ground state totally antisymmetric, a more serious objection is that nonrelativistic quark model calculations give radial nodes which should be observable in nucleon form factors, though these may be at higher momentum transfers than experiment has so far reached. The way out of the difficulty is to have parastatics of order 3 for quarks (Green, 1953; Greenberg and Messiah, 1965), so that, essentially each 3-quark function is made of three different kinds of quarks, obeying Fermi statistics. In that case the wave function, eq. (6-65), for physical para-quarks can have any statistics, because it is still to be multiplied by a para-quark label function, and it is only that latter product which needs to be antisymmetric. In particular, function (6-65) can be totally symmetric, giving the symmetric quark model for $f_{SU3} \, f_S \, f_L$ with symmetric radial function.

6.5. THE L-EXCITATION QUARK MODEL

The $(1s)^2 (1p)$ cannot form a totally symmetrical function with either the $\{56\}$ or the $\{20\}$.

Linear combinations of the wave functions, eq. (6-67), can be formed so that the orbital angular momentum $L = 1$ [displayed as a vector in eqs. (6-66) and (6-67)] combines with the quark *spin* of the $\{70\}$, the multiplets for which are displayed along with the SU3 multiplets in eqs. (6-57) and (6-58), to form states of total angular momentum J. The notation $\{M\}^N(L)_J$ is used to denote these multiplets where M, N are the dimensions of the SU3 and spin multiplets respectively and L is the quark orbital angular momentum (denoted spectroscopically by $S, P, D, \ldots$). It is easily seen that these multiplets are:

$$\{10\}\left({}^2P_{\tfrac{3}{2}}, {}^2P_{\tfrac{1}{2}}\right), \ \{8\}\left({}^4P_{\tfrac{5}{2}}, {}^4P_{\tfrac{3}{2}}, {}^4P_{\tfrac{1}{2}}, {}^2P_{\tfrac{3}{2}}, {}^2P_{\tfrac{1}{2}}\right), \ \{1\}\left({}^2P_{\tfrac{3}{2}}, {}^2P_{\tfrac{1}{2}}\right).$$

(6-68a)

Fig. 6-2. The $\{70\}$ 1^- supermultiplet of the L-excitation quark model. The $\{SU3\}^{2S+1}L$ multiplets are shown, split into the J^P submultiplets. The nucleonic state corresponding to each submultiplet is indicated, in pion-nucleon scattering notation. (Candidates for the SU3 singlet states are also indicated). On assigning the masses of the nucleonic states from Table 5-1, and in particular the lower s_{11} and d_{13} to the $\{8\}^2P$ multiplet, there is an evident mass ordering in the $\{SU3\}^{2S+1}L$ multiplets, which is that shown in the figure and also illustrated in Fig. 6-4. The ordering of the J^P submultiplets, within the $\{SU3\}^{2S+1}L$ multiplets, is subject to experimental uncertainty.

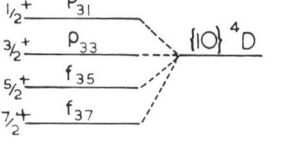
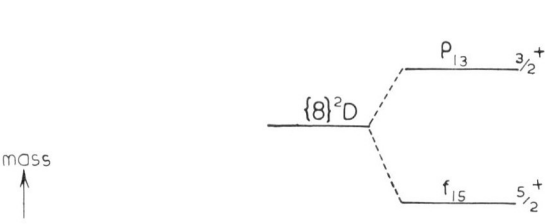

Fig. 6-3. The $\{56\}\, 2^+$ supermultiplet of the L-excitation quark model. As in Fig. 6-2, the ordering of the $\{SU3\}^{2S+1}L$ multiplets is taken from the known experimental masses of Table 5-1. The ordering of the J^P submultiplets within $\{10\}\, ^4D$ is not experimentally significant.

Because of the quark orbital angular momentum these are all negative parity states and the N^* members of the above are in pion-nucleon scattering notation, respectively:

$$d_{33},\ s_{31},\ d_{15},\ d_{13},\ s_{11},\ d_{13},\ s_{11},\ -,\ -\ . \tag{6-68b}$$

From Table 5-1 we see that there are resonances or strong interactions corresponding to all these states. The states for eq. (6-68) are shown in Fig. 6-2.

If one now considers as the next excited states the harmonic oscillator configurations $(1s)^2 (1d)$, $(1s)(1p)^2$ and $(1s)^2 (2s)$, taking care to exclude spurious center of mass motion (Faiman and Hendry, 1968) we find the following totally symmetric combinations:

$$\begin{array}{llllll} \text{SU6 representation} & \{56\} & \{56\} & \{70\} & \{70\} & \{20\} \\ L,\ P & =\ 2^+ & 0^+ & 2^+ & 0^+ & 1^+ \end{array} \tag{6-69}$$

The rotationally excited $\{56\}\, L^P = 2^+$ contains the following positive parity multiplets, shown in Fig. 6-3 with the nucleonic members in pion-nucleon scattering notation listed below:

6.5. THE L-EXCITATION QUARK MODEL

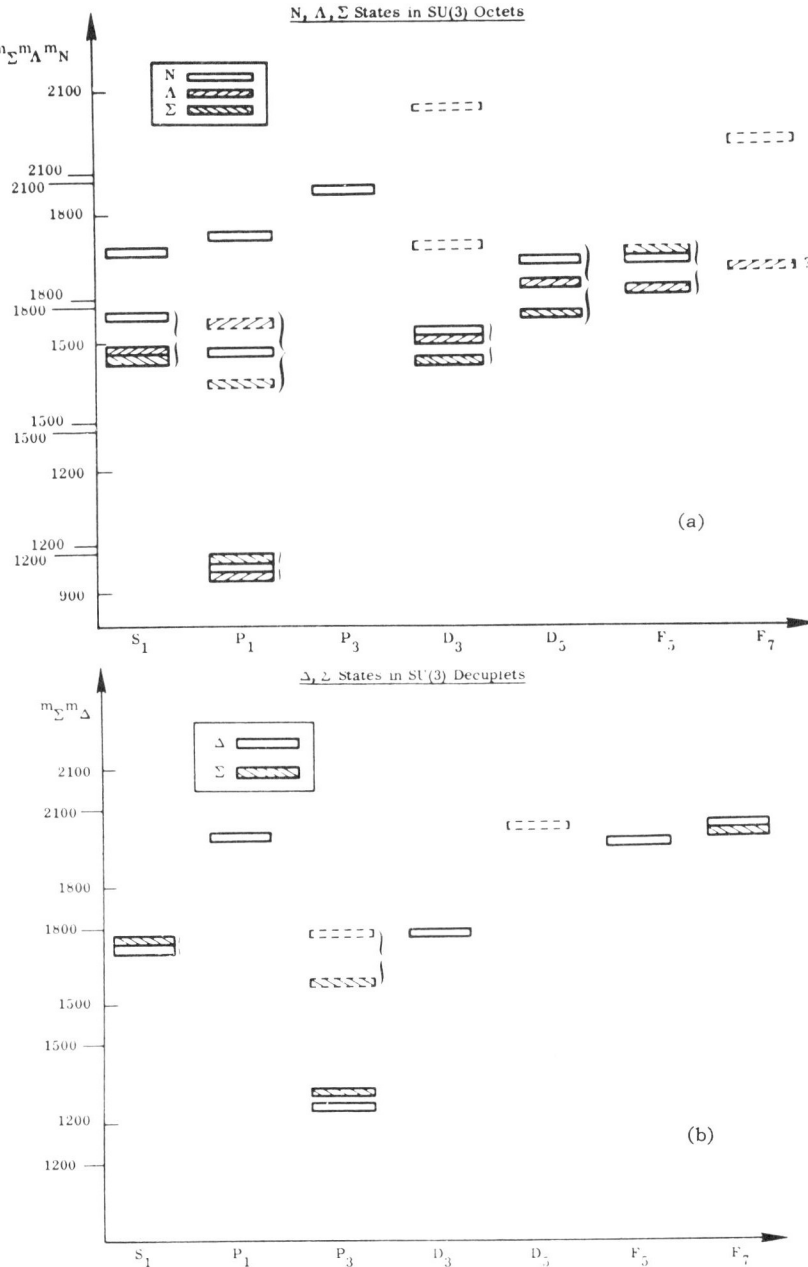

Fig. 6-4. (a) N, Λ, and Σ excited states assigned to SU3 octets – the ideal world in which the N-Λ-Σ mass splitting in all octets is identical would correspond to coincident boxes for the corresponding states; (b) Δ and Σ excited states assigned to SU3 decuplets, similarly to (a).

$$\{10\} \left({}^4D_{\frac{7}{2}},\ {}^4D_{\frac{5}{2}},\ {}^4D_{\frac{3}{2}},\ {}^4D_{\frac{1}{2}} \right),\ \{8\} \left({}^2D_{\frac{5}{2}},\ {}^2D_{\frac{3}{2}} \right) \qquad (6\text{-}70a)$$

$$f_{37},\ f_{35},\ p_{33},\ p_{31},\qquad\qquad f_{15},\ p_{13}\ . \qquad (6\text{-}70b)$$

Comparison with Table 5-1 shows that there are resonances or strong interactions which can be put into correspondence with all these states. The supermultiplet $\{56\}\ 0+$ of eq. (6-69) may contain the $J^P = \frac{1}{2}^+ \ p_{11}(1460)$ resonance, and also the $p_{33}(1690)$ — if that is not assigned to the $\{56\}\ 2^+$. Higher nucleonic states may be assigned to higher orbital excitations as discussed below in section 6.7.

By whatever mechanism the symmetry may arise, we see that the existing lower-energy nucleonic strong interacting states may be classifiable into the following $\{SU6\}\ L^P$ supermultiplets:

$$\{56\}\ 0^+,\ \{56\}\ 0^{+'},\ \{70\}\ 1^-,\ \{56\}\ 2^+ \qquad (6\text{-}71)$$

where the listing is in ascending order of median mass. The existing evidence from strange particle states is that there are SU3 multiplets corresponding in J^P and mass to the N^* states of Table 5-1, eq. (6-68b) and eq. (6-70b), and thus to the multiplets of eqs. (6-68a) and (6-70a) or the supermultiplets of eq. (6-71). This is shown diagramatically in Fig. 6-4.

Mesonic States in the L-Excitation Model

Higher spin mesonic states are formed by considering the quark-anti-quark pair as a two-particle system having orbital angular momentum L. The quark-anti-quark mesonic wave function is then:

$$f\ (\text{meson}) = f\ (\text{SU3})\ f_S\ (\text{spin})\ f_L\ (\text{orbital}) \qquad (6\text{-}72)$$

where f (SU3) is a nonet function given by eq. (6-47) or eq. (6-49) [symmetrical or antisymmetrical under interchange $1 \leftrightarrow 2$ as in eq. (6-63)],

6.5. THE L-EXCITATION QUARK MODEL

f_S(spin) is a quark spin function with $S = 0$ [eq. (6-61)] or $S = 1$ [eq. (6-62)], and f_L(orbital) is an orbital wave function with orbital angular momentum L. As in section 6.4, we take the total wave function antisymmetric (corresponding to anticommuting quark-anti-quark operators) and so for the neutral $Y = 0$ states, the charge conjugation parity C is given by:

$$C = (-1)^{L+S} \tag{6-72}$$

and for a $Y = 0$ substate with isospin I, the G-parity is then:

$$G = C(-1)^I \tag{6-73}$$

and the parity P of any state is:

$$P = -(-1)^L . \tag{6-74}$$

Since the spin, J, of the meson is given by $J = L + S, \ldots, |L-S|$ and $S = 0$ or 1, it follows that:

$$P = C \text{ if } P = (-1)^J . \tag{6-75}$$

This quark model selection rule that for *natural parity* states $(0^+, 1^-, 2^+ \ldots)$ $P = C$ appears to hold experimentally. In Table 6-3 we show mesons which may correspond (Dalitz, 1968; Particle Data Group, 1972) to $L = 1$ excitations.

$^{2S+1}(L)_J$	J^{PC}	$I = 1$ $Y = 0$	$\frac{1}{2}$ ± 1	0 0	0 0
$^3P_2$	2^{++}	$A_2(1310)$	$K^*(1419)$	$f(1260)$	$f'(1514)$
$^3P_1$	1^{++}	$A_1(1070)$	$K^*(1240)$	$D'(?)$	$D(1285)$
$^3P_0$	0^{++}	?	$K\pi(1100)$	?	$S^*(1060)$
$^1P_1$	1^{+-}	$B(1235)$	$K^*(1320)$	?	?

Table 6-3. Each of the established mesons (except those of Table 6-2) with mass below about 1500 MeV is assigned to the nonet appropriate to the evidence on its spin parity.

6.6. SU3 COUPLINGS AND N^* DECAYS

We wish to find baryon-baryon-meson couplings that express the SU3 invariance, as nucleon-nucleon-meson couplings like eq. (6-9a, b) express isospin invariance. To do this one must know how multiplets transform under SU3. In the case of octets it is easiest to do this by considering the meson octet functions, eq. (6-49), which, using eq. (6-27) and q_1, q_2, $q_3 = p, n, \lambda$ we can express as linear combinations of:

$$\bar{q}\lambda_i q \equiv \bar{q}_j^{(1)}(\lambda_i)_{jk} q_k(2):$$

$I = \frac{1}{2}, Y = 1 \quad \frac{1}{2}\bar{q}(\lambda_6 - i\lambda_7)q, \frac{1}{2}\bar{q}(\lambda_4 - i\lambda_5)q$

$I = \frac{1}{2}, Y = -1 \quad \frac{1}{2}\bar{q}(\lambda_4 + i\lambda_5)q, \frac{1}{2}\bar{q}(\lambda_6 + i\lambda_7)q$

$I = 1, Y = 0 \quad \frac{1}{2}\bar{q}(\lambda_1 + i\lambda_2)q, \frac{1}{2}\bar{q}\lambda_3 q, \frac{1}{2}\bar{q}(\lambda_1 - i\lambda_2)q$

$I = 0, Y = 0 \quad \frac{1}{2}\bar{q}\lambda_8 q .$ \hfill (6-76)

In the case of the pseudoscalar octet and the baryon octet the correspondence with the above expressions is immediately made, and if M_i and B_i are those combinations of the meson and baryon octets respectively that transform like $\bar{q}\lambda_i q$, we have:

$K_0 = \frac{1}{2}(M_6 - iM_7), K^+ = \frac{1}{2}(M_4 - iM_5); N = \frac{1}{2}(B_6 - iB_7), P = \frac{1}{2}(B_4 - iB_5)$

$K^- = -\frac{1}{2}(M_4 + iM_5), \overline{K^0} = \frac{1}{2}(M_6 + iM_7); \Xi^- = \frac{1}{2}(B_4 + iB_5), \Xi^0 = \frac{1}{2}(B_6 + iB_7)$

$\pi^- = \frac{1}{2}(M_1 + iM_2), \pi^0 = \frac{1}{\sqrt{2}}M_3, \Sigma^- = \frac{1}{2}(B_1 + iB_2), \Sigma^0 = \frac{1}{\sqrt{2}}B_3 ,$

$\pi^+ = -\frac{1}{2}(M_1 - iM_2); \qquad\qquad \Sigma^+ = -\frac{1}{2}(B_1 - iB_2)$

$\eta = \frac{1}{\sqrt{2}}M_8; \qquad\qquad \Lambda = \frac{1}{\sqrt{2}}B_8 .$ \hfill (6-77)

6.6. SU3 COUPLINGS AND N^* DECAYS

From eqs. (6-39)-(6-42) we find that under infinitesimal (generator) transformations:

$$\bar{q}\lambda_j q \to \bar{q}\lambda_j q + i\bar{q}[\lambda_j, \lambda_i] q \Delta a_i . \qquad (6\text{-}78)$$

To obtain eq. (6-78) we have used the fact that the λ_ν are hermitian. Using the fact (section 6.3) that $[\lambda_j, \lambda_i] = 2 i f_{jik}\lambda_k$, we find that:

$$\bar{q}\lambda_j q \to U_{jk}\bar{q}\lambda_k q \qquad (6\text{-}79)$$

where $U = 1 + i f_i(2\Delta a_i)$, $(f_i)_{jk} = -i f_{ijk}$. In other words in the octet representation, where M_i or B_i or $\bar{q}\lambda_i q$ ($i = 1, \ldots, 8$) form a base multiplet, the generators are represented by the 8×8 matrices f_i, whose matrix elements are given by the structure constants f_{ijk} of eq. (6-26) according to $(f_i)_{jk} = -i f_{ijk}$. It is easily verified that $(F_i =) f_i$ obey the commutation relations of eq. (6-25); the analogous property for SU2 is given by eq. (6-10).

An alternative but more difficult method of deriving the infinitesimal transformations of B_i, (M_i) would have been to consider the transformations of the baryon quark functions, eq. (6-38).

The anti-baryons to eq. (6-38), following the anti-quarks, transform by the complex conjugate or contragredient transformation. For the antibaryons B_i, $i = 1, \ldots, 8$ to transform by the contragradient complex conjugate transformation to B_i we must have:

$$\bar{N} = \tfrac{1}{2}(\bar{B}_6 + i \bar{B}_7), \quad \bar{P} = \tfrac{1}{2}(\bar{B}_4 + i \bar{B}_5)$$

$$\bar{\Xi}^- = \tfrac{1}{2}(\bar{B}_4 - i \bar{B}_5), \quad \bar{\Xi}^0 = \tfrac{1}{2}(\bar{B}_6 + i \bar{B}_7)$$

$$\bar{\Sigma}^- = \tfrac{1}{2}(\bar{B}_1 - i \bar{B}_2), \quad \bar{\Sigma}^0 = \tfrac{1}{\sqrt{2}} \bar{B}_3, \quad \bar{\Sigma}^+ = -\tfrac{1}{2}(\bar{B}_1 + i\bar{B}_2)$$

$$\bar{\Lambda} = \tfrac{1}{\sqrt{2}} \bar{B}_8 . \qquad (6\text{-}80)$$

F and D Couplings of Octets

The baryon-baryon-meson coupling can in general be written:

$$\bar{B}_i (\theta_k)_{ij} B_j M_k . \tag{6-81}$$

By applying the infinitesimal transformation, eq. (6-79), to B, M, and $\bar{B}$, (since U is real) that the condition for SU3 invariance is:

$$[f_i, \theta_j] = i f_{ijk} \theta_k \tag{6-82}$$

and the two sets of matrices θ_i, $i = 1, \ldots, 8$ that satisfy eq. (6-82) are (Gell-Mann, 1961):

$$\theta_i = f_i, \quad \theta_i = d_i \tag{6-83}$$

$$(f_i)_{jk} = -i f_{ijk} \tag{6-84a}$$

$$(d_i)_{jk} = d_{ijk} . \tag{6-84b}$$

The f_{ijk} are the antisymmetric structure constants given by eq. (6-26) and also listed in Table 6-4. The d_{ijk} are symmetric under permutations of ijk are and given in Table 6-4. The most general coupling is:

$$F(\bar{B} f_i B) M_i + D(\bar{B} d_i B) M_i . \tag{6-85}$$

The information in the above expression is best displayed by dividing by $F + D$, using the ratio $a' = F/(F + D)$ and isospin multiplets to obtain:

$$(\bar{N}\tau N)\cdot\pi - (1-2a')(\bar{\Xi}\tau\Xi)\cdot\pi + \frac{2}{\sqrt{3}}(1-a')(\bar{\Lambda}\Sigma+\bar{\Sigma}\Lambda)\cdot\pi - 2a'i\{\bar{\Sigma}\times\Sigma\}\cdot\pi$$

$$+ \frac{1}{\sqrt{3}}(4a'-1)(\bar{N}N)\eta - \frac{1}{\sqrt{3}}(1+2a')(\bar{\Xi}\Xi)\eta - \frac{2}{\sqrt{3}}(1-a')(\bar{\Lambda}\Lambda)\eta$$

$$+ \frac{2}{\sqrt{3}}(1-a')(\bar{\Sigma}\cdot\Sigma)\eta - \frac{1}{\sqrt{3}}(1+2a')\{(\bar{N}K)\Lambda + \bar{\Lambda}(\bar{K}N)\}$$

$$+ \frac{1}{\sqrt{3}}(4a'-1)\{(\bar{\Xi}K_c)\Lambda + \bar{\Lambda}(\bar{K}_c\Xi)\} + (1-2a')\{\bar{\Sigma}\cdot(\bar{K}\tau N)$$

$$+ (\bar{N}\tau K)\cdot\Sigma\} - \{\bar{\Sigma}\cdot(\bar{K}_c\tau\Xi) + (\bar{\Xi}\tau K_c)\cdot\Sigma\} \tag{6-86}$$

6.6. SU3 COUPLINGS AND N^* DECAYS

ijk	f_{ijk}	ijk	d_{ijk}
123	1	118	$1/\sqrt{3}$
147	$\frac{1}{2}$	228	$1/\sqrt{3}$
156	$-\frac{1}{2}$	338	$1/\sqrt{3}$
246	$\frac{1}{2}$	146	$\frac{1}{2}$
257	$\frac{1}{2}$	157	$\frac{1}{2}$
345	$\frac{1}{2}$	256	$\frac{1}{2}$
367	$-\frac{1}{2}$	344	$\frac{1}{2}$
458	$\sqrt{3}/2$	355	$\frac{1}{2}$
678	$\sqrt{3}/2$	247	$-\frac{1}{2}$
		366	$-\frac{1}{2}$
		377	$-\frac{1}{2}$
		448	$-1/2\sqrt{3}$
		558	$-1/2\sqrt{3}$
		668	$-1/2\sqrt{3}$
		778	$-1/2\sqrt{3}$
		888	$-1/\sqrt{3}$

Table 6-4. Nonzero elements of f_{ijk} and d_{ijk}. The f_{ijk} are odd, and the d_{ijk} even, under permutations of ijk.

where we have used scalars, vectors and two-component spinors in isospin space defined by:

$$N = \begin{pmatrix} P \\ N \end{pmatrix}, \bar{N} = (\bar{P}\,\bar{N}), \Xi = \begin{pmatrix} \Xi^0 \\ \Xi^- \end{pmatrix}, \bar{\Xi} = (\bar{\Xi}^0\,\bar{\Xi}^-)$$

$$K = \begin{pmatrix} K^+ \\ K^0 \end{pmatrix} \bar{K} = (K^-\,\bar{K}_0)\; K_c = \begin{pmatrix} \bar{K}^0 \\ -K^- \end{pmatrix} \bar{K}_c = (K^0\,-K^+)$$

$$\boldsymbol{\pi} = (\pi_1, \pi_2, \pi_3);\; \pi_1 = \frac{1}{\sqrt{2}}(-\pi^+ + \pi^-),\; \pi_2 = \frac{-i}{\sqrt{2}}(\pi^+ + \pi^-),\; \pi_3 = \pi_0$$

$$\boldsymbol{\Sigma} = (\Sigma_1, \Sigma_2, \Sigma_3);\; \Sigma_1 = \frac{1}{\sqrt{2}}(-\Sigma^+ + \Sigma^-),\; \Sigma_2 = \frac{-i}{\sqrt{2}}(\Sigma^+ + \Sigma^-),\; \Sigma_3 = \Sigma_0$$

$$\bar{\boldsymbol{\Sigma}} = (\bar{\Sigma}_1, \bar{\Sigma}_2, \bar{\Sigma}_3)\; \bar{\Sigma}_1 = \frac{1}{\sqrt{2}}(-\bar{\Sigma}^+ + \bar{\Sigma}^-),\; \bar{\Sigma}_2 = \frac{+i}{\sqrt{2}}(\bar{\Sigma}^+ + \bar{\Sigma}^-),\; \bar{\Sigma}_3 = \bar{\Sigma}_0 \quad (6\text{-}87)$$

For example:

$$i(\bar{\boldsymbol{\Sigma}} \times \boldsymbol{\Sigma}) \cdot \boldsymbol{\pi} = (\bar{\Sigma}^+ \pi^+ - \bar{\Sigma}^- \pi^-)\Sigma^0 + \bar{\Sigma}_0(\Sigma^- \pi^+ - \Sigma^+ \pi^-) + (\bar{\Sigma}^- \Sigma^- - \bar{\Sigma}^+ \Sigma^+)\pi^0.$$

In eqs. (6-86) and (6-87) we have written η as the isospin singlet member of the pseudoscalar octet. It should be remembered that this is not necessarily the physical η, which though predominantly in an SU3 octet, seems to contain an admixture of SU3 singlet. In terms of physical particles one should substitute for η in eq. (6-86), $a\eta + bX^0$. Since the admixture is small ($b \ll a$) one can usually ignore it, in this case of the pseudoscalar octet.

The coefficients of the terms in eq. (6-86) give, when multiplied by the isospin Clebsch-Gordan coefficients contained in the terms, Clebsch-Gordan coefficients for octet-octet coupling to the octet. SU3 Clebsch-Gordan coefficients have been evaluated (de Swart, 1963; Particle Data Group, 1972) for a large range of SU3 multiplets. We quote here the results for the octet-octet coupling coefficients to the decuplet in a similar form to eq. (6-86) above:

$$-\sqrt{\frac{1}{5}}\bar{\Delta}[N\pi] + \sqrt{\frac{1}{5}}\bar{\Delta}[\Sigma K] + \bar{Y}_1^* \cdot \{-\frac{1}{2\sqrt{10}}(\bar{K}\tau N) - \frac{1}{2\sqrt{10}}(\bar{K}_c \tau \Xi)$$

$$+ \frac{i}{2\sqrt{10}}\boldsymbol{\Sigma}\times\boldsymbol{\pi} - \frac{1}{2}\sqrt{\frac{3}{10}}\boldsymbol{\Sigma}\eta + \frac{1}{2}\sqrt{\frac{3}{10}}\Lambda\boldsymbol{\pi}\} + (\bar{\Xi}^*\{\frac{1}{2}\sqrt{\frac{1}{15}}\tau\,\Xi)\cdot\boldsymbol{\pi} - \frac{1}{2}\sqrt{\frac{1}{15}}\tau\,K_c)\cdot\boldsymbol{\Sigma}$$

$$+ \frac{1}{2\sqrt{5}}\Xi)\eta - \frac{1}{2\sqrt{5}}K_c)\Lambda\} - \sqrt{\frac{1}{20}}\bar{\Omega}^-(\Xi\,K_c) \quad (6\text{-}88)$$

6.6. SU3 COUPLINGS AND N^* DECAYS

where $\bar{\Delta} = [\bar{\Delta}^{++}, \bar{\Delta}^{+}, \bar{\Delta}^{0}, \bar{\Delta}^{-}]$, and:

$$[N\pi] = \begin{bmatrix} P\,\pi^+ \\ \sqrt{\frac{2}{3}}\,P\,\pi^0 + \sqrt{\frac{1}{3}}\,N\,\pi^+ \\ \sqrt{\frac{1}{3}}\,P\,\pi^- + \sqrt{\frac{2}{3}}\,N\,\pi^0 \\ N\,\pi^- \end{bmatrix}$$

$$\overline{Y_1^*} = ((\bar{Y}_1^*)_1, (\bar{Y}_1^*)_2, (\bar{Y}_1^*)_3); \quad (Y_1^*)_1 = -\frac{1}{\sqrt{2}}(\overline{Y_1^{*+}} + \overline{Y_1^{*-}}), \quad (Y_1^*)_2$$

$$= -\frac{i}{\sqrt{2}}(\overline{Y_1^{*+}} - \overline{Y_1^{*-}}), \quad (Y_1^*)_3 = \overline{Y_1^{*0}}$$

$$\bar{\Xi}^* = (\overline{\Xi^{*0}}, \overline{\Xi^{*-}}) \ .$$

The notation is as for the three-octet case, with round brackets such as $(\Xi\,\tau\,K)$ denoting a two-component isospinor product, which is an isovector or isoscalar according as τ is present or absent.

The expressions (6-86) and (6-88) have been written in terms of particular octets and a particular decuplet. They are of course perfectly general and express octet-octet-octet and decuplet-octet-octet couplings for any octets and decuplets on suitable substitution. For example to find the couplings involving the pseudovector mesons we perform the substitution:

$$\pi \to \rho, \ K \to K^*, \ \bar{K} \to \bar{K}^* \quad \text{and} \quad \eta \to V(8)$$

where $V(8)$ is the octet isospin singlet vector meson and is a combination of the physical ω and ϕ roughly according to [eq. (6-49) ff.]:

$$V(8) \simeq \frac{1}{\sqrt{3}}\omega - \sqrt{\frac{2}{3}}\phi \ .$$

We may define baryon-baryon-meson coupling constants, $g_{\Xi\Xi\pi}$, $g_{\Lambda\Sigma\pi}$, $g_{NN\eta}\ldots$, in analogy to the nucleon-nucleon-pion coupling constant g. Then if the strong interactions were SU3 symmetric, the meson field sources and hence the coupling constants would be in the ratios given by the expression, eq. (6-86); for example $g : g_{\Xi\pi} : g_{\Lambda\Sigma\pi} : \ldots = 1 : -(1-2a') : \frac{2}{\sqrt{3}}(1-a') : \ldots$. We already know that SU3 is broken since the masses in an SU3 multiplet are not equal — indeed for the pseudoscalar octet are very unequal because of the pion. Consequently we expect some deviation from SU3.

In some classification and symmetry schemes which subsume SU3, such as SU6, $a' = F/(F+D)$ is determined. For example in the quark model, the basic interaction for the emission of a meson by a quark is:

$$\sum_{\nu=1}^{8} g'_\nu \, \bar{q} \lambda_\nu \, (\vec{\sigma} \vec{k}) q \, M_\nu$$

where $\vec{k}$ is the meson momentum and q is a six-component quark spinor.[11] In this interaction, illustrated in Fig. 6-4(a), the meson is envisaged as a particle rather than a $q\bar{q}$ pair; this does not affect the SU3 and spin properties.

In the quark model the assumption of SU3 symmetry of interactions is not necessary. The interaction is made SU3 invariant by imposing the condition $g'_\nu = g'$ for all ν. (Isospin and charge conjugation invariance only require $g'_1 = g'_2 = g'_3$ and $g'_4 = g'_5 = g'_6 = g'_7$.)

The interaction is used in the form:

$$g' \sum_{j=1}^{3} \bar{q}(j) \lambda_\nu \, (\vec{\sigma} \vec{k}) q(j) \, M_\nu$$

as a transition operator for the transition $B_\mu \to B_\rho M_\nu$, where B_μ and B_ρ are the initial and final baryons. Consequently the transition probability is proportional to:

---

[11] In the non-relativistic quark model there may be additional recoil correction terms added to this interaction; such additional terms occur in the relativistic model of Feynman, Kislinger and Ravndal (1971) and are sometimes important. The use of the basic interaction given here, with $g'_\nu = g'$, is equivalent to the collinear SU6$_W$ symmetry applied to particle decays (Lipkin and Meshkov, 1965, 1966).

6.6. SU3 COUPLINGS AND N^* DECAYS

$$g'\left(\overline{B}_\mu \left| \sum_{j=1}^{3} \lambda_\nu^{(j)} \sigma_3^{(j)} \right| B_\rho \right) \qquad (6\text{-}89)$$

where B_μ and B_ρ are quark baryon wave functions such as eq. (6-56), and $\lambda_\nu^{(j)}$ and $\sigma_3^{(j)}$ are the Gell-Mann and Pauli matrices, eq. (6-27), sandwiched between the jth quark in B_μ and the jth quark in B_ρ; a frame in which the process is collinear along the 3-axis has been taken with the quark spins quantized in that direction. As only one coupling constant g' is involved, independent of μ, ρ, or ν, it is evident that the ratio $\alpha' = F/(F+D)$ is determined. The result is $\alpha' = 0.4$ or $\alpha = D/(D+F) = 0.6$. Though the quark model, even with the (unnecessary) restriction of SU3 symmetry, is not equivalent to SU6 or SU6$_W$ symmetry of elementary particles, in this case the result is the same.[11,12] The above argument is equivalent to a method for deriving SU6$_W$ Clebsch-Gordan coefficients [given by eq. (6-89)].

In the interaction of the (meta-)stable baryons, one multiplies eq. (6-86) by the pion-nucleon coupling constant g and then part of eq. (6-86) is:

$$g N \vec{\tau} N \vec{\pi} + g_{N\Lambda K} \{(\overline{N}K)\Lambda + \overline{\Lambda}(KN)\} + g_{N\Sigma K} \{\vec{\Sigma}\cdot(K\vec{\tau} N) + (N\vec{\tau} K)\cdot\vec{\Sigma}\}$$
$$(6\text{-}90\text{a})$$

where the ΛK and ΣK coupling constants are given by:

$$g_{N\Lambda k} = -\frac{g}{\sqrt{3}}(1+2\alpha') = -\frac{g}{\sqrt{3}}(3-2\alpha); \quad g_{N\Sigma K} = g(1-2\alpha') = g(2\alpha-1).$$
$$(6\text{-}90\text{b})$$

For $\alpha = 0.6$ we get the values:

$$\frac{g^2_{N\Lambda K}}{4\pi} = \frac{25}{27}\frac{g^2}{4\pi}; \quad \frac{g^2_{N\Sigma K}}{4\pi} = \frac{1}{9}\frac{g^2}{4\pi}. \qquad (6\text{-}91)$$

---

[12] A similar calculation gives the $D/(D+F)$ ratio for decay of the [8,2] and [8,4] of a {70} of SU6 into stable baryon + pseudoscalar meson as 0.375 and 1.5 respectively. All these quark model calculations are independent of the orbital angular momentum assigned to the SU6 multiplets.

These coupling constants can be found in principle from forward dispersion relation analysis of low energy kaon-nucleon scattering experiments, but the experiments are not yet accurate enough to permit an accurate evaluation, and so to check the validity of the SU3 prediction eq. (6-90) and the SU6 prediction eq. (6-91).

It is evident that the decays of unstable octets can be checked for consistency with SU3 using eq. (6-89). Such checks have been reviewed by Tripp (1968) and by Levi-Setti (1969), including the simpler cases of singlet and decuplet [eq. (6-87)] decay which do not involve a $F/(D+F)$ ratio. The results seem to be qualitatively consistent with SU3.

Our particular interest is in the nucleonic resonances of which the $I = \frac{3}{2}$ are decuplet members and the $I = \frac{1}{2}$ are octet members. The $N^*_{\frac{3}{2}}$ resonances have, according to SU3, equal probability of decay into $N\pi$ or $K\Sigma$ states. The $N^*_{\frac{1}{2}}$ resonances should decay, according to SU3, by:

$$(N^{*+}_{\frac{1}{2}} \to P\pi^0 + \sqrt{2}N\pi^-) : (N^{*+}_{\frac{1}{2}} \to P\eta) : (N^{*+}_{\frac{1}{2}} \to K^+\Lambda) : (N^{*+}_{\frac{1}{2}} \to K^+\Sigma^0 + \sqrt{2}K^0\Sigma^-)$$

$$= 1 : \frac{1}{3}(4a-1)^2 : \frac{1}{3}(1+2a)^2 : (1-2a)^2 \quad . \tag{6-92}$$

The nonrelativistic real quark model, in its more specific forms such as the harmonic oscillator model, gives specific values for the partial width for N^* decays into $N\pi$ states. The partial width expressions are in terms of a quark(p or n)-pion coupling constant, and in the case of the harmonic oscillator, the spring constant already determined from the level spacing. In this model all the $N^* \to N\pi$ partial widths are determined in terms of one multiplicative parameter, the quark-pion coupling. The authors (Faiman and Hendry, 1968) claim success for the calculations, though it is difficult to assess the significance in view of the large uncertainty on the partial widths discussed in Chapter 5.

6.6. SU3 COUPLINGS AND N^* DECAYS

SU3 Breaking: Mass Formulae

To complete our limited discussion of SU3 we must mention this topic briefly.

The (first) decuplet states are evenly spaced in mass with a mass difference of ~ 147 MeV. This could possibly be attributed to a one-quark operator which caused a λ quark to be worth 147 MeV more of energy than a p or n quark. Thinking of the (first) baryon octet in the same way, we see that the extra energy associated with a λ quark there would be $(M(\Xi) - M(N))/2 \sim 190$ MeV, which would then predict the Σ and Λ states to lie together at a mass $(M(\Xi) + M(N))/2 \sim 1130$ MeV. Although the mean mass for Λ and Σ is quite close to this prediction, the mass splitting $M(\Sigma) - M(\Lambda) \sim 80$ MeV is a significant discrepancy. On this model it is necessary to introduce two-quark operators (Dalitz, 1967a).

The problem may be approached in another way that is due to Gell-Mann and also to Okubo (Gell-Mann and Ne'eman, 1964). If we consider an SU3 baryon multiplet, dimension m with members $B_i (i = 1, 2 \ldots , m)$ we may write the masses as:

$$M(B_i) = <B_i | \mathfrak{M}_0 + \mathfrak{M}_1 + \ldots | B_i> \qquad (6\text{-}93)$$

where $\mathfrak{M}_0$ is an SU3 invariant mass operator, giving rise to equal masses in the multiplet, and $\mathfrak{M}_1 \ldots$ breaks SU3, giving rise to mass splitting within the multiplet. For strong interactions strangeness is preserved and isospin nearly preserved, as evinced by the near equality of masses within an isospin multiplet. So we take $\mathfrak{M}_1$ to be an operator which, while breaking SU3, preserves isospin and hypercharge, and thus gives the mass differences between isospin multiplets (further terms $\mathfrak{M}_2, \mathfrak{M}_3 \ldots$ could be taken to break isospin, strangeness ...). The term $\mathfrak{M}_1$ is an $I = 0$, $Y = 0$ operator, and the simplest hypothesis is that it is the 8th component of an SU3 octet. (It could also belong to a {27} or a higher representation, though none of these representations has evidently appeared as a resonance multiplet.) We know that if B is an eight-component spinor of a baryon octet, then $\bar{B} f_\lambda B$ and $\bar{B} d_\lambda B$ ($\lambda = 1, 2 \ldots 8$) are the only quantities transforming as octets that can be formed. Consequently $\mathfrak{M}_1$ is a combination of f_8 and d_8, and thus the mass differences between isospin multiplets within an octet are given by:

$$\Delta M(B_i) = a B_i^+ f_8 B_i + b B_i^+ d_8 B_i \, . \tag{6-94}$$

Elimination of the constants a and b gives the Gell-Mann-Okubo octet mass formula which is satisfied to within 1% of either side:

$$\tfrac{1}{2} M(N) + \tfrac{1}{2} M(\Xi) = \tfrac{3}{4} M(\Lambda) + \tfrac{1}{4} M(\Sigma) \, . \tag{6-95}$$

More generally it can be shown (Okubo, 1962), on the assumption of octet only mass breaking, that in an SU3 multiplet:

$$M = M_0 + M_1 Y + M_2 \left(I(I+1) - \tfrac{1}{4} Y^2 \right) \, . \tag{6-96}$$

This formula gives rise to the baryon octet relation, eq. (6-95). For a decuplet we know that $I = 1 + \tfrac{1}{2} Y$ leading to $M = M_0 + M_1 Y$, and the *equal spacing rule* for decuplet masses, which as we saw above is well satisfied for the first decuplet with $\Delta M \sim 147$ MeV. The discovery (Barnes et al., 1964) of the Ω^- particle where predicted (Gell-Mann, 1962) confirmed SU3 as a good, though broken, symmetry.

It seems more appropriate for the mesons, which obey the Klein-Gordon equation where mass occurs squared, to use $mass^2$ in the above formulae. For the pseudoscalar octet we then obtain, since $M(\bar{K}) = M(K)$:

$$M(K)^2 = \tfrac{1}{4} (3 M(\eta)^2 + M(\pi)^2) \tag{6-97}$$

which is only satisfied to 10%. Perhaps the reason for this is that η is not a pure octet particle but mixes with the X^0. For the vector octet:

$$M(K^*)^2 = \tfrac{1}{4} (3 M(V(8))^2 + M(\rho)^2) \tag{6-98}$$

where $V(8)$ is the isoscalar component of the vector octet. Substitution of the masses of the K^* and the ρ leads to $M(V(8)) \simeq 930$ MeV, which is intermediate between the masses of the ω and ϕ. We conclude that these nonet members are not pure singlet or octet. If $V(1)$ is the SU3 singlet member of the nonet, we write:

$$\phi = V(1) \sin\theta - V(8) \cos\theta$$
$$\omega = V(1) \cos\theta + V(8) \sin\theta \ . \qquad (6\text{-}99)$$

The mixing angle θ can be estimated from ϕ, ω decay widths or by diagonalizing the mass matrix (Glashow, 1963; Dashen and Sharp, 1964). It is found that $\theta \simeq \pm 40°$. We note that the value corresponding to $\phi = \bar{\lambda}\lambda$, $\omega = \frac{1}{2}(\bar{p}p + \bar{n}n)$ is $\theta = 35°$.

6.7. ELECTROMAGNETIC INTERACTIONS AND U-SPIN

We saw in section 6.3 (eqs. (6-34)) that the charge operator acting on the quark triplet is:

$$Q/e = \frac{1}{2}\left(\lambda_3 + \frac{1}{3}\lambda_8\right) = \begin{pmatrix} \frac{2}{3} & 0 & 0 \\ 0 & -\frac{1}{3} & 0 \\ 0 & 0 & -\frac{1}{3} \end{pmatrix} \qquad (6\text{-}100)$$

Electromagnetic interactions in SU3 are treated by assuming that they all transform like $F_3 + \frac{1}{\sqrt{3}} F_8$, and so, acting on a quark triplet, are given by eq. (6-100). [For example, in terms of a quark model, quarks may have a magnetic moment as well as a charge, and the SU3 assumption is that the magnetic moment is also proportional to eq. (6-100). Of course, the quarks of this section are mathematical quarks.] The term F_3 (or λ_3) is the third component of an isovector, and F_8 (or λ_8) is an isoscalar; so the electromagnetic interaction breaks both isospin symmetry and SU3 symmetry.

The hypercharge and isospin operators acting on a quark triplet are:

$$Y = \begin{pmatrix} \frac{1}{3} & 0 & 0 \\ 0 & \frac{1}{3} & 0 \\ 0 & 0 & -\frac{2}{3} \end{pmatrix}, \quad I_1 = \frac{1}{2}\lambda_1 = \frac{1}{2}\begin{pmatrix} 0 & 1 & 0 \\ 1 & 0 & 0 \\ 0 & 0 & 0 \end{pmatrix}, \quad I_2 = \frac{1}{2}\lambda_2$$

$$= \begin{pmatrix} 0 & -i & 0 \\ i & 0 & 0 \\ 0 & 0 & 0 \end{pmatrix}, \quad I_3 = \frac{1}{2}\lambda_3 = \frac{1}{2}\begin{pmatrix} 1 & 0 & 0 \\ 0 & -1 & 0 \\ 0 & 0 & 0 \end{pmatrix}. \quad (6\text{-}101)$$

The hypercharge and isospin operators treat the p and n quarks as a doublet and the λ as a singlet. Since SU3 symmetry is symmetry between p, n, and λ quarks, the equivalent operators with n and λ treated as a doublet and p as a singlet have the same SU3 status. These are obviously:

$$E = -Q/e = -\begin{pmatrix} \frac{2}{3} & 0 & 0 \\ 0 & -\frac{1}{3} & 0 \\ 0 & 0 & -\frac{1}{3} \end{pmatrix} \quad U_1 = \frac{1}{2}\lambda_6 = \frac{1}{2}\begin{pmatrix} 0 & 0 & 0 \\ 0 & 0 & 1 \\ 0 & 1 & 0 \end{pmatrix}, \quad U_2 = \frac{1}{2}\lambda_7$$

$$= \frac{1}{2}\begin{pmatrix} 0 & 0 & 0 \\ 0 & 0 & -i \\ 0 & i & 0 \end{pmatrix} \quad U_3 = \frac{1}{4}(\sqrt{3}\,\lambda_8 - \lambda_3) = \frac{1}{2}\begin{pmatrix} 0 & 0 & 0 \\ 0 & 1 & 0 \\ 0 & 0 & -1 \end{pmatrix}. \quad (6\text{-}102)$$

Any SU3 multiplet must have the same number and type of U-spin submultiplets as of I-spin submultiplets. Term E is a U-spin scalar, since it commutes with U_1, U_2, and U_3, and so all U-spin multiplets have the same charge; they are marked in Fig. 6-1 for a decuplet and an octet. The U-spin multiplets for the octet are (p, Σ^+), (Σ^-, Ξ^-), $(n, \frac{1}{2}(\sqrt{3}\,\Lambda - \Sigma^0), \Xi^0)$, $\frac{1}{2}(\Lambda + \sqrt{3}\,\Sigma^0)$. [The combinations of Λ and Σ^0 that are U-spin singlets and triplets can be found from combination of the wave functions of eqs. (6-38) or (6-49), remembering that I-spin $\to U$-spin corresponds to p, n, $\lambda \to n$, λ, p. It is also obvious from the expression

6.8. ELECTROMAGNETIC DECAYS OF BARYON RESONANCES

for U_3.] The mass differences of particles of different charge within an isospin multiplet are usually attributed to the interaction of the usual electromagnetic forces. In that case the electromagnetic increments, which we denote by δM, are generated by a U-spin scalar (namely $Q/e = -E$), so that:

$$\delta M(P) = \delta M(\Sigma^+), \quad \delta M(N) = \delta M(\Xi^0), \quad \delta M(\Sigma^-) = \delta M(\Xi^-)$$

from which we derive:

$$\delta M(\Xi^-) - \delta M(\Xi^0) = \delta M(P) - \delta M(N) + \delta M(\Sigma^-) - \delta M(\Sigma^+) .$$

In this relation only differences between particles in the same isomultiplet occur, so we can substitute the masses themselves, giving rise to the *Coleman-Glashow* (1961) *relation*:

$$(M(\Xi^-) - M(\Xi^0)) = (M(P) - M(N)) + (M(\Sigma^-) - M(\Sigma^+)) .$$

Experimentally $M(P) - M(N) = -1.29$ MeV, $M(\Sigma^-) - M(\Sigma^+) = 7.92 \pm .13$ MeV, $M(\Xi^-) - M(\Xi)^0 = 6.6 \pm 0.7$ MeV, giving a rather good agreement with the Coleman-Glashow relation. (Particle Data Group, 1972.)

Directly from the fact that the electromagnetic interaction is a U-spin scalar, there are the magnetic moment relations $\mu(\Sigma^+) = \mu(P)$, $\mu(\Xi^0) = \mu(N)$, $\mu(\Xi^-) = \mu(\Sigma^-)$. It can further be shown (Coleman and Glashow, 1961) that $\mu(\Lambda) = \frac{1}{2}\mu(N) = -\mu(\Sigma^0)$, $\mu(\Sigma^-) = -\mu(P) - \mu(N)$. Experimentally $\mu(N) = -1.913$, $\mu(\Lambda) = -0.67 \pm .06$, $\mu(P) = 2.793$, $\mu(\Sigma^+) = 2.59 \pm 0.46$.

6.8. ELECTROMAGNETIC DECAYS OF BARYON RESONANCES

Among the principal experimental foundations of modern physics is atomic spectroscopy, through which atomic structure is revealed by the electromagnetic excitation and decay of atomic energy levels. In nuclear physics too the electromagnetic excitation or decay of nuclear energy levels is very important. It may be hoped that analogous experiments in elementary particle physics will be equally revealing of particle structure as the nuclear physics experiments are of nuclear structure. To understand this hope let us consider the experimental situation on the nucleonic resonances which we consider under the categories of $N^*_{\frac{1}{2}}$ (isospin $\frac{1}{2}$, SU3 octet) and $N^*_{\frac{3}{2}}$ (isospin $\frac{3}{2}$, SU3 decuplet). The information we have on these resonances as largely covered in Chapter 5 is their mass, total width Γ, and partial width Γ_π for decay into πN. The information on other strong decay modes is small and likely, in general, to remain so (except for the possibility in the future of information on $N^* \to K\Lambda$ or $K\Sigma$). However, events of the type $\gamma N \to \pi N$ in the resonance region largely proceed through resonance formation: $\gamma N \to N^* \to \pi N$. So by use of the already existing information on $\Gamma_\pi = \Gamma(N^* \to \pi N)$ one can deduce, though with some uncertainties of principle and practice, $\Gamma_\gamma = \Gamma(N^* \to \gamma N)$ (Gourdin and Salin, 1962; Chau, Dombey, and Moorhouse, 1967; Walker, 1969; Dalitz and Sutherland, 1966). Now the photon has spin 1 with, for the real photon, possible helicities ± 1; consequently in the decay $N^* \to \gamma + N$, the γN system can either have total helicity $(\pm)\frac{1}{2}$ or $(\pm)\frac{3}{2}$ leading to two possible decay modes into γN, equivalent of course to linear superpositions of the electric and magnetic multipole decay modes (cf. Chapter 2). Additionally the photon can convey either zero or unit isospin giving the possible decays:

$$N^*_{\frac{3}{2}} \to \gamma(I=1) + N; \ N^*_{\frac{1}{2}} \to \gamma(I=0) + N; \ N^*_{\frac{1}{2}} \to \gamma(I=1) + N \ .$$

Thus there are two possible decay modes $N^*_{\frac{3}{2}} \to \gamma + N$ with two measurable widths $\Gamma(N^*_{\frac{3}{2}} \to \gamma N)$ and four possible decay modes $N^*_{\frac{1}{2}} \to \gamma + N$ with four measurable widths $\Gamma(N^*_{\frac{1}{2}} \to \gamma N)$.

6.8. ELECTROMAGNETIC DECAYS OF BARYON RESONANCES

This represents a proportionately large potential addition to our data on resonance states. Any theory of the elementary particles should explain these numbers (to some approximation), and accordingly, they are a guide in the search for a correct theory. We shall illustrate the potential importance of radiative widths by considering the L-excitation quark model, which we choose for two reasons. First, being a composite model with explicit wave functions, it is, in a sense, a complete, though relativistically unsatisfactory, theory; certain symmetries, such as $SU6_W$, also give some predictions in radiative decay widths but they are not as complete as those of the quark model. Second, it happens that the radiative widths, as presently known, are generally in fairly good agreement with the quark model calculations, so that our discussion can illustrate a productive relationship of theory and experiment, which, presumably, should also happen in a more satisfactory theory. We shall explain the results by using the L-excitation quark model in its nonrelativistic form, without possible relativistic corrections (discussed by Copley, Karl, and Obryk, 1969) to the most primitive nonrelativistic interaction Hamiltonian. Feynman, Kislinger, and Ravndal (1971) have attempted a relativistic L-excitation quark model, still however containing some unsatisfactory features, which differs little from the naive nonrelativistic model in predicted radiative widths, though some other electromagnetic results are significantly different (Ravndal, 1971).

Electromagnetic Interaction in the Nonrelativistic Quark Model

The simplest nonrelativistic electromagnetic Hamiltonian for transitions between baryons each containing three quarks is:

$$H_\gamma = \sum_{j=1}^{3} \left[\frac{e_j}{2 M_Q} (2\vec{A} \cdot \vec{p_j}) - i\mu_j \, \vec{\sigma}^{(j)} \cdot \vec{k} \times \vec{A} \right] \qquad (6\text{-}103)$$

where A is the electromagnetic field and k the photon momentum, $\sigma^{(j)}$ and $p^{(j)}$ are the spin and momentum operators acting on the jth quark; M_Q is the quark mass and e_j, μ_j are the charge and magnetic moment of the jth quark. These latter are given (cf. section 6.7) by the matrices sandwiched between appropriate quark spinors q_j:

$$e_j = \bar{q}_j e \begin{pmatrix} \frac{2}{3} & 0 & 0 \\ 0 & -\frac{1}{3} & 0 \\ 0 & 0 & -\frac{1}{3} \end{pmatrix} q_j, \mu_j = \bar{q}_j \mu \begin{pmatrix} \frac{2}{3} & 0 & 0 \\ 0 & -\frac{1}{3} & 0 \\ 0 & 0 & -\frac{1}{3} \end{pmatrix} q_j \qquad (6\text{-}104)$$

so that for example $e_j = \frac{2}{3} e$ when j is a proton quark. It is an immediate consequence of eq. (6-104) that sandwiching eq. (6-103) between the nucleon states of section 6.4 gives:

$$\mu_P = \mu \quad \text{and} \quad \mu_N = -\frac{2}{3}\mu \qquad (6\text{-}105)$$

and consequently:

$$\mu_P/\mu_N = -\frac{3}{2} \qquad (6\text{-}106)$$

in close accord with the experimental ratio of -1.46. The result, eq. (6-106) was first obtained by Beg, Lee, and Pais (1964) using SU6 symmetry; the above quark model derivation, of course, does *not* require SU6 symmetric interactions. [The SU6 symmetry was first proposed by Gursey and Radicati (1964) and by Sakita (1964).] Because of eq. (6-105) the quark scale moment μ of eq. (6-104) is taken to be equal to μ_P:

$$\mu \equiv \mu_P . \qquad (6\text{-}107)$$

$M1$ Selection Rule and the Radiative Width of the p_{33} (1236)

Having just dealt with what can be regarded as the transition $N \to N + \gamma$, we can now consider another transition within the $\{56\}$ namely $\Delta \to N + \gamma$. Since the Δ (1236) has spin-parity $\frac{3^+}{2}$ and the nucleon has spin-parity $\frac{1^+}{2}$, the multipole moment transitions involved are electric quadrupole, $E2$, or magnetic dipole, $M1$. As discussed in Chapter 2, section 2.4,[13]

---

[13] In Chapter 2 the CGLN amplitudes $E_{\ell+}$, $M_{\ell+}$ proportional to $E\ell$ and $M\ell$ amplitudes were mainly used; in the $\Delta \to N+\gamma$ transition, $E2 \sim E_{1+}$ and $M1 \sim M_{1+}$.

6.8. ELECTROMAGNETIC DECAYS OF BARYON RESONANCES

the $E2$ transition almost vanishes. Becchi and Morpurgo (1965) noted that in the quark model there is no $E2$ transition: the electric interaction with the quarks, the first term on the R.H.S. of eq. (6-103), gives no contribution since the Δ has quark spin $S = \frac{3}{2}$ and the N has quark spin $S = \frac{1}{2}$; the interaction with the magnetic moment of the quarks might also give a contribution to $E2$ transitions, but that $E2$ operator is a second-rank tensor formed from products $\vec{\sigma} Y_2^m(\vec{v}^j)$ whose matrix element vanishes for $L = 0$ to $L = 0$ transitions.

To obtain the magnetic moments of the nucleons, eq. (6-105), one sandwiched the operator:

$$M_z = \sum_{j=1}^{3} \mu_j \sigma_z^{(j)} \qquad (6\text{-}108)$$

between nucleon wave functions. The same magnetic dipole operator, sandwiched between the Δ^+ and the proton, P, with m denoting z component of angular momentum, gives the matrix element:

$$M = \left(P, m = \frac{1}{2} |M_z| \Delta^+, m, = \frac{1}{2} \right) \qquad (6\text{-}109)$$

proportional to the magnetic dipole transition matrix element $M1$ (or M_{1+}). If we insert eq. (6-56) for the left-hand wave function of eq. (6-109) and for the right-hand wave function insert (cf. section 6.4):

$$\left| \Delta^+, m = \frac{1}{2} \right) = \frac{1}{3}(p(1)\,p(2)\,n(3) + p(1)\,n(2)\,p(3) + n(1)\,p(2)\,p(3))\,(\alpha(1)\,\alpha(2)\,\beta(3)$$
$$+ \alpha(1)\,\beta(2)\,\alpha(3) + \beta(1)\,\alpha(2)\,\alpha(3))$$

a simple evaluation gives:

$$M = \frac{2\sqrt{2}}{3} \mu_P \ . \qquad (6\text{-}110)$$

From analysis of experiments Dalitz and Sutherland (1966) find the value:

$$M^{\exp} = (1.28 \pm 0.02) \frac{2\sqrt{2}}{3} \mu_P \ . \qquad (6\text{-}111)$$

However, in calculating M in a physical nonrelativistic quark model there is the additional spatial form factor $G(k^2)$ multiplying the theoretical expression, eq. (6-110):

$$G(k^2) = \int \phi_\Delta^*(r_1, r_2, r_3) \phi_N(r_1, r_2, r_3) e^{i\vec{k} \cdot \vec{r}_1} d^3r_1 d^3r_2 d^3r_3$$

which Dalitz and Sutherland estimate to be 0.79 (neglecting any difference in the ϕ_Δ and ϕ_N wave functions). The comparison with experiment may then be given be expressing M^{exp} as:

$$M^{exp} = (1.62)(0.79 \times \tfrac{2\sqrt{2}}{3} \mu_p) \qquad (6\text{-}112)$$

where the second bracket represents the calculated quantity. The error is unstated here because of the uncertainty in the estimation of the form factor, but we should say that there is a 50% difference between theory and experiment.

Selection Rules and Widths of the Excited States

There are some interesting selection rules on the radiative decays of the excited states of the L-excitation quark model (Copley, Karl, and Obryk, 1969; Moorhouse, 1966). It is desirable to introduce a simple notation for the decay amplitudes. Since the photon has helicity $\lambda_1 = \pm 1$ and the nucleon has helicity $\lambda_2 = \pm\frac{1}{2}$, the total helicity of the γN system can take on the values $\lambda = \lambda_1 - \lambda_2 = \pm\frac{1}{2}, \pm\frac{3}{2}$. Now take the center of mass system of the decaying N^* and quantize the N^* spin m_z along the z direction, the direction of the outgoing particles. Then there are four possible decay processes $|N^*, m_z\rangle \to |\gamma, \lambda_1\rangle + |N, \lambda_2\rangle$ since $m_z = \lambda_1 - \lambda_2 = \pm\frac{1}{2}, \pm\frac{3}{2}$; by rotation through $180°$ about an axis perpendicular to the z-axis combined with space reflection, we see that there are only two independent decay amplitudes which we denote by $A_{\frac{1}{2}}, A_{\frac{3}{2}}$ having respectively $|\lambda| = \frac{1}{2}, \frac{3}{2}$ (the normalization of these amplitudes is defined below in the section on formalism, together with their linear relation to the electric and magnetic multipole decay amplitudes).

6.8. ELECTROMAGNETIC DECAYS OF BARYON RESONANCES

There are two general selection rules, one involving 56-plets and the other 70-plets:

I. *The helicity $\frac{3}{2}$ amplitude, $A_{\frac{3}{2}}$, for the photoexcitation off neutrons of $I = \frac{1}{2}$ resonances belonging to 56-plets vanishes identically.* The derivation of this selection rule is simple. In the center of mass system we take the γ to have spin +1 along the z direction. Then for the helicity $\frac{3}{2}$ amplitude we have to start from a nucleon state with quark spin projection $+\frac{1}{2}$; consequently only the orbital term in H_γ is effective since the magnetic moment term would lead to quark spin $\frac{3}{2}$ which is not possible for an $I = \frac{1}{2}$ resonance belonging to a 56-plet. Further, for these transitions, the orbital term is proportional in spin-unitary spin space to the charge operator and this gives zero for neutral particles. (The addition of a spin-orbit term to H_γ would invalidate this selection rule.)

Observed resonances classified as above are the $f_{15}(1690)$ and the $p_{13}(1850)$, both members, in the L-excitation quark model, of the 56, $L = 2^+$ multiplet. Evidence on the photoexcitation of the p_{13} is not yet available; the existing evidence on the f_{15} supports the selection rule (Walker, 1969b; Moorhouse, 1973).

II. *The photoexcitation off protons of resonances belonging to the [8,4] submultiplet of a 70-plet vanishes.* Since such an excitation changes the quark spin from $S = \frac{1}{2}$ to $S = \frac{3}{2}$, it must take place via the magnetic moment term in H_γ. Explicit calculation shows that this vanishes when sandwiched between the spin-unitary spin wave functions of an [8, 4] charge +1 state of the {70} and an [8, 2] charge +1 state of the {56}. It does *not* vanish for neutron states. (The spin-unitary spin wave functions of the relevant states have been given in sections 6.4 and 6.5 and the calculation is elementary.)

The observed resonances usually assigned to the [8, 4] of the {70} $L = 1^-$ are the $d_{15}(1670)$, $d_{13}(1700)$ and $s_{11}(1700)$; the two latter may mix with the $d_{13}(1520)$ and $s_{11}(1530)$, usually assigned to the [8, 2] of the {70} $L = 1^-$.

There is little evidence on this selection rule for the d_{15}. If it proves correct, it may be a useful guide on the amount of mixing between the [8, 2] and the [8, 4] of the 70-plet. If correct, to some approximation, these selection rules would be a remarkable verification of the L-excitation quark model, the resonance assignments therein, and a simple form of the effective electromagnetic interaction. If they significantly failed, then one would presumably look in the first place, to relativistic effects including more complicated effective Hamiltonians.

It is interesting that the selection rules I and II depend on the spin-unitary spin structure of the wave functions and not on the details of the spatial wave functions. It is perhaps even more interesting that there is a resonance photoproduction feature which is explained by a calculation involving details of the quark spatial wave functions. It appears that the $A_{\frac{1}{2}}$ amplitude for the photoexcitation of the $d_{13}(1520)$ off protons, and probably also for the $A_{\frac{1}{2}}$ photoexcitation of the $f_{15}(1690)$ off protons, are small compared to the appropriate $A_{\frac{3}{2}}$ amplitude. Copley, Karl, and Obryk (1969) show that using harmonic oscillator spatial wave functions one obtains:

$$A_{\frac{1}{2}}(d_{13}^+) = i\sqrt{\frac{\pi}{k}}\frac{2}{3} a\, e^{-k^2/6a^2}\left\{\frac{\mu k^2}{a^2} - \frac{e}{2M_Q}\right\} \qquad (6\text{-}113a)$$

$$A_{\frac{1}{2}}(f_{15}^+) = -\sqrt{\frac{2}{5}\pi k}\,\frac{2}{3} e^{-k^2/6a^2}\left\{\frac{\mu k^2}{2a^2} - \frac{e}{2M_Q}\right\} \qquad (6\text{-}113b)$$

a is the spring constant; k is the magnitude of the photon momentum in the rest system of the decaying resonance, and

$$k^2(d_{13}) = .22 \text{ GeV}^2; \quad \frac{1}{2}k^2(f_{15}) = .17 \text{ GeV}^2$$

6.8. ELECTROMAGNETIC DECAYS OF BARYON RESONANCES

Copley, Karl and Obryk find suitably small values of eq. (6-113), for $a^2 = 0.17$ GeV^2 and $M_Q = 3.8$ GeV/c^2, which are quite close to "normal" quark model values for these parameters when explicit calculations involving spatial wave functions are made.

There is also an impressive non-trivial agreement between the quark model and experiment in the *sign* of the resonance electromagnetic couplings — determined experimentally and theoretically relative to the Born approximation. (Walker, 1969b; Feynman et al., 1971; Moorhouse, 1973.)

Formalism for Resonance Photoexcitation

Let A_λ be the transition matrix element for a resonance of total spin J and spin component λ along the z-direction decaying in its rest system into a photon in the +z-direction and a nucleon in the −z-direction, so that $\lambda = \lambda_1 - \lambda_2$ where λ_1 and λ_2 are the photon and nucleon helicities respectively and the states have the normalization of eq. (2-3). By parity and rotational invariances, as explained above, $A_\lambda =$ (phase factor) $A_{-\lambda}$. This definition of A_λ is general, but applying it to the L-excitation symmetric quark model gives:

$$A_\lambda = \left(N; M_Z = -\lambda_2 \left| 3 \sqrt{\frac{4\pi}{2k}} e^{ikz_3} \left\{ \frac{e_3}{2M_Q} 2(\vec{\epsilon} \cdot \vec{p}^{(3)}) - i\mu_3 \vec{\sigma}^{(3)} \cdot \vec{k} \times \vec{\epsilon} \right\} \right| N^*; M_Z = \lambda_1 \right) \quad (6\text{-}114)$$

where $\epsilon = \frac{1}{\sqrt{2}} (1, +i, 0)$ for $\lambda_1 = \pm 1$, and where we have taken advantage of the overall symmetry of the wave functions to use only the third quark in the Hamiltonian. (We have selected quark 3, rather than quarks 1 or 2, for most convenient use with the particular spin-unitary spin waves constructed previously in this chapter.)

Now consider a resonance of spin J and spin projection M along some fixed direction which we denote as the Z-axis. Then the amplitude for decay into γN along a direction at angle (θ, ϕ) to the Z-axis, is, by superposition:

$$A_M^J(\theta, \phi) = \sum_{\lambda=-\frac{3}{2}}^{\frac{3}{2}} A_\lambda \, d_{\lambda M}^J(\theta) \, e^{i(\lambda - M)\phi} \quad . \quad (6\text{-}115)$$

The transition matrix element, eq. (6-115), is related to the partial decay width, Γ_γ, by the formula:

$$\Gamma_\gamma = 2\pi \int |A_M^J(\theta,\phi)|^2 \frac{k^2 dk d\Omega}{(2\pi)^3} \frac{M_N}{E_N} \delta(k+E_N-M_R)$$

$$= \frac{k^2}{4\pi} \frac{M_N}{M_R} \frac{8}{2J+1} \left\{ |A_{\frac{3}{2}}|^2 + |A_{\frac{1}{2}}|^2 \right\} \qquad (6\text{-}116)$$

where M_N and M_R are the nucleon and resonance masses respectively. Ignoring nonresonant background within the partial wave of the resonance, the contribution of the resonance to the integrated single pion photoproduction cross section is

$$\sigma_T = (2J+1) \frac{\pi}{k^2} \frac{\Gamma_\gamma \Gamma_\pi}{\Gamma^2}$$

$$= \frac{M_N}{M_R} \frac{\Gamma_\pi}{\Gamma} \frac{2}{\Gamma} \left\{ |A_{\frac{3}{2}}|^2 + |A_{\frac{1}{2}}|^2 \right\} \qquad (6\text{-}117)$$

where Γ_π is the partial width for the resonance decay into πN. Equation (6-117) is the form appropriate to photoproduction of pions of all charges through excitation of a resonance of given J^P, I. For photoproduction of a pion of a particular type we must multiply eq. (6-117) by an isospin factor; for example for photoproduction of π^0 the factor is $\frac{1}{3}$ or $\frac{2}{3}$ for an isospin state of $I = \frac{3}{2}$ or $\frac{1}{2}$ respectively.

We can connect the above resonance photoexcitation amplitudes with the general formalism for pion photoproduction outlined in Chapter 2, by using the formulae given in Appendix B.

REFERENCES

V. E. Barnes, et al., Phys. Rev. Letters 12, 204 (1964).

M. A. Beg, B. W. Lee and A. Pais, Phys. Rev. Letters 13, 514 (1964).

C. Becchi and G. Morpurgo, Phys. Letters 17, 352 (1965).

A. Y. Chau, N. Dombey and R. G. Moorhouse, Phys. Rev. 163, 1632 (1967).

S. Coleman and S. L. Glashow, Phys. Rev. Letters 6, 423 (1961).

L. A. Copley, G. Karl and E. Obryk, Nucl. Phys. B13, 303; Phys. Letters 29B, 117 (1969).

R. H. Dalitz, Topical Conference on Meson Spectroscopy, University of Pennsylvania (1965a).

R. H. Dalitz, Les Houches Summer School, in *High Energy Physics* (Gordon and Breach, New York, 1966). p. 253 (1965b).

R. H. Dalitz, Proc. Oxford Int. Conf. on Elementary Particles (Rutherford High Energy Laboratory, Chilton, January, 1966).

R. H. Dalitz, Proc. XIII Int. Conf. on High Energy Physics (University of California Press, Berkeley, 1967).

R. H. Dalitz, In Proc. Second Hawaii Topical Conference in Particle Physics (University of Hawaii Press, 1968), p. 327.

R. H. Dalitz and D. G. Sutherland, Phys. Rev. 146, 1180 (1966).

R. F. Dashen and D. H. Sharp, Phys. Rev. 133, 1585 (1964).

J. J. de Swart, Rev. Mod. Phys. 35, 916 (1963); this paper is also reprinted in the *The Eightfold Way*, M. Gell-Mann and Y. Ne'eman (Benjamin, New York, 1964).

D. Dorfan, J. Eades, L. Lederman, W. Lee and C. Ting, Phys. Rev. Letters 14, 999 (1965).

D. Faiman and A. W. Hendry, Phys. Rev. 173, 1720 (1968).

R. P. Feynman, M. Kislinger and F. Ravndal, Phys. Rev. 103, 2706 (1971).

M. Gell-Mann, California Institute of Technology Report CTSL-20 (1961); this paper is also reprinted in the *The Eightfold Way*, M. Gell-Mann and Y. Ne'eman (Benjamin, New York, 1964).

M. Gell-Mann, Proc. Int. Conf. High Energy Phys. (CERN, 1962), p. 805.

S. L. Glashow, Phys. Rev. Letters 11, 48 (1963).

M. Gourdin and Ph. Salin, Nuovo Cimento 27, 193 (1962).

H. S. Green, Phys. Rev. 90, 270 (1953).

O. W. Greenberg, Phys. Rev. Letters 13, 598 (1964).

O. W. Greenberg and A. Messiah, Phys. Rev. 138, B1155 (1965).

O. W. Greenberg and S. Resnikoff, Phys. Rev. 172, 1850 (1968).

F. Gursey and L. A. Radicati, Phys. Rev. Letters 13, 173 (1964).

M. Hammermesh, *Group Theory* (Addison Wesley, London, 1962).

H. Harari, Proc. Vienna Conf. on High Energy Physics (1968) CERN (1969).

R. Levi-Setti, Proc. Lund Int. Conf. on High Energy Physics (1969).

H. Lipkin and S. Meshkov, Phys. Rev. Letters 14, 670 (1965).

H. Lipkin and S. Meshkov, Phys. Rev. 143, 1269 (1966).

A. N. Mitra, Nuovo Cimento A 56, 1164 (1968).

A. N. Mitra and M. Ross, Phys. Rev. 158, 1630 (1967).

R. G. Moorhouse, Phys. Rev. Letters 16, 772 (1966).

R. G. Moorhouse, Proc. of XVI Inter. Conf. on High Energy Physics (ed. A. Roberts, National Accelerator Laboratory, 1973).

S. Okubo, Prog. Theoret. Phys. (Kyoto) 27, 949 (1962).

Particle Data Group, Phys. Letters, April 1972.

F. Ravndal, Phys. Rev. D4 (1971).

S. Sakata, Prog. Theor. Phys. (Kyoto) 16, 686 (1956).

B. Sakita, Phys. Rev. 136, B1756 (1964).

R. D. Tripp, Proc. XIV Int. Conf. on High Energy Physics (CERN, Geneva, 1968), p. 173.

R. L. Walker, Phys. Rev. 182, 1729 (1969a).

R. L. Walker, 4th International Symposium on Electron and Photon Interactions at High Energies, Daresbury, 1969 (Science Research Council, Daresbury, 1969b).

H. Weyl, *The Classical Groups* (Princeton University Press, Princeton, 1946).

CHAPTER 7

CURRENT ALGEBRA AND SUM RULES

In the last chapter we were concerned with particles and resonances as multiplets (that is, bases of representations) of symmetry groups SU2, SU3, or SU6 and with their interactions as approximately invariant under SU2, SU3, or SU6. However these symmetries may also play a part in elementary particle physics not as a group, but as an algebra of equal-time commutations of *currents* (Gell-Mann, 1962). When one sandwiches such a relation between state vectors, and inserts a complete sum over states between the currents or other operators of the commutator, one obtains a *sum rule* that may enable one to utilize the *current algebra*. In evaluation of a sum rule the approximation known usually as the partially conserved axial current (PCAC) and sometimes as pole dominance of the divergence of the axial vector current (PDDAVC), which relates the divergence of the axial vector current to the pion field, is often used. Other techniques in current algebra sum rules may give relations like $\int \text{Im } T(\nu) \, d\nu = 0$, where $T(\nu)$ is, for example, a scattering amplitude and ν an energy variable. Relations like these, known as *superconvergence relations*, may be formulated independently of current algebra, being a consequence of analyticity and dispersion relations in $T(\nu)$, together with a superconvergence condition such as $\nu T(\nu) \to 0$ as $\nu \to \infty$.

Sum rules and superconvergence relations have a wide range of applications. Our interests here are in the narrower field of those involving the pion-nucleon interaction, particularly where the assumptions of current algebra and PCAC predicate quantities or relations in the pion-nucleon interactions. The Born approximation, with a certain derivative coupling of the pion field, gives results equivalent to PCAC. In section 7.1 we derive this Born approximation, since it is a simple view of some of the results that can be obtained in a more complicated way. In subsequent sections we discuss PCAC and current algebra, leading to the derivation of the s-wave pion-nucleon scattering lengths. [We refer generally, for this chapter, to Adler and Dashen (1968).]

7.1. BORN APPROXIMATION

In the Heisenberg picture in which the field operations vary and the state vectors remain constant, the pion and nucleon fields $\phi_i(x)$ and $\psi(x)$ respectively obey the equations:

$$(\Box + M_\pi^2)\, \phi_i(x) = j_i(x) \tag{7-1}$$

$$\left(i\gamma^\mu \frac{\partial}{\partial x^\mu} - M_N\right) \psi(x) = X(x) \tag{7-2}$$

where the source functions $j_i(x)$ and $X(x)$ are functions of ϕ_i and ψ. In eq. (7-1) $i = 1, 2, 3$ is the isospin index of the pion field; in eq. (7-2) $\psi = \begin{pmatrix} \psi_1 \\ \psi_2 \end{pmatrix}$, where ψ_1 is the four-component spinor of the proton field and ψ_2 is the 4-component spinor of the neutron field. The condition that j_i and X are functions of ϕ_i and ψ, together with isospin invariance, Lorentz invariance with space reversal invariance (parity conservation), limits the possible forms of j_i and X. The simplest form is:

$$j_i(x) = ig\, \bar{\psi}(x)\, \gamma_5\, \tau_i\, \psi(x) + \text{C.T.} \tag{7-3}$$

$$X(x) = ig\, \gamma_5\, \tau_i\, \psi(x)\, \phi_i(x) + \text{C.T.} \tag{7-4}$$

7.1. BORN APPROXIMATION

where C.T. denotes possible counter terms which operate when calculations are made in non-lowest-order perturbation theory, essentially to preserve the definitions of the coupling constant g and the masses implicit in eqs. (7-1), (7-2) and the first terms of eqs. (7-3), (7-4). We will use the lowest-order perturbation theory so that they will not enter. (In any case they have a formal rather than an operational significance, since higher-order perturbation calculations are mostly easily performed using rules for S-matrix power series expansion in g in which the counter terms do not enter explicitly.) The coupling constant g enters in accord with the Watson-Lepore definition, namely:

$$<p_2|j_i(0)|p_1> = i\, g(s)\, \frac{M_N}{(2\pi)^3}\, \bar{u}(p_2)\, \gamma_5\, \tau_i\, u(p_1)$$

$$g \equiv g(M_\pi^2) \tag{7-5}$$

where $s = (p_1-p_2)^2$. The first terms of eqs. (7-3), (7-4) are derivable from an interaction part of the Lagrangian density given by:

$$\mathcal{L}_I(x) = ig\, \bar{\psi}(x)\, \gamma_5\, \tau_i\, \psi(x)\, \phi_i(x)\ . \tag{7-6}$$

We will now evaluate to lowest order in g, the pion-nucleon scattering S-matrix, or equivalently the A and B amplitudes. Necessary definitions are:

$$S_{FI} \equiv <p_2\, q_2^{(\text{out})}|p_1\, q_1^{(\text{in})}> \tag{7-7}$$

$$S_{FI} = \delta_{FI} + 2i\, \delta(p_1+q_1-p_2-q_2)\, T \tag{7-8}$$

$$T = \frac{M_N}{16\pi^2}\, \bar{u}(p_2) \left[A + \frac{1}{2}(\slashed{q}_1+\slashed{q}_2)\, B\right] u(p_1)\ . \tag{7-9}$$

In eqs. (7-7)-(7-9) the 4-momenta of the initial and final nucleon are p_1 and p_2 respectively, and of the initial and final pion q_1 and q_2 respectively.

In eq. (7-7) we apply the usual reduction technique (Bjorken and Drell, 1965) to the reduction of the pions obtaining:

$$S_{FI} = \delta_{FI} + \frac{i^2}{2(2\pi)^3} \iint d^4x\, d^4y <p_2|T\{j_b(x)\, j_a(y)\}|p_1> \exp(+iq_2 x - iq_1 y)$$
(7-10)

where a and b are the isospin indices of the initial and final pions respectively. In eq. (7-10) we substitute the first term of eq. (7-3) for the pion source:

$$S_{FI} = \delta_{FI} + \frac{g^2}{2(2\pi)^3} \iint d^4x\, d^4y <p_2|T\{\bar{\psi}(x)\, \gamma_5\, \tau_b\, \psi(x)\, \bar{\psi}(y)\, \gamma_5\, \tau_a\, \psi(y)\}|p_1>$$

$$\exp(+iq_2 x - iq_1 y).$$
(7-11)

To obtain the result to lowest order in g^2 it is sufficient to evaluate the integral to zero order in g. This is equivalent to treating the nucleon fields and state vectors as those of free particles and gives:

$$S_{FI} = \delta_{FI}$$

$$+ \frac{g^2 M_N}{2(2\pi)^6} \iint d^4x\, d^4y\, \bar{u}(p_2) <0|[\tau_b\, \gamma_5\, T\{\psi(x)\, \bar{\psi}(y)\}\, \tau_a\, \gamma_5\, \exp(+ip_2 x - ip_1 y)$$

$$+ \tau_a\, \gamma_5\, T\{\psi(y)\, \bar{\psi}(x)\}\, \tau_b\, \gamma_5\, \exp(+ip_2 y - ip_1 x)]|0> u(p_1)\, \exp(+iq_2 x - iq_1 y)$$

$$= \delta_{FI} - \frac{g^2 M_N}{4(2\pi)^6} \iint d^4x\, d^4y\, \bar{u}(p_2) <0|[\tau_b\, \gamma_5\, S_F(x-y)\, \tau_a\, \gamma_5\, \exp(+ip_2 x - ip_1 y)$$

$$+ \tau_a\, \gamma_5\, S_F(y-x)\, \tau_b\, \gamma_5\, \exp(+ip_2 y - ip_1 x)]|0> u(p_1)\, \exp(+iq_2 x - iq_1 y)$$
(7-12)

where S_F is the free field, otherwise zero order, nucleon Feynman propagator given by:

$$S_F(x) = -2i \int d^4p\, \frac{\not{p} + M_N}{p^2 - M_N^2 + i\epsilon}\, \frac{e^{-ipx}}{(2\pi)^4}.$$
(7-13)

Substitution of eq. (7-13) into eq. (7-12), performing the integrations, and using the definitions $s = (p_1 + q_1)^2$, $u = (p_1 - q_2)^2$ yields:

7.1. BORN APPROXIMATION

$$S_{FI} = \delta_{FI} + 2i\,\delta(p_2+q_2-p_1-q_1)\,\frac{g^2 M_N}{16\pi^2}\,\bar{u}(p_2)\left[\tau_b\gamma_5\,\frac{(\not{p}_1+\not{q}_1+M_N)}{s-M_N^2}\,\tau_a\gamma_5\right.$$

$$\left. + \tau_a\gamma_5\,\frac{(\not{p}_1-\not{q}_2+M_N)}{u-M_N^2}\,\tau_b\gamma_5\right]u(p_1)\,. \tag{7-14}$$

On using $(\not{p}_1-M_N)u(p_1) = 0$, $\bar{u}(p_2)(\not{p}_2-M_N) = 0$, this reduces to:

$$S_{FI} = \delta_{FI} + 2i\,\delta(p_2+q_2-p_1-q_1)\,\frac{g^2 M_N}{16\pi^2}\,\bar{u}(p_2)\tfrac{1}{2}(\not{q}_1+\not{q}_2)\left(\frac{\tau_b\tau_a}{M_N^2-s} - \frac{\tau_a\tau_b}{M_N^2-u}\right)u(p_1)\,.$$

$$\tag{7-15}$$

Comparison with eqs. (7-8)-(7-9) gives:

$$A = 0 \tag{7-16a}$$

$$B = g^2\left(\frac{\tau_b\tau_a}{M_N^2-s} - \frac{\tau_a\tau_b}{M_N^2-u}\right). \tag{7-16b}$$

The (+) and (−) amplitudes are defined by:

$$B = B^{(+)}\tfrac{1}{2}[\tau_a,\tau_b]_+ + B^{(-)}\tfrac{1}{2}[\tau_b,\tau_a] = B^{(+)}\delta_{ab} + B^{(-)}\tfrac{1}{2}[\tau_b,\tau_a] \tag{7-17}$$

and similarly for A. Equation (7-16) yields:

$$A^{(+)} = A^{(-)} = 0\,. \tag{7-18a}$$

$$B^{(+)} = g^2\left(\frac{1}{M_N^2-s} - \frac{1}{M_N^2-u}\right),\quad B^{(-)} = g^2\left(\frac{1}{M_N^2-s} + \frac{1}{M_N^2-u}\right). \tag{7-18b}$$

We recognize in eq. (7-18) the single particle contributions to the dispersion relations, eq. (3-67) for the $A^{(\pm)}$ and $B^{(\pm)}$ amplitudes. We see that the definition, eq. (7-5), of the coupling constant g is equivalent to the definition of g^2 as the residue of the pole at $s = M_N^2$ of the analytic function $B^{(+)}(s,t)$ or $B^{(-)}(s,t)$. There are no contributions, other than the Born term, eq. (7-18b), to these pole terms. The corresponding Feynman diagrams are illustrated in Fig. 7-1; Fig. 7-1(a) corresponds to the pole terms at $s = M_N^2$, and Fig. 7-1(b) to the pole terms at $u = M_N^2$.

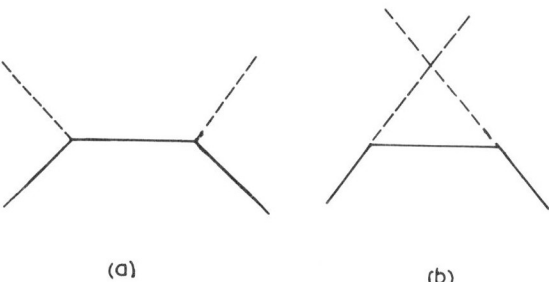

Fig. 7-1. The Born amplitudes in pion-nucleon scattering.

We may ask whether these pole terms approximate to the observed low energy scattering. For this comparison we wish to find the contribution of eq. (7-18) to the s-wave scattering lengths, which is found by evaluation at $t = 0$, $s = (M_N+M_\pi)^2$, $u = (M_N-M_\pi)^2$, giving:

$$B^+(s,t)\Big|_{\substack{s=(M_N+M_\pi)^2 \\ t=0}} = -\frac{g^2}{M_N M_\pi}\left(1+O\left(\frac{M_\pi^2}{M_N^2}\right)\right)$$

$$B^-(s,t)\Big|_{\substack{s=(M_N+M_\pi)^2 \\ t=0}} = \frac{g^2}{2 M_N^2}\left(1+O\left(\frac{M_\pi^2}{M_N^2}\right)\right) \qquad (7\text{-}19)$$

From formulae given below (eq. (7-98) ff.) the s-wave scattering lengths are given by:

$$\tfrac{1}{3}(a_1-a_3) = \frac{1}{4\pi\left(1+\frac{M_\pi}{M_N}\right)}(A^{(-)}(s,t)+M_\pi B^{(-)}(s,t))\Big|_{\substack{s=(M_N+M_\pi)^2 \\ t=0}}$$

$$= \frac{g^2}{4\pi}\frac{M_\pi}{2 M_N^2\left(1+\frac{M_\pi}{M_N}\right)} \qquad (7\text{-}20a)$$

7.1. BORN APPROXIMATION

$$\tfrac{1}{3}(a_1+2a_3) = \frac{1}{4\pi\left(1+\frac{M_\pi}{M_N}\right)}(A^{(+)}(s,t)+M_\pi B^{(+)}(s,t))\Big|_{\substack{s=(M_N+M_\pi)^2\\t=0}}$$

$$= -\frac{g^2}{4\pi}\frac{1}{(M_N+M_\pi)} \quad . \tag{7-20b}$$

Since experimentally $a_1 - a_3 \simeq 0.3\, M_\pi^{-1}$ and $a_1 + 2a_3 < 0.1\, M_\pi^{-1}$, the agreement is reasonable for the (−) amplitudes (isospin 1 exchange) but bad for the (+) amplitudes (isospin 0 exchange). It is interesting to see whether the use of the next most simple interaction, the derivative interaction, improves the above results. We replace eq. (7-6) by the following interaction part of the Lagrangian density:

$$\mathcal{L}_I(x) = \sqrt{4\pi}\,\frac{f}{M_\pi}\,\bar{\psi}(x)\,\gamma_5\,\gamma_\mu\,\tau_i\,\psi(x)\,\frac{\partial\phi_i(x)}{\partial x_\mu} \quad . \tag{7-21}$$

The use of the new $\mathcal{L}_I(x)$ leads to extra factors $-iq_1{}^\mu\gamma_\mu$ and $iq_2{}^\mu\gamma_\mu$, so that eq. (7-12) becomes:

$$S_{FI} = \delta_{FI}$$
$$+ 4\pi\,\frac{f^2}{M_\pi^2}\,\frac{M_N}{4(2\pi)^6}\iint d^4x\, d^4y\, \bar{u}(p_2)[\tau_b\gamma_5\slashed{q}_2\, S_F(x-y)\tau_a\gamma_5\slashed{q}_1\exp(+ip_2 x - i p_1 y)$$
$$+ \tau_a\gamma_5\slashed{q}_1\, S_F(y-x)\tau_b\gamma_5\slashed{q}_2\exp(+ip_2 y - i p_1 y)]\, u(p_1)\exp(+iq_2 x - i q_1 y)$$

$$= \delta_{FI} - 2i\,\delta(p_2+q_2-p_1-q_1)\,\frac{f^2}{M_\pi^2}\,4\pi\,\frac{M_N}{16\pi^2}\,\bar{u}(p_2)\Bigg[\tau_b\slashed{q}_2\,\frac{(\slashed{p}_1+\slashed{q}_1-M_N)}{s-M_N^2}\,\tau_a\slashed{q}_1$$
$$+ \tau_a\slashed{q}_1\left(\frac{\slashed{p}_1-\slashed{q}_2-M_N}{u-M_N^2}\right)\slashed{q}_2\,\tau_b\Bigg]u(p_1) \quad . \tag{7-22}$$

Using $(\slashed{p}_1-M_N)u(p_1) = 0 = \bar{u}(p_2)(\slashed{p}_2-M_N)$ and the identity:

$$\slashed{A}\slashed{B} + \slashed{B}\slashed{A} = 2AB \tag{7-23}$$

leads to:

$$S_{FI} = \delta_{FI} + 2i\,\delta(p_2+q_2-p_1-q_1)\,\frac{f^2}{M_\pi^2}\,4\pi\,\frac{M_N}{16\pi^2}\,\bar{u}(p_2)\left[2M_N[\tau_a,\tau_b]_+ \right.$$

$$\left. + \frac{1}{2}(\slashed{q}_1+\slashed{q}_2)\left\{4M_N^2\left(\frac{\tau_b\tau_a}{M_N^2-s} - \frac{\tau_a\tau_b}{M_N^2-u}\right) + \tau_a\tau_b - \tau_b\tau_a\right\}\right]u(p_1).$$

(7-24)

Comparison of eq. (7-24) with eqs. (7-8)-(7-9) yields:

$$A^{(+)} = 4\pi\,\frac{f^2}{M_\pi^2}\,4M_N, \quad A^{(-)} = 0 \tag{7-25a}$$

$$B^{(+)} = 4\pi\,\frac{f^2}{M_\pi^2}\,4M_N^2\left(\frac{1}{M_N^2-s} - \frac{1}{M_N^2-u}\right)$$

$$B^{(-)} = 4\pi\,\frac{f^2}{M_\pi^2}\left\{4M_N^2\left(\frac{1}{M_N^2-s} + \frac{1}{M_N^2-u}\right) - 2\right\}. \tag{7-25b}$$

Defining the coupling constants by the pole residues requires, as is seen on comparison of eqs. (7-25b) and (7-18b):

$$f^2 \equiv \frac{g^2}{4\pi}\left(\frac{M_\pi}{2M_N}\right)^2. \tag{7-26}$$

The pion-nucleon coupling constant can be determined by forward dispersion relations, as discussed in Chapter 3, to within about 6%. Convenient round numbers, not exactly related by eq. (7-26), are:

$$f^2 = 0.08,\quad \frac{g^2}{4\pi} = 15. \tag{7-27}$$

Equations (7-25) can be expressed in terms of g^2 as:

$$A^{(+)} = \frac{g^2}{M_N},\quad A^{(-)} = 0 \tag{7-28a}$$

$$B^{(+)} = g^2\left(\frac{1}{M_N^2-s} - \frac{1}{M_N^2-u}\right),\quad B^{(-)} = g^2\left(\frac{1}{M_N^2-s} - \frac{1}{M_N^2-u}\right) - \frac{g^2}{2M_N^2}.$$

(7-28b)

7.1. BORN APPROXIMATION

As discussed in section 7.3 below, the value of $A^{(+)}$ is verified from dispersion relations and for small momentum pions to 10%. We will see that eq. (7-28a) (with one pion of zero mass, and both pion momenta tending to zero) can be derived using PCAC; it is then known as the *Adler consistency condition*. Equation (7-28) also gives:

$$(A^{(+)}(s,t) + M_\pi B^{(+)}(s,t))\big|_{\substack{s=(M_N+M_\pi)^2 \\ t=0}} \simeq 0 \ .$$

Using (7-20b) above we see that this accords with the experimental result:

$$a_1 + 2a_3 = 0 \ .$$

On the other hand we get from the (−) amplitudes of eq. (7-20b) that $a_1 - a_3 = 0$, which is not in accord with experiment. If we add a term:

$$\mathcal{L}_I = -\left(\frac{f}{M_\pi}\right)^2 \bar{\psi}\gamma^\mu \vec{\tau} \cdot \vec{\phi} \times \frac{\partial \vec{\phi}}{\partial x^\mu} \psi \tag{7-29}$$

to the Lagrangian density, then the (+) result is unaltered but the (−) result becomes:

$$a_1 - a_3 = \frac{g^2}{4\pi} \frac{3M_\pi}{2M_N^2 \left(1 + \frac{M_\pi}{M_N}\right)}$$

$$\simeq 0.36 \, M_\pi^{-1} \tag{7-30}$$

in approximate agreement with experiment.

[The total interaction Lagrangian, eq. (7-21) + eq. (7-29), can be obtained by a transformation of the pseudoscalar interaction Lagrangian in which the interaction term is eq. (7-6), with a similar transformation to that given by Schweber (1955). There is an additional interaction term which results from this transformation whose inclusion would restore the results, eq. (7-18).]

7.2. MESONS AND CURRENTS

The hadronic currents with which elementary particle physics is commonly concerned are the electromagnetic current and the weak interaction current(s). In this book we have encountered the electromagnetic current, $j_\mu(x)$, as a superposition of an isovector part and an isoscalar part, and in the last chapter (see for instance section 6.8) we discussed the reasons for regarding it as the superposition:

$$j_\mu(x) = j_{3,\mu}(x) + \frac{1}{\sqrt{3}} j_{8,\mu}(x) \qquad (7\text{-}31)$$

where $j_{3,\mu}$ and $j_{8,\mu}$ transform as the third and eighth components respectively of an octet and are, of course, the third components of an isovector, and an isoscalar term. The electromagnetic current is conserved:

$$\frac{\partial j_\mu(x)}{\partial x_\mu} = 0 \qquad (7\text{-}32)$$

which leads to zero renormalization of the charge by strong interactions, and shows why the charge on the proton need not be different from that on the electron, despite the fact that the proton is subject immediately to the strong interactions while the electron is not (see Bernstein, 1968).

The weak interaction hadronic current consists of a vector (negative parity) and an axial vector (positive parity) part (for an account of weak interactions see Marshak, Ryan and Riazzudin, 1969):

$$J_\mu(x) = J_\mu^V(x) + J_\mu^A(x) \ . \qquad (7\text{-}33)$$

More exactly one defines currents $J_{i,\mu}^V(x)$, $J_{i,\mu}^A(x)$ $i = 1,\ldots,8$ which transform as an octet, but only some components of the octet appear in the weak interactions. There is no evidence for neutral currents in the weak interaction, and it seems that the current appears as:

$$J_\mu^{V,A}(x) = \left(J_{1,\mu}^{V,A}(x) \pm i J_{2,\mu}^{V,A}(x)\right) \cos\theta_c + \left(J_{4,\mu}^{V,A} \pm i J_{5,\mu}^{V,A}\right) \sin\theta_c \qquad (7\text{-}34)$$

7.2. MESONS AND CURRENTS

where θ_C is the Cabibbo angle and is of magnitude $\sin\theta_C \simeq 0.21$. Terms $J_4 \pm i J_5$ are strangeness changing charge raising and lowering operators (as may easily be seen by reference to the last chapter and considering $\lambda_4 \pm i \lambda_5$ acting on a quark triplet) and $J_1 \pm i J_2$ are non-strangeness changing charge raising and lowering operators. These latter give for instance, the neutron-proton transitions of a β-decay.

The leptonic parts of the weak current J_μ^L is coupled to itself by a coupling constant G and to the hadronic vector part by G_V so that:

$$H_{weak} = \frac{G}{\sqrt{2}} J_\mu^L J^{L\mu\dagger} + \ldots \frac{G_V}{\sqrt{2}} J_\mu^V J^{L\mu\dagger} + \ldots + h.c.$$

Experiment (on comparison of μ-decay with β-decay) shows that G_V and G are equal to within about 2% leading, since $\cos\theta_C \simeq 1$, to the hypothesis of no renormalization by the strong interactions of the non-strangeness changing vector current $J_{1,\mu}^V(x) \pm i J_{2,\mu}^V(x)$. This leads to the identification of $J_{1,\mu}^V(x)$, $J_{2,\mu}^V(x)$ with the first two components of *conserved* isospin current[1] so that $\frac{\partial}{\partial x_\mu}\{J_{1,\mu}(x) \pm i J_{2,\mu}(x)\} = 0$. Also, because the $J_{i,\mu}^V(x)$ $i = 1, \ldots, 8$ are postulated as an octet, $J_{3,\mu}^V(x)$ is the third component of isospin current, and $J_{8,\mu}^V(x)$ is the hypercharge current. The latter current is conserved since hypercharge (as well as isospin) is conserved in strong interactions. So one has:

$$\frac{\partial J_{i,\mu}^V(x)}{\partial x_\mu} = 0 \qquad i = 1, 2, 3, 8 \; . \tag{7-35}$$

---

[1] Since isospin is conserved in strong interactions, there exists an isospin current $\vec{J}_\mu(x)$ obeying $\frac{\partial \vec{J}_\mu}{\partial x_\mu} = 0$. The term $\vec{J}_\mu$ is a vector in isospin space; for example, in a world of pions and nucleons only, the isospin current is:

$$\vec{J}_\mu(x) = \{\overline{\psi}(x) \gamma_\mu \frac{\vec{\tau}}{2} \psi(x) + \vec{\phi}(x) \times \frac{\partial}{\partial x^\mu} \vec{\phi}(x)\}$$

and the isospin operator is:

$$\vec{I} = \int d^3x \, \vec{J}_0(x) \; .$$

Since the electromagnetic current is conserved, $j_{3,\mu}$ and $j_{8,\mu}$ are a third component of the isospin current and the hypercharge current respectively, so that:

$$j_{3,\mu}(x) \equiv J_{3,\mu}^{V}(x), \quad j_{8,\mu}(x) \equiv J_{8,\mu}^{V}(x) \ .$$

If SU3 were a perfect symmetry, $J_{i,\mu}^{V}(x)$ $i = 4, 5, 6, 7$ would be conserved, but SU3 breaking, isospin conserving, strong interactions lead to nonconservation of these currents. Of course eq. (7-35) only holds in the limit of isospin invariance, which we know is broken (by electromagnetic forces) to a few percent in the strong interactions. The identification of the non-strangeness changing weak vector current with the isospin current, with the consequence eq. (7-5) is known as the *conserved vector current* hypothesis or *C.V.C.* (Feynman and Gell-Mann, 1958).

Partially Conserved Axial Current and the Goldberger-Treiman Relation

If we consider the β-decay of neutron into proton, then a model of the hadronic current is:

$$\bar{\psi}(x) \gamma_\mu (1+g_A \gamma_5) \frac{\tau_1 \pm i \tau_2}{2} \psi(x) \tag{7-36}$$

where $\psi(x)$ is the nucleon spinor function (with 2×4 components) $g_A = G_A/G_V$, with G_A the axial vector and G_V the vector coupling constants. We can extract from this a model of the axial vector current as $g_A \bar{\psi}(x) \gamma_\mu \gamma_5 \frac{\tau_i}{2} \psi(x)$. It is evident that, under the invariant transformations of strong interactions (Lorentz and isospin transformations and P, C, and T), $\frac{\partial}{\partial x_\mu} \bar{\psi}(x) \gamma_\mu \gamma_5 \tau_i \psi(x)$ transforms as the pion field $\phi_i(x)$. Since the model and the original have the same strong interaction transformation properties, the divergence of the axial vector current transforms like the pion field:

$$\frac{\partial}{\partial x_\mu} J_{i,\mu}^{A}(x) \sim \phi_i(x) \ . \tag{7-37}$$

7.2. MESONS AND CURRENTS

We are, accordingly, at liberty to define the pion interpolating field, apart from a normalization constant, to be the divergence of the axial vector current:

$$\frac{\partial}{\partial x_\mu} J_{i,\mu}^A(x) = c \phi_i(x) \ . \tag{7-38a}$$

The parity and Lorentz transformation properties of J_μ^A show that:

$$<P, p_2 | J_{1,\mu}^A(0) + i J_{2,\mu}^A(0) | N, p_1>$$

$$= \frac{M_N}{(2\pi)^3} \bar{u}_P(p_2) \left(\gamma_\mu g_A(q^2) + q_\mu h_A(q^2) \right) \gamma_5 u_N(p_1) \tag{7-39}$$

where p, p_2 and N, p_1 denote proton and neutron states of momenta p_2 and p_1 respectively, and $q = p_2 - p_1$. The nucleon axial-vector coupling constant is then *defined* as:

$$g_A = g_A(M_\pi^2) \tag{7-40}$$

while the pion-nucleon coupling constant is defined through the pion source $j_i(x)$:

$$<p_2 | j_i(0) | p_1> = \frac{i M_N}{(2\pi)^3} \bar{u}(p_2) \gamma_5 \tau_i u(p_1) g(q^2)$$

$$g \equiv g(0) \ . \tag{7-41}$$

Since $j_i(x) \equiv (\Box + M_\pi^2) \phi_i(x)$, eq. (7-41) gives:

$$<p_2 | \phi_i(0) | p_1> = \frac{i M_N}{(2\pi)^3} \frac{g(q^2)}{M_\pi^2 - q^2} \bar{u}(p_2) \gamma_5 \tau_i u(p_1) \ . \tag{7-42}$$

Equations (7-38) and (7-39) give:

$$c <p_2 | \phi_1 + i \phi_2 | p_1> = \frac{i M_N}{(2\pi)^3} \bar{u}_N(p_2) \left(2M g_A(q^2) + q^2 h_A(q^2) \right) \gamma_5 u_N(p_1) \tag{7-43}$$

and letting $q \to 0$ and comparing with eq. (7-42), we find:

$$c = \frac{M_N M_\pi^2 g_A}{g(0)} . \tag{7-44}$$

We can now find c independently, using eq. (7-38), by relating it to the pion decay coupling constant. The decay $\pi^+ \to \mu^+ + \nu_\mu$ is mediated by the weak interaction $\frac{G}{\sqrt{2}} J_\mu^L(x) J_\mu^\dagger(x) + \text{h.c.}$, the matrix element of the hadronic vector current vanishing. The constant f_π is defined at $q^2 = M_\pi^2$ by:

$$\sqrt{2} <\pi^+, q | J_{1,\mu}^A(x) + i J_{2,\mu}^A(x) | 0> = \frac{iq_\mu f_\pi}{M_\pi^2} (2\pi)^{-\frac{3}{2}} e^{iqx} \tag{7-45}$$

and one then finds the decay width for $\pi^+ \to \mu^+ + \nu_\mu$ as:

$$\Gamma_\pi = \frac{f_\pi^2}{8\pi} G^2 \cos^2\theta_c \frac{M_\mu^2}{M_\pi^2} \left(1 - \frac{M_\mu^2}{M_\pi^2}\right)^2 . \tag{7-46}$$

However operating on eq. (7-45) with $\frac{\partial}{\partial x_\mu}$ gives:

$$c \sqrt{2} <\pi^+, q | \phi_1(0) + i \phi_2(0) | 0> = \frac{f_\pi}{(2\pi)^{\frac{3}{2}}}$$

or:

$$c = f_\pi / \sqrt{2} . \tag{7-47}$$

The two expressions for c, eqs. (7-44) and (7-47), give:

$$f_\pi = \frac{\sqrt{2} M_N M_\pi^2 g_A}{g(0)} . \tag{7-48}$$

If we postulate that the matrix element eq. (7-41) is a slowly varying function of q^2 so that $g(0) \simeq g(M_\pi^2)$, we obtain the *Goldberger-Treiman* relation:

$$f_\pi \simeq \frac{\sqrt{2} M_N M_\pi^2 g_A}{g} . \tag{7-49}$$

7.3. SOFT PIONS

Equation (7-46) and measurement of Γ_π gives the experimental value of $f_\pi = 0.96\, M_\pi^3$, while the right-hand side of eq. (7-49) is $0.83\, M_\pi^3$.

The partially conserved axial vector current hypothesis (PCAC) consists in the definition eq. (7-38) (not in itself having physical content) *together with* the hypothesis of slow variation of matrix elements of the pion current [such as eq. (7-41)] in the vicinity of $q^2 = M_\pi^2$. As we have just seen, PCAC is used to derive the Goldberger-Treiman relation. Equation (7-38a) is sometimes quoted in the form:

$$\frac{\partial}{\partial x_\mu} J_{i,\mu}^A(x) = \frac{M_N M_\pi^2 g_A}{g} \phi_i(x) \ . \tag{7-38b}$$

Another way of deriving the Goldberger-Treiman relation is to postulate that the matrix element $<p_2|\frac{\partial}{\partial x_\mu} J_{i,\mu}(0)|p_1>$ satisfies an unsubtracted dispersion relation in q^2, which in the interval $0 < q^2 < M_\pi^2$ is pion pole dominated. From this alternative viewpoint, the hypothesis is sometimes known as pion pole dominance of the divergence of the axial vector current: PDDAVC.

The Goldberger-Treiman relations is correct to within 10% and this can be taken as a first estimate of the error in PCAC in other situations where it is used.

7.3. SOFT PIONS

In the next section (7.4) we shall develop current algebra and use it in section 7.5 to derive s-wave pion-nucleon scattering lengths in the approximation of zero 4-momentum (implying zero mass) pions. First we must derive in this section rules which relate the matrix element for any strong interaction process with the matrix element for the corresponding process in which an additional zero-mass, zero-energy pion is emitted or absorbed.

Pions tending to this zero-mass, zero-energy limit are known as *soft pions*. The matrix elements thus found for soft pions are used to approximate matrix elements for real pions of small 4-momentum.

Let a be any incoming system of strongly interacting particles, total 4-momentum p_I, which can make a transition to an outgoing system β, total 4-momentum p_F. Let $\mathfrak{M}(a \to \beta)$ be the matrix element for this process defined in terms of the S-matrix element, $S_{\beta a}$, as:

$$S_{\beta a} = \langle \beta(p_F) \text{ out} | a(p_I) \rangle = \delta_{\beta a} + i(2\pi)^4 \delta(p_F - p_I) \mathfrak{M}(a \to \beta) . \qquad (7\text{-}50)$$

We wish to find the modification to $\mathfrak{M}(a \to \beta)$ that gives the matrix element $\mathfrak{M}(a \to \beta + \text{zero-mass, zero-momentum pion})$. We commence the derivation by using eq. (7-38a) to find:

$$q^\lambda \langle \beta(p_F) \text{ out} | J_{i,\lambda}^A(0) | a(p_I) \text{ in} \rangle = i c \langle \beta(p_F) \text{ out} | \phi_i(0) | a(p_I) \text{ in} \rangle$$

$$= i \frac{M_N \, g_A}{g(0)} \frac{M_\pi^2}{M_\pi^2 - q^2} \langle \beta(p_F) \text{ out} | j_i(0) | a(p_I) \text{ in} \rangle \qquad (7\text{-}51)$$

where $q = -p_F + p_I$, and $j_i(0)$ is the source of the pion field ϕ. Now $\lim_{p_F \to p_I} \langle \beta(p_F) \text{ out} | j_i(0) | a(p_I) \text{ in} \rangle$ is the matrix element for $a \to \beta + $ (zero-mass, zero-energy pion) and is in general nonzero [the value of the limit may depend on the direction of $\lim (p_F - p_I)$]. So $\langle \beta(p_F) \text{ out} | J_{i,\lambda}^A(0) | a(p_I) \text{ in} \rangle$ must contain pole terms which go as $1/q$ in order that eq. (7-51) be nonzero in the limit $q \to 0$. In order to find the residue at that pole and thus evaluate the right-hand side of eq. (7-51) we list the ways in which the pole can arise.

The Matrix Elements $\langle \beta(p_F) \text{ out} | J_{i,\lambda}^A(0) | a(p_I) \text{ in} \rangle$

First, no pole can come from attachment to an internal line — that is a line inside the blob of Fig. 7-2(a). For in that case $\langle \beta | J^A | a \rangle$ differs from the corresponding $\langle \beta | a \rangle$ (we have omitted labels from the matrix elements for simplicity) by having two internal propagators instead of one with (in the limit $q \to 0$) the same 4-momentum. The propagator cannot

7.3. SOFT PIONS 301

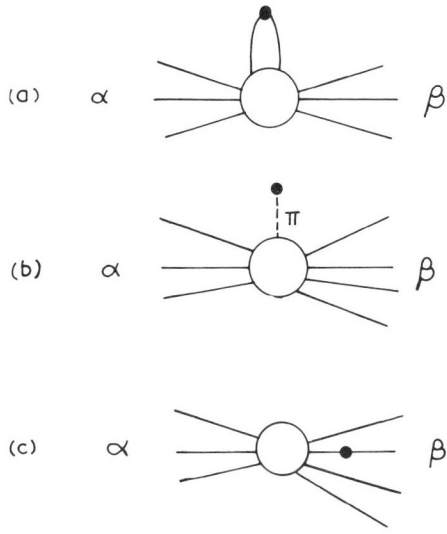

Fig. 7-2. Methods of joining the proper vertex of $J_{i,\lambda}^{A}$, represented by a heavy dot.

give rise to an infinity, otherwise the original matrix element $<a|\beta>$ in which the insertion is made would be singular, and we assume we are not considering processes $a \to \beta$ which have infinite matrix elements. [For further details see Adler (1965b).] Consequently such insertions give rise to terms $q^\lambda <\beta(p_F)$ out $|J_{i,\lambda}^A(0)|a(p_I)$ in$>$ of order q. Another type of vertex insertion is at the end of an external pion line, as shown in Fig. 7-2(b) and gives rise to a contribution to eq. (7-51) equal to:

$$<\beta(p_F) \text{ out } |j_\ell|a(p_I) \text{ in}> \frac{1}{q^2 - M_\pi^2} <\pi_{\ell,q}|q^\lambda J_{i,\lambda}^A(0)|0> .$$

Using eq. (7-38a) as in eq. (7-45) ff. to evaluate the last factor, we obtain:

$$\frac{q^2}{q^2 - M_\pi^2} \frac{M_N \, g_A}{g(0)} <\beta(p_F) \text{ out } |j_i|a(p_I) \text{ in}> . \qquad (7\text{-}52)$$

Equation (7-52) is of order q^2 and may be neglected.

Finally we consider the vertex insertion into an external line, as in Fig. 7-2(c), which does give a contribution. G-parity prevents such an insertion into a pion line or any pseudoscalar meson line, but it is allowed in a nucleon line. Let p be the 4-momentum of the final nucleon line; then the graph with the insertion gives a contribution (including a pion normalization factor of $\frac{1}{\sqrt{2}} (2\pi)^{-\frac{3}{2}}$):

$$\frac{g_A \sqrt{M_N}}{\sqrt{2}\,(2\pi)^3} \bar{u}(p)\, \gamma_\lambda\, \gamma_5 \frac{\tau_i}{2} \frac{1}{\not{p}-\not{q}-M_N}\, \mathfrak{M}^c \sim \frac{g_A \sqrt{M_N}}{\sqrt{2}\,(2\pi)^3} \bar{u}(p)\, \gamma_\lambda\, \gamma_5 \frac{\tau_i}{2} \frac{\not{p}+M_N}{-2pq}\, \mathfrak{M}^c \tag{7-53}$$

where $\mathfrak{M}^c$ is defined in terms of $\mathfrak{M}(\alpha \to \beta)$ [eq. (7-50)] by:

$$-i\,\mathfrak{M}(\alpha \to \beta) = \sqrt{\frac{M_N}{(2\pi)^3}}\, \bar{u}(p)\, \mathfrak{M}^c . \tag{7-54}$$

Letting $q^2 \to 0$ and using $p^2 - M_N^2 = 0$ gives the right-hand side of eq. (7-53); we note that it contains the required pole. From eqs. (7-51) and (7-53) we see that as $q \to 0$ the contribution to $<\beta(p_F)$ out $|j_i(0)|a(p_I)$ in$>$ from the insertion on a β line as in Fig. 7.2(c) is obtained by the substitution in $<\beta(p_F)$ out $|a(p_I)$ in$>$:

$$\bar{u}(p) \to (2\,(2\pi)^3)^{-\frac{1}{2}} \frac{g(0)}{M_N}\, \bar{u}(p)\, \not{q}\, \gamma_5 \frac{\tau_i}{2} \frac{\not{p}+M_N}{-2pq} . \tag{7-55}$$

The result of the substitution of eq. (7-54) in the matrix element $<\beta(p_F)$ out $|a(p_I)$ in$>$ is just the matrix element for the emission of a soft pion from the external, final state nucleon line of momentum p. The preceding discussion shows that soft pions are emitted only from external (nucleon or baryon) lines, as the matrix elements for other emissions are $O(q)$. The emission of a soft pion from an incoming line of momentum p is given by the substitution:

$$u(p) \to (2\,(2\pi)^3)^{-\frac{1}{2}} \frac{g(0)}{M_N} \frac{\tau_i}{2} \frac{\not{p}+M}{-2pq}\, \gamma_5\, \not{q}\, u(p) . \tag{7-56}$$

7.3. SOFT PIONS

The sum of substitutions, eqs. (7-55) and (7-56), in the matrix element $<\beta(p_F)$ out $|a(p_I)$ in$>$ gives the total matrix element for $<\beta(p_F)$ out $|j_i(0)|a(p_I)$ in$>$ corresponding to the emission of a soft pion: $a \to \beta +$ (soft pion).

Adler Consistency Condition

We shall illustrate these soft pion emission or absorption rules by considering the case when $a = N$, $\beta = \pi + N$ and deriving an important *consistency condition* due to Adler (1965a). We consider the absorption of a soft pion so that the ultimate process is $N +$ (soft π) $\to N + \pi$ or pion-nucleon scattering with the initial pion soft, as illustrated in Fig. 7.3.

We first recollect some of the formalism of pion-nucleon scattering. Let the 4-momenta of the initial and final nucleon be p_1 and p_2 respectively and of the initial and final pion be q_1 and q_2 respectively. The S-matrix is given by eqs. (7-7)-(7-9), where A and B are Lorentz scalar functions of two independent kinematic invariants, for example ν and t. Terms A and B are tensors in the isospin indices a and b of the initial and final pion:

$$A = A^{(+)} \delta_{ab} + A^{(-)} \tfrac{1}{2}[\tau_a, \tau_b] \tag{7-57}$$

$$B = B^{(+)} \delta_{ab} + B^{(-)} \tfrac{1}{2}[\tau_a, \tau_b] . \tag{7-58}$$

Terms $A^{(+)}$ and $A^{(-)}$ are given in terms of amplitudes $A^{\tfrac{1}{2}}$ and $A^{\tfrac{3}{2}}$ for total isospin $\tfrac{1}{2}$ and $\tfrac{3}{2}$ respectively by:

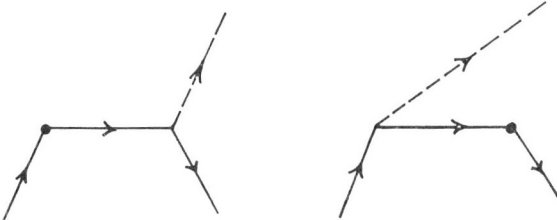

Fig. 7-3. Generalized Born amplitudes for pion-nucleon scattering with absorption of a soft pion. The proper vertex of $J^A_{i,\lambda}$ is represented by a heavy dot.

$$A^{(+)} = \frac{1}{3} A^{\frac{1}{2}} + \frac{2}{3} A^{\frac{3}{2}}, \ A^{(-)} = \frac{1}{3} A^{\frac{1}{2}} - \frac{1}{3} A^{\frac{3}{2}} \tag{7-59}$$

and similarly for B.

We now apply the substitutions eqs. (7-55) and (7-56) to the case where $\alpha = N$, $\beta = \pi + N$, to produce the matrix element for πN scattering with the initial pion soft. In this case:

$$\mathcal{M}(\alpha \to \beta) = \frac{g M_N}{\sqrt{2}} (2\pi)^{-\frac{9}{2}} \bar{u}(p_2) \tau_b \gamma_5 u(p_1) \tag{7-60}$$

is just the matrix element for the pion-nucleon vertex with the initial (final) nucleon having 4-momentum p_1 (p_2) and the pion having isospin index b. From eqs. (7-55) and (7-56) we find that, if π^a denotes a soft pion with isospin index a and $q = q_1$:

$$\mathcal{M}(\pi^a + N \to \pi^b + N) = \frac{M_N}{2 (2\pi)^6} \bar{u}(p_2) \, (A + \frac{1}{2} (\not{q}_1 + \not{q}_2) B) \, u(p_1)$$

$$= \frac{M_N}{2 (2\pi)^6} \bar{u}(p_2) \left\{ \left[\frac{g(0) \not{q} \gamma_5 \tau_a}{2 M_N} \frac{\not{p}_2 + M_N}{-2p_2 q} \right] [g \gamma_5 \tau_b] \right.$$

$$\left. + [g \gamma_5 \tau_b] \left[\frac{\not{p}_1 + M_N}{-2p_1 q} \frac{\tau_a \gamma_5 \not{q} \, g(0)}{2 M_N} \right] \right\} u(p_1)$$

$$= \frac{M_N}{2 (2\pi)^6} g(0) g \, \bar{u}(p_2) \left\{ \frac{\delta_{ab}}{M_N} - \not{q} \left(\frac{\tau_b \tau_a}{s - M_N^2} - \frac{\tau_a \tau_b}{u - M_N^2} \right) \right\} u(p_1). \tag{7-61}$$

In writing down the last line of eq. (7-61) we have used:

$$s = (p_1 + q)^2 = M_N^2 + 2p_1 q + q^2, \ u = (p_2 - q)^2 = M_N^2 - 2p_2 q + q^2$$

with $q^2 = 0$.

Comparison of eq. (7-61) and eqs. (7-57), (7-58) shows that:

7.3. SOFT PIONS

$$A^{(+)}(\nu = 0, q^2 = 0) = \frac{g(0)\,\bar{g}}{M_N} \qquad (7\text{-}62)$$

$$A^{(-)}(\nu = 0, q^2 = 0) = 0 \,. \qquad (7\text{-}63)$$

Equation (7-63) is automatically satisfied since the $A^{(-)}$ amplitude is odd under the crossing $s \leftrightarrow u$, which is equivalent to $\nu \leftrightarrow -\nu$ since $\nu = (s-u)/4M$. Equation (7-62) is a nontrivial relation depending on PCAC, known as the Adler consistency condition (see section 7.1).

The Born approximation with the basic pion-nucleon vertex coupling $\bar{g}\,\gamma_5\,\tau$ is given in terms of the A and B amplitudes by eq. (7-16):

$$A^{(+)} = A^{(-)} = 0$$

$$B = +\bar{g}^2 \left(\frac{\tau_b \tau_a}{M_N^2 - s} - \frac{\tau_a \tau_b}{M_N^2 - u} \right) . \qquad (7\text{-}64)$$

It is seen that eq. (7-64) agrees with the expression obtained from eq. (7-61) for B on putting $g(0) = \bar{g}$. (The Adler consistency condition is not derivable from the Born terms in γ_5 theory, but is in derivative coupling.)

A numerical check of the Adler consistency condition may be attempted. For this it is best expressed as:

$$\left. \frac{A^{(+)}(\nu=0, q^2)}{K(q^2)} \right|_{q^2=0} = \frac{\bar{g}^2}{M_N} \qquad (7\text{-}65)$$

where $K(q^2) = g(q^2)/\bar{g}$; $(\bar{g} \equiv g(M_\pi^2))$. One may evaluate the pion on mass shell amplitude $\left. \dfrac{A^{(+)}(\nu=0, q^2)}{K(q^2)} \right|_{q^2=M_\pi^2}$ by fixed momentum dispersion relations (the point $\nu = 0$ is unphysical, so that a direct experimental comparison is not possible), with subtraction constants depending on the scattering lengths in the various partial waves. Adler (1965a) finds that:

$$\left. \frac{A^{(+)}(\nu=0, q^2)}{K(q^2)} \right|_{q^2=M_\pi^2} = 30 M_\pi^{-1} \,. \qquad (7\text{-}66)$$

We express eq. (7-65) through the equivalent pseudovector coupling constant.

$$f^2 = \left(\frac{M_\pi}{2M_N}\right)^2 \frac{g^2}{4\pi} \qquad (7\text{-}67)$$

The pion-nucleon coupling constant is usually quoted in terms of f^2 and may be assumed to be within 6% of:

$$f^2 = 0.08 \qquad (7\text{-}68)$$

giving:

$$\frac{g^2}{M_N} = 27\, M_\pi^{-1} \,. \qquad (7\text{-}69)$$

Comparison of eqs. (7-69) and (7-66) is quite satisfactorily within the error expected — which is due to PCAC (10%), value of f^2 (6%), and dispersion relation evaluation (up to 10%). The Adler condition, eq. (7-65), requires an off-mass-shell extrapolation from eq. (7-66), which is estimated (Adler 1965a) to give a correction of $\sim -0.5\, M_\pi^{-1}$ to eq. (7-66). This is a small correction, but in the direction of bettering the agreement with eq. (7-69).

7.4. CURRENT ALGEBRA

The three components of the isospin operator I are given by:

$$I_i = \int J_{i,0}^V(x)\, d^3x \qquad i = 1, 2, 3 \qquad (7\text{-}70)$$

and are conserved quantities (corresponding to the conserved isospin current).[2] They obey the commutation rules:

---

[2] The vanishing of $\frac{\partial I}{\partial x_0}$, when sandwiched between any two Heisenberg representation state vectors which vanish asymptotically in spatial directions, is an

7.4. CURRENT ALGEBRA

$$[I_i, I_j] = i\,\epsilon_{ijk}\,I_k \tag{7-71}$$

where $\epsilon_{ijk} = +1, -1$ or 0 according as $i\,j\,k$ is an odd, even, or no permutation of 1 2 3. Terms I_1, I_2, I_3 are a set of operators which are closed under commutation (meaning that the commutator of any two of the set is a linear combination of operators of that set). Such a system is called a *Lie algebra* and the ϵ_{ijk} are the *structure constants* of the algebra as stated in section 6.1 of Chapter 6. If we consider the set of infinitesimal unitary operators $1 + i\Delta_j I_j$, where Δ_j is infinitesimal, and all possible products of these, we obtain a continuous group of unitary transformations. We refer to eq. (7-71) as the *algebra* of the *group*. The group in this case is SU2. We know that, if H is the Hamiltonian of strong interactions, then (approximately) $[I_i, H] = 0$, corresponding to the fact that (approximately) I_i is conserved. The relations $[I_i, H] = 0$ mean that H is invariant under the isospin group SU2. We now wish to consider the situation when we have an *algebra* of physical operators, but H is no longer invariant under transformations of the group. Define operators which are, in analogy to eq. (7-70), the spatial integral of the time component of the *axial* vector current:

$$D_i = \int J^A_{i,0}(x)\,d^3x \qquad i = 1, 2, 3 \tag{7-72}$$

$$\frac{\partial}{\partial x_0} D_i + \int \frac{\partial}{\partial x_k} J^A_{i,k}(x)\,d^3x \neq 0\,. \tag{7-73}$$

From eq. (7-73) D_i is not a constant of the motion and does not commute with the Hamiltonian. However D_i is an isovector [since $J^A_{i,0}(x)$ have

---

immediate consequence of the current conservation $\dfrac{\partial J^V_{i,\mu}(x)}{\partial x_\mu} = 0$. The vanishing of the operator $\dfrac{\partial I}{\partial x_0}$ follows if the states, for example, are the in-states or the out-states used in defining the S-matrix.

been formed so as to transform like a vector under isospin space rotations] and thus obeys[3] the commutation relations:

$$[I_i, D_j] = i\epsilon_{ijk} D_k . \tag{7-74}$$

Although one can perfectly well obtain sum rules from eq. (7-71) or eq. (7-74), it is the essence of current algebra to postulate a commutator of $\vec{D}$ with itself which makes $\vec{I}$ and $\vec{D}$ a closed system under commutation. (This makes the current algebra a Lie algebra.) If one had a world consisting just of free nucleons, then:

$$I_i = \frac{1}{2} \int d^3x\, \bar{\psi}(\vec{x},t)\, \tau_i \psi(\vec{x},t),$$

$$D_i(t) = \frac{1}{2} \int d^3x\, \bar{\psi}(\vec{x},t)\, \gamma_5 \tau_i \psi(\vec{x},t)$$

$$[D_i(t), D_j(t)] = \frac{1}{4} \int d^3x\, d^3y\, (\bar{\psi}(\vec{x},t)\, \gamma_5 \tau_i \{\psi(\vec{x},t), \bar{\psi}(\vec{y},t)\}\gamma_5 \tau_j \psi$$

$$- \bar{\psi}^+(\vec{y},t)\, \gamma_5 \tau_j \{\psi(\vec{y},t), \bar{\psi}(\vec{x},t)\}\gamma_5 \tau_i \psi(\vec{x},t))$$

$$= \frac{1}{4} \int d^3x\, \bar{\psi}(\vec{x},t)\, (\gamma_5 \tau_i \gamma_5 \tau_j - \gamma_5 \tau_j \gamma_5 \tau_i)\, \psi(\vec{x},t)$$

$$= \frac{1}{4} \int d^3x\, \bar{\psi}(\vec{x},t)\, 2i\epsilon_{ijk}\tau_k \psi(\vec{x},t) = i\epsilon_{ijk} I_k . \tag{7-75}$$

Consequently on this free field model we complete eqs. (7-71) and (7-74) by adding eq. (7-75):

$$[I_i, I_j] = i\epsilon_{ijk} I_k$$

$$[I_i, D_j(t)] = i\epsilon_{ijk} D_k(t) \tag{7-76}$$

$$[D_i(t), D_j(t)] = i\epsilon_{ijk} I_k$$

---

[3] That these commutation relations are a consequence of the vector transformation of D_i in isospin space can be seen from the fact that infinitesimal isospin rotations of D_i are given by $D_i \rightarrow U^{-1} D_i U$, where $U = 1 + i\Delta_j I_j$.

7.4. CURRENT ALGEBRA

which is a closed system of *equal-time* commutation relations. Current algebra postulates that these *equal-time* commutation relations hold not only for the model but also for the true original D_k. If we form the combinations:

$$\vec{I}^+ = \tfrac{1}{2}(\vec{I}+\vec{D}(t)), \quad \vec{I}^- = \tfrac{1}{2}(\vec{I}-\vec{D}(t)) \tag{7-77}$$

we find the commutation relations:

$$[I_i^\pm, I_j^\pm] = i\,\epsilon_{ijk}\, I_k^\pm$$

$$[I_i^-, I_j^+] = 0\;. \tag{7-78}$$

Equation (7-46) corresponds to two independent angular momenta or isovectors, and I^+ and I^- each generate an SU2 algebra. So the commutation relations, eq. (7-76) or eq. (7-78) are said to form an algebra of SU2 ⊗ SU2. However as we noted previously, D does not commute with the Hamiltonian, which is consequently not invariant under this SU2 × SU2 group.

The algebra is sometimes distinguished as the *chiral* algebra of SU2 ⊗ SU2. The *chirality operator* is defined for action upon a fermion state of spin $\tfrac{1}{2}$. In the Weyl representation $\gamma_5 = \begin{pmatrix} 1 & 0 \\ 0 & -1 \end{pmatrix}$, where 1 is a 2 × 2 unit matrix and $\tfrac{1}{2}(1+\gamma_5) = \begin{pmatrix} 1 & 0 \\ 0 & 0 \end{pmatrix}$, $\tfrac{1}{2}(1-\gamma_5) = \begin{pmatrix} 0 & 0 \\ 0 & 1 \end{pmatrix}$. In the model for $I^\pm$, I^+ contains $\bar\psi \tfrac{1}{2}(1+\gamma_5)\psi$ and I^- contains $\bar\psi \tfrac{1}{2}(1-\gamma_5)\psi$, so that I^+ only annihilates fermions of chirality +1, creates fermions of chirality −1, annihilates anti-fermions of chirality −1, and creates anti-fermions of chirality +1, while I^- has opposite chiralities.

The algebra of eq. (7-78) is the least restrictive version of current algebra. The hypothesis can be extended from the integrated current and time components to the time components themselves (again using the free field model as a guide):

$$\left[J^V_{i,0}(x), J^V_{j,0}(y)\right]_{x_0=y_0} = i\,\epsilon_{ijk}\,J^V_{k,0}(x)\,\delta^3(\vec{x}-\vec{y})$$

$$\left[J^V_{i,0}(x), J^A_{j,0}(y)\right]_{x_0=y_0} = i\,\epsilon_{ijk}\,J^A_{k,0}(x)\,\delta^3(\vec{x}-\vec{y})$$

$$\left[J^A_{i,0}(x), J^A_{j,0}(y)\right]_{x_0=y_0} = i\,\epsilon_{ijk}\,J^V_{k,0}(x)\,\delta^3(\vec{x}-\vec{y}) . \tag{7-79}$$

If one goes further and introduces spatial components of currents, then one finds:

$$\left[J^V_{i,0}(x), J^V_{j,\mu}(y)\right]_{x_0=y_0} = i\,\epsilon_{ijk}\,J^V_{k,\mu}(x)\,\delta^3(\vec{x}-\vec{y}) + \text{S.T.}$$

$$\left[J^V_{i,0}(x), J^A_{j,\mu}(y)\right]_{x_0=y_0} = i\,\epsilon_{ijk}\,J^A_{k,\mu}(x)\,\delta^3(\vec{x}-\vec{y}) + \text{S.T.}$$

$$\left[J^A_{i,0}(x), J^A_{j,\mu}(y)\right]_{x_0=y_0} = i\,\epsilon_{ijk}\,J^V_{k,\mu}(x)\,\delta^3(\vec{x}-\vec{y}) + \text{S.T.} \tag{7-80}$$

where S.T. stands for "Schwinger terms" which are terms involving derivatives of $\delta^3(\vec{x}-\vec{y})$, and must necessarily occur on the right-hand side of eq. (7-80) (Schwinger, 1959).

Using PCAC and the commutator:

$$[D_1 + i D_2, D_1 - i D_2] = 2 I_3 . \tag{7-81}$$

Adler (1965c) and Weisberger (1965) found a sum rule for $|g_A| = |G_A/G_V|$:

$$1 - \frac{1}{g_A^2} = \frac{4 M_N^2}{g(0)^2}\frac{1}{\pi}\int_{M_N+M_\pi}^{\infty} \frac{W\,dW}{W^2 - M_N^2}\,[\sigma_0^+(W) - \sigma_0^-(W)] \tag{7-82}$$

where, according to our usual notation, M_N and M_π are the nucleon and pion masses, $g(s)$ is the πNN form factor so that $g(M_\pi^2) = g$, the pion nucleon coupling constant, and $\sigma_0^\pm$ are the total cross sections for scattering of a zero mass $\pi^\pm$ on a proton, at center of mass energy W. On extrapolation through an $(energy)^2 = M_\pi^2$, one finds an agreement (in magnitude) of g_A from eq. (7-82) with the experimental value of about -1.21 to $\sim 10\%$.

7.5. S-WAVE PION-NUCLEON SCATTERING LENGTHS

The chiral algebra of $SU2 \otimes SU2$ can be generalized to the chiral algebra of $SU3 \otimes SU3$. The commutation relations, eqs. (7-76), (7-79), and (7-80) all have their analogue simply obtained by substituting f_{ijk} for ϵ_{ijk}. For example, the analogue of eq. (7-79) is for i, j, and k in the range $1, 2, \ldots, 8$:

$$\left[J_{i,0}^V(x), J_{j,0}^V(y)\right]_{x_0=y_0} = i\, f_{ijk}\, J_{k,0}^V(x)\, \delta^3(\vec{x}-\vec{y})$$

$$\left[J_{i,0}^V(x), J_{j,0}^A(y)\right]_{x_0=y_0} = i\, f_{ijk}\, J_{k,0}^A\, \delta^3(\vec{x}-\vec{y})$$

$$\left[J_{i,0}^A(x), J_{j,0}^A(y)\right]_{x_0=y_0} = i\, f_{ijk}\, J_{k,0}^V\, \delta^3(\vec{x}-\vec{y}) \ . \qquad (7\text{-}83)$$

One does not necessarily expect the algebra of $SU3 \otimes SU3$ to be so good as the algebra of $SU2 \otimes SU2$. Also the application of the analogue of PCAC in the strangeness changing case involves extrapolations from real to zero K-meson mass.

7.5. S-WAVE PION-NUCLEON SCATTERING LENGTHS

Using PCAC and the current algebra relations of eq. (7-48), a value for the s-wave pion-nucleon scattering lengths can be derived which is in good agreement with the values found from experiment. We outline the derivation in this section, using the notation that $p_1\, q_1$ and $p_2\, q_2$ are the 4-momenta of the initial and final nucleon pion respectively, and that the pions have isospin indices a and b respectively. The usual definitions eqs. (7-8) and (7-9) give the A and B amplitudes in terms of the S-matrix element S_{FI} where:

$$S_{FI} \equiv \langle p_2\, q_2\ (\text{out}) | p_1\, q_1\ (\text{in}) \rangle \ . \qquad (7\text{-}84)$$

By the usual reduction technique:

$<p_2\ q_2\ (\text{out})|p_1\ q_1\ (\text{in})> = \delta_{FI}$

$$+ \frac{i^2}{2(2\pi)^3}(q_1^2-M_\pi^2)(q_2^2-M_\pi^2)\int d^4x\, d^4y$$

$<p_2|T\{\phi_b(x)\,\phi_a(y)\}|p_1> \exp(i\,q_2 x - i\,q_1 y)$

(7-85)

where $T\{\phi_b(x)\phi_a(y)\}$ is a time-ordered product.

In eq. (7-85) we substitute for ϕ_b and ϕ_a using the PCAC relation eq. (7-38); later in this derivation we shall use the second part of the PCAC hypothesis, namely the slow variation of matrix elements such as eq. (7-85) with q_1^2 and q_2^2 in the neighborhood of M_π^2:

$$<p_2\ q_2\ (\text{out})|p_1\ q_1\ (\text{in})> = \delta_{FI} + \frac{i^2(q_1^2-M_\pi^2)(q_2^2-M_\pi^2)}{2((2\pi)^3)M_N^2\,M_\pi^4\,g_A^2}\,g^2\int d^4x\,d^4y$$

$$<p_2|T\left\{\frac{\partial J_{b,\mu}(x)}{\partial x_\mu}\frac{\partial J_{a,\nu}(y)}{\partial y_\nu}\right\}|p_1>\exp(-i\,q_1 y)\exp(i\,q_2 x)\,. \quad (7\text{-}86)$$

The time-ordered product is defined as:

$$T\{A(x)\,B(y)\} = \theta(x-y)\,A(x)\,B(y) + \theta(y-x)\,B(x)\,A(y) \quad (7\text{-}87)$$

where:

$$\theta(x) = 1 \qquad x_0 > 0$$

$$\theta(x) = 0 \qquad x_0 < 0\,. \quad (7\text{-}88)$$

The definition eq. (7-88) entails that:

$$\frac{\partial}{\partial x_i}\theta(x) = 0 \qquad i = 1,\,2,\,3$$

$$\frac{\partial}{\partial x_0}\theta(x) = \delta(x_0)\,. \quad (7\text{-}89)$$

7.5. S-WAVE PION-NUCLEON SCATTERING LENGTHS

Using eq. (7-89) and integrating by parts, one obtains for the integrand in eq. (7-86):

$$\left\{-\delta(x_0-y_0)<p_2|\left[J^A_{b,0}(x),\frac{\partial}{\partial y_\nu}J^A_{a,\nu}(y)\right]|p_1> - iq_2^\mu \delta(x_0-y_0)\right.$$

$$\left.<p_2|\left[J^A_{a,0}(y),J^A_{b,\mu}(x)\right]|p_1> + q_2^\mu q_1^\nu <p_2|T\{J^A_{b,\mu}(x),J^A_{a,\nu}(y)\}|p_1>\right\}. \quad (7\text{-}90)$$

It turns out that the first and last terms of expression (7-90) do not contribute, which we shall comment on at the end of the derivation. We use eq. (7-80) to obtain as the contribution to eq. (7-86) from the middle term of (7-90):

$$\frac{i^2(q_1^2-M_\pi^2)(q_2^2-M_\pi^2)g^2}{2(2\pi)^3 M_N^2 M_\pi^4 g_A^2} \int d^4x\, d^4y\, (iq_2^\mu) \exp(-iq_1 y)\exp(+iq_2 x)$$

$$<p_2|(\delta^4(x-y)\,i\,\epsilon_{abc}\,J^V_{c,\mu}(x) + \delta(x_0-y_0)\,(S.T.))|p_1> . \quad (7\text{-}91)$$

The Schwinger terms [S.T. in (7-91)] involve gradients of δ-functions which on integration by parts can be neglected in the limit $q_1 \to q_2 \to 0$ (which we are about to take), since they are of higher order in q. On letting $q_1 \to q_2 \to 0$ the term of first order in q in eq. (7-91) is:

$$\frac{g^2}{2(2\pi)^3 M_N^2 g_A^2} \int d^4x\, q_2^\mu <p_2|\epsilon_{abc}\,J^V_{c,\mu}(x)|p_1> \exp(i(q_2-q_1)x)$$

$$= \frac{g^2}{2(2\pi)^3 M_N^2 g_A^2} (2\pi)^4 \delta^4(p_2+q_2-p_1-q_1) q_2^\mu <p_2|\epsilon_{abc}\,J^V_{c,\mu}(0)|p_1>$$

$$= \frac{g^2}{M_N^2 g_A^2} (2\pi)^4 \delta^4(p_F-p_I)\frac{M_N}{2(2\pi)^6} q_2^\mu \bar{u}(p_2)\gamma_\mu \frac{1}{2}\tau_c \epsilon_{abc}\,u(p_1)$$

$$= i(2\pi)^4 \delta^4(p_F-p_I)\frac{M_N}{2(2\pi)^6}\frac{g^2}{M_N^2 g_A^2} \bar{u}(p_2)\frac{1}{4}[\tau_b,\tau_a]\not{q}_2\, u(p_1) . \quad (7\text{-}92)$$

In the limit $q_2 \to q_1 \to 0$, it follows that $p_2 \to p_1$ and $\bar{u}(p_2) q_{2\mu} \gamma^\mu u(p_1) M_N = p^\mu q_{2\mu} = \frac{1}{2} p_1(q_1+q_2)$ to first order in q. Since $\nu = (s-u)/4M_N = ((p_1+q_1)^2 - (p_1-q_2)^2)/4M_N = \frac{1}{2M_N} p_1(q_1+q_2)$, we can express eq. (7-92) to first order in q as:

$$i(2\pi)^4 \delta(p_F - p_I) \frac{M_N}{2(2\pi)^6} \frac{g^2}{M_N^2 g_A^2} \nu X_2 \frac{1}{4}[\tau_b, \tau_a] X_1 \qquad (7\text{-}93)$$

where X_1 and X_2 are respectively the initial and final isospinors of the nucleon. Equation (7-93) is the contribution to the S-matrix eq. (7-85) to first order in q as $q_2 \to q_1 \to 0$. We can also express the same quantity in terms of the A and B amplitudes using eqs. (7-57)-(7-58) and eqs. (7-7)-(7-9):

$$S_{FI} = \delta_{FI} + i(2\pi)^4 \delta(p_F - p_I) \frac{M_N}{2(2\pi)^6} \bar{u}(p_2) \Big\{ (A^{(+)}(\nu, t)$$
$$+ \frac{1}{2}(\slashed{q}_1 + \slashed{q}_2) B^{(+)}(\nu, t)) \frac{1}{2}[\tau_a, \tau_b]_+ + (A^{(-)}(\nu, t)$$
$$+ \frac{1}{2}(\slashed{q}_1 + \slashed{q}_2) B^{(-)}(\nu, t)) \frac{1}{2}[\tau_b, \tau_a] \Big\} u(p_1) \, . \qquad (7\text{-}94)$$

We first compare the odd isospin or $(-)$ amplitudes in eqs. (7-61) and (7-62). We are comparing at $t = 0$ since $q_1 = q_2 \to 0$ and $\nu \to 0$. The term $A^{(-)}(\nu, t)$ is odd under crossing, $\nu \leftrightarrow -\nu$, so we write:

$$A^{(-)}(\nu, 0) = \nu A^{(-)\prime} + O(\nu^3) \, . \qquad (7\text{-}95)$$

So to first order in q the isospin odd part of eq. (7-62) becomes:

$$i(2\pi)^4 \delta(p_F - p_I) \frac{M_N}{2(2\pi)^6} \nu (A^{(-)\prime} + B^{(-)}(0,0)) X_2 \frac{1}{2}[\tau_b, \tau_a] X_1 \, . \qquad (7\text{-}96)$$

Comparison of eqs. (7-63) and (7-61) gives:

$$A^{(-)\prime} + B^{(-)}(0,0) = \frac{1}{2} \frac{g^2}{M_N^2 g_A^2} \, . \qquad (7\text{-}97)$$

7.5. S-WAVE PION-NUCLEON SCATTERING LENGTHS

To establish a connection with scattering lengths we return to the on-mass-shell physical amplitudes. We refer the reader to sections 2.2 and 3.4 for some of the subsequent formulae and parameters. The forward scattering amplitudes are given in terms of the A and B amplitudes by:

$$f^{(\pm)}(\nu) = \frac{M_N}{4\pi W} (A^{(\pm)}(\nu, 0) + \nu B^{(\pm)}(\nu, 0)) \qquad (7\text{-}98)$$

and if a_1, a_3 are the $T = \frac{1}{2}, \frac{3}{2}$ s-wave scattering lengths:

$$f^{(-)}(M_\pi) = \frac{1}{3}(a_1 - a_3) \qquad (7\text{-}99\text{a})$$

$$f^{(+)}(M_\pi) = \frac{1}{3}(a_1 + 2a_3) \;. \qquad (7\text{-}99\text{b})$$

We now approximate eq. (7-98), for the $(-)$ amplitudes, for ν between 0 and M_π by:

$$f^{(\pm)}(\nu) \simeq \frac{M_N}{4\pi W} \nu(A^{(-)'} + B^{(-)}(0,0)) \;. \qquad (7\text{-}100)$$

Using eqs. (7-96) and (7-99a) gives at $\nu = M_\pi$:

$$\frac{1}{4\pi} M_\pi \frac{1}{2} \frac{g^2}{M_N^2 \, g_A^2} \simeq \left(1 + \frac{M_\pi}{M_N}\right) \frac{1}{3}(a_1 - a_3) \;. \qquad (7\text{-}101)$$

Using $g_A = 1.2$, $\frac{g^2}{4\pi} = 15$, we find that:

$$a_1 - a_3 \simeq 0.25 \, M_\pi^{-1} \;. \qquad (7\text{-}102)$$

We now consider the $(+)$ amplitudes. As is evident in the derivation of the Adler consistency condition [for example, eq. (7-62)] $B^{(+)}$ has a pole in ν; it is of course odd in ν; $A^{(+)}$ is even in ν. So the expansion of eq. (7-98) in the region $0 \lesssim \nu \lesssim M_\pi$ is, for the $(+)$ amplitudes:

$$f^{(+)}(\nu) = \frac{M_N}{4\pi W}(A^{(+)}(0,0) + B^{(+)'}) + 0(\nu^2) \qquad (7\text{-}103)$$

when $B^{(+)'}$ is the residue of the pole. Comparison with eq. (7-93) which has no isospin even part gives:

$$A^{(+)}(0,0) + B^{(+)'} = 0 \qquad (7\text{-}104)$$

$$a_1 + 2a_3 = 0 \ . \qquad (7\text{-}105)$$

Equations (7-102 and (7-105) give:

$$a_1 \simeq 0.17\ M_\pi^{-1} \qquad a_3 \simeq -0.08\ M_\pi^{-1} \ . \qquad (7\text{-}106)$$

These current algebra scattering lengths (with their inherent 10% error from PCAC) are in agreement with estimates from low energy scattering and forward dispersion relations as given in Chapter 3. Hamilton (1966), for example, gives:

$$a_1 - a_3 = 0.271 \pm 0.007, \quad a_1 + 2a_3 = -0.002 \pm 0.008 \ .$$

We must now mention why the first and third terms of expression (7-90) do not contribute. The Adler self-consistency argument (section 7.2) shows that expression (7-90) must vanish, except for poles coming from propagators, when $q_2 = 0$, $q_1^2 = m^2$. By current algebra relations, eq. (7-80), the first term of expression (7-90) contains:

$$< p_2 |\{i\ \sigma_{ab}(x)\ \delta^4(x-y) + \text{S.T.}\}| p_1 >$$

where $\sigma_{ab}(x)$ is a scalar field. This contains no pole terms and, from invariance arguments, no contributing linear terms. Consequently it gives a negligible contribution in the present linear approximation. The last term of eq. (7-58) can contain pole terms from propagators, but they are not present for s-waves. For a more detailed discussion we refer the reader to Weinberg (1966b).

REFERENCES

S. L. Adler, Phys. Rev. 137, B1022 (1965a).

S. L. Adler, Phys. Rev. 139, B1638 (1965b).

S. L. Adler, Phys. Rev. 140, 736 (1965c).

S. L. Adler and R. F. Dashen, *Current Algebras* (W. A. Benjamin, Inc., New York, 1968).

J. Bernstein, *Elementary Particles and Their Currents* (W. H. Freeman and Co., San Francisco, 1968).

J. Bjorken and S. Drell, *Relativistic Quantum Fields* (McGraw Hill, New York, 1965)

R. P. Feynman and M. Gell-Mann, Phys. Rev. 109, 193 (1958).

M. Gell-Mann, Phys. Rev. 125, 1067 (1962).

J. Hamilton, Phys. Letters 20, 687 (1966).

R. E. Marshak, Riazzudin and C. P. Ryan, *Theory of Weak Interactions in Particle Physics* (Wiley-Interscience, New York, 1969).

S. S. Schweber, *Mesons and Fields* (Row, Peterson and Co., Evanston, 1955), Vol. I: Fields.

J. Schwinger, Phys. Rev. Letters 3, 296 (1959).

S. Weinberg, Phys. Rev. Letters 16, 879 (1966a).

S. Weinberg, Phys. Rev. Letters 17, 616 (1966b).

W. Weisberger, Phys. Rev. 143, 1302 (1965).

CHAPTER 8

SCATTERING AT HIGHER ENERGIES

8.1. ELASTIC SCATTERING AT LABORATORY MOMENTA $\gtrsim$ 1 GeV

The Diffraction Peak

One of the features of elementary particle scattering is that for center of mass momenta greater than 0.5 GeV/c the differential cross sections for the observed elastic scatterings have a marked peak at forward scattering angles in the center of mass system. As collision energies increase, the forward peak increasingly dominates the elastic cross section, the greater part of which is contained in a decreasingly small forward angle (with the exception that when the particles are alike, as in proton-proton scattering the differential cross section is necessarily symmetric between forward and backward directions). The forward, or *diffraction*, peaking is shown for the case of $\pi-p$ scattering in Figs. 8-1 and 8-2.

At high energies the colliding particles have a large probability of producing other particles, and the diffraction peak will now be shown to be a necessary concomitant of this inelastic scattering, or absorption. We shall conduct our demonstration for the scattering of spin 0 by spin $\frac{1}{2}$ particles (having pion or kaon scattering on protons in mind), but the argument may be generalized to other spins.

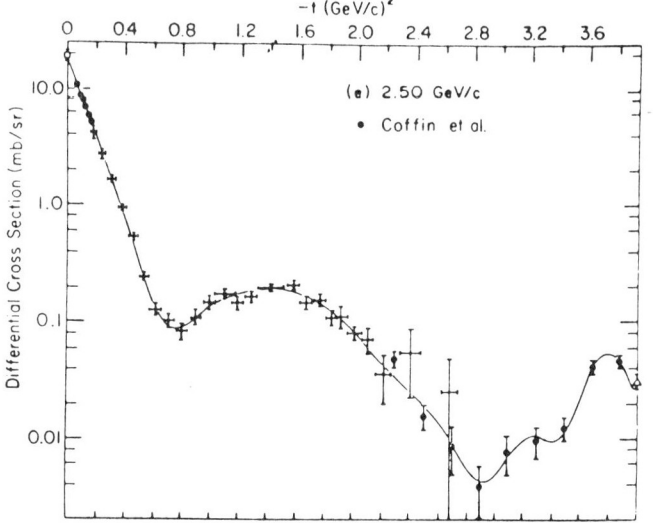

Fig. 8-1. $\pi^- p$ elastic differential cross section at 2.5 GeV/c pion laboratory momentum.

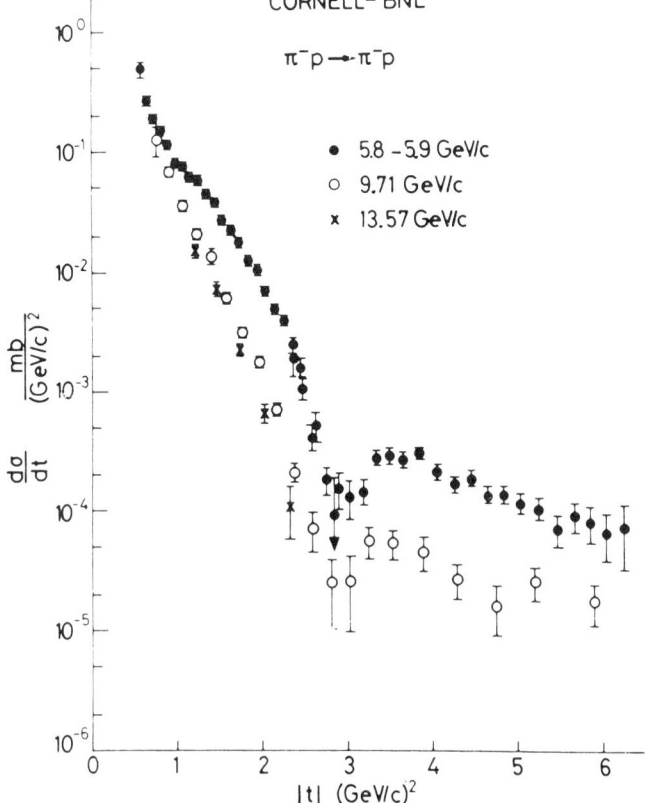

Fig. 8-2. $\pi^- p$ elastic differential cross section between 6.0 GeV/c and 13.5 GeV/c pion laboratory momentum.

8.1. ELASTIC SCATTERING

We consider the scattering through an angle θ in the center of mass system where each particle has 3-momentum of magnitude q. Then, as shown in Chapter 2, eqs. (2-52)-(2-54) the non-spin flip and spin flip amplitudes may be written (where $f_{\ell\pm}$ are the partial wave amplitudes for $j = \ell \pm \frac{1}{2}$):

$$f(\theta) = \sum_{\ell=0}^{\infty} \{(\ell+1) f_{\ell+} + \ell\, f_{\ell-}\} P_\ell(\cos\theta) \tag{8-1}$$

$$g(\theta) = \sum_{\ell=1}^{\infty} \{f_{\ell+} - f_{\ell-}\} \sin\theta\, P_\ell{}'(\cos\theta) \tag{8-2}$$

and the elastic differential cross section is:

$$\frac{d\sigma}{d\Omega} = |f(\theta)|^2 + |g(\theta)|^2 \,. \tag{8-3}$$

We first note that the spin flip amplitude $g(\theta)$ is small at small forward angles [formally because of the factor $\sin\theta$ in eq. (8-2), corresponding to the fact that there is no spin flip at $\theta = 0°$, $180°$ by conservation of angular momentum]. Consequently it does not contribute significantly to the diffraction peak, and we can concentrate our attention on $f(\theta)$. Now the spin-nonflip amplitude $f(\theta)$ is in a certain sense the pure elastic scattering amplitude since it is the scattering amplitude, $F_{aa}(\theta)$, from channel a to channel a where the quantum numbers a include the spin[1]:

$$f(\theta) \equiv F_{aa}(\theta) \tag{8-4}$$

and the *optical theorem* [see eq. (2-29)] applies to the purely elastic scattering amplitude:

$$\frac{4\pi}{q} \operatorname{Im} f(0) = \sigma_{\text{total}} \,. \tag{8-5}$$

---

[1] It may be preferable to regard the helicity nonflip amplitude, $f_{++}(\theta) = f(\theta) \cos\frac{\theta}{2} + g(\theta) \sin\frac{\theta}{2}$, as the pure elastic scattering amplitude, Terms $f(\theta)$ and $f_{++}(\theta)$ are equal in the forward direction, and almost the same in the near forward direction comprising the diffraction peak.

We may express $f(\theta)$ as:

$$f(\theta) = \sum_{\ell=0}^{\infty} \frac{1}{2iq} (2\ell+1)(\eta_\ell e^{2i\delta_\ell} - 1) P_\ell(\cos\theta) \tag{8-6}$$

and, from subtracting $\int |f(\theta)|^2 d\Omega$ from eq. (8-5):

$$\sigma(\text{inelastic + spin flip}) = \sum_{\ell=0}^{\infty} \frac{\pi}{q^2} (2\ell+1)(1-\eta_\ell^2) \tag{8-7}$$

with the unitarity requirement that the absorption parameters η_ℓ satisfy[2]:

$$\eta_\ell^2 \leq 1 \ . \tag{8-8}$$

From eq. (8-6):

$$\text{Im } f(\theta) = \sum_{\ell=0}^{\infty} \frac{(2\ell+1)}{2q} (1 - \eta_\ell \cos 2\delta_\ell) P_\ell(\cos\theta) \ . \tag{8-9}$$

$$\text{Re } f(\theta) = \sum_\ell \frac{2\ell+1}{2q} \eta_\ell \sin 2\delta_\ell P_\ell(\cos\theta) \ . \tag{8-10}$$

For high energy scattering a large number of partial wave amplitudes are nonzero; that is, a large number of the coefficients of P_ℓ in eq. (8-9) and eq. (8-10) are nonzero. This has an immediate consequence for Im $f(\theta)$. Since we see from eq. (8-9) that all the coefficients of $P_\ell(\cos\theta)$ are positive or zero, and we note that $P_\ell(\cos\theta)|_{\theta=0} = 1$ while at non-forward angles $P_\ell(\cos\theta)$ are of varying sign (for instance $P_\ell(\cos\theta)|_{\theta=\pi} = (-1)^\ell$), Im $f(\theta)$ has a maximum in forward directions, which becomes more pronounced at higher energies as more nonzero partial waves occur in the sum in eq. (8-9). If furthermore absorption is relatively large so that $\eta_\ell^2 \ll 1$ for a large number of partial waves we see that Im $f(\theta)$ will be

---

[2] Equation (8-8) also follows from eq. (8-1), which gives:

$$\eta_\ell e^{2i\delta_\ell} = \frac{\ell+1}{2\ell+1} \eta_{\ell+} e^{2i\delta_{\ell+}} + \frac{\ell}{2\ell+1} \eta_{\ell-} e^{2i\delta_{\ell-}}$$

together with the usual unitarity condition $\eta_{\ell\pm}^2 \leq 1$.

8.1. ELASTIC SCATTERING

larger than Re $f(\theta)$. This happens both because of the smallness of η_ℓ in the coefficients of P_ℓ in eqs. (8-9) and (8-10) and because of the varying sign of $\eta_\ell \sin 2\delta_\ell$ in eq. (8-10). For the latter reason the dominance of Im $f(\theta)$ over Re $f(\theta)$ is to be particularly expected in the forward direction and we expect $\frac{\text{Re } f(0)}{\text{Im } f(0)} \ll 1$. This is verified by experiment but so far has not been found to reach the sometimes surmised zero limit, except perhaps in $\pi^- p$ scattering at 24 GeV (Foley et al., 1963a,b, 1965). More generally:

$$\text{Im } f(0) \gg \text{Re } f(\theta) \,. \tag{8-11}$$

The above arguments show the reason why (for the high energy, absorptive situation envisaged) $|f(\theta)|^2$ has a forward peak, giving rise to a forward peak in the cross section in eq. (8-3). The key principal used is *unitarity*, which relates the elastic to the inelastic cross section through the parameter η_ℓ.

We shall now make the above arguments more quantitative through the use of a definite model, and we shall examine how far the results of the model agree with the experimental data.

The Black Sphere Model

The physical idea of the black sphere model is that elastic scattering in a high energy collision is characterized by a distance R, the radius of a black sphere. If the wave functions of the two particles are at a greater distance than R, then there is no interaction; if they are within R, then there is maximum absorption into inelastic channels.

We consider the scattering of spin 0 by spin $\frac{1}{2}$ particles and the totally elastic (that is non-spin flip) amplitude:

$$f(\theta) = \sum_{\ell=0}^{\infty} (2\ell+1) f_\ell P_\ell(\cos\theta) \tag{8-12}$$

where from eq. (8-1) (q is the center of mass momentum):

$$(2\ell+1)f_\ell = (\ell+1) f_{\ell_+} + \ell\, f_{\ell_-}$$

$$f_{\ell\pm} = \frac{1}{2iq}\left(\eta_{\ell\pm}\, e^{2i\delta_{\ell\pm}} - 1\right) .\tag{8-13}$$

In the black sphere model we either have no interaction, when $\eta_{\ell\pm} = 1$, $\delta_{\ell\pm} = 0$, or maximum absorption when from the fact that:

$$\sigma_{\ell\pm}^{abs} = \frac{\pi}{q^2}\left(\ell + \frac{1}{2} \pm \frac{1}{2}\right)\left(1 - \eta_{\ell\pm}^2\right)\tag{8-14}$$

we must have $\eta_{\ell\pm} = 0$. Consequently we have:

$$\text{no interaction: } f_\ell = 0$$
$$\text{maximum absorption: } f_\ell = \frac{i}{2q} .\tag{8-15}$$

If we consider a collision with orbital angular momentum ℓ, center of mass momentum q, we can define the *impact parameter* b by:

$$\ell + \frac{1}{2} = qb .\tag{8-16}$$

Semiclassically b is the closest distance of approach of the colliding particles of angular momentum ℓ. [For a detailed discussion of scattering based on the impact parameter representation, see, for example, Blankenbecler and Goldberger (1964).] Quantum mechanically, when as here we are considering plane wave scattering, b is some measure of a mean distance of the component of the plane wave of orbital angular momentum ℓ.

Using eqs. (8-15), (8-16) we can now more exactly formulate the black sphere approximation as:

$$b > R : f_\ell = 0$$
$$b < R : f_\ell = \frac{i}{2q}\tag{8-17}$$

and using eqs. (8-16), (8-17) we approximate eq. (8-12) by:

8.1. ELASTIC SCATTERING

$$f(\theta) = iq \int_0^R J_0(qb\sin\theta)\, b\, db \, . \tag{8-18}$$

The approximation eq. (8-18) depends on the energy being great enough so that there are a "large" number of partial waves such that $b < R$ and so the replacement of the summation by an integration is valid. It is also valid only for angles such that (Watson, 1958, p. 158):

$$qb\sin\theta \lesssim 1 \tag{8-19}$$

when:

$$P_\ell(\cos\theta) \simeq J_0(qb\sin\theta) \, . \tag{8-20}$$

We can perform the integration in eq. (8-18) to obtain (Watson, 1958, p. 373):

$$f(\theta) = \frac{iR J_1(qR\sin\theta)}{\sin\theta} \, . \tag{8-21}$$

From eq. (8-2) we see that the black sphere model implies that $g(\theta) = 0$, and consequently from eqs. (8-3) and (8-21) that the differential cross section is given by:

$$\frac{d\sigma}{d\Omega} = R^2 \left| \frac{J_1(qR\sin\theta)}{\sin\theta} \right|^2 \, . \tag{8-22}$$

$\left|\frac{J_1(x)}{x}\right|^2$ has a maximum at $x = 0$, zeros at $x = 3.82, 7.01, \ldots$, with successively lower maxima between the zeros. These are the characteristic features of *diffraction scattering*. From the fact that eq. (8-22) has zeros which the data does not possess, we see at once that such a simple picture as the black sphere model cannot be a completely correct representation of elementary particle interaction. However we may readily see that the minima in the data occur at approximately the positions in the model. From Figs. 8-1 and 8-2, and the discussion in section 8.8 below, the $\pi^+ p$ elastic scattering data has minima (or, at high energy, kinks) at $-t \simeq 0.8$ and $-t \simeq 2.8$ (GeV/c)^2, where $t = -2q^2(1-\cos\theta)$ is the invariant momentum transfer squared. At high energy $x^2 = q^2 R^2 \sin^2\theta \simeq -tR^2$. The forward point and the first two minima in the model are at $x^2 = 0, 14.7, 49.1$ and in the data at $-t = 0, 0.8, 2.8$. We see that these numbers are fairly consistent and can be fit with a radius, $R \sim 0.8\text{-}0.9$ F. Also it is easy to see, using eqs. (8-12), (8-14), and (8-17) that the cross sections integrated over angles are:

$$\sigma_{\text{elastic}} = \pi R^2, \quad \sigma_{\text{absorption}} = \pi R^2, \quad \sigma_{\text{total}} = 2\pi R^2 \quad . \quad (8\text{-}23)$$

The black sphere model can be generalized, as for example by Kozlowsky and Dar (1966), who use a 4-parameter optical model to fit $\pi^- p$ elastic scattering between 2 and 12 GeV/c incident laboratory momentum. Their sphere is characterized by a radius, edge diffusivity, transparency, and a parameter giving a real part of the amplitude. They find rather constant values of the parameters except for the edge diffusivity; for example the radius decreases from 0.84 F at 2 GeV/c to 0.81 F at 12 GeV/c (incident laboratory momentum).

One may obtain an idea of how valid is the neglect of $g(\theta)$ in the high energy differential cross section by considering the polarization P given by:

$$P = \frac{2 \operatorname{Im} (f(\theta) \, g^*(\theta))}{|f(\theta)|^2 + |g(\theta)|^2} \quad . \quad (8\text{-}24)$$

For example in $\pi^- p$ scattering at 6 GeV/c incident laboratory momentum (1.61 GeV/c center of mass momentum) $|P| \lesssim 0.15$ from $10°$ to $30°$ and takes similar values at higher energies. There is, of course, an unknown phase factor in eq. (8-24), but we may take the consistent smallness of P as evidence that the neglect of $|g|^2$ at high energies and small angles is justified.

Not only the $\pi^+ p$ and $\pi^- p$ but also the $K^- p$ differential cross section exhibits a minimum at $-t = 0.8$, in all cases fading into a kink at incident laboratory momenta above about 4 or 5 GeV/c. There is a more persistent minimum in $\bar{p} p$ elastic scattering at $-t = 0.6$, but there are no minima in pp or $K^+ p$. These two latter systems, unlike $\pi^+ p, \pi^- p,$

8.1. ELASTIC SCATTERING

K^-p, or pp, do not contain obvious resonances. We shall comment further on this difference between exotic and nonexotic channels below, but now we pass on to the parametrization of elastic scattering at higher energies.

Forward Cross Sections and the Impact Parameter Representation

Some features of experimental results at energies from 6 to 20 GeV are illustrated in Fig. 8-3 plotted as a function of the invariant momentum transfer $t = -2q^2(1-\cos\theta)$. We see that the differential cross section falls several decades in the forward direction. To a good approximation

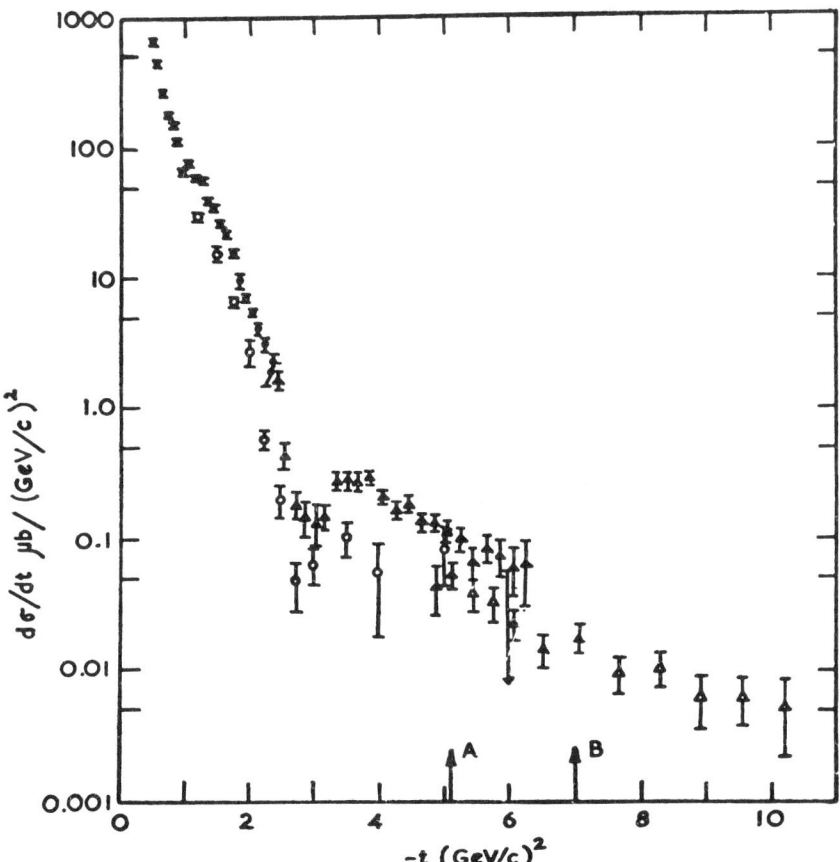

Fig. 8-3. Pion-nucleon elastic differential cross sections in the range $0.2 < -t < 1.0$ $(GeV/c)^2$ and for laboratory momenta between 3.5 and 13.5 GeV/c.

one can parametrize the forward high energy data[3] in scattering of *all* elementary particles for $1.0\ (\text{GeV}/c)^2 > -t > 0.2\ (\text{GeV}/c)^2$ by the expression [first given by Serber (1964)]:

$$\frac{d\sigma}{dt} = \left(\frac{d\sigma}{dt}\right)_{t=0} e^{a_1 t + a_2 t^2} . \tag{8-27}$$

At 13 GeV/c the following values are found[3]

$$pp: \frac{d\sigma}{dt} = 104 \pm 4\ \text{mb}/(\text{GeV}/c)^2,\ a_1 = 10.03 \pm 0.28\ (\text{GeV}/c)^{-1}$$

$$a_2 = 2.19 \pm 0.38\ (\text{GeV}/c)^{-4} .$$

$$\pi^+ p: \left(\frac{d\sigma}{dt}\right)_{t=0} = 37.8 \pm 1.8\ \text{mb}/(\text{GeV}/c)^2,\ a_1 = 8.93 \pm 0.27\ (\text{GeV}/c)^{-2},$$

$$a_2 = 2.36 \pm 0.34\ (\text{GeV}/c)^{-4} .$$

$$\pi^- p: \left(\frac{d\sigma}{dt}\right)_{t=0} = 42.4 \pm 1.6\ \text{mb}/(\text{GeV}/c)^2,\ a_1 = 9.71 \pm 0.26\ (\text{GeV}/c)^{-2},$$

$$a_2 = 3.02 \pm 0.35\ (\text{GeV}/c)^{-4} .$$

$$K^+ p: \left(\frac{d\sigma}{dt}\right)_{t=0} = 22.0 \pm 1.4,\ a_1 = 6.93 \pm 0.38\ (\text{GeV}/c)^{-2},$$

$$a_2 = 1.18 \pm 0.45\ (\text{GeV}/c)^{-4} .$$

The variation in a_1, a_2 in a momentum range 6-17 GeV/c is not significant. We use these results to illustrate the use of the *impact parameter representation* (Blankenbecler and Goldberger, 1964).

To put the above numbers in terms of amplitudes we reasonably assume, as discussed above, that the imaginary part of the nonflip amplitude dominates the diffraction peak. This enables us to obtain the imaginary part of the non-flip amplitude as essentially the square root of the

---

[3] The range of energy considered here is for incident momenta up to 20 (GeV/c). For the data see for example Foley et al., (1963a,b). The parameters are from Gasiorowitz (1967, p. 48).

8.1. ELASTIC SCATTERING

differential cross section. In the *impact parameter* representation we write for the imaginary part of the non-flip scattering amplitude in an obvious generalization of eq. (8-18):

$$f(t)(\equiv f(\theta)) = 2q^2 \int_0^\infty F(b)\, J_0(b\sqrt{-t})\, b\, db \qquad (8\text{-}28)$$

where $b = (\ell + \tfrac{1}{2})/q$ is the impact parameter and $F(b)$ the non-spin flip partial wave amplitude corresponding to impact parameter b. We may invert eq. (8-28) to get, where $x = \sqrt{-t}$:

$$F(b) = \frac{1}{2q^2} \int_0^\infty J_0(bx)\, x f(-x^2)\, dx \, . \qquad (8\text{-}29)$$

In eq. (8-29) we assume the form:

$$f(-x^2) = f(0)\, e^{-a_1 x^2/2} \qquad (8\text{-}30)$$

corresponding to eq. (8-27) with neglect of the a_2 term. Then:

$$F(b) = \frac{1}{2q^2\, a_1}\, F(0)\, e^{-b^2/2a_1} \, . \qquad (8\text{-}31)$$

The magnitude of $F(b)$ is a measure of the interaction (mainly absorptive) at a distance b of approach. Equation (8-31) thus represents a diffuse sphere with absorption strongest at the center and falling off as a Gaussian with a width $\sqrt{2a_1}$. For πp collisions of 12 GeV/c, $\sqrt{2a_1} \simeq .85$ fermis; for $K^+ p$ collisions at 12 GeV/c, $\sqrt{2a_1} \simeq .75$ fermis, corresponding to a lack of long range absorption in the $K^+ p$ system, which is also very evident in low energy collisions. The perturbative effect of the factor $e^{a_2 t^2/2}$ in the cross section is to weight the absorption more towards the center of the interaction sphere.

The Eikonal Approximation

We have discussed high energy scattering using certain rather explicit and simple types of absorptive potential. It is useful to have a method which

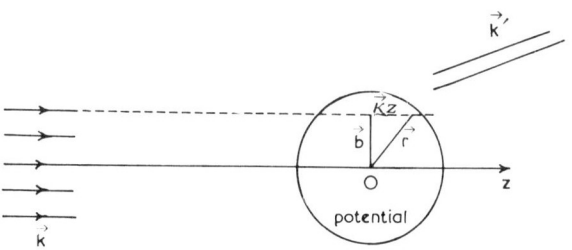

Fig. 8-4. Some features of the eikonal approximation to small angle potential scattering in a potential $V(r)$. The incident particle has momentum $\vec{k}$, in the z-direction, and $\hat{\kappa}$ is a unit vector in that direction. The dotted line parallel to the z-axis is the "approximate path" of the particle through the potential and the perpendicular vector b is the impact parameter. The position vector of a point on the "path" is given by $\vec{r} = \vec{b} + \hat{\kappa}z$.

gives a simple, though approximate, expression for the scattering amplitude with more general types of potential. Such a method, appropriate for high energy scattering at small angles, is the *eikonal approximation*. The physical idea of the eikonal approximation is that the incident particle, viewed semiclassically, takes a path approximately in a straight line parallel to the incident direction through the potential. This is illustrated in Fig. 8-4 and the scattering amplitude is given by:

$$f(\vec{k}', \vec{k}) = \frac{k}{2\pi i} \int e^{i(\vec{k}-\vec{k}')\cdot\vec{b}} \left\{ \exp\left[-\frac{i}{v}\int_{-\infty}^{\infty} V(\vec{b}+\hat{\kappa}z)\,dz\right] - 1 \right\} d^2b \quad (8\text{-}32)$$

where v is the velocity of the incident particle. If a is the radius of the potential, then eq. (8-32) holds[14] for energies and scattering angle θ such that:

$$ka \gg 1 \quad \text{and} \quad \theta^2\,ka \ll 1 \ . \quad (8\text{-}33)$$

---

[4] Glauber (1958) gives a derivation and discussion of eq. (8-32); if $\bar{V}$ is an average magnitude of the potential, the condition $\bar{V}/k \ll 1$ is imposed. Arnold (1967) discusses the application of the eikonal approximation to elementary particle scattering.

8.1. ELASTIC SCATTERING

In the eikonal approximation the scattering amplitude acquires a phase shift from each portion of the path, as is evident in eq. (8-32). We may write:

$$S(\vec{b}) = e^{2i\Delta(\vec{b})} \tag{8-34}$$

$$\Delta(\vec{b}) = -\frac{1}{2v}\int_{-\infty}^{\infty} V(\vec{b} + \vec{\kappa}z)\,dz \quad . \tag{8-35}$$

The term $S(\vec{b})$ is the S-matrix for impact parameter $\vec{b}$. The term $\Delta(\vec{b})$ is the corresponding phase shift, in general complex with its imaginary part given by the imaginary, that is absorptive, part of the potential.

In elementary particle scattering, rigorous arguments using potential theory as a basis cannot be made. However it seems plausible [see, for example, Arnold (1967)] that the eikonal approximation may be used in many elementary particle models, subject to the conditions of eq. (8-33). Corresponding to eq. (8-32) we write the spin-nonflip amplitude of pion-nucleon scattering, for example, as:

$$f(\theta) = iq_1\, a(\vec{q}) \tag{8-36}$$

$$a(\vec{q}) = \frac{1}{2\pi}\int d^2b\, e^{i\vec{q}\cdot\vec{b}}(1 - S(\vec{b})) \tag{8-37}$$

where $\vec{q} = \vec{q}_1 - \vec{q}_2$, $\vec{q}_1$ and $\vec{q}_2$ being the initial and final pion momenta respectively, and $S(\vec{b}) = e^{2i\Delta(\vec{b})}$, $\Delta(\vec{b})$ complex, is the S-matrix for impact parameter $\vec{b}$. [If we put $S(\vec{b}) = 1$, $b > R$ and $S(\vec{b}) = 0$, $b < R$ we regain eq. (8-18).] The term $\Delta(\vec{b})$ is to be calculated as a sum of contributions along the path shown in Fig. 8-4, and in the case of a potential model is given by eq. (8-35).

Composite Models and Multiple Scattering

In considering the black sphere model we saw how high energy scattering leads to the idea of elementary particles as extended objects, as is also evident in the form factors found by probing with electromagnetic radiation.

Such a composite structure arises naturally in the quark model where the nucleon structure observed is due to the extension of the 3-quark wave function. Whether or not this be true, any view of the nucleon as an extended object means that we can consider scattering as arising from a series of component scatters on different parts of the nucleon. This view can be given more definite form by using the eikonal approximation, as follows.

Let us consider the scattering of two elementary particles, which we shall call the pion and the nucleon. We first consider the pion to be non-composite and the nucleon to be composed of N fixed scattering centers located at $r_j = \hat{\kappa} z_j + s_j$, where s_j is the transverse coordinate vector. The incident pion accumulates phase from each of the scatterers [one may consider, for example, eq. (8-32) to see how this comes about] so that:

$$\Delta(\vec{b}) = \sum_{j=1} \Delta_j(\vec{b} - \vec{s}_j) \tag{8-38}$$

where $\Delta_j(\vec{b} - \vec{s}_j)$ a function of $\vec{b} - \vec{s}_j$ is the phase accumulated at impact parameter $\vec{b}$ due to the jth scatterer. The S-matrix is:

$$S(\vec{b}) = \prod_{j=1}^{N} \exp[2i \Delta_j(\vec{b} - \vec{s}_j)] . \tag{8-39}$$

We can express $S(\vec{b})$ more explicitly as a product of scatterings by the individual centers. Define the amplitude for scattering by the jth center:

$$a_j(\vec{q}) = \frac{1}{2\pi} \int d^2 b \, e^{i \vec{q} \cdot \vec{b}} \left(1 - e^{2i \Delta_j(\vec{b})}\right) .$$

Noting that only the x-y plane (Fig. 8-4) projection of q enters, we may define the two-dimensional Fourier transform:

$$\tilde{a}_j(\vec{b}) = \frac{1}{2\pi} \int d^2 q \, e^{-i \vec{q} \cdot \vec{b}} a_j(\vec{q}) \tag{8-40}$$

to give:

8.1. ELASTIC SCATTERING

$$\exp[2i\Delta_j(\vec{b}-\vec{s}_j)] = 1 - \tilde{a}_j(\vec{b}-\vec{s}_j)$$

$$S(\vec{b}) = \prod_{j=1}^{N} [1 - \tilde{a}_j(\vec{b}-\vec{s}_j)] . \tag{8-41}$$

In the expansion of this product the terms linear in $\tilde{a}$ give the single scatters, bilinear the double scatters, and so on. We can consider a more realistic case than that of fixed scattering centers, by considering the jth scattering center to have a distribution $\rho_j(r_j)$ in space so that the S-matrix for impact parameter $\vec{b}$ and with the composite scatterer remaining unchanged in its internal structure is:

$$<i|S(\vec{b})|i> = \prod_{j=1}^{N} \int d^3 r_j \, \rho_j(r_j) [1 - \tilde{a}_j(\vec{b}-\vec{s}_j)] \tag{8-42}$$

where $r_j = (\vec{s}_j, z_j)$.

In a realistic quark model for pion-nucleon (or any elementary particle) scattering both particles are composite. If we label the scatterers (quarks) from one particle by $j = 1, 2 \ldots N$ and the scatterers (quarks) from the other particle by $j' = 1, \ldots M$, we shall have a scattering amplitude $a_{jj'}(\vec{q})$ for momentum transfer $\vec{q}$ between the two quarks, and we define a two-dimensional Fourier transform:

$$\tilde{a}_{jj'}(\vec{b}) = \frac{1}{2\pi} \int d^2 \vec{q} \, e^{-i\vec{q}\cdot\vec{b}} \, a_{jj'}(\vec{q}) \tag{8-43}$$

leading to an S-matrix for fixed scatterers analogous to eq. (8-41):

$$S(\vec{b}) = \prod_{j=1}^{N} \prod_{j'=1}^{M} [1 - \tilde{a}_{jj'}(\vec{b}+\vec{s}_{j'}-\vec{s}_j)] . \tag{8-44}$$

For pion-nucleon elastic scattering in the realistic quark model, $N = 3$, $M = 2$ and we would take expectation values of eq. (8-44) between particles in their quark model ground states. In the quark model or composite models generally the observed diffraction minima and maxima, or kinks, are due to interference between successive orders of multiple scattering[5] as given by the expansion of eq. (8-44). For example, the most forward dip and rise, or at higher energies kink, in πN scattering is a region of interference between single and double scattering.

Continuum Approximation for Elementary Particle Matter

Instead of considering a small number of scatterers in each elementary particle, as in the quark model, one may consider a continuous distribution of matter in one or both of the particles. First we regard just one of the particles as composite, and return to eq. (8-42), where, if all the scatterers are the same, we find:

$$<i|S(\vec{b})|i> = \prod_{j=1}^{N} \int d^3 r_j \, \rho_j(\mathbf{r}_j) \, [1 - \tilde{a}_j(\vec{b} - \vec{s}_j)]$$

$$= \left[1 - \int d^2 s \, dz \, \rho(\vec{s}, z) \frac{1}{2\pi} \int d^2 q \, e^{-i \vec{q}(\vec{b} - \vec{s})} \, a(\vec{q}) \right]^N$$

As N becomes large, the scattering amplitude $a(\vec{q})$ of each individual scatterer becomes small, linearly with N, so that we write:

$$a(q) = A(\vec{q})/N \ .$$

Provided that the spatial extension of $\tilde{a}(\vec{b})$ is small compared to the distance in which $\rho(\vec{s}, z)$ changes appreciably we can replace $a(\vec{q})$ by $a(0)$ to obtain:

$$<i|S(\vec{b})|i> = \left(1 - \frac{2\pi A(0)}{N} D(\vec{b}) \right)^N \tag{8-45}$$

where the two-dimensional density $D(\vec{b})$ is a measure of the interacting matter encountered by the incident particles passing through the system at impact parameter b:

---

[5] These aspects of quark model scattering are discussed for example by Dean (1968) and Pagnamenta (1970).

8.1. ELASTIC SCATTERING

$$D(\vec{b}) = \int dz\, \rho(\vec{b}, z) \,. \tag{8-46}$$

As $N \to \infty$ we find that:

$$\left.\begin{array}{c} <i|S(\vec{b})|i> = \exp(-2\pi A(0) D(\vec{b})) \\ \text{or} \\ \Delta(\vec{b}) = i\pi A(0) D(\vec{b}) \end{array}\right\} \tag{8-47}$$

which gives us back the optical model discussed previously.

For both particles, A and B say, composite we start with eq. (8-44) and similarly proceed to the limit of infinite M and N obtaining:

$$\Delta(\vec{b}) = K_{AB} \int d^2\vec{b}'\, D_A(\vec{b}-\vec{b}')\, D_B(\vec{b}') \tag{8-48}$$

where $D_A(\vec{b})$ and $D_B(\vec{b})$ are the two-dimensional densities of interacting matter [eq. (8-46)] for particles A and B respectively, and K_{AB} is a complex interaction constant of the material of A with that of B.

It is a very reasonable hypothesis to take the extension of a hadronic particle in space, as given by its electromagnetic form factors as a measure of the density of interacting matter:

$$D(\vec{b}) = \frac{1}{2\pi} \int d^2\vec{q}\, e^{-i\vec{q}\cdot\vec{b}}\, F_{em}(\vec{q}^2) \,.$$

This approach has been used with success in high proton-proton scattering up to momentum transfers $|t| = 20$ (GeV/c)^2 (Wu and Yang, 1965; Byers and Yang 1966; Chou and Yang, 1967, 1968a,b).

8.2. TOTAL CROSS SECTIONS AND THE POMERANCHUK THEOREM

Total Cross Sections

Measurements of elementary particle total cross sections up to laboratory momentum of 60 GeV/c (Denisov et al., 1971) show a trend for cross sections to become constant at high energies (with the exceptions of K^+p and $\bar{p}p$ scattering). Indeed for the case where information is available, namely pp scattering, this may persist out to the higher energies of cosmic ray data (Perkins, 1960).

It is interesting to see the effect of an assumption of bounded range. Let us suppose that there is a radius R outside which there is no effective interaction. We know from eqs. (8-5), (8-6) that:

$$\sigma_{\text{total}}(q) = \frac{2\pi}{q^2} \sum_{\ell=0}^{\infty} (2\ell+1) \, \text{Re}\left(1 - \eta_\ell \, e^{2i\delta_\ell}\right).$$

The impact parameter $b = \frac{\ell}{q}$ measures the distance of approach, and thus our assumption is that for $b > R$ there is no interaction, and $\eta_\ell \, e^{2i\delta_\ell} - 1 = 0$, and:

$$\sigma_{\text{total}}(q) = \frac{2\pi}{q^2} \sum_{\ell=0}^{qR} (2\ell+1) \, \text{Re}\left(1 - \eta_\ell \, e^{2i\delta_\ell}\right) \leq 2\pi R^2. \quad (8\text{-}49)$$

If the range is bounded, that is, R in eq. (8-49) is independent of the energy E of the incident particle, then σ_{total} is bounded and it may seem plausible that it tends to a constant (though of course it may oscillate finitely).

Term R may not be bounded but may be an increasing function $R(E)$ of E. If the target particle has a probability density distribution:

$$P(b) = P_0 \exp(-b/R_0)$$

at impact parameter b, and the probability of interaction is bounded above by E^N (N fixed), then the probability of interaction at impact parameter b and energy E will satisfy:

8.2. TOTAL CROSS SECTIONS

$$P(E, b) < P_0 \, E^N \exp(-b/R_0)$$

which is negligible for $b > R(E)$ where:

$$R(E) = NR_0 \ln E$$

giving from eq. (8-49) the *Froissart bound*:

$$\sigma_{total} < C \ln^2 E \qquad (8-50)$$

where C is a constant.

The bound, eq. (8-50), was first deduced (Froissart, 1961) assuming the Mandelstam representation, and was later deduced by Martin (1966) assuming the axioms of quantum field theory.

Pomeranchuk Theorem

The Pomeranchuk theorem states that if $\sigma_+(E)$ and $\sigma_-(E)$ are the total cross sections at laboratory energy E for incident particle and incident anti-particle on the same target, then $\sigma_+(\infty) = \sigma_-(\infty)$ or:

$$\sigma_{total}(A+B \to A+B) = \sigma_{total}(\bar{A}+B \to \bar{A}+B) \, .$$

The original proof of Pomeranchuk (1958) is based on the assumption of constant cross section at large energies; other authors (Amati et al., 1960; Weinberg, 1961) have proved the theorem under less restrictive conditions; for example Weinberg's proof replaces the condition of constancy by the requirement that $\sigma_+(E) - \sigma_-(E)$ changes sign at most a finite number of times. All proofs make the assumption that the imaginary part is greater than the real part at high energies.

We shall now illustrate Pomeranchuk's argument (which was originally given for pp and $\bar{p}p$ scattering) by the case of $\pi^+ p$ and $\pi^- p$ scattering. Let k, ω be the laboratory momentum, energy of the pion, and $\omega_N = \dfrac{M_\pi^2}{2M_N}$ the laboratory energy at the nucleon pole. If $F^\pm(\omega)$ and $\sigma^\pm(\omega)$ denote the $\pi^\pm$ forward amplitudes and total cross sections in the laboratory system, then we define:

$$F(\omega^2) \equiv \frac{F^-(\omega) - F^+(\omega)}{2\omega} \equiv \frac{F^-(\omega) - F^-(-\omega)}{2\omega} \tag{8-51}$$

the last relation following from crossing. Since [from eq. (8-5), the optical theorem being of the same form in the laboratory as in the center of mass system]:

$$\text{Im } F^{\pm}(\omega) = \frac{k}{4\pi} \sigma^{\pm}(\omega) \tag{8-52}$$

and since we are adopting the Pomeranchuk assumption that $\sigma^{\pm}(\omega)$ tend to constants at large energy, then we can write a subtracted dispersion relation for $F(\omega^2)$ (see section 3.4):

$$\text{Re}(F(\omega^2) - F(0)) = \frac{f^2}{\omega_N^2} \frac{\omega^2}{\omega^2 - \omega_N^2}$$

$$+ \frac{\omega^2}{4\pi^2} P \int_{M_\pi}^{\infty} \frac{k' d\omega'}{\omega'^2(\omega'^2 - \omega^2)} (\sigma^-(\omega') - \sigma^+(\omega')) . \tag{8-53}$$

Let ϵ be a number such that for $\omega > \epsilon$, $\sigma_{\pm}(\omega) = \sigma_{\pm}(\infty) + O\left(\frac{1}{\omega}\right)$ and divide the range of integration into two halves $m < \omega' < \epsilon$ and $\epsilon < \omega' < \infty$. Then we find that for $\omega \gg \epsilon$:

$$\text{Re } F(\omega^2) = -(\sigma^-(\infty) - \sigma^+(\infty)) \frac{1}{4\pi^2} \ell n \left(\frac{\omega}{\epsilon}\right) + 0(1) . \tag{8-54}$$

Now from eq. (8-52) and the Pomeranchuk assumption Im $F(\omega^2)$ is bounded at large energy, and from the discussion in section 8.1 above we expect Re $F(\omega^2)$ < Im $F(\omega^2)$. The only way that these assumptions can be reconciled with eq. (8-54) is by setting:

$$\sigma^-(\infty) = \sigma^+(\infty) . \tag{8-55}$$

This is the Pomeranchuk theorem for the particular case of $\pi^{\pm} p$ scattering.

We see from Fig. 8-5 that the $\pi^{\pm} p$ total cross sections have more nearly fulfilled the Pomeranchuk theorem at 60 GeV than the $K^{\pm} p$ or the pp, $\bar{p}p$ total cross sections. In neither of these two latter cases

8.2. TOTAL CROSS SECTIONS

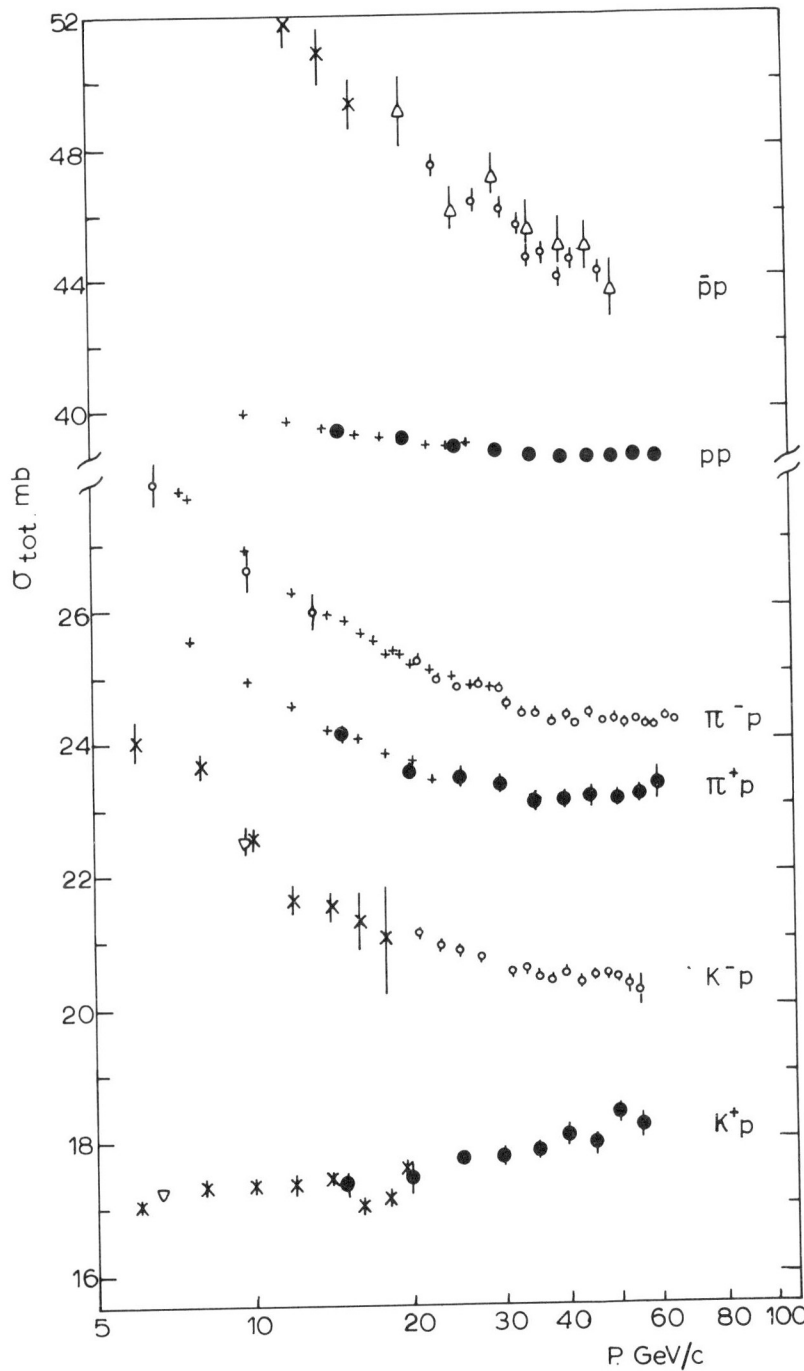

Fig. 8-5. Total cross sections 6 to 60 GeV/c (Denisov et al., 1971).

would one expect it to have been fulfilled in terms of the original Pomeranchuk proof, because one of the conditions for validity — namely constancy of the cross sections (or even difference of cross sections) — has not been satisfied.

8.3. PARTICLE EXCHANGE AND REGGE EXCHANGE

In section 8.1 we discussed the interpretation of the high energy elastic scattering data by the optical model, and consistently with that model but independently of its special features we interpreted the forward peak as diffraction scattering associated with the occurrence of a fairly large number of partial waves in the amplitude. We believe that consideration of the exchange of virtual particles may give a more fundamental interpretation of scattering, perhaps most obviously relevant and fruitful in *non-diffractive* inelastic two body reactions such as $\pi^- p \to \pi^0 n$. Such *non-diffractive* scattering is obviously also a component in elastic scattering, as we see immediately from the relations, eq. (2-87), between the elastic $\pi^{\pm} p$ scattering amplitudes and the charge exchange scattering amplitude.

The role of particle exchange is suggested not only by considerations of the range of nuclear forces, such as those of Yukawa in predicting the pion, and the accepted theory of the dominance of pion exchange for high angular momentum in the nucleon-nucleon interaction,[6] but also by the empirical observation of Morrison (1966) that cross sections behave with energy like:

$$\sigma = \sigma_0 \, q_L^{-n} \qquad (8\text{-}56)$$

---

[6] This has been used (MacGregor, Arndt, and Wright, 1968) to obtain a value of the pion-nucleon coupling constant not significantly in disagreement with that obtained from pion-nucleon scattering.

8.3. PARTICLE EXCHANGE AND REGGE EXCHANGE

where q_L is the laboratory momentum and n depends on the quantum numbers allowed to be exchanged in the particular scattering process: (a) $n \sim 0$, vacuum exchange; (b) $n \sim 2$, charge or isospin exchange; (c) $n \sim 2.5$, strangeness exchange; (d) $n \sim 3\text{-}4$, baryon number exchange. Low energy pion-nucleon scattering [see Chapter 9 and, for example, Hamilton (1964)] can be interpreted as mainly due to the Born term (nucleon exchange), N^* exchange, 2π-exchange, and 3π-exchange; nucleon-nucleon scattering up to 300 MeV can be explained by pseudoscalar and vector meson exchange and 2π-exchange (Scotti and Wong, 1965). These results are obtained by using matrix elements illustrated by diagrams like Fig. 8-6(a), giving rise to diagrams like Fig. 8-6(b).

While rescattering of diagrams like Fig. 8-6(a) is necessary to explain lower energy scattering, it is even more essential in high energy scattering. The first reason is that any pole term, or sum of pole terms, like Fig. 8-6(a) necessarily produces a real amplitude, while in high energy scattering the amplitude is dominantly imaginary. Secondly the exchange of vector, or higher spin mesons, in an amplitude represented by Fig. 8-6(a) gives a too large high energy cross section. We illustrate this result by the scattering of pseudoscalar mesons:

If the exchanged particle is a vector (spin 1) meson, then the scattering amplitude is:

$$A(s,t) \sim g^2 \frac{(p_1+p_3)^\mu \, g_{\mu\nu} \, (p_2+p_4)^\nu}{t-m^2} \tag{8-57}$$

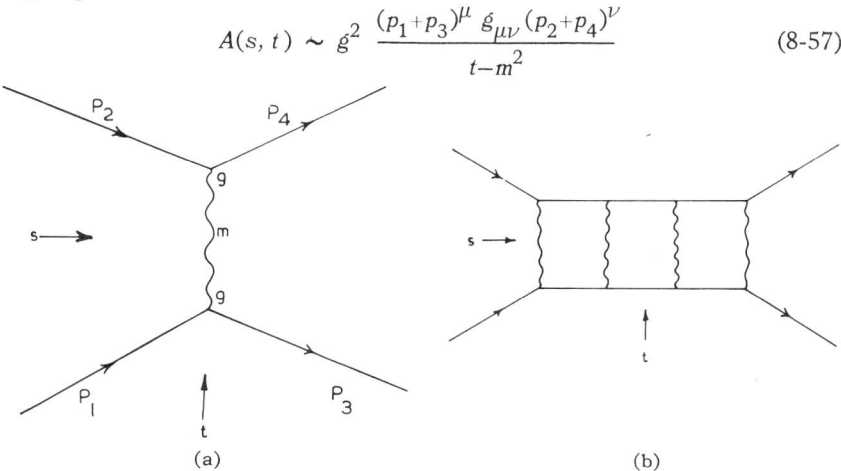

Fig. 8-6. (a) The exchange of a particle of mass m with coupling constant g in an elastic scattering process; (b) repeated scatterings of the type illustrated in (a).

and thus for large s:

$$A(s, t) \sim g^2 \frac{s}{t-m^2} \; ; \; \frac{d\sigma}{dt} \sim \left|\frac{g^2}{t-m^2}\right|^2 . \tag{8-58}$$

If the meson exchanged is a tensor (spin 2), then there are more momentum factors in the numerator (in order to cancel the tensor subscripts), and:

$$A(s, t) \sim \frac{g^2 s^2}{t-m^2} , \; \frac{d\sigma}{dt} \sim \left|\frac{g^2 s}{t-m^2}\right|^2 . \tag{8-59}$$

In general the exchange of a particle of integer spin J gives an amplitude

$$\sim g^2 \frac{P_J(\cos\theta_t)}{(t-m^2)} \sim \frac{g^2 s^J}{(t-m^2)}$$

where θ_t is the *t-channel* scattering angle. Thus:

$$\frac{d\sigma}{dt} \sim \frac{g^4 s^{2J-2}}{|t-m^2|^2} .$$

We see the difficulty that cross sections would be predicted to arise with energy if the exchange of any single higher spin particle were dominant. Indeed in the case of pion-nucleon charge exchange scattering:

$$\pi^- + p \to \pi^0 + n$$

the lowest spin particle that can be exchanged is the ρ and eq. (8-58) is certainly in disagreement with the experimental data (see section 8.4).

Thus no simple sum of amplitudes like eq. (8-6a) is going to explain high energy scattering. Also in the high energy case, where a shorter range part of the force comes in with corresponding contributions from the exchange of more massive virtual systems, we would be extremely fortunate if we needed to consider only the exchange of particles like the π or ρ. We might a priori expect graphs like, for example, Fig. 8-7 to contribute to high energy elastic scattering. The Regge hypothesis solves

8.3. PARTICLE EXCHANGE AND REGGE EXCHANGE

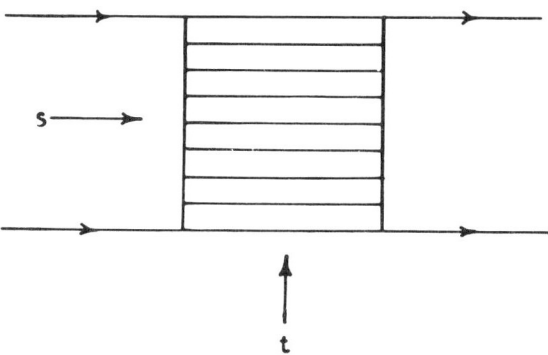

Fig. 8-7. A multiperipheral contribution to elastic scattering in the s-channel (Fubini, 1964).

both the technical difficulty of the reality and possible divergence of non-rescattered exchange amplitudes and the difficulty of incorporating, in principle, graphs like Fig. 8-7. The Regge pole hypothesis replaces particle or pole exchange that has a contribution to the amplitude such as:

$$A(s,t) \sim g^2 \frac{s}{t-m^2} \qquad (8\text{-}60)$$

by Regge pole exchange with a contribution to the amplitude:

$$A(s,t) \sim \beta(t) \frac{1 \pm \exp(-i\pi a(t))}{\sin \pi a(t)} s^{a(t)} \qquad (8\text{-}61)$$

where $a(t)$ is a *Regge trajectory* and $a(m^2) = 1$ (if the Regge trajectory corresponds to a physical particle of spin 1). We note that eq. (8-61) has a pole at $t = m^2$, that for suitable behavior of $a(t)$ it gives well-behaved cross sections for large s in contrast with eqs. (8-58), (8-59), and that it is a complex amplitude.

It is conceptually possible for a sum of amplitudes like eq. (8-61) to dominate the high energy scattering amplitude, and this conjecture was strongly investigated both experimentally and theoretically during the period 1960-1967. Difficulties have appeared which render this hypothesis unattractive and almost untenable, and some of these will be discussed below. However, the success of Regge poles in generally giving a first order description of the energy dependence of cross sections leaves them as an important component of high energy phenomenology. We discuss the Regge pole formalism and the role of Regge poles in the next few sections.

8.4. t-CHANNEL REGGE POLES

First we need to introduce t-channel Regge poles into the pion-nucleon scattering formalism, and we do this by expanding the two invariant amplitudes $A(s,t)$, $B(s,t)$, for large s, as a sum of Regge poles. It is convenient to define:

$$A' = A + \frac{\nu B}{1 - t/4M_N^2} . \qquad (8\text{-}62)$$

The amplitudes $F^{(\pm)}(\nu)$ defined by eq. (3-79) are equal to $\frac{1}{4\pi} A'^{(\pm)}(\nu, t=0)$. We expand A' and B in t-channel partial wave amplitudes using eq. (9-28) to obtain:

$$A'^{(\pm)}(s, t) = \sum_j \frac{8\pi}{\bar{p}^2} (p\bar{p})^j \left(j + \frac{1}{2}\right) f_+^{(\pm)j}(t) \, P_j(z)$$

$$B^{(\pm)}(s, t) = \sum_j 8\pi (p\bar{p})^{j-1} P'_j(z) \, \overline{f}_-^{(\pm)j}(t) \qquad (8\text{-}63)$$

where z is the cosine of the scattering angle in the t-channel, p and $\bar{p}$ are the center of mass momentum in the $\pi\pi$ and $N\bar{N}$ systems respectively, and f_+^j and f_-^j are defined in eq. (9-26) with:

$$\overline{f}_-^{(\pm)j}(t) = \frac{j + \frac{1}{2}}{\sqrt{j(j+1)}} f_-^{(\pm)j}(t) \qquad (8\text{-}64)$$

and are expressed in terms of the A and B amplitudes by:

8.4. t-CHANNEL REGGE POLES

$$f_+^{(\pm)j}(t) = \frac{1}{8\pi} \left\{ \frac{\bar{p}^2}{(p\bar{p})^j} A_j'^{(\pm)} \right\}$$

$$\bar{f}_-^{(\pm)j}(t) = \frac{1}{16\pi(p\bar{p})^{j-1}} \left\{ B_{j-1}^{(\pm)} - B_{j+1}^{(\pm)} \right\}$$

(8-65)

where:

$$A_j'^{(\pm)} = \int_{-1}^{1} dz \, A'^{(\pm)}(s,t) \, P_j(z), \quad B_j^{(\pm)} = \int_{-1}^{1} dz \, B^{(\pm)}(s,t) \, P_j(z) \,. \quad (8\text{-}67)$$

We introduce the fixed-t dispersion relations such as eq. (4-59) and use crossing of the pions ($s \leftrightarrow u$) to find, on using $s = 2p\bar{p}z - p^2 - \bar{p}^2$:

$$f_+^{(\pm)j}(t) = [1 \pm (-1)^j] \, f_+^{(\pm)}(j,t) \quad \text{where}$$

$$f_+^{(\pm)}(j,t) = \frac{\bar{p}^2}{8\pi^2} \int ds' \left[a_1^{(\pm)}(s',t) + \frac{s'+p^2+\bar{p}^2}{2\bar{p}^2} M_N \, b_1^{(\pm)}(s',t) \right] Q_j\left(\frac{s'+p^2+\bar{p}^2}{2p\bar{p}}\right)$$

(8-68)

$$\bar{f}_-^{(\pm)j}(t) = [1 \pm (-1)^j] \, \bar{f}_-^{(\pm)}(j,t) \quad \text{where}$$

$$\bar{f}_-^{(\pm)}(j,t) = \frac{1}{16\pi^2} \int ds' \, b_1^{(\pm)}(s',t) \left[Q_{j-1}\left(\frac{s'+p^2+\bar{p}^2}{2p\bar{p}}\right) - Q_{j+1}\left(\frac{s'+p^2+\bar{p}^2}{2p\bar{p}}\right) \right].$$

(8-69)

We see that $f_+^{(\pm)}(j,t)$ and $\bar{f}_-^{(\pm)}(j,t)$ define analytic continuations that [being without the factor $(-1)^j$ that has unsuitable behavior (see for example Squires, 1963; Collins and Squires, 1968) for large j] are suitable for making Sommerfeld-Watson transforms. Using the fact that $P_j(z) = (-1)^j P_j(-z)$ for j integer, we find that:

$$A'^{(\pm)}(s,t) = \sum_{j=1}^{\infty} \frac{8\pi}{p^2} \left[(j+\tfrac{1}{2}) \, f_+^{(\pm)}(j,t)(P_j(z) \pm P_j(-z)) \right]$$

$$B^{(\pm)}(s,t) = \sum_{j=1}^{\infty} 8\pi \, (p\bar{p})^{j-1} \, \bar{f}_-^{(\pm)}(j,t)(P_j'(z) \mp P_j'(-z)) \,. \quad (8\text{-}70)$$

We make on eq. (8-70) Sommerfeld-Watson transforms, which give, as in Chapter 4:

$$A'^{(\pm)}(s,t) = \sum_i C_i^{(\pm)}(t) \frac{P_{a_i^{(\pm)}(t)}(+z) \pm P_{a_i^{(\pm)}(t)}(-z)}{\sin \pi \, a_i^{(\pm)}(t)}$$

+ background integral (8-71a)

$$B^{(\pm)}(s,t) = \sum_i D_i^{(\pm)}(t) \frac{P'_{a_i^{(\pm)}(t)}(+z) \mp P'_{a_i^{(\pm)}(t)}(-z)}{\sin \pi \, a_i^{(\pm)}(t)}$$

+ background integral . (8-71b)

Here C and D are essentially residues at the Regge poles, but also have absorbed factors from eq. (8-70), while $a_i^{(+)}(t)$ and $a_i^{(-)}(t)$ come from $f^{(+)}$ and $f^{(-)}$ functions respectively and are known as Regge trajectories of positive and negative signature respectively. For positive signature the trajectory corresponds to $j^P = 0^+$, 2^+, 4^+... mesons in the $\pi\pi \leftrightarrow N\bar{N}$ channel (0^-, 2^-, 4^-... mesons being forbidden by parity conservation to couple to two pions). This is readily seen from eq. (8-70) where $j = $ odd makes no contribution to the positive signature amplitude. Similarly for negative signature the trajectory corresponds to $j^P = 1^-$, 3^-, 5^-... mesons in the $\pi\pi \leftrightarrow N\bar{N}$ channel. From Bose statistics, mesons coupled to two pions with even (odd) spin have $I = 0(1)$, giving the fact that positive signature trajectories have $I = 0$ and negative have $I = 1$. [Signature can be defined as the relative sign between $P_a(z)$ and $P_a(-z)$ in eq. (8-71a) or equivalently between 1 and $\exp(-i\pi a)$ in eq. (8-72).]

For positive $t \geq 4M_N^2$ and negative $s (= 2 p\bar{p}z - p^2 - \bar{p}^2, -1 \leq z \leq 1)$, eq. (8-71) can give the amplitudes for scattering in the t-channel, the denominators $\sin \pi \, a_i^{(\pm)}(t)$ giving rise to resonances in the scattering as shown in Chapter 4. By analytic continuation, we can consider scattering in the s channel, $\pi N \to \pi N$, in which t is negative. As also discussed in Chapter 4, for large s we neglect the background and take the asymptotic form of $P_a(\pm z)$, giving:

8.5. THE ρ-REGGE POLE AND CHARGE EXCHANGE SCATTERING

$$A'^{(\pm)}(s,t) \sim \sum_i C_i^{(\pm)}(t) \frac{1 \pm \exp(-i\pi a_i^{(\pm)}(t))}{\sin \pi a_i^{(\pm)}(t)} \left(\frac{s}{s_0}\right)^{a_i^{(\pm)}(t)} \quad (8\text{-}72\text{a})$$

$$B^{(\pm)}(s,t) \sim \sum_i a_i^{(\pm)}(t) D_i^{(\pm)}(t) \frac{1 \pm \exp(-i\pi a_i^{(\pm)}(t))}{\sin \pi a_i^{(\pm)}(t)} \left(\frac{s}{s_0}\right)^{a_i^{(\pm)}(t)-1}. \quad (8\text{-}72\text{b})$$

It is common to use this asymptotic expression even down to $s \sim 2(\text{GeV}/c)^2$ and to take $s_0 (=2\ p\bar{p})$ as a constant "scale factor."

Using eq. (8-72) it is readily seen, for example from eq. (8-75) below, that the characteristic energy dependence of the cross section induced by a single Regge pole $a(t)$ is given by:

$$\frac{d\sigma}{dt} \propto s^{2a(t)-2}$$

or: $\quad (8\text{-}73)$

$$\frac{d\sigma}{d\Omega} \propto s^{2a(t)-1}.$$

One of the purposes of the next section is to illustrate the characteristics of the exchange of a single Regge pole by viewing, as far as is possible, the phenomena in $\pi^- p \to \pi^0 n$ as due to the exchange of a single Regge trajectory, the ρ.

8.5. THE ρ-REGGE POLE AND CHARGE EXCHANGE SCATTERING

The process of πp charge exchange scattering, $\pi^- p \to \pi^0 n$, is one of the principal proving grounds of theories of high energy inelastic two body reactions. Reactions with charge, or isospin, or strangeness exchange, have the important property of only containing the *nondiffractive* part of elementary particle scattering, while high energy elastic scattering necessarily contains a (usually dominant) diffractive component. It is to such reactions

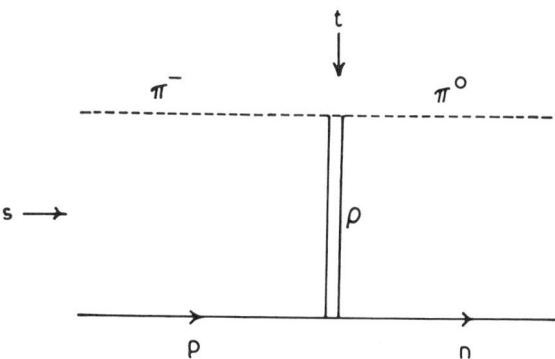

Fig. 8-8. The exchange of a ρ-meson, or Reggeized ρ-meson, in the t-channel giving rise to charge exchange scattering in the s-channel.

that we look for important clues on the nature of the nondiffractive scattering, and particularly we look to $\pi^- p \to \pi^0 n$, which is the best studied among meson-baryon reactions (though the data is significantly less accurate than in $\pi^+ p$ elastic scattering, mainly because of technical difficulties of π^0 detection). Another attraction of πp charge exchange is that because of quantum number limitations the known mesons that can be exchanged (see Fig. 8-8) are the ρ and ρ' (Bingham et al., 1972) of which the ρ' is relatively weakly coupled. This leads naturally to a one Regge pole theory of high energy forward charge exchange scattering, and insofar as this viewpoint is correct, and thus Regge pole theory is here unencumbered by complications, the process is a test of the pure Regge pole hypothesis. We shall see that while there are great successes, at least some modification is necessary, and in sections to come we shall see that one success (the explanation of the dip at $t = -0.6$) is probably more complicated than simple Regge pole theory would suggest.

As discussed in Chapter 2 the center of mass differential cross section for pion-nucleon scattering is:

$$\frac{d\sigma}{d\Omega} = |f(\theta)|^2 + |g(\theta)|^2 + 2(\vec{P}_1 \cdot \vec{n})\,\text{Im}\,(f(\theta)\,g^*(\theta)) \qquad (8\text{-}74)$$

8.5. THE ρ-REGGE POLE AND CHARGE EXCHANGE SCATTERING

where $\vec{n} = \vec{q}_1 \times \vec{q}_2 / |\vec{q}_1 \times \vec{q}_2|$ is the normal to the scattering plane, and $\vec{P}_1$ is the polarization of the target nucleons. Equation (8-74) can be expressed in terms of the A' and B amplitudes, using eqs. (3-63) and (3-66), as:

$$\frac{d\sigma}{dt} = \frac{1}{\pi s}\left(\frac{M_N}{4q}\right)^2 \left\{ (1 - t/4M_N^2)|A'|^2 + (t/4M_N^2)\frac{q_L^2 - st/4M_N^2}{1 - t/4M_N^2}|B|^2 \right\}$$

$$+ \frac{\sin\theta}{16\pi\sqrt{s}} (\vec{n} \cdot \vec{P}_1) \operatorname{Im}(A'B^*) \tag{8-75}$$

where $q^2 = (s - M_N^2 - M_\pi^2)^2/4M_N^2 - M_\pi^2$, is the square of the laboratory 3-momentum.

For charge exchange scattering the amplitudes A' and B in eq. (8-75) are $-\sqrt{2}A'^{(-)}$ and $-\sqrt{2}B^{(-)}$. The high energy behavior should be given by eqs. (8-73) and (8-74) with nonstrange mesonic Regge trajectories $a_I^{(-)}(t)$ having isospin $I = 1$ and G-parity $= +1$. As discussed above, such trajectories comprise particles having $J^P = 1^-, 3^-, 5^-\ldots$. The ρ-meson is the only strongly coupled 1^- particle known, so we write:

$$A'^{(-)} = C_\rho(t) \frac{1 - \exp(-i\pi a_\rho(t))}{\sin \pi a_\rho(t)} \left(\frac{s}{s_0}\right)^{a_\rho(t)} \tag{8-76a}$$

$$B^{(-)} = a_\rho(t) D_\rho(t) \frac{1 - \exp(-i\pi a_\rho(t))}{\sin \pi a_\rho(t)} \left(\frac{s}{s_0}\right)^{a_\rho(t)-1} \tag{8-76b}$$

where $a_\rho(t)$ is the ρ-meson trajectory. Substitution of eq. (8-76) in eq. (8-75) gives the following results:

(i) **Shrinkage**

$$\frac{d\sigma}{dt}(s, t) = F(t) \left(\frac{s}{s_0}\right)^{2a_\rho(t)-2} \tag{8-77}$$

This formula gives *shrinkage*; that is, for fixed t the differential cross section shrinks as s increases if $a_\rho(t) < 0$, and the rate of shrinkage is given by the value of $a_\rho(t)$. This at the same time verifies the Regge hypothesis, which predicts such exponential shrinkage and gives a means for determining the ρ-trajectory for $t < 0$. It turns out that a linear trajectory fits the data; a fit in the range $0 > -t > 1$ (GeV/c)^2 is:

$$a_\rho(t) = 0.57 + 0.91\, t \tag{8-78}$$

which extrapolates closely to the ρ-meson position at $\alpha = 1$, $t = 0.56$ (GeV/c)^2, thus confirming the connection of Regge trajectories with particles.

(ii) **Phase Rule**

The term $a_\rho(0)$ according to eq. (8-78) is roughly $\frac{1}{2}$ and so the real and imaginary parts of the forward amplitude are roughly equal. The imaginary part can be found from the optical theorem together with isospin invariance. Using eq. (3-81) and the fact that $A'^{(\pm)}(\nu, t=0) = 4\pi F^{(\pm)}(\nu)$ we find that:

$$\frac{1}{q_L} \text{Im}\, A'^{(\text{c.e.})}(t=0) = \frac{1}{\sqrt{2}} (\sigma_{\text{tot}}(\pi^+ P) - \sigma_{\text{tot}}(\pi^- p)) . \tag{8-79}$$

The real part to within a sign can be found from the observed charge exchange forward cross section, and is found to agree with the phase rule.

(iii) **Dip at** $t = -0.6$

The contribution of B to the cross section must vanish at $t = 0$ and a strong B term thus explains the rise near $t = 0$. However, from eq. (8-76b), B vanishes at $\alpha(t) = 0$ corresponding to $t = -0.6$, which is just where a dip is observed in the differential cross section. This point is further discussed in section 8.6 below, dealing with factorization.

The points (i), (ii), and (iii) support the description of high energy (2-18 GeV/c) πp charge exchange by a single Regge ρ-exchange. However, this model has A' and B in phase according to eq. (8-76) and consequently zero polarization [eq. (8-58)], and experiment shows appreciable polarization at momenta greater than 5 GeV/c (Bonamy et al., 1968). This polarization can be produced by not too large additions to the amplitude consistent with the maintenance of Regge ρ-exchange as the dominant effect and (i), (ii), and (iii) above.

The required modification may be produced by invoking other Regge trajectories of the same quantum numbers as the ρ, or by the effects of absorptive scattering subsequent to the exchange of the Regge ρ. These possibilities are discussed further in sections 8.7 and 8.8 below, and we shall see that an absorptive mechanism can provide an alternative explanation for dip to the one discussed above and in the next section.

8.6. FACTORIZATION AND ZEROS OF REGGE POLE RESIDUES

The vanishing of the B amplitude is a particular case of a general feature of Regge poles which is best described using helicity amplitudes. Consider a scattering process in the s-channel

$$1 + 3 \to 2 + 4$$

which is achieved by t-channel exchange as shown in Fig. 8-9. Then the t-channel helicity amplitudes are given by eq. (2-136). In the factor $d^j_{\lambda\mu}(\theta)$, $\theta = \theta_t$ is the t-channel scattering angle, j is the spin of the exchanged particle or system, and $\lambda = \lambda_1 - \lambda_2$, $\mu = \lambda_3 - \lambda_4$. The terms λ_1, λ_2, λ_3, and λ_4 are the t-channel helicities of particles 1, 2, 3, and 4 respectively. If an elementary particle is exchanged in Fig. 8-9, we see, on considering the center of momentum system at each vertex, that we must have $|\lambda| \leq j$ and $|\mu| \leq j$; $d^j_{\lambda\mu}(\theta)$ vanishes when this condition is not satisfied.

Now suppose that in Fig. 8-9 a Regge pole is being exchanged so that $j = \alpha(t)$. Then at integer (or $\frac{1}{2}$-odd integer if the Regge pole is a fermion) values of α, one attaches the label *sense* or *nonsense* to the lower vertex, *sense* if $|\lambda| \leq \alpha$ and *nonsense* if $|\lambda| > \alpha$; similarly for the upper vertex. The whole amplitude is classified in three categories: (a) a sense-sense amplitude T_{ss} if both vertices are sense; (b) a sense-nonsense amplitude T_{sn} if one vertex is sense and the other nonsense; (c) a nonsense-

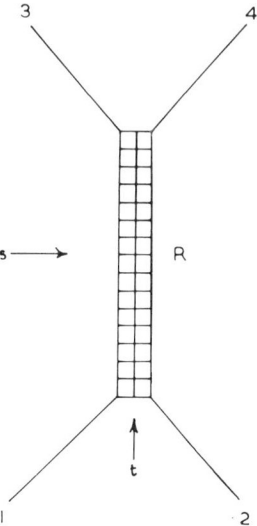

Fig. 8-9. The *t*-channel exchange of a Regge pole.

nonsense amplitude T_{nn} if both vertices are nonsense. In the case of pion-nucleon scattering where particles 3 and 4 are pions, then at $a = 0$ the top vertex is always sense while the bottom vertex is sense for spin-nonflip and nonsense for spin flip.

The behavior of the three classes of amplitude near an integer (or $\frac{1}{2}$ - odd integer as the case may be) value of a, $a = j_1$ say, is given by:

$$T_{ss} \propto \beta_{ss} \Sigma_a s^a, \; T_{sn} \propto (a-j_1)^{\frac{1}{2}} \beta_{sn} \Sigma_a s^a, \; T_{nn} \propto (a-j_1) \beta_{nn} \Sigma_a s^a \; . \quad (8\text{-}80)$$

In eq. (8-80) the β's are the Regge pole residues corresponding to C and D in eq. (8-76) and Σ_a are the signature factors:

$$\Sigma_a = \frac{1 \pm \exp(-i\pi a)}{\sin \pi a} .$$

Equation (8-80) is obtained by taking the Sommerfeld-Watson transform of eq. (2-136). In eq. (2-136) the $d^j_{\lambda\mu}(z)$, where $z = \cos\theta_t$, are analytic functions of z except for possible singularities at $z = \pm 1$, and can be

8.6. FACTORIZATION AND ZEROS OF REGGE POLE RESIDUES

analytically continued in $j \to a$ so that for Re $a > -\frac{1}{2}$, $d_{\lambda\mu}^{a}(z) \sim z^a$ for z large. We thus obtain the expression eq. (8-80) where the factors $(a-j_1)^{\frac{1}{2}}$ and $(a-j_1)$ come from the $d_{\lambda\mu}^{a}(z)$ (Collins and Squires, 1968, p. 101).

We also need the *factorization condition* on Regge pole residues. We are led to this if we assume that the Regge pole residue is a product of a factor due to the vertex 12R and a factor due to the vertex 34R (Fig. 8-9). In addition to Fig. 8-9 we can consider processes in which *both* vertices are 12R or *both* 34R, which leads to:

$$(\beta_{sn})^2 = \beta_{ss} \beta_{nn} . \qquad (8-81)$$

We now consider the behavior of the β's and the consequent behavior of the T's near $a(t) = j_1$ in the region of t and s corresponding to physical scattering in the s-channel. If j_1 is a *right signature point*, that is such that Σ_a has a pole, the β's must be such so as to cancel this pole. (Such a pole is called a *ghost state* and the means of its elimination is termed a *ghost-killing mechanism*.) At a *wrong signature point*, that is such that Σ_a has no pole, such a cancellation is not necessary. Also there is no reason to expect a branch-point singularity of the form $(a(t)-j_1)^{\frac{1}{2}}$ in an amplitude, nor is one ever observed. Thus β_{sn} contains a factor $(a-j_1)^{(2n+1)/2}$, n integer, leading from eq. (8-81), to corresponding integral powers of $(a-j_1)$ in β_{ss} or β_{nn} or both. From these considerations a number of related behaviors of T_{ss}, T_{sn}, and T_{nn} at $a \sim j_1$ arise. The principal ones are classified and named as in Table 8-1.

We see now that the zero in the B-amplitude charge exchange scattering is not a unique consequence of Regge pole theory, but because of the freedom in the residue functions is just one of a number of possible behaviors. It is termed a nonsense choosing mechanism at a wrong signature point of the ρ-trajectory.

	Name	T_{ss}	T_{sn}	T_{nn}
Right signature	(i) Choosing sense mechanism	finite nonzero	$(a(t)-j_1)$	$(a(t)-j_1)^2$
	(ii) Choosing nonsense (Gell-Mann) mechanism	finite nonzero	finite nonzero	finite nonzero
	(iii) Noncompensating† or choosing nonsense mechanism	$(a(t)-j_1)$	$(a(t)-j_1)$	$(a(t)-j_1)$
Wrong signature	(i) Choosing sense mechanism	finite nonzero	$(a(t)-j_1)$	$(a(t)-j_1)^2$
	(ii) Choosing nonsense (Gell-Mann) mechanism.	$(a(t)-j_1)$	$(a(t)-j_1)$	$(a(t)-j_1)$

Table 8-1. Some possible, named, behaviors of transition amplitudes near integer (meson trajectories) or $\frac{1}{2}$-odd integer (baryon trajectories) values of $a(t)$.

† The mechanism is named noncompensating in contrast to the finite nonzero nonsense choosing behavior (ii) where poles in the nonasymptotic term are supposed to be removed by compensating trajectories.

8.7. ELASTIC SCATTERING, REGGE POLES, AND ABSORPTION

P, P' and ω Trajectories

The πN charge exchange scattering discussed in section 8.5 necessarily involved the t-channel exchange of isospin, leading to the exchange of a Regge trajectory (the ρ-trajectory) with isospin $I = 1$. In elastic scattering we can have t-channel exchange of trajectories with the quantum numbers of the vacuum: $I = 0, P = +1, C = +1, B = Y = 0, C_2 = +1$. The exchange of such a Regge pole alone will give particle-particle and particle-antiparticle cross sections equal as in the Pomeranchuk theorem discussed in section 8.2 above. Such trajectories are the *Pomeranchuk trajectory* or

8.7. ELASTIC SCATTERING, REGGE POLES, AND ABSORPTION

Pomeron or *P* trajectory, and also the *P'* trajectory. Also allowed is the exchange of the ω trajectory with $T = 0$, $P = -1$, $C = -1$.

The Pomeron is hypothesized to have positive signature, to maintain a finite cross section at $t = 0$ without sharp variation of the residue function. The *P'* has likewise positive signature, containing the $f^0(1260)$, which is a $J^P = 2^+$ meson. The ω and ρ meson trajectories have negative signature since they contain the $J^P = 1^-$ ω and ρ mesons.

The Pomeron is a trajectory needed in a pure Regge pole picture of high energy scattering, but which has, so far as is known, no physical particles on it. It is only used for $t < 0$. From the equations of section 8.4 we see that asymptotically:

$$\frac{d\sigma}{dt}(s,t) = F(t)\left(\frac{s}{s_0}\right)^{2\alpha(t)-2} + \ldots$$

where α is the leading trajectory (that is the trajectory of greatest α), and also that:

$$\sigma_T(s) \sim \left(\frac{s}{s_0}\right)^{\alpha(t)-1} + \ldots$$

so that insofar as total cross sections are constant at high energies, the Regge pole picture demands that $\alpha(0) = 1$ for the leading trajectory. Without calling in aid the *t*-dependence of the residue functions, $C(t)$ and $D(t)$ of eq. (8-76), a positive slope of Regge trajectories with

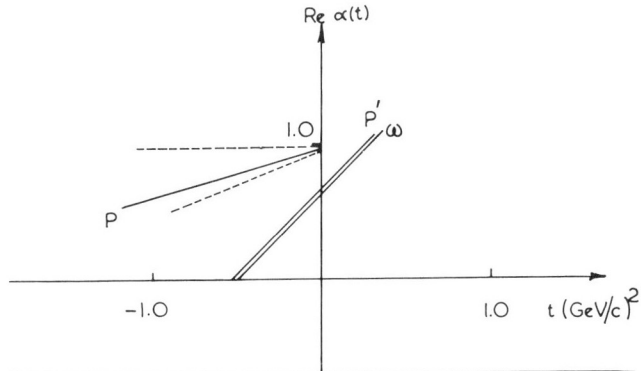

Fig. 8-10. Schematic diagram of *P*, *P'* and ω trajectories. The likely limits on the *P* trajectory are shown by dotted lines.

$a(0) \leq 1$, as in Fig. 8-10, obviously gives a forward peak in $d\sigma/dt$. That $\pi^{\pm}N$ high energy elastic scattering up to 20 GeV/c laboratory momentum (and not in the backward direction where u-channel poles would be important) can be well parametrized by a superposition of the P, P', ω, and ρ Regge poles using the formulae of eq. (8-76) has been shown by Philips and Rarita (1965). However there are difficulties with any pure Regge pole model, as will be discussed below.

The Crossover Effect

If the differential cross sections for particle-particle elastic scattering $(AB \to AB)$ and particle-antiparticle elastic scattering $(\bar{A}B \to \bar{A}B)$ are plotted together as a function of t, it is usually observed experimentally that the two curves cross over each other for some small value of t in the range $-.10$ to $-.15$ $(\text{GeV}/c)^2$. This phenomenon is called the *crossover effect* and the value of t at which the cross sections are equal is called the crossover point. The crossover effect occurs for example when AB is $\pi^+ p$, $K^+ p$, pp and $\bar{A}B$ is, correspondingly, $\pi^- p$, $K^- p$, $\bar{p}p$. With this identification $\frac{d\sigma}{dt}(\bar{A}B) - \frac{d\sigma}{dt}(AB)$ is positive for small values of t, and negative for t larger than the crossover value. Now the $\bar{A}B$ systems as just defined couple to a greater number of direct channels than the AB systems: $\pi^- p$ and $K^- p$ both have two isospin states as against one for $\pi^+ p$ and $K^+ p$; and the $K^+ p$ has exotic quantum numbers; the $\bar{p}p$ couples via baryon annihilation to the multimeson systems. So the fact that the differential elastic cross section for $\bar{A}B$ is greater at $t=0$ than that for AB and more sharply peaked finds its most natural explanation in a diffractive effect associated with the larger number of channels open to the $\bar{A}B$ system. Consequently an explanation of the crossover effect in terms of cross-channel exchanges such as Regge poles is complicated.

For example in the pp, $\bar{p}p$ case explanations in a pure Regge pole model require the residue of the ω trajectory to vanish at $t \sim -.15$. The factorization theorem then implies a dip in $d\sigma/dt$ at this t-value for all reactions dominated by the ω-trajectory, whereas no sign of such a dip is

8.7. ELASTIC SCATTERING, REGGE POLES, AND ABSORPTION

found in $\pi p \to \rho N$ or $\gamma p \to \pi^0 p$, which are both assumed to be dominated by ω exchange. Difficulties also appear in the $\pi^+ p$, $\pi^- p$ crossover with a pure Regge pole model, which probably need an explanation involving conspiracy (Leader, 1968) between various poles, or the introduction of cuts in the angular momentum plane.

Diffractive Features of Elastic πN Scattering

In both $\pi^+ p$ and $\pi^- p$ elastic scattering there are dips at constant values of t, namely $t = -0.8$ $(\text{GeV}/c)^2$ and $t \sim -3$ $(\text{GeV}/c)^2$. The dip at $t \sim -1$, corresponding on a diffraction picture to the first diffraction minimum, is very evident at a pion laboratory momentum of 2.0 GeV/c, and at higher energies as in Fig. 8-10 becomes a kink or shoulder. For

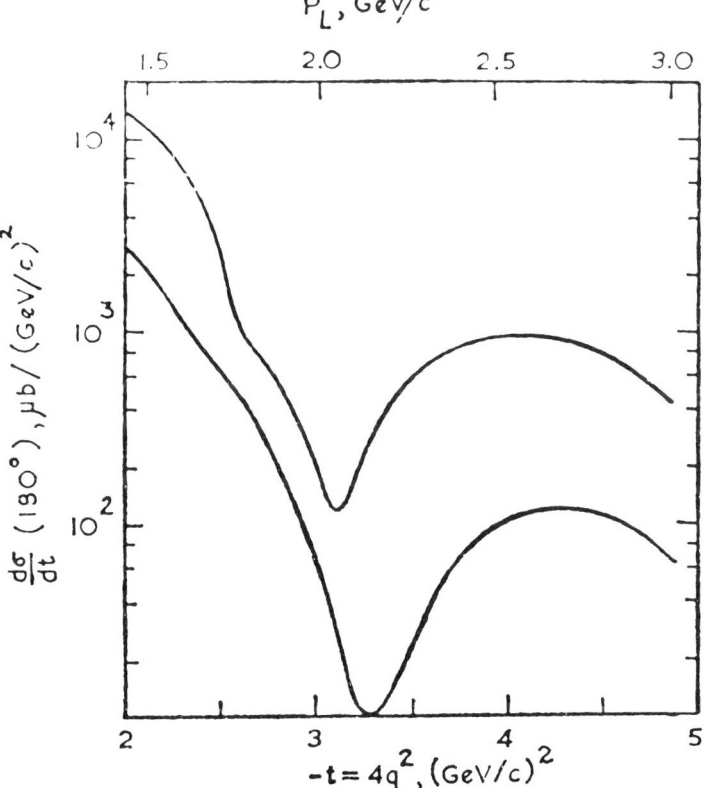

Fig. 8-11. Differential cross sections at 180° for $\pi^+ p$ scattering (upper curve) and $\pi^- p$ scattering (lower curve).

scattering with pion laboratory energy > 2 GeV there is an evident dip (Fig. 8-11) in $\pi^- p$ and $\pi^+ p$ at $t \sim -3$ $(GeV/c)^2$. Below 2 GeV incident energy, $t = -3$ is in an unphysical region of the scattering amplitude corresponding to $\cos\theta < -1$. The movement of this dip in the angular distribution from $\cos\theta < -1$ for $q_L < 2$ GeV/c to $\cos\theta > -1$ for $q_L > 2$ GeV/c corresponds to the minimum in the 180° differential cross section at $q_L = 2$ GeV/c shown in Fig. 8-10 (Booth, 1968). Since the dip moves through 180° where the spin-flip amplitude $g(\theta)$ vanishes, it must be associated with the spin-nonflip amplitude $f(\theta)$. A simple picture here would be of the high energy scattering as predominantly spin-nonflip diffraction scattering with diffraction minima at $t = -0.8$ and -3 $(GeV/c)^2$.

This and the crossover effect [as well as other considerations from, for example, charged pion photoproduction[7] where $\frac{d\sigma}{dt} \propto s^{-2}$ for a large range of $-t$ extending from 0 to at least 1.0 $(GeV/c)^2$] make desirable a modification of a pure Regge picture in the direction of the "classical" diffraction picture of section 8.1. Such a modification can be achieved for example by the introduction of absorptive scattering after Regge pole exchange, or by multiple Regge pole exchange. There is an intimate connection here with the *J-plane cuts*.

It is important that at higher energies (incident laboratory momenta > 6 GeV/c) the diffraction minima in πN elastic scattering tend to become kinks and to eventually disappear. The same thing happens to the minima in $K^- p$ and pp elastic scattering, whereas the differential cross section in the exotic channel[8] $K^+ p$ has no dips or bumps as a function of t from incident energy 1 GeV/c to the highest energies

---

[7] See Richter (1967); Harari (1967; 1969), Dombey (1969); Barbour, Malone, and Moorhouse (1971).

[8] An exotic channel is one not having the quantum numbers of 3-quarks or a quark-anti-quark pair (Chapter 6).

8.7. ELASTIC SCATTERING, REGGE POLES, AND ABSORPTION

available. This leads to the diffractive scattering being regarded as composed of two parts: (a) the Pomeron part contributing to all elastic scattering processes, and having no dips or bumps as a function of t; and (b) the non-Pomeron part (in Regge pole terms associated with the P', ω, etc., trajectories) which becomes small compared to (a) as energy increases and only contributes to nonexotic channels (such as πN, $\bar{K}N$, $\bar{N}N$). This theory is discussed further in the next section, 8.8.

Absorption and J-Plane Cuts. Dips in Inelastic Processes

In eqs. (8-62)-(8-70) we expressed the πN scattering amplitudes $A^{(\pm)}(s,t)$ and $B^{(\pm)}(s,t)$ in terms of the amplitudes $f_+^{(\pm)}(j,t)$ and, $f_-^{(\pm)}(j,t)$ respectively. These latter amplitudes are generalizations of the helicity amplitudes of eq. (9-26) to complex j, and are analytic functions of j in the right-hand j-plane (Re $j > -\frac{1}{2}$) with the exception of certain singularities. We have discussed at some length in this chapter the singularities which are j-plane poles given by $j = \alpha_\rho(t)$, $j = \alpha_\omega(t)$, and so on. It is also mathematically possible and physically likely that there are branch-point singularities, or j-plane cuts, in the right-hand j-plane. [Mandelstam (1963 a,b) showed that with Feynman diagram results there was reason to expect j-plane cuts.]. A cut obtruding from the left-hand into the right-hand j-plane will give a contribution to an amplitude, say $A^{(+)}(s,t)$, of the form:

$$A'^{(+)}(s,t) \propto \int^{\alpha_c(t)} dj \left(\frac{s}{s_0}\right)^j \text{disc } f_+^{(+)}(j,t) + \cdots$$

where disc $f_+^{(+)}(j,t)$ is the discontinuity of $f_+^{(+)}(j,t)$ across the cut and $\alpha_c(t)$ is the end point of the cut. If the discontinuity behaves as $(\alpha_c(t)-j)^n$ at the end point, then it is easy to show that the large s behavior, due to the cut, is:

$$\lim_{s \to \infty} A'^{(+)}(s,t) \propto s^{\alpha_c(t)} \left[\ln\left(\frac{s}{s_0}\right) - \frac{i\pi}{2}\right]^{-n-1} + \cdots$$

j-plane cuts can be generated by the exchange of two Regge poles. As shown by Collins and Squires (1968) in the simple case of the exchange of Regge poles described by two linear Regge trajectories $a_1(t) = a_1(0) + a_1'(0) t$, $a_2(t) = a_2(0) + a_2'(0) t$, the term a_c is given by:

$$a_c(t) = a_1(0) + a_2(0) - 1 + \frac{a_1'(0) a_2'(0)}{a_1'(0) + a_2'(0)} t \ .$$

At the end of the last subsection we mentioned that the Regge pole picture could be modified in the direction of the "classical" diffraction picture by allowing multiple Regge pole exchange. We have now seen that this involves the generation of Regge cuts. One way of carrying the program into effect is by making the Regge pole exchange the basic scattering process in the eikonal approximation, so that the full scattering amplitude incorporates multi-Regge exchanges. A paper by Arnold and Blackmon (1968), following the method of Arnold (1967), illustrates the implementation of this approach in the case of πN scattering. The method in principle solves the difficulty of the observed polarization in high energy πN charge exchange scattering and also the problem of *dips*, which we just mention briefly now, preparatory to an extended coverage in section 8.8. *Dips* are observed at fixed t-values (such as $t = -0.6$ in πN charge exchange scattering) in some inelastic processes but not in others. In a pure Regge pole theory, with factorization of Regge pole residues, dips in some processes require dips in others, and an insoluble problem of consistency with the experiments is posed. However, the addition of the impurity of absorptive scattering solves the problem, because absorptive scattering can be used to fill in and level up unwanted dips; the dips that remain are due to the nonsense wrong signature zeros. The importance given to absorption is even greater in the strong cut Reggeized absorption model of Henyey et al., (1969), where absorption is associated with inelastic intermediate states and where there are no nonsense wrong signature zeros of Regge pole amplitudes, but the dips are due to the destructive interference between the poles and the cut (amplitude $M = M^{\text{poles}} + M^{\text{cut}}$), so that the dips are diffraction minima (Kelly et al., 1970; Ross, Henyey, and Kane 1970; Kane et al., 1970).

8.8. THE s-CHANNEL, DUALITY, AND PERIPHERALITY

We started this chapter by discussing high energy scattering in terms of partial wave amplitudes, unitarity, the optical model, and other concepts, all in terms of s-channel quantum numbers. After that we turned our attention to the t-channel and Regge poles, but we found that a necessary modification of Regge pole ideas lay in the direction of absorption, an s-channel concept. The time has come to return to the s-channel viewpoint, and configuration space, but now to consider simultaneously the t-channel viewpoint using the ideas of duality.

Duality. Historically this concept arose through the use of the finite energy sum rules and this work will be discussed in Chapter 9. In the qualitative aspect of duality discussed here, it is suitable to start with the preliminary observation that some observed scattering channels have prominent resonance formation at lower energies and some do not. Those that do are reactions initiated by the nonexotic channels πN, $\overline{K}N$ and $\overline{N}N$ and those that do not are the exotic channels KN and pp. Also it is just the elastic scattering in the nonexotic channels that contains a dip and bump structure (dying away for momenta $\gtrsim 6$ GeV/c) in the differential cross section while the exotic channels are structureless. These considerations lead to the hypothesis that one can divide the imaginary part of an elastic scattering amplitude into two parts:

$$\text{Im } A(s,t) = P(s,t) + R(s,t) \qquad (8\text{-}82)$$

where $P(s,t)$ has no dips or bumps as a function of t, and where such a part is present in both exotic and nonexotic amplitudes, while $R(s,t)$ has dips and bumps as a function of t, is present only in nonexotic amplitudes and contributes a decreasing proportion of the amplitude with increasing energy. Since R only contributes to those amplitudes that have prominent resonances at low energies, it is natural to identify this as a contribution from resonances; at the same time Regge poles or Regge poles + cuts are successful in describing the high energy amplitude, and (cf. Chapter 9) high energy Regge amplitudes, extrapolated to low energies, average low energy cross sections, including resonance-dominated ones such as πp charge exchange. Considerations like these have led to the hypothesis of *Harari-Freund duality* (Harari, 1968, Freund, 1968), which we express symbolically as follows:

$$\begin{array}{cc} \text{Im (amplitude)} \ equals & \text{background} + \text{resonances} \\ & \updownarrow \qquad\qquad \updownarrow \\ or\ equals & \text{Pomeron} + \text{Regge poles (+ cuts)}. \end{array} \qquad (8\text{-}83)$$

Equation (8-83) expresses the hypothesis that we can regard any two-body reaction amplitude as composed of background and resonances but equally well as Pomeron and Regge poles (+ cuts), and that there is an equivalence between the Regge and the resonance description — though one is simple and more appropriate at low energies and one is simpler and more appropriate at high energies. We should note that the vanishing of R for exotic reactions involves the exchange degeneracy of Regge trajectories (cf. Chapter 9). It is important that eq. (8-83) contains a conceptual separation of the Pomeron from ordinary Regge poles: the Pomeron is *dual* to the background (and only contributes to elastic scattering) while the Regge poles (+ cuts) are *dual* to the resonances. In eq. (8-82) R is mnemonic for Regge *or* resonance as is P for Pomeron or background (sound shifted).

In support of some of the consequences of eqs. (8-82) and (8-83) we quote the following facts:

8.8. THE s-CHANNEL, DUALITY, AND PERIPHERALITY

(a) Harari and Zarmi (1969) show that the partial wave projections of πN elastic scattering amplitudes with definite t-channel isospin, $I_t = 1$, are resonance dominated while those with $I_t = 0$ show imaginary nonresonant background, corresponding to the Pomeron of eq. (8-82) or eq. (8-83) having $I_t = 0$.

(b) As shown in Fig. 8-5 the K^+p and pp total cross sections are rather constant with energy over the range 5-60 GeV/c laboratory momentum while the nonexotic total cross sections markedly decrease (the contrast is even more marked if the range 2-5 GeV/c is also considered). This is in accord with the vanishing of R with increasing energy in nonexotic channels and its nonoccurrence in exotic channels. [There is an interesting slight increase in $\sigma_{total}(K^+p)$.]

(c) The optical theorem and isospin conservation shows that $\text{Im}(K^+n \rightarrow K^0p)_{t=0} \propto \sigma_{tot}(K^+p) - \sigma_{tot}(K^+n)$ while experimentally $\sigma_{tot}(K^+p) \simeq \sigma_{tot}(K^+n)$. Since $K^+n \rightarrow K^0p$ has no resonances *and* no Pomeron, its imaginary part *should* be zero by eqs. (8-82), (8-83).

The Pomeron and s-Channel Helicity. So far in this section we have not mentioned spin. In pion-nucleon scattering for example there are two spin amplitudes which may be taken as the helicity flip and helicity nonflip amplitudes, so that in the πN case eq. (8-82) is a symbolic representation of two equations describing two amplitudes. Now *every* elastic scattering process has a helicity nonflip amplitude, and it may be regarded as *the* pure elastic amplitude, since in a sense there is no change of quantum numbers other than those of particle position. There are some experimental indications that "the Pomeron conserves helicity," in other words that $P(s,t)$ in eq. (8-82) is zero except where the amplitude has zero helicity flip (Gilman, Pumplin, Schwimmer, and Stodolsky, 1970).

Peripheral and Nonperipheral Resonances. If a sum of resonances is an adequate description of the term $R(s,t)$ in eq. (8-82), then this sum must reproduce the dips and bumps at fixed t and all energies (such as the dip at $t = -0.6$ in $\pi^-p - \pi^0n$) that occur so luxuriantly in inelastic scattering.

Dolen, Horn, and Schmid (1968) pointed out that the prominent, peripheral resonances *individually* have this property with respect to dips in πN scattering from the empirical fact that the s-channel contribution of each prominent N^* resonance exhibits zeros at approximately fixed t-values. Harari (1971) has emphasized that although this is not the only possible way that fixed-t features could arise when viewed through the s-channel amplitudes (for instance it could arise from a conspiracy of s-channel amplitudes as in the Veneziano model), yet it is the simplest and would also seem to require a dominance of peripheral partial waves in $R(s,t)$. To investigate this point Davier and Harari (1971) have approximately evaluated $R(s,t)$ from the data on K^+p and K^-p scattering. On the assumptions that the Pomeron is nearly all imaginary and helicity nonflip and that R is small compared to the Pomeron, one finds from eq. (8-82) that:

$$\left(\frac{d\sigma}{dt}\right)_{\text{elastic}} = P^2 + 2P \cdot R_{\Delta\lambda=0} \tag{8-84}$$

where $R_{\Delta\lambda=0}$ is the helicity nonflip part of $R(s,t)$. With the assumption that P is the same for K^+p and K^-p scattering, one thus obtains for the imaginary part of the helicity nonflip resonance (or Regge) contribution to K^-p scattering:

$$R_{\Delta\lambda=0}(s,t) = \frac{\frac{d\sigma}{dt}(K^-p) - \frac{d\sigma}{dt}(K^+p)}{2\sqrt{\frac{d\sigma}{dt}(K^+p)}} . \tag{8-85}$$

At $p_{\text{lab}} = 5$ GeV/c Davier and Harari find that a good fit is:

$$R_{\Delta\lambda=0}(t) = A \exp(Bt) J_0(r\sqrt{-t}) \tag{8-86}$$

where $A = 1.6 \; mb^{\frac{1}{2}} \; \text{GeV}^{-1}$, $B = 1.3 \; \text{GeV}^{-2}$, $r = 4.8 \; \text{GeV}^{-1} = 0.95 \; fm$. The projection of eq. (8-86) into the impact parameter representation shows a strongly peripheral form, sharply peaked at an impact parameter of $\sim 0.9 \; fm$ (corresponding, at that energy, to $J \sim {}^{11}/2$). By contrast the corresponding impact parameter representation of P shows that almost all the scattering comes from impact parameters $b < 0.5 \; fm$.

8.8. THE s-CHANNEL, DUALITY, AND PERIPHERALITY

The calculation of Davier and Harari suggests that the non-Pomeron part of the scattering amplitude is dominated by peripheral (probably resonant) partial waves, at any rate in the helicity nonflip part. Let us now consider the mathematical consequences of a peripheral $R(s,t)$ for general helicity flip.

In the process $a + b \to c + d$ where the (s-channel) helicities of particles a, b, c, d are respectively λ_a, λ_b, λ_c, λ_d, and where we are considering forward hemisphere scattering in the center of mass with particles a and c forward, the definition of the total *helicity flip*, $\Delta\lambda$, is:

$$\Delta\lambda = (\lambda_a - \lambda_b) - (\lambda_c - \lambda_d) . \tag{8-87}$$

Analogously to eq. (8-12) we write for the imaginary part of the non-Pomeron amplitude of helicity flip $\Delta\lambda$:

$$R_{\Delta\lambda}(s,t) = \sum_j (2j+1) \, I^j(s) \, d^j_{\lambda_a-\lambda_b, \lambda_c-\lambda_d}(\theta) \tag{8-88}$$

where I^j is the imaginary part of the corresponding partial wave amplitude of spin j (and appropriate helicity), and θ is the scattering angle (cf. Chapter 2). As was done in section 8.1, we transfer to the impact parameter representation using a small angle Bessel function approximation for the d^j:

$$R_{\Delta\lambda}(s,t) = 2q^2 \int b\,db\, I(s,b) \, J_{\Delta\lambda}(b\sqrt{-t}) \tag{8-89}$$

where q is the center of mass momentum. If the distribution were so peripheral that $I(s,b)$ were a δ-function in b, then eq. (8-89) would give:

$$R_{\Delta\lambda}(t) = A \, J_{\Delta\lambda}(r\sqrt{-t}) \tag{8-90}$$

where A is independent of t. A merely sharply peaked distribution in b for $I(s,b)$ rather than a δ-function will lead to a form for R as a function of t which has the characteristics, such as zeros and minima, exhibited in eq. (8-90), and which thus may be written:

$$R_{\Lambda\lambda}(t) \simeq A(t) J_{\Lambda\lambda}(r\sqrt{-t}) \qquad (8\text{-}91)$$

where $A(t)$ is a smooth monotonic function of t and r will be in the region 0.8 to 1.0 fm. Equation (8-86) is an example of such a function.

The form eq. (8-82) with eq. (8-91) immediately explains the "crossover" effect in $\pi^{\pm}p$ elastic scattering as due to the zero of J_0 near $t = -0.2$.

If we accept duality in the form outlined previously in this section, then $R_{\Lambda\lambda}(s,t)$ can also be obtained from Regge poles + cuts (where the cuts probably are Regge pole × Pomeron). Ross, Henyey, and Kane (1970) came to very similar expressions to eq. (8-91) using their strong cut Regge absorption model; in their calculations explicit expressions for the real part were also obtained as well as an explicit, rather Regge pole-like, s-dependence.

Finally, we note that the dominance of R by peripheral resonances agrees well with the L-excitation quark model without radial excitation. Since in this case the spin J of higher mass resonances is obtained by adding L to quark spin $S = \frac{1}{2}$ or $\frac{3}{2}$, the resonances are necessarily peripheral.

Dips. The form eq. (8-91) has some immediate consequences which can be compared with experiment. The first is the form eq. (8-82) for elastic scattering which as already noted gives a featureless cross section for K^+p or pp forward scattering but features for K^-p, $\pi^{\pm}p$, or $\bar{p}p$ elastic scattering. In addition $J_0(r\sqrt{-t})$ has a zero at $t \sim -0.2$ followed by a negative minimum at $t \simeq -0.6$. One should therefore expect a dip in the elastic cross sections at about this value, or somewhat above because of the fast-falling Pomeron in $\pi^{\pm}p$ elastic scattering. We have previously noted the occurrence of this dip (disappearing at higher energies) at $t \simeq -0.8$, and of course there is a dip in K^-p to which eq. (8-86) constitutes a fit. In the $\bar{p}p$ case the dip is at $t \simeq -0.5$.

In the case of inelastic scattering the situation is much more complicated. We have:

8.8. THE s-CHANNEL, DUALITY, AND PERIPHERALITY

$$\left(\frac{d\sigma}{dt}\right)_{\text{inel}} \propto \sum_{\lambda_a, \lambda_b, \lambda_c, \lambda_d} |A_{\lambda_a, \lambda_b, \lambda_c, \lambda_d}(s,t)|^2 \tag{8-92}$$

where $A_{\lambda_a, \lambda_b, \lambda_c, \lambda_d}(s,t)$ is a helicity amplitude, whose imaginary part is $R_{\Delta\lambda}(s,t)$ where $\Delta\lambda = |(\lambda_a - \lambda_b) - (\lambda_c - \lambda_d)|$. For πN charge exchange, for example, eq. (8-92) reduces to:

$$\left(\frac{d\sigma}{dt}\right)_{\text{inel}} \propto \sum_{\Delta\lambda = 0,1} |A_{\Delta\lambda}(s,t)|^2 . \tag{8-93}$$

The term $J_1(r\sqrt{-t})$, which occurs in the imaginary part of $A_{\Delta\lambda=1}(s,t)$ has a zero at $t \simeq -0.6$, which is the value of t at which many inelastic processes (including πN charge exchange) have dips, but where many do not. The only simple way of achieving a dip at $t \simeq -0.6$ is through the major contribution to the cross section coming from $A_{\Delta\lambda=1}$ and more specifically from its imaginary part $R_{\Delta\lambda=1}$, though this depends on the phase relation between the real and imaginary part. In the specific case of $\pi^- p \to \pi^0 n$ it is highly probable that $A_{\Delta\lambda=1}$ is larger than $A_{\Delta\lambda=0}$ from the shoulder in $d\sigma/dt$ at $t \simeq 0$ (the contribution from $A_{\Delta\lambda=1}$ vanishes at $t = 0$). This immediately gives the observed dip in $d\sigma/dt$ at $t \simeq -0.6$, since from the phase relations discussed below, one expects a small contribution from the real part of $A_{\Delta\lambda=1}$ at $t \simeq -0.6$.

In general one might expect spin flip at the baryon vertex in ρ-exchange processes (like $\pi^- p \to \pi^0 n$), since vector dominance arguments indicate a larger magnetic than electric ρNN coupling. Using arguments of this nature on t-channel exchanges, Harari (1969, 1971c) finds agreement with the occurrence or nonoccurrence of dips in all fifteen cases in which the contributions of vector and/or tensor meson exchange can be isolated.

Phase Relations and Polarization. Given an amplitude $A(\nu,t)$ with well-defined crossing $(\nu \leftrightarrow -\nu)$ properties and with the property, for fixed t:

$$A(\nu,t) \to \nu^{\alpha(t)} \quad \text{as} \quad \nu \to \infty \tag{8-94}$$

then by an application of the Phragmen-Lindeloff theorem (Khuri and Kinoshita, 1965):

$$\text{Re } A(\nu,t) \to \nu^{a(t)} \tan \frac{\pi a(t)}{2} \quad \text{(crossing-odd)} \quad (8\text{-}95a)$$

$$\text{Re } A(\nu,t) \to \nu^{a(t)} \cot \frac{\pi a(t)}{2} \quad \text{(crossing-even)} . \quad (8\text{-}95b)$$

Equation (8-94) may contain logarithmic terms without affecting the results. [If a behavior like eq. (8-94) is observed, it may be interpreted as due to Regge poles, or poles and cuts.] It must be emphasized that eq. (8-95) is an asymptotic relation and so must only be used with due caution and precaution. If, however, we assume it, we can consider the polarization $\mathcal{P}$ in $\pi^{\pm}p$ and $K^{\pm}p$ elastic scattering which is given by:

$$\mathcal{P} \frac{d\sigma}{dt} = \text{Im}\left[A^*_{\Delta\lambda=0} A_{\Delta\lambda=1}\right]$$

which using Pomeron dominance of $A_{\Delta\lambda=0}$ gives:

$$\mathcal{P} \frac{d\sigma}{dt} \sim P(t) \text{ Re } A_{\Delta\lambda=1} . \quad (8\text{-}96)$$

Using eq. (8-96), the phase relations eq. (8-95), and eq. (8-91) for $\text{Im } A_{\Delta\lambda=1}$, we find, where $+$ and $-$ refer to positive and negative meson scattering:

$$\left(\mathcal{P} \frac{d\sigma}{dt}\right)_+ + \left(\mathcal{P} \frac{d\sigma}{dt}\right)_- \sim P(t) A(t) J_1(r\sqrt{-t}) \cot \frac{\pi a(t)}{2} \quad (8\text{-}97a)$$

$$\left(\mathcal{P} \frac{d\sigma}{dt}\right)_+ - \left(\mathcal{P} \frac{d\sigma}{dt}\right)_- \sim P(t) A(t) J_1(r\sqrt{-t}) \tan \frac{\pi a(t)}{2} \quad (8\text{-}97b)$$

In eq. (8-97) $P(t) A(t)$ is a featureless function, so the polarization combinations should follow the features of $J_1(r\sqrt{-t}) \tan \frac{\pi a(t)}{2}$ or $J_1(r\sqrt{-t}) \cot \frac{\pi a(t)}{2}$. In fact they do; in particular the combination $\left(\mathcal{P} \frac{d\sigma}{dt}\right)_- - \left(\mathcal{P} \frac{d\sigma}{dt}\right)_-$ in both πp and Kp scattering shows two zeros near $t = -0.6$.

We may point out another success of the phase relation in the Harari peripheral picture: $\pi^- p \to \eta^0 n$ has no dip at $t = -0.6$, although like $\pi^- p \to \pi^0 n$ the differential cross section also has a shoulder near $t = 0$ indicating $\Delta\lambda = 1$ dominance. However, $A_{\Delta\lambda=1}(\pi^- p \to \eta^0 n)$ is *crossing even* and the large real part [eq. (8-97a)] fills in the J_1 dip near $t = -0.6$, where $\alpha(t) \sim 0$. (Obviously the scattering amplitude or the polarization cannot be infinite when $\alpha = 0$; the infinity in $\cot \frac{\pi\alpha(t)}{2}$ must be cancelled by an appropriate zero, and there is one at hand in $J_1(r\sqrt{-t})$; from this point of view a correlation between $\alpha_{\text{effective}}$ and the dip position seems necessary.)

8.9. BACKWARD PION-NUCLEON SCATTERING, u-CHANNEL EXCHANGE AND NUCLEONIC REGGE TRAJECTORIES

The relationship between the s- and u-channels in pion-nucleon scattering is illustrated in Fig. 8-10. The exchanged particles in the u-channel must necessarily be baryons. The value of u in terms of the center of mass energy, s, and scattering angle θ is:

$$u = -2q^2(1 + \cos\theta) + (M_N^2 - M_\pi^2)^2/s \qquad (8\text{-}98)$$

where q is the center of mass momentum given by eq. (2-20). We see that for large s (say $s \gg M_N^2$) and in the forward direction, $\cos\theta \sim 1$, u is large and negative, while in the backward direction, $\cos\theta \sim -1$, u can be small. Now the particles and resonances which can be exchanged in the u-channel, as in Fig. 8-12, give rise to poles at $\operatorname{Re} u > 0$, the lowest u value being that for the nucleon exchange itself at $u = M^2$ [which is given explicitly in the Born approximation for the A and B amplitudes by eq. (7-18)]. These singularities should be much more evident in backward s-channel scattering where they are nearer to physical

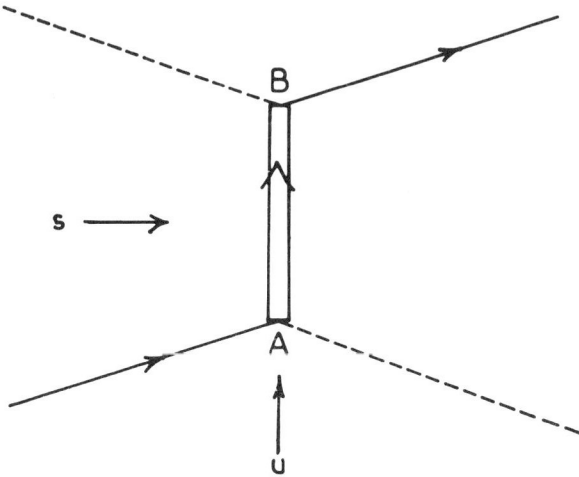

Fig. 8-12. Diagram of a contribution to s-channel pion-nucleon scattering from u-channel exchange.

u values than in forward s-channel scattering. The situation is contrariwise for the t-channel, meson, exchanges which have been discussed in section 8.4 above. We expect forward scattering to be dominated by t-channel meson exchange and backward scattering by u-channel baryon exchange.

From the existence of nucleonic resonances, as revealed in pion-nucleon scattering and total cross sections and as discussed in Chapter 5, there are at least three nucleonic Regge trajectories, two of which appear established as linear in u for $M_N^2 < u < 10$ (GeV/c)^2. These are:

$\Delta(p_{33}(1))$ trajectory: $J^P = \frac{3+}{2}, \frac{7+}{2}, \frac{11+}{2}; \ldots$; signature $-$:

$$\text{Re } a(\sqrt{u}) = 0.15 + 0.90\, u \tag{8-99}$$

N_α(nucleon) trajectory: $J^P = \frac{1+}{2}, \frac{5+}{2}, \ldots$; signature $+$:

$$\text{Re } a(\sqrt{u}) = -0.39 + 1.01\, u \tag{8-100}$$

8.9. BACKWARD PION-NUCLEON SCATTERING

N_γ trajectory: $J^P = \frac{3-}{2}, \frac{7-}{2}, \frac{11-}{2}, \ldots$; signature $-$:

$$\text{Re } a(\sqrt{u}) = -0.90 + 0.92\, u \ . \tag{8-101}$$

The above trajectories are expected to contribute to s-channel scattering when u is less than zero. It is conjectured, with some success, as we shall see below, that one may simply continue the linear formulae (8-99)-(8-101) to negative u. We now develop the formalism for u-channel Regge poles. For this purpose we consider the invariant amplitudes A and B as functions of the two independent variables s and u — that is, $A(s,u)$ and $B(s,u)$. The nucleon energy in the s-channel center of mass system is given by:

$$E^s = (s + M_N^2 - M_\pi^2)/2W \tag{8-102}$$

and is an odd function of $W = \sqrt{s}$ so that from eq. (3-63):

$$f_1^s(\sqrt{s}, u) = \frac{E^s + M_N}{8\pi\sqrt{s}}\, [A(s,u) + (\sqrt{s} - M_N)\, B(s,u)] \tag{8-103}$$

$$f_2^s(\sqrt{s}, u) = -f_1^s(-\sqrt{s}, u) \ . \tag{8-104}$$

Here f_1^s, f_2^s are just the usual functions f_1 and f_2 of eq. (3-63) which we have here labelled with s because we are going to define the corresponding functions for u-channel scattering which we will label with u. The corresponding u-channel amplitude to eq. (8-103) is:

$$f_1^u(\sqrt{u}, s) = \frac{E^u + M_N}{8\pi\sqrt{u}}\, [A(u,s) + (\sqrt{u} - M_N)\, B(u,s)] \ . \tag{8-105}$$

We consider $\pi^- p$ scattering in the s-channel, the amplitudes being denoted by the suffix $-$. Then under crossing $s \leftrightarrow u$, $\pi^- \leftrightarrow \pi^+$ and, since the $\pi^+ p$ system has isospin $\frac{3}{2}$, we have the crossing relations:

$$A_-(s,u) = A^{\frac{3}{2}}(u,s), \quad B_-(s,u) = -B^{\frac{3}{2}}(u,s) \tag{8-106}$$

which from eqs. (8-103)-(8-105) leads to:

$$f_1^S(\sqrt{s},u)_- = \frac{E^S+M_N}{2\sqrt{s}}\left[C(\sqrt{u},\sqrt{s})\,f_1^{u\frac{3}{2}}(\sqrt{u},s) + C(-\sqrt{u},\sqrt{s})\,f_1^{u\frac{3}{2}}(-\sqrt{u},s)\right]$$

where: (8-107)

$$C(\sqrt{u},\sqrt{s}) = (\sqrt{u}-\sqrt{s}+2M)/(E^u+M_N) \,. \qquad (8\text{-}108)$$

For π^+p and charge exchange scattering the kinematic factors are the same but the crossing relations lead to different isospin relations:

$$f_1^S(\sqrt{s},u)_+ = \frac{E^S+M_N}{2\sqrt{s}}\Bigg[C(\sqrt{u},\sqrt{s})\left(\tfrac{1}{3}f_1^{u\frac{3}{2}}(\sqrt{u},s) + \tfrac{2}{3}f_1^{u\frac{1}{2}}(\sqrt{u},s)\right)$$

$$+ C(-\sqrt{u},s)\left(\tfrac{1}{3}f_1^{u\frac{3}{2}}(-\sqrt{u},s) + \tfrac{2}{3}f_1^{u\frac{1}{2}}(-\sqrt{u},s)\right)\Bigg] \qquad (8\text{-}109)$$

$$f_1^S(\sqrt{s},u)_{C.E.} = \frac{E^S+M_N}{2\sqrt{s}}\Bigg[C(\sqrt{u},\sqrt{s})(f_1^{u\frac{3}{2}}(\sqrt{u},s) - f_1^{u\frac{1}{2}}(\sqrt{u},s))$$

$$+ C(-\sqrt{u},s)(f_1^{u\frac{3}{2}}(-\sqrt{u},s) - f_1^{u\frac{1}{2}}(-\sqrt{u},s))\Bigg]\frac{\sqrt{2}}{3}. \qquad (8\text{-}110)$$

The amplitudes in eqs. (8-107), (8-109), and (8-110) for π^+p, π^-p, and charge exchange scattering respectively have been expressed in terms of u-channel amplitudes. As discussed below these may be expressed in terms of Regge poles. In terms of the trajectories mentioned above, the u-channel Regge pole contribution to the amplitudes may be written (Barger and Cline, 1967):

$$f_1^{u\frac{3}{2}} = \frac{E^u+M_N}{\sqrt{u}}\gamma_\Delta(\sqrt{u})\,\frac{1-i\exp(-i\pi\alpha_\Delta(\sqrt{u}))}{\cos(\pi\alpha_\Delta(\sqrt{u}))}\left(\frac{s-M_N^2}{s_0}\right)^{\alpha_\Delta(\sqrt{u})-\frac{1}{2}} \qquad (8\text{-}111)$$

8.9. BACKWARD PION-NUCLEON SCATTERING

$$f_1^{u\frac{1}{2}} = \frac{E^u_+ M_N}{\sqrt{u}} \left\{ \gamma_\alpha(\sqrt{u}) \, \frac{1 + i \exp(-i\pi a_\alpha(\sqrt{u}))}{\cos(\pi a_\alpha(\sqrt{u}))} \left(\frac{s - M_N^2}{s_0}\right)^{a_\alpha(\sqrt{u}) - \frac{1}{2}} \right.$$

$$\left. + \gamma_\gamma(\sqrt{u}) \, \frac{1 - i \exp(-i\pi a_\gamma(\sqrt{u}))}{\cos(\pi a_\gamma(\sqrt{u}))} \left(\frac{s - M_N^2}{s_0}\right)^{a_\gamma(\sqrt{u}) - \frac{1}{2}} \right\}. \qquad (8\text{-}112)$$

We may write this result a little more transparently by inserting eqs. (8-111) and (8-112) in eqs. (8-103)-(8-110) to find the contributions of a typical Regge pole to the A and B amplitudes. Eliminating some non-asymptotic factors, we write the result (Berger and Fox, 1971) as:

$$A = \frac{1}{2} \pi \, \frac{[1 + \tau i \, e^{-i\pi\alpha}]}{\Gamma\left(\frac{1}{2} + \alpha\right) \cos \pi\alpha} \, \gamma_A(u) \left(\frac{s}{s_0}\right)^{\alpha - \frac{1}{2}} \qquad (8\text{-}113\text{a})$$

$$B = \frac{1}{2} \pi \, \frac{[1 + \tau i \, e^{-i\pi\alpha}]}{\Gamma\left(\frac{1}{2} + \alpha\right) \cos \pi\alpha} \, \gamma_B(u) \left(\frac{s}{s_0}\right)^{\alpha - \frac{1}{2}} \qquad (8\text{-}113\text{b})$$

where τ is the signature and the $\Gamma\left(\frac{1}{2} + \alpha\right)$ ensures that nonsense wrong signature zeros are present (unless there are implicit poles in the $\gamma_A(u)$, $\gamma_B(u)$). Equation (8-113) immediately gives for the cross section $d\sigma/du$ at fixed u due to one Regge pole:

$$\frac{d\sigma}{du} \propto \left(\frac{s}{s_0}\right)^{2\alpha(u) - 2}. \qquad (8\text{-}114)$$

It is an empirical observation that high energy $\pi^{\pm} p$ backward scattering at $180°$ has the property that $P^3_{\text{lab}} \left(\frac{d\sigma}{du}\right)_{180°}$ is very approximately constant. We note that the dominant trajectory is the N_α given by eq. (8-100), so that $a_N(0) = -0.4$, giving:

$$\left(\frac{d\sigma}{du}\right)_{180°} = \left(\frac{s}{s_0}\right)^{-2.8}. \qquad (8\text{-}115)$$

Since at high energies $P_{lab} \propto s$, this again exhibits, but now in the backward direction, the effectiveness of Regge poles in predicting energy dependence.

We may also note that the well-observed dip in the $\pi^{\pm}p$ cross sections occurs at a constant value $u = -0.2$ (GeV/c)^2 over a large range of energy. Now $\alpha_\alpha = -\frac{1}{2}$ is a wrong signature point for the nucleon trajectory, with a nonsense-nonsense transition. If there is a nonsense wrong signature zero we see that the residue function must vanish when:

$$\alpha_\alpha(u) + \frac{1}{2} = 0 . \tag{8-116}$$

This occurs near $u = -0.2$, explaining the dip. On comparing eqs. (8-99) and (8-101) we see that α_Δ is greater than α_γ by ~ 1 and thus at $u = -0.2$ and large s the scattering amplitude should be dominated by Δ trajectory exchange. The Clebsch-Gordan isospin factors for pure isospin $\frac{3}{2}$ u-channel exchange then give a ratio 1:9 for $\pi^+p:\pi^-p$ scattering. This ratio is satisfied at $u = -0.2$ (and not at other u-values). Thus Regge pole theory, with the factorization of Regge pole residues, gives an explanation of some prominent facts in pion-nucleon backward scattering.

On the other hand we may note the absence of a dip in π^-p scattering at $u = -1.8$ (GeV/c)^2 which would be given by a linear Δ trajectory; this might be explained (inter alia) by deviations from linearity for larger negative values of u.

Chu and Hendry (1971) have particularly emphasized the applicability of the s-channel absorptive and peripheral ideas discussed in section 8.8 to the backward direction. They obtain the dip at $u \simeq -0.2$ from the s-channel picture, using the same peripheral partial waves that can generate fixed-t structure.

8.10. u-CHANNEL REGGE POLES AND DAUGHTER TRAJECTORIES

Some difficulties arise in the Reggeization of eqs. (8-107), (8-103), (8-104). Suppose we make the standard partial wave expansion, but in this case in the channel where $\sqrt{u}$ is the energy:

$$f_1^u(\sqrt{u}, s) = \sum_{\ell=0}^{\infty} (f_{\ell+}^u(\sqrt{u}) P'_{\ell+1}(Z_u) - f_{\ell-}^u(\sqrt{u}) P'_{\ell-1}(Z_u)) \quad (8\text{-}117)$$

where $f_{\ell\pm}^u(\sqrt{u})$ are u-channel partial waves and Z_u is the cosine of the u-channel center of mass scattering angle, given by:

$$Z_u = 1 + t/2q_u^2 = 1 + \frac{2M^2 + 2m^2 - u - s}{2q_u^2} \quad (8\text{-}118)$$

where q_u is the center of mass momentum. If we perform a Sommerfeld-Watson transform, the angular functions become $P'_{a+\frac{1}{2}}(Z_u)$ and from eq. (8-118) and eqs. (8-107)-(8-110) we appear to get $s^{a-\frac{1}{2}}$ as the asymptotic dependence on s as in eq. (8-111) or eq. (8-112). This is indeed the case if we maintain a fixed nonzero value of u (for example the value $u = -0.2$ discussed in the previous section). For then:

$$q_u^2 = \frac{1}{4} \left\{ u - (2M^2 + 2m^2) + \frac{(M^2 - m^2)^2}{u} \right\} \quad (8\text{-}119)$$

is also finite. But at $u = 0$, q_u^2 and thus Z_u is undefined and in the exact backward direction, $\theta_s = 180°$, $u = \frac{(M^2 - m^2)^2}{s} \to 0$ as $s \to \infty$. Consequently it is not immediately clear that we can apply eqs. (8-111) and (8-112) at $\theta_s = 180°$.

However a careful treatment of the analytic properties of the scattering amplitudes near $u = 0$ shows that the large s asymptotic behavior of eqs. (8-111) and (8-112) holds in the backward direction, $\theta_s = 180°$. It has been found that a necessary concomitant is the existence of *daughter trajectories* where, if $a(u)$ is the primary Regge trajectory, the nth daughter trajectory, $a_n(u)$, is such that:

$$a_n(0) = a(0) - n$$

for $n = 1, 2, 3, \ldots$. The daughter trajectories lie in a sequence, each member one unit of angular momentum lower than the preceding one at $u = 0$.[9]

The existence of daughter trajectories of a parent trajectory R shown in Fig. 8-13, follows if either $m_1 \neq m_3$ or $m_2 \neq m_4$ or both. In pion nucleon scattering if R is a t-channel Regge pole then $m_2 = m_4$ and $m_1 = m_3$ so the existence of a daughter to the ρ-trajectory, for example, is not implied. However ρ-trajectories are exchanged in other processes, implying ρ daughters. It should be remembered that daughter trajectories are not necessarily parallel to the parent, and so physical particles lying on daughter trajectories are not necessarily implied.

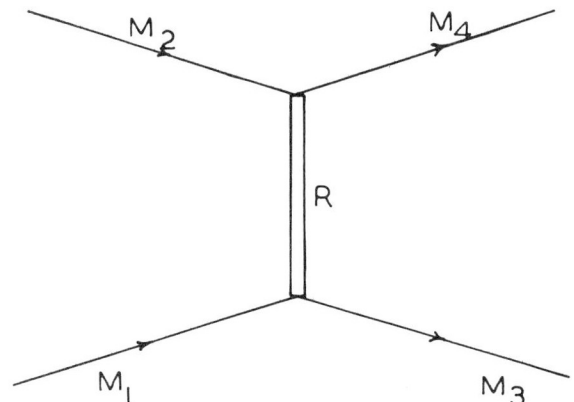

Fig. 8-13. Regge pole scattering of unequal mass particles.

---

[9] For a review containing references to the original work we refer the reader to Bertocchi (1968).

REFERENCES

D. Amati, M. Fierz and V. Glasser, Phys. Rev. Letters 4, 89 (1960).

R. C. Arnold, Phys. Rev. 153, 1523 (1967).

R. C. Arnold and M. L. Blackmon, Phys. Rev. 176, 2082 (1968).

I. Barbour, W. Malone and R. G. Moorhouse, Phys. Rev. 4, 1521 (1971).

V. D. Barger and D. Cline, Phys. Rev. 155, 1792 (1967).

E. L. Berger and G. C. Fox, Nucl. Phys. B26, 1 (1971).

L. Bertocchi, *Proc. 1967 Heidelberg Int. Conf. on Elementary Particles*, ed. H. Filtuth (North-Holland, Amsterdam, 1968).

H. Bingham, W. B. Fretter, W. J. Podolsky, M. S. Rabin, A. H. Rosenfeld, G. Smadja, G. P. Yost, J. Ballam, G. B. Chadwick, Y. Eisenberg, E. Kogan, K. C. Moffeit, P. Seyboth, I. O. Skillicorn, H. Spitzer and G. Wolf, Phys. Letters, 41B, 635 (1972).

R. Blankenbecler and M. L. Goldberger, Phys. Rev. 126, 766 (1964).

P. Bonamy, P. Borgeaud, P. Falk-Vairant, O. Guisan, P. Sondereger, C. Caverzasio, J. P. Guillaud, J. Schneider, M. Yvert, I. Manneli, F. Sergiampetri and L. Vincelli, Phys. Letters 23, 501 (1968).

N. E. Booth, Phys. Rev. Letters 21, 465 (1968).

N. Byers and C. N. Yang, Phys. Rev. 142, 976 (1966).

T. T. Chou and C. N. Yang, *High Energy Physics and Nuclear Structure*, ed. G. Alexander (North Holland Press, Amsterdam, 1967), p. 348.

T. T. Chou and C. N. Yang, Phys. Rev. 170, 1591 (1968a).

T. T. Chou and C. N. Yang, Phys. Rev. 175, 1382 (1968b).

S. Y. Chu and A. W. Hendry, unpublished.

P. D. B. Collins and E. J. Squires, *Regge Poles in Particle Physics* (Springer, Berlin, 1968).

M. Davier and H. Harari, Phys. Letters 35B, 239 (1971).

N. W. Dean, Nucl. Phys. B4, 534 (1968).

S. P. Denisov, S. V. Donskov, Yu. P. Gorin, A. I. Petrukhin, Yu D. Prokoshkin, D. A. Stoyanova, J. V. Allaby and G. Giacomelli, Phys. Letters 36B, 415 (1971).

R. Dolen, D. Horn and C. Schmid, Phys. Rev. 166, 1768 (1968).

N. Dombey, Phys. Letters 30B, 646 (1969).

K. J. Foley, S. J. Lindenbaum, W. A. Love, S. Ozaki, J. J. Russell and L. C. L. Yuan, Phys. Rev. Letters 11, 425 (1963a).

K. J. Foley, S. J. Lindenbaum, W. A. Love, S. Ozaki, J. J. Russell and L. C. L. Yuan, Phys. Rev. Letters 11, 503 (1963b).

K. J. Foley, R. S. Gilmore, S. T. Lindenbaum, W. A. Love, S. Ozaki, E. H. Witten, R. Yamada and L. C. L. Yuan, Phys. Rev. Letters 15, 45 (1965).

P. G. O. Freund, Phys. Rev. Letters 20, 235 (1968).

M. Froissart, Phys. Rev. 123, 1053 (1961).

S. Fubini, *Strong Interactions and High Energy Physics*, ed. R. G. Moorhouse (Oliver and Boyd, Edinburgh, 1964).

W. Galbraith, E. W. Jenkins, T. F. Kycia, B. A. Leontic, R. H. Phillips, A. L. Read and R. Rubinstein, Phys. Rev. 138, 13913 (1965).

S. Gasiorowicz, *Elementary Particle Physics* (John Wiley and Sons, New York, 1967).

F. J. Gilman, J. Pumplin, A. Schwimmer and L. Stodolsky, Phys. Letters 31B, 387 (1970).

REFERENCES

R. G. Glauber, *Lectures in Theoretical Physics*, ed. E. Brittin and L. G. Dunham (Interscience Publishers, New York, 1958).

J. Hamilton, *Strong Interactions and High Energy Physics*, ed. R. G. Moorhouse (Oliver and Boyd, Edinburgh 1964), p. 281.

H. Harari, Proc. 1967 Stanford Electron-Photon Symposium, p. 309 (1967).

H. Harari, Phys. Rev. Letters 20, 1395 (1968).

H. Harari, Proc. Daresbury Int. Conf. on Electron and Photon Interactions at High Energies, ed. D. W. Braben, Daresbury Nuclear Physics Laboratory (Daresbury, 1969).

H. Harari and Y. Zarmi, Phys. Rev. 187, 2230 (1969).

H. Harari, Annals Phys. 63, 432 (1971a).

H. Harari, Phys. Rev. Letters 26, 14000 (1971b).

H. Harari, *Hadronic Interactions of Electrons and Photons*, ed. J. Cumming and H. Osborn, (Academic Press, London, 1971). (1971c).

F. Henyey, G. L. Kane, J. Pumplin, M. Ross, Phys. Rev. 182, 1579 (1969).

G. L. Kane, F. Henyey, D. R. Richards, M. Ross and G. Williamson, Phys. Rev. Letters 25, 1519 (1970).

R. L. Kelly, G. L. Kane and G. Henyey, Phys. Rev. Letters 24, 1511 (1970).

N. N. Khuri and T. Kinoshita, Phys. Rev. 137B, 720 (1965).

B. Kozlowsky and A. Dar, Phys. Letters 20, 311 (1966).

E. Leader, Phys. Rev. 166, 1549 (1968).

M. H. MacGregor, R. A. Arndt and R. M. Wright, Phys. Rev. 169, 1128 (1968).

S. Mandelstam, Nuovo Cimento 30, 1127 (1963a).

S. Mandelstam, Nuovo Cimento 30, 1148 (1963b).

A. Martin, Nuovo Cimento 42, 930 ibid 44, 1219 (1966).

D. Morrison, Phys. Letters 22, 528 (1966).

A. Pagnamenta, Phys. Rev. 2, 1150 (1970).

D. H. Perkins, Progress in Elementary Particle and Cosmic Ray Physics, V, 297 (1960).

R. J. N. Phillips and W. Rarita, Phys. Rev. 139, B1336 (1965).

I. Ia. Pomeranchuk, J. Exp. Theoret. Phys. (U.S.S.R.) 34, 725 and (Soviet Phys.) JEPT 34, 499 (1958).

B. Richter, Proc. 1967 Stanford Electron-Photon Symposium, p. 309 (1967).

M. Ross, F. S. Henyey and G. L. Kane, Nucl. Phys. B23, 269 (1970).

A. Scotti and D. Y. Wong, Phys. Rev. 138, B145 (1965).

R. Serber, Phys. Rev. Letters 13, 32 (1964).

E. J. Squires, *Complex Angular Momentum and Particle Physics* (W. A. Benjamin, New York, 1963).

G. N. Watson, *Theory of Bessel Functions* (Cambridge University Press, Cambridge, 1958).

S. Weinberg, Phys. Rev. 124, 2049 (1961).

T. T. Wu and C. N. Yang, Phys. Rev. 137, B708 (1965).

CHAPTER 9

PION-NUCLEON DYNAMICS

9.1. INTRODUCTION

The ultimate goal towards which the study of elementary particles strives is the provision of an adequate dynamical theory. At first, following the ideas of Yukawa, attempts were made to describe the pion nucleon system in terms of a Lagrangian field theory,[1] similar in structure to the successful theory of quantum electrodynamics. Qualitatively the strong attractive forces between nucleons can be understood in terms of the exchange of

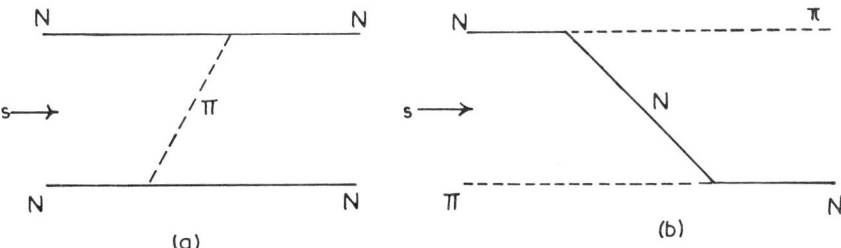

Fig. 9-1. Single particle exchange diagrams: (a) pion exchange in nucleon-nucleon scattering; (b) nucleon exchange in pion-nucleon scattering.

---

[1] The field theory of the pion nucleon system has been described in a book by Hamilton (1959).

(virtual) pions, in lowest order pictured in the Feynman diagram of Fig. 9-1a, while the p_{33} resonance in pion-nucleon scattering can be attributed to nucleon exchange (Fig. 9-1b). The quantitative implementation of these ideas has generally miscarried, because the strength of the interaction is such that perturbation series for the scattering amplitude fail to converge. However attempts have been made to avoid the use of perturbative methods, for example in the Tamm-Dancoff approach (a variant of the Hartree-Fock method), but these approximations have been largely abandoned, because it has become apparent that a new and more promising theoretical framework can be found by making use of the analytic properties of the scattering amplitude, discussed in Chapters 3 and 4. It does not, of course, follow that an approach from a field theoretical standpoint cannot succeed; recently it has been shown that a method based on Padé approximates leads to promising results, and we shall return briefly to this point at a later stage.

From the discussion given in Chapter 4, it is clear that from a knowledge of the double spectral function, appearing in the Mandelstam representation [eq. (4-53)], the scattering amplitudes for each of a set of two-particle collisions related by crossing can be calculated, provided that the number of subtractions required in forming dispersion relations can be determined. If it is assumed that the only singularities of the scattering amplitude are those generated from the bound state and resonance poles by unitarity (the principle of maximum analyticity of the first kind), it is possible to attempt to compute the double spectral function from a knowledge of the position and coupling strengths of the observed particles. If a continuation of the scattering amplitude into the complex angular momentum (J) plane encounters only isolated singularities (maximum analyticity of the second kind), the question of subtractions can be settled, and the partial waves for smaller values of the angular momentum can be determined uniquely. The partial wave amplitudes calculated from the double spectral function should then contain the same set of poles representing bound states and resonances as those employed in the construction of the

9.1. INTRODUCTION

double spectral function. This is a strong self-consistency condition, so strong that the hypothesis has been made that this condition is sufficient to determine the number and position of the poles representing particles, uniquely. This is the celebrated "bootstrap" hypothesis of Chew and Mandelstam (1960). In practical calculations it was hoped that it would be possible to generate, by an iterative procedure, the resonant states in some limited sector of the complete system of elementary particles. It was shown in Chapter 4 and will be shown in more detail below that the singularities in the scattering amplitude nearest the physical region, which are due to the exchange of particles of low mass, generate the forces of longest range. In turn, the long range forces are chiefly responsible for scattering at low energies, and for the generation of low mass resonances. It might be reasonable to suppose that the low energy region of a given two-particle scattering process could be treated self-consistently, or, if this were not so, the forces of shorter range could be represented in terms of some small numbers of parameters. In accordance with these ideas, the two pion system has been extensively studied, to see whether the ρ meson (which appears as a resonance in the $I = 1$, $J = 1$, pion-pion scattering amplitude) can be generated. In this case, the force between the pions is itself due to the exchange of the ρ meson and an iterative solution of the crossing, unitarity, and dispersion equations should determine the mass and coupling strength of the ρ.

The inherent difficulty of the bootstrap approach is that, while the general hypothesis may well be correct, in practice it may not be possible to isolate a subsystem of particles; for example, the forces that determine the ρ meson may be due to the exchange of not only the ρ meson itself but, also, of more massive particles. The most extensive and careful calculations on the pion-pion system have been made by Collins and Johnson (1969), who succeeded in generating self-consistent P (Pomeranchon) and ρ Regge trajectories. The calculated parameters of the ρ meson are in harmony with experiment, but the output Regge trajectories do not show the continuous rise with energy that is apparently indicated by experiment (see Chapter 8). This work involved numerical computation at the limit of the capacity of the latest generation of computers, but unfortunately the many simplifications proposed in the past were demonstrated not to be valid. For this reason, it is difficult to contemplate the extension of the work to include other channels coupled to the pion-pion system or to extensions to other more complicated systems.

In this situation a much less ambitious approach is open, which casts light on the underlying physical assumptions of the "bootstrap hypothesis" without performing the full iterative calculation. It is possible to consider the exchange of known particles or resonances of mass < 1 GeV/c^2 and to ask whether the forces so obtained are attractive, and large enough in the correct channels, to generate the observed resonances at about the correct masses. In this approach there is no attempt at an iterative bootstrap, and we must be careful not to try to apply the method where forces of shorter range than those due to the exchange of the particles in question are important. This approach has been developed by Donnachie and Hamilton (1965) in a discussion of the pion-nucleon system, and in the next section we shall describe results of this and related work.

9.2. THE ANALYSIS OF LOW ENERGY π-N SCATTERING BY PARTIAL WAVE DISPERSION RELATIONS

In this section, we shall describe in more detail how the dispersion relations of eq. (4-72) can be used to calculate the partial wave amplitudes $f_{\ell\pm}^I(s)$ in the physical region. To do this we need to know the discontinuities across the "left-hand" or unphysical cuts. As we shall, in practice, not be able to calculate all the discontinuities, it is necessary to consider which discontinuities are most important. Quite generally, it is expected that an amplitude at a certain value of s will be determined mainly by the singularities lying nearest to s. To take an example, if the

9.2. THE ANALYSIS OF LOW ENERGY π-N SCATTERING

left-hand cuts in $f_\ell(s)$ were replaced by a series of poles at positions $-s_i$ with residues R_i, and further if the amplitude in the physical region were small and elastic ($R(s) = 1$ in eq. (4-67)), so that on the right hand cut Im $f_\ell(s) \ll$ Re $f_\ell(s)$, then we would have:

$$\text{Re } f_i(s) \approx \sum_i \frac{R_i}{(s+s_i)} \,. \tag{9-1}$$

Consider the variation of $f_\ell(s)$ in some interval $s' < s < s''$ of the real axis. It is clear that distant poles (for which $s_i \gg s''$) provide a contribution to $f_\ell(s)$ which is constant or slowly varying, while nearby poles provide a rapidly varying contribution. The situation will be most favorable at low energies, because at high energies (large s) there will be relatively less difference between the poles situated at the nearer parts of the left-hand cut and those further away. Using the N/D method, it can be shown that this picture remains valid for the larger amplitudes, for which we cannot neglect Im f_ℓ in comparison with Re f_ℓ on the right-hand cut.

On this basis, we would expect that it would be essential to obtain the discontinuities across the cuts nearest to the physical region (see Fig. 4-8) when calculating the πN scattering amplitudes at low energies, if necessary representing the effects of distant singularities by a constant or by a pole. Fortunately, as we shall see, it is just the nearer parts of the left-hand cuts that can be calculated.

The Range of Interaction and the Peripheral Method

In section 4.1 we noted that in the case of potential scattering the left-hand cut discontinuities could be directly related to the potential. As an example, we shall calculate the discontinuity arising from the term $f_B(t)$ [eq. (4-12)], which we found to be equal to the Fourier transform of the potential. From eqs. (4-2), (4-9), and (4-21) and using the relation:

$$\frac{1}{2} \int_{-1}^{+1} d\mu \, P_\ell(\mu) \frac{1}{\lambda^2 + 2s(1-\mu)} = \frac{1}{2s} Q_\ell\left(1 + \frac{\lambda^2}{2s}\right) \tag{9-2}$$

where Q_ℓ is the Legendre function of the second kind, the contribution to $f_\ell(s)$ from $f_B(t)$ is:

$$\int_{m^2}^{\infty} d\lambda \, \frac{\sigma(\lambda)}{2s} \, Q_\ell\left(1 + \frac{\lambda^2}{2s}\right). \tag{9-3}$$

The function $Q(x)$ for positive integral values of ℓ is analytic in the x plane cut from $x = -1$ to $x = +1$, so that $Q_\ell(1+\lambda^2/2s)$ is analytic in s, with a cut in the interval $-\infty < s < -\lambda^2/4$. Looking at eq. (9-3), one sees that the nearest cut to the physical region $(s > 0)$, comes from the smallest value of λ, that is, from the potential of longest range. Farther away from the physical region contributions arise from potentials of shorter range (larger λ). The same can be shown to be true for the complete left-hand cut discontinuity, so that there is a correlation between the distance along the left-hand cut and the range of the contributing potentials. It is possible to make a similar statement for elementary particle scattering. In the notation employed in section 4.3, the three pion-nucleon channels connected by crossing were labeled channels I, II, and III, where the channel I is the process $\pi + N \to \pi + N$, with the physical region $s > (M_\pi + M_N)^2$, $t < 0$, channel II is $\pi + \pi \to N + N$ with the physical region $t > 4M_\pi^2$, $s < 0$, while channel III is the crossed $\pi + N \to \pi + N$ channel, with physical region $u > (M_\pi + M_N)^2$, $t < 0$. When a system of mass M is exchanged in channels II or III, this gives rise to an effective potential in channel I. In the static approximation the potential will have the shape of

Table 9-1. Values of q^2 at points on the circle cut in the pion-nucleon partial wave scattering amplitudes, with the maximum values t and the range of the corresponding exchange force, R.

θ	q^2	t_{max}	R
	(Units of M_π^2)		$\times 10^{-13}$ cm
$\pm 10°$	-1.34	5.3	4.7
$\pm 30°$	-3.96	15.8	0.72
$\pm 50°$	-8.89	35.6	0.35

θ = Angle defined in Fig. 4-10.

9.2. THE ANALYSIS OF LOW ENERGY π-N SCATTERING

Table 9-2. Values of q^2 at points along the cut in the interval $0 \leq s < (M_N - M_\pi)^2$ in the pion-nucleon partial wave scattering amplitude, together with the maximum values of u and the corresponding range of the exchange forces.

s	q^2 (units of M_π^2)	u_{max}	R $\times 10^{-13}$ cm
32.7	0	59.6	7.8
25.0	2.67	78.0	1.56
15.0	13.17	130	0.39
10.0	28.17	195	0.23
0			

a Yukawa well and be of range $\sim (1/M)$. As the nearest parts of the left-hand cuts in channel I to the physical region correspond to small values of t and u, it follows that these discontinuities represent the forces of longest range in the system. For example in Table 9-1 the maximum values of t which contribute to various points on the circular cut are shown, and larger values of t, corresponding to interactions of shorter range, become increasingly important as we move around the circle away from the physical region. On the real axis, the nearest singularity to the physical region arises from the "short cut" which is centered about $s = M_N^2 = 45.2$ in the interval $43.2 < s < 47.2$, where we have used units such that $M_\pi = 1$ ($M_\pi = 139.6$ MeV). This cut is due to nucleon exchange in channel III and represents an interaction of range $R = 1.4 \times 10^{-13}$ cm.[2] The next nearest singularity is the cut in the interval $0 \leq s \leq (M_N - M_\pi)^2$ arising from channel III. The maximum values of u that contribute and the corresponding ranges of the effective interactions are shown in Table 9-2. The back of the circle cut and the cuts with $s < 0$ represent forces of very short range.

---

[2] We adopt the definition of the range of the exchange force given by Hamilton et al., (1962a), that $R = \frac{\hbar}{Mc} (2 - \sqrt{t}/M_\pi)^{-1}$ for t-channel exchanges and $R = \frac{\hbar}{M_\pi c} \left(1 + \frac{M_N}{M_\pi} - \frac{\sqrt{u}}{M_\pi}\right)^{-1}$ for u-channel exchanges.

The short-range effects, corresponding to distant singularities, contribute slowly varying terms to the amplitude in the physical region and can often be represented approximately by a pole situated at a point distant from the physical region; for partial waves with $\ell > 0$ it is possible to suppress the short-range effects by using the function $F_{\ell\pm}^I(s) = q^{-2\ell} f_{\ell\pm}^I(s)$, in place of $f_{\ell\pm}^I(s)$ (Donnachie et al., 1964). The new function possesses the same cuts and poles as $f_{\ell\pm}^I(s)$, and no additive singularity is introduced when q vanishes at the physical threshold $s = (M_N + M_\pi)^2$, since $f_{\ell\pm}^I(s)$ is proportional to $q^{2\ell}$ at this point. The momentum q becomes large along the unphysical cuts away from the physical region (this is illustrated in Tables 9-1 and 9-2), and to the first approximation the short range forces can be neglected for $\ell > 0$.

It is then reasonable to discuss low energy πN scattering in terms of the singularities along the real axis for $s > 0$ and along the front of the circular cut alone. We shall look at the various contributions in turn.

Nucleon Exchange and the Chew-Low Plot

We shall start by evaluating the contribution to the amplitudes $F_{\ell\pm}^I$ from the short cut in the interval [eq. (4-65b)], which arises from the long range part of nucleon exchange. Looking at eqs. (4-55) and (4-61), we see that we require the projections on the various angular momentum states of $C(I)(M_N^2 - u)^{-1}$, where $C(\frac{3}{2}) = -2g^2$ and $C(\frac{1}{2}) = +g^2$. These projections are [from eqs. (3-63) and (4-60)], $^B f_{\ell\pm}^I$ where:

$$^B f_{\ell\pm}^I = \frac{C(I)}{2} \int_{-1}^{+1} d\mu \, \frac{(E+M_N)(W-M_N) P_\ell(\mu) + (E-M_N)(W+M_N) P_{\ell+1}(\mu)}{8\pi W (M_N^2 - u)} \quad (9\text{-}4)$$

and where $u = -s + 2M_N^2 + 2M_\pi^2 + 2q^2(1-\mu)$. Using the integral of eq. (9-2), we find that:

$$^B f_{\ell\pm}^I = \frac{C(I)(-1)^\ell}{16\pi W \, q^2} [(E+M_N)(W-M_N) Q_\ell(Z) - (E-M_N)(W+M_N) Q_{\ell+1}(Z)] \quad (9\text{-}5)$$

where $Z = (s - M_N^2 - 2M_\pi^2 - 2q^2)/2q^2$.

9.2. THE ANALYSIS OF LOW ENERGY π-N SCATTERING

This expression can be calculated numerically, although it contains some short range terms which must be subtracted first, but a useful approximation can be made for low energy scattering (when q is of order M_π) since in that case Z is large of order $\left(\frac{M_N}{M_\pi}\right)^2 = 45$, and we may use just the leading term in the asymptotic expansion of $Q_\ell(Z)$. This is (Morse and Feshbach, 1953):

$$Q_\ell(Z) \sim \sqrt{\pi} \, \frac{\Gamma(\ell+1)}{\Gamma\left(\ell+\frac{3}{2}\right)} \, (2Z)^{-\ell-1} \tag{9-6}$$

To the same order, we can write:

$$Z \approx \frac{(s-M_N^2)}{2q^2}, \quad W - M_N \approx \frac{(s-M_N^2)}{2M_N}, \quad E - M_N \approx \frac{q^2}{2M_N}$$

and short calculation gives:

$$B_{f_{\ell+}}^I = \frac{C(I)(-1)^\ell \, \Gamma(\ell+1)}{16\sqrt{\pi} \, M_N \Gamma\left(\ell+\frac{3}{2}\right)} \left(\frac{q^2}{(s-M_N^2)}\right)^\ell ; \quad B_{f_{\ell-}}^I = -\frac{1}{2\ell} B_{f_{\ell+}}^I . \tag{9-7}$$

In this approximation, the s wave amplitudes $B_{f_{0+}}^I$ are constant, and it follows from our previous discussion that these represent short range effects, and are a contribution to the cut from $s = -0$ to $s = -\infty$. There is no s wave contribution, in this approximation, to the short cut. The contributions for $\ell > 0$ take the form of poles at $s = M_N^2$, the center of the short cut. These are manifestly all long range contributions, but as $q^2/(s-M_N^2)$ is of order $(q^2/2M_N)$ for small q, it follows that, at low energies, nucleon exchange is important only in the p waves, ($\ell=1$). In channel I, the Born pole to order M_π/M_N contributes only in the P_{11} state, the contribution being:

$$\frac{-3 \, g^2 \, q^2}{8\pi \, M_N^2 \, (s-M_N^2)} . \tag{9-8}$$

On dividing by q^2 and adding in this contribution, the total contributions to the four p waves, $F_{1\pm}^I = q^{-2} f_{1\pm}^I$, from the nucleon poles are:

$$B_{F_{1+}^{\frac{3}{2}}}(s) = \frac{g^2}{6\pi M_N} \frac{1}{(s-M_N^2)} \qquad B_{F_{1-}^{\frac{3}{2}}}(s) = \frac{-g^2}{12\pi M_N} \frac{1}{(s-M_N^2)}$$

$$B_{F_{1+}^{\frac{1}{2}}}(s) = -\frac{g^2}{12\pi M_N} \frac{1}{(s-M_N^2)} \qquad B_{F_{1-}^{\frac{1}{2}}}(s) = -\frac{g^2}{3\pi M_N} \frac{1}{(s-M_N^2)}. \quad (9\text{-}9)$$

It is evident from eq. (9-1) that a pole in the scattering amplitude gives rise to a phase shift which is of the same sign as the residue of the pole. It follows that in the p_{33} state, nucleon exchange provides an attractive force (positive phase shift), while in the p_{31}, p_{13} waves the effective force is repulsive. It is now interesting to see whether the attraction in the p_{33} state due to nucleon exchange is, by itself, sufficient to account for the $N^*(1236)$ resonance in that wave. The N/D method may be used to calculate the amplitude $F_{1+}^{\frac{3}{2}}(s)$, by setting $N(s)$ equal to $B_{F_{1+}^{\frac{3}{2}}}(s)$, so that:

$$N(s) = \frac{g^2}{6\pi M_N} \frac{1}{(s-M_N^2)}. \qquad (9\text{-}10)$$

The denominator function $D(s)$ [from eqs. (4-76), (4-74c) and (4-77)] is then given by:

$$\text{Re } D(s) = 1 - \frac{1}{\pi} (s-M_N^2) P \int_{s_0}^{\infty} \frac{g^2 \, q^3(s') \, R(s')}{6\pi M_N (s'-M_N^2)^2 (s'-s)} \, ds' \qquad (9\text{-}11)$$

where the subtraction has been made at $s = M_N^2$, so that $D(M_N^2) = 1$.

Below a π meson laboratory kinetic energy of 300 MeV, inelastic scattering is unimportant, and $\eta_{\ell\pm}^I = 1$. The real phase shift $\delta_{1+}^{\frac{3}{2}}$ is found from:

$$q^3 \cot \delta_{1+}^{\frac{3}{2}} = \text{Re} \left[F_{1+}^{\frac{3}{2}}(s) \right]^{-1} = \frac{\text{Re } D(s)}{N(s)} = (s-M_N^2) \left[\frac{6\pi M_N}{g^2} - (s-M_N^2) J(s) \right]$$

$$(9\text{-}12)$$

9.2. THE ANALYSIS OF LOW ENERGY π-N SCATTERING

where:

$$J(s) = \frac{1}{\pi} \int_{s_0}^{\infty} \frac{q^3(s') R(s')}{(s'-M_N)^2 (s'-s)} ds' . \quad (9\text{-}13)$$

To order (M_π/M_N), $q \approx q_L$ and $(s-M_N^2) \approx 2M_N \omega$ where ω is the total laboratory energy of the π-meson [cf. eq. (3-21)]. Making these approximations and expressing g^2 in terms of the effective pseudovector coupling f^2 by eq. (3-85), the formula for $\cot \delta_{1+}^{\frac{3}{2}}$ becomes:

$$\left(\frac{q^3}{\omega}\right) \cot \delta_{1+}^{\frac{3}{2}} = \frac{3M_\pi^2}{4f^2} - 4M_N^2 \omega J(s) . \quad (9\text{-}14)$$

Before proceeding further, it is necessary to know the inelasticity function $R(s)$. Below 300 MeV, $R(s) \simeq 1$ and, as a first approximation, we may put $R(s) = 1$ for all s. Using $f^2 = 0.08$, the value found from the forward dispersion relation analysis, $\delta_{1+}^{\frac{3}{2}}$ can then be calculated from eq. (9-14). It is found that $\left(\frac{q^3}{\omega}\right) \cot \delta_{1+}^{\frac{3}{2}}$ is positive at very low energies and falls linearly, passing through zero near 400 MeV, at which energy $\delta_{1+}^{\frac{3}{2}} = \frac{\pi}{2}$ and there is a resonance. Qualitatively this is an encouraging result, but quantitatively it is poor, since the resonance position is in fact near 200 MeV. The disagreement is partly due to the effect of the omitted left-hand cuts, but also to the approximation in which $R(s)$ is set equal to unity. In fact, the result is quite sensitive to the value of $R(s)$ in the high energy region, since integral $J(s)$ converges slowly. Because of this, the N/D method can provide accurate results for the small partial waves (provided all the long range forces are included) as in these waves Im $f_\ell \ll$ Re f_ℓ and the integral over the right-hand cut in the dispersion relation can be neglected. We can also expect to predict which amplitudes can be resonant at low energies, even if we cannot determine the position and width of the resonance.

As $J(s)$ is a slowly varying function of s, an effective range formula (Chew and Low, 1956) is obtained by approximating $J(s)$ by a constant. In this case, eq. (9-14) can be written in the form:

$$\left(\frac{q^3}{\omega}\right) \cot \delta_{1+}^{\frac{3}{2}} = \frac{3M_\pi^2}{4f^2}\left(1 - \frac{\omega}{\omega_R}\right) \tag{9-15}$$

where ω_R is the total π-meson laboratory energy at resonance. The plot of $(q^3/\omega) \cot \delta_{1+}$ against ω is a straight line known as a Chew-Low plot. If f^2 and ω_R are determined to fit the experimental data, a very good fit is found with $f^2 = 0.087$, which is in excellent agreement with the more accurate value given by forward dispersion relations.

N^* Exchange

It is now necessary to look at the other long range interactions to see what role they play relative to nucleon exchange. We have seen that the cut in the interval $0 < s < (M_N - M_\pi)$ is connected with π-N scattering in channel III. For a given s, the range of values of u (the square of the center of mass energy in channel III) can be calculated using:

$$u = 2M_N^2 + 2M_\pi^2 - s + 2q^2(1 - \cos\theta) . \tag{9-16}$$

We have already noted, in Table 9-2, the maximum values of u at various points along the cut, and we have seen that the nearer parts of the cut depend on the low energy scattering in channel III. The further parts are suppressed by the factors $(1/q^{2\ell})$. Each partial wave in channel I receives contributions along the cut from all partial waves in the crossed channel. This is because the relation between the scattering angle in channel III, θ_u, and that in channel I, θ, is nonlinear:

$$q^2(s)(1 - \cos\theta) = q^2(u)(1 - \cos\theta_u)$$

where $q^2(u)$ is the center of mass momentum in channel III. As $\cos\theta$ varies from $+1$ to -1, $\cos\theta_u$ spans the same interval, and this implies

9.2. THE ANALYSIS OF LOW ENERGY π-N SCATTERING

that the partial wave expansion is within its radius of convergence. As at low energies the p_{33} wave is resonant, this wave is expected to dominate the discontinuity in each partial wave in channel I. We shall refer to this as N^* exchange. Detailed calculations (Hamilton et al., 1962b; Donnachie et al., 1964) confirm that N^* exchange is dominant for $\ell = 1$ and $\ell = 2$. For $\ell = 3$ the s waves in channel III provide a comparable contribution, but in this case both contributions are very small and can be ignored.

The calculations are carried out as follows. First we notice that the crossing relation, eq. (4-56), together with eq. (4-61) shows that both isotopic spin amplitudes in channel III contribute to the discontinuity in channel I. We easily find that (cf. the footnote on page 143):

$$A^I(s,t) = \sum_{I'} M_{II'} A^{I'}(u,t)$$

$$B^I(u,t) = -\sum_{I'} M_{II'} B^{I'}(u,t) \qquad (9\text{-}17)$$

where M, the crossing matrix, is (with $I = \tfrac{1}{2}$ first):

$$M = \begin{pmatrix} -\tfrac{1}{3} & \tfrac{4}{3} \\ \tfrac{2}{3} & \tfrac{1}{3} \end{pmatrix}. \qquad (9\text{-}18)$$

From the partial wave amplitudes in channel III, we can calculate Im $A^I(u,t)$ and Im $B^I(u,t)$ and then Im $A^I(s,t)$, Im $B^I(s,t)$ using eq. (9-17). The discontinuity across the cut in the partial wave amplitude is then $2i$ Im $f^I_{\ell\pm}(s)$ where Im $f^I_{\ell\pm}(s)$ is obtained by projection from Im $A^I(s,t)$, Im $B^I(s,t)$. This is, of course, a numerical calculation, but it is interesting to compute N^* exchange in a static approximation, where terms of order (M_π/M_N) are neglected. In this limit, the laboratory and center of mass system coincide ($q \approx q_L$) and:

$$k_1 = (\vec{q}_b, \omega) \qquad k_2 = (\vec{q}_a, \omega)$$

$$q = |\vec{q}_a| = |\vec{q}_b|$$

where k_1 and k_2 are the four-momentum vectors of the incident and scattered mesons [cf. eq. (3-10)]. Under crossing $k_1 \to -k_2$, $k_2 \to -k_1$ so that $\omega \to -\omega$, while $\cos\theta \equiv (\vec{q}_a \cdot \vec{q}_b / q^2)$ is unaltered. This shows that in this limit, crossing connects partial waves of the same order and N^* exchange contributes only to p waves in channel I. The relation between s and u also simplifies in this limit, because in channel III (positive energies):

$$u \approx M_N^2 + 2M_N\omega \tag{9-19a}$$

while in channel I, (negative energies) we have:

$$s \simeq M_N^2 - 2M_N\omega \ . \tag{9-19b}$$

Ignoring the isotopic spins of the particles, we can calculate the scattering amplitude in channel III that arises from partial waves of order ℓ. It is:

$$f(\theta,\phi) = [(\ell+1)f_{\ell_+}(\omega) + \ell f_{\ell_-}(\omega)] P_\ell(\cos\theta) - i(f_{\ell_-}(\omega) - f_{\ell_+}(\omega))\sin\theta\, P'_\ell(\cos\theta)(\vec{\sigma}\cdot\vec{n}) . \tag{9-20}$$

Under crossing $\vec{n} = (\vec{q}_\alpha \times \vec{q}_\beta)/q^2 \to -\vec{n}$, so we must have that:

$$(\ell+1)f_{\ell_+}(\omega) + \ell f_{\ell_-}(\omega) = (\ell+1)f_{\ell_+}(-\omega) + \ell f_{\ell_-}(-\omega)$$

and:

$$f_{\ell_-}(\omega) - f_{\ell_+}(\omega) = -(f_{\ell_-}(-\omega) - f_{\ell_+}(-\omega))$$

from which:

$$f_{\ell_+}(-\omega) = \left(\frac{1}{2\ell+1}\right)[f_{\ell_+}(\omega) + 2\ell f_{\ell_-}(\omega)]$$

$$f_{\ell_-}(-\omega) = \left(\frac{1}{2\ell+1}\right)[(2\ell+2) f_{\ell_+}(\omega) - f_{\ell_-}(\omega)] \ . \tag{9-21}$$

9.2. THE ANALYSIS OF LOW ENERGY π-N SCATTERING

From these general relations, we specialize to N^* exchange by keeping only $f_{1+}^{\frac{3}{2}}(\omega)$ on the right-hand side of eq. (9-21) and restoring the isotopic spin crossing matrix M, so that:

$$f_{1+}^I(-\omega) = \frac{1}{3} M_{I\frac{3}{2}} f_1^{\frac{3}{2}}(\omega)$$

$$f_{1-}^I(-\omega) = \frac{4}{3} M_{I\frac{3}{2}} f_1^{\frac{3}{2}}(\omega) \ . \qquad (9\text{-}22)$$

Remembering (Chapter 4) that the continuation is from the upper side of the cut in channel III ($u+i\epsilon$) to the lower side of the crossed cut ($s-i\epsilon$), we have finally:

$$\text{Im } f_{1\pm}^I(s) = N_\pm(I) \text{ Im } f_1^{\frac{3}{2}}(2M_N^2 - s) \qquad (9\text{-}23)$$

where we have restored the variable s and where:

$$N_+\left(\frac{3}{2}\right) = -\frac{1}{9} \qquad N_-\left(\frac{3}{2}\right) = -\frac{4}{9}$$

$$N_+\left(\frac{1}{2}\right) = -\frac{4}{9} \qquad N_-\left(\frac{1}{2}\right) = -\frac{16}{9} \ . \qquad (9\text{-}24)$$

On the basis of this approximation, we expect N^* exchange to be only very important in p states. The effective potential is attractive in each of the p states, but is very weak in the p_{33} state and quite strong in the p_{11} state. This is borne out by exact calculation. Following Donnachie et al., (1964), we write the dispersion relation, eq. (4-72), as:

$$\text{Re } F_{\ell\pm}^I(s) = \mathcal{F}_{\ell\pm}^I(s) + \frac{1}{\pi} \int_{s_0}^\infty \frac{\text{Im } F_{\ell+}^I(s')ds'}{(s'-s)} \qquad (9\text{-}25)$$

where $\mathcal{F}_{\ell\pm}^I(s)$ is the contribution from the integrals over all the left-hand cuts, then the exact contributions from nucleon and N^* exchange are compared at threshold, $s = s_0$, in Table (9-3) for the p states.

Table 9-3. The nucleon and N^* contributions to the p wave pion-nucleon scattering amplitudes, scattering amplitudes, at threshold.

Partial Wave	N-exchange	N^* exchange
P_{11}	0.026	0.026
P_{13}	−0.052	0.006
P_{31}	−0.052	0.006
P_{33}	0.104	0.001

Units are $\hbar = c = M_\pi = 1$. (From Hamilton et al., 1962b.)

ρ and σ Exchange and the Circle Cut

We have already seen that the circle cut is associated with the channel II reaction $\pi + \pi \to N + \bar{N}$. Table 9-1 showed that the front of the circle is connected with small values of t, the square of the center of mass energy in channel II. The lowest normal threshold in this channel is at $t = 4M_\pi^2$, which is far below the physical threshold at $4M_N^2$. Nevertheless, we shall see that a partial wave expansion is possible in this channel which converges in the region of interest. As each π-meson has unit isotopic spin, the π-π system can have isotopic spin $I = 0$, 1, or 2. However, the final N-$\bar{N}$ state can only be in the states with $I = 0$ or $I = 1$, and these are the only states that concern us here. By the Pauli principle, the $I = 0$, π-π system must be in an even angular momentum state with $J = 0, 2, 4, \ldots$ and the $I = 1$ system in an odd state with $J = 1, 3, \ldots$.

The states with even J are symmetrical under interchange of the π-mesons, that is, under the interchange $s \leftrightarrow u$ and must be connected with the amplitudes $A^{(+)}(s,t)$ and $B^{(+)}(s,t)$, while the $I = 1$ states with odd J are odd under the interchange $s \leftrightarrow u$ and are connected with $A^{(-)}(s,t)$ and $B^{(-)}(s,t)$ (see section 8-4).

9.2. THE ANALYSIS OF LOW ENERGY π-N SCATTERING

To define partial wave amplitudes for the reaction $\pi + \pi \to N + \bar{N}$ we may use the helicity formalism developed in Chapters 2 and 3. As in the case discussed on page 73, there are four elements of the reduced S-matrix, of which, by parity conservation, only two are independent. These are S^j_{++} ($= S^j_{--}$) and S^j_{+-} ($= S^j_{-+}$) where $\pm$ denotes production of a nucleon or antinucleon with helicities $\pm\frac{1}{2}$. Frazer and Fulco (1960) have shown that the partial wave amplitudes $f^j_\pm(t)$ defined by:

$$f^j_+(t) = \frac{E}{(p\bar{p})^j} S^j_{++}$$

$$f^j_-(t) = \frac{1}{(p\bar{p})^j} S^j_{+-} \qquad (9\text{-}26)$$

have simple analytic properties. Here p and $\bar{p}$ are the center of mass momenta in the π-π and N-$\bar{N}$ systems and:

$$t = 4(M_\pi^2 + p^2) = 4(M_N^2 + \bar{p}^2) \qquad (9\text{-}27)$$

and E is the center of mass nucleon energy:

$$t = 4E^2 \;.$$

The expansion of the amplitudes $A^{(\pm)}(s,t)$ and $B^{(\pm)}(s,t)$ terms of the $f^j_\pm$ is (Frazer and Fulco, 1960):

$$A^{(\pm)}(s,t) = \sum_j \frac{4\pi}{p^2} (2j+1)(p\bar{p})^j \left[f^j_+(t) P_j(x) - \frac{M_N x}{\sqrt{j(j+1)}} P'_j(x) f^j_-(t) \right]$$

$$B^{(\pm)}(s,t) = \sum_j \frac{4\pi(2j+1)}{\sqrt{j(j+1)}} (p\bar{p})^{j-1} P'_j(x) f^j_-(t) \qquad (9\text{-}28)$$

where for $A^{(+)}, B^{(+)}$ the sum runs over even j, and for $A^{(-)}, B^{(-)}$ over odd j. The discontinuity in the channel I amplitudes Im $f_{\ell\pm}$ can be calculated from Im $A^{(\pm)}$ and Im $B^{(\pm)}$, which are determined by eqs. (9-28). However, as we are below the physical N-$\bar{N}$, threshold $\bar{p}$ must be replaced by $i|\bar{p}|$ in these equations. In (9-28), $x = \cos\Phi$, where Φ is the angle of scattering in channel II. We have:

$$x = \cos \Phi = \frac{s - |\bar{p}| + p^2}{2i|\bar{p}|p} \qquad (9\text{-}29)$$

and this is complex for values of s and t on the circle. This means that Im $A^{(\pm)}$ and Im $B^{(\pm)}$ are complex quantities and also that the partial wave expansions do not always converge. However Frazer and Fulco (1960) have shown that round the front of the circle the expansion does converge up to an angle $|\phi| = 66°$, in which region $t \leq 26 M_\pi^2$. As beyond this region the effective potentials are in any case of rather short range, this is not too serious.

The analytic structure of the partial wave amplitudes $f_\pm^j(t)$ can be investigated in a similar way to the amplitudes in channel I. They of course possess the cut in t from the threshold at $t = 4 M_\pi^2$ and in addition possess a left-hand cut along the real axis from $t = t_1$ where $t_1 = 4 M_\pi^2 \left(1 - \frac{M_\pi^2}{4 M_N^2}\right)$ to $t = -\infty$. The discontinuity across the nearest part of the left-hand cut is determined by the N and N^* exchange in channels I and III. Although the right-hand cut is below the physical threshold, it is possible to continue the unitarity equation to determine the discontinuity across the cut. Between $t = 4 M_\pi^2$ and $t = 16 M_\pi^2$ (the four pion threshold) only two π-meson intermediate states are possible,[3] and the unitarity condition reads:

$$\text{Im } f_\pm^j(t) = c \{f_\pm^j(t)\}^* y_j(t) \qquad (9\text{-}30)$$

where c is a real constant and $y_j(t)$ is the amplitude for the elastic scattering of two π-mesons. As usual $y_j(t)$ can be expressed in terms of a real phase shift δ_j, and we set:

$$y_j(t) = \frac{\sqrt{t}}{p} e^{i \delta_j} \sin \delta_j \; .$$

---

[3] States of three pions are forbidden by G parity invariance.

9.2. THE ANALYSIS OF LOW ENERGY π-N SCATTERING

The unitarity equation shows that $f_{\pm}^j(t)$ has the same phase as $y_j(t)$; that is, we have:

$$f_{\pm}^j(t) = |f_{\pm}^j(t)| e^{i\delta j} . \qquad (9\text{-}31)$$

The magnitude of the circle contribution thus depends on the π-π phase shift. If this were zero, Im $f_{\pm}^j$ would vanish and there would be no contribution from channel II in channel I.

We are now in a position to discuss which amplitudes should give important contributions to the long range forces. From studying the final state interactions in processes such as π-meson production in π-N collisions:

$$\pi + N \to \pi + \pi + N$$

it is known that there is a resonance in the π-π system at 740 MeV. This occurs in the $I = 1$, $j = 1$ state, and is known as the ρ meson. Indirect evidence confirming this has also come from the study of nucleon form factors discussed in Chapter 11. A direct measurement of the parameters of the ρ is now possible by an experiment in which the reaction:

$$e^+ + e^- \to \rho^0 \to \pi^+ + \pi^-$$

is observed. The measured width is 110 MeV. The value of t at resonance is about $28\, M_\pi^2$, and the ρ begins to contribute to the circle cut for $|\phi| > 40°$. Although this is quite far from the physical region, it provides a most important contribution to the phase shifts that we shall be discussing. A further resonance the f^0 has been identified in the π-π system, in the $I = 0$, $J = 2$ state, but as the resonance energy is 1250 MeV, it only contributes to the short range interactions in the π-N system.

There is also evidence of a strong interaction in the $I = 0$, $J = 0$ π-π system, for example from a study of the reaction (Abashian et al., 1963):

$$p + d \to He^3 + \pi + \pi$$

from studies of final state interactions in $\pi^- + p$ scattering (Samois et al., 1962; Scharenguivel et al., 1969; Deinet et al., 1969), and in η and K^+

decay (Brown and Singer, 1964). Although the results are difficult to interpret, it is agreed that the interaction in this state is strong and attractive. It seems there is a broad effect, the σ, at ~ 600 MeV and a possible resonance, ϵ, at ~ 1000 MeV (Flatte et al., 1972).

π-N s-Wave Scattering

The contribution to πN scattering from the exchange of the s-wave pair of π-mesons and from ρ exchange has been analyzed by Hamilton and his collaborators (Hamilton et al., 1962b). As the s-wave π-π phase shift is not known experimentally, these workers first used the experimental values of the πN s-wave phase shifts to calculate the π-π wave parameters, which then could be used for the prediction of the phase shifts for $\ell > 0$ in the π-N system. The contribution to the dispersion relation from all the left-hand cuts $\mathcal{F}_{\ell\pm}^I(s)$ can be written as for $1 = 0$:

$$\mathcal{F}_{0+}^I(s) = \overline{\mathcal{F}}_{0+}^I(s) + \Delta_0^I(s) \tag{9-32}$$

where $\overline{\mathcal{F}}_{0+}(s)$ is the contribution from N and N^* exchange, which, as we have discussed earlier, is known. It is then possible, using the experimental values of Re f_{0+}^I and Im f_{0+}^I, to calculate the quantity $\Delta_0^I(s)$ which is called the discrepancy. From Δ_0^I we can find $\Delta_0^{(\pm)}(s)$, the discrepancy in the even and odd amplitudes. Then $\Delta_0^{(+)}$ will, apart from short range effects, contain contributions from the $I = 0$, $j = 0$, $\pi + \pi \to N + \overline{N}$, state, while $\Delta_0^{(-)}$ will be connected with ρ exchange in the $I = 1$, $j = 1$ state.

The term Δ_{0+} is well-fitted by a parametrization:

$$\Delta_0^{(+)}(s) = \frac{c^{(+)}}{s_1 - s} + \frac{c^{(+)}}{s_1^t - s} + \frac{b^{(+)}}{\overline{s} - s} \tag{9-33}$$

where s_1 is on the circle at $\phi = 26°$ and $\overline{s} = -20$. The last term in eq. (9-33) represents short range effects, which are strong and repulsive, and the first two terms represent the π-π attraction in the $j = 0$ state.

9.2. THE ANALYSIS OF LOW ENERGY π-N SCATTERING

In the same way $\Delta_0^{(-)}$ can be fitted by a form:

$$\Delta_0^{(-)}(s) = \frac{2c^{(-)}(s - \text{Re } \bar{s})}{(s - \text{Re } s_1)^2 + (\text{Im } s_1)^2} + \frac{b^{(-)}}{\bar{s} - s} \tag{9-34}$$

the short range interaction in this case being much smaller.

This work has been extended and revised by Nielsen et al., (1970), making use of the extensive phase shift analyses of πN scattering data up to 2 GeV and using Regge pole high energy extrapolations in the evaluation of the dispersion relation integrals. The results are in reasonable accord with those of the earlier work. The interpretation of the discrepancy functions in terms of $\pi\pi$ scattering essentially involves the analytic continuation of the discrepancy functions away from the real axis. The specific parametrizations, eqs. (9-33) and (9-34), represent a particular ansatz for doing this. Systematic methods of analytic continuation suitable for such applications have been discussed by Cutkosky and Deo (1968a,b) and Ciulli (1969a,b), who have paid due attention to the problems of the convergence of the procedures and to the errors involved.[4]

The s-Wave π-π Phases

To calculate the π-π phase shift $\delta_0(t)$ it is necessary to assume that the unitarity relation, eq. (9-31), holds for $t > 16M_\pi^2$. This is probably a good approximation unless inelastic π-meson processes are exceptionally strong. If the π-π amplitude $y_0(t)$ is written in N/D form:

$$y_0(t) = N(t)/D(t) \tag{9-35}$$

then, D, which has the physical cut, has the phase $(-\delta_0)$. The function $g(t) = D(t) f_+^0(t)$, then is real from $t = 4M_\pi^2$ to $t = \infty$. It has no right-hand cut, but has the left-hand cut from $t = t_1$ to $t = -\infty$. Writing a dispersion relation, we have:

---

[4] A useful review has been given by Hamilton (1971).

$$g(t) = D(t) f_+^0(t) = \frac{1}{\pi} \int_{-\infty}^{t_1} \frac{D(t') \text{Im } f_+^0(t') \, dt'}{(t'-t)} . \qquad (9\text{-}36)$$

The discrepancy analysis gives a measure of $f_+^0(t)$ on the right-hand cut, while Im $f_+^0(t)$ on the left-hand cut can be calculated from the known channel II and III phase shifts, so that eq. (9-36) is an integral equation for $D(t)$. In practice subtractions must be made, but the subtraction constants can also be found from πN scattering data. Finally the π-π amplitude can be expressed in the effective range form:

$$\frac{p}{\sqrt{p^2+1}} \cot \delta_0 = \frac{1}{a_0} + \frac{1}{2} r \, p^2 \qquad (9\text{-}37)$$

with $a_0 = 1.3 \pm 0.4 \left(\frac{1}{M_\pi}\right)$ (Hamilton et al., 1962b). In the most recent analyses, the phase shift δ_0 is found to rise to about $50°$ near $t = 10 M_\pi^2$, and values consistent with this can be obtained by methods based on an analysis of backward π-N scattering (Nielsen et al., 1970). Pion production experiments, (Scharenguivel et al., 1969), $\pi + N \to \pi + \pi + N$, on the other hand appear to indicate a π-π resonance, known as the σ resonance near $t = 25 M_\pi^2$, while studies of the reactions $\pi^+ p \to \Delta^{++} \pi^+ \pi^-$ or $\Delta^{++} K^+ K^-$ (Flatte et al., 1972) indicate a broad σ effect at ~ 600 MeV and a resonance (the ϵ) at 990 MeV.

A similar analysis of $\Delta_0^{(-)}$ provides parameters which are consistent with what is known about nucleon form factors (Chapter 11) and with the ρ-meson parameters. It should be noted that if $\Delta_0^{(-)}$ were zero, low energy π-N s-wave scattering would be the same in both isotopic spin states, because the N^* exchange contribution to $f_0^I(s)$ is small. As the short range contribution to $\Delta_0^{(-)}$ is also small, ρ exchange is responsible for the large splitting observed.

The Prediction of π-N Phase Shifts

Using the peripheral method to minimize the effects of short range forces, Donnachie et al., (1964) and Donnachie and Hamilton (1965a) have made

9.2. THE ANALYSIS OF LOW ENERGY π-N SCATTERING

predictions of the p, d, and f, π-N partial waves, using the long range forces given by N, N^*, σ, and ρ exchange, where the symbol σ is used to denote a strong π-π s-wave interaction. These predictions are expected to be good at energies for which inelastic scattering is small, in the partial wave concerned. The lowest genuine two-body reaction threshold is given by the reaction:

$$\pi + N \to \eta + N \tag{9-38}$$

at a π meson laboratory kinetic energy of $T = 558$ MeV. This threshold is associated with the s_{11} amplitude (see Chapter 8). Single meson production is possible above 170 MeV, but this occurs primarily by the pseudo-two-body reactions:

$$\pi + N \to \sigma + N$$
$$\pi + N \to \pi + N^* . \tag{9-39}$$

The first of these reactions becomes important at 300 MeV. If the σ-N system is produced at low energies in a state of zero angular momentum, the π-N state concerned must be the p_{11}, and in fact this partial wave becomes very inelastic above 300 MeV. The second reaction occurs above 330 MeV in the d_{33} and d_{35}, π-N state, but is not important below 550 MeV. We would then expect reasonable results up to energies of the order 500 MeV except for the p_{11} wave.

From eq. (9-7), it is seen that N exchange is most important in the $I = \frac{3}{2}$ states and that it is attractive in the p_{33} and f_{37} waves, repulsive in d_{35}. N^* exchange is, as expected, most important in the $\ell = 1$ states and gives an important attractive contribution to p_{11}. σ exchange, which is of importance at low energies, contributes an equal attraction in all states with the same value of ℓ, but ρ exchange differentiates between states with the same ℓ. It is large and attractive in the states with $I = \frac{1}{2}$ and $j = \ell - \frac{1}{2}$, that is, in the p_{11}, d_{13}, f_{15} states. To illustrate these points, the different contributions are shown in the case of d waves in Fig. 9-2.

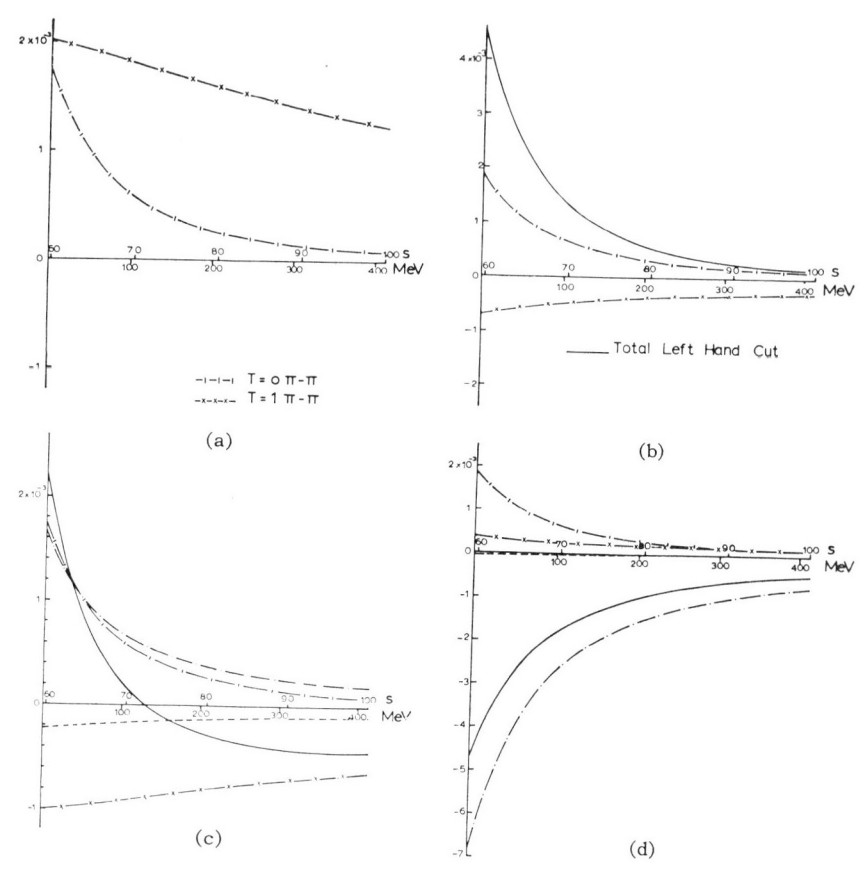

Fig. 9-2. The contribution of various exchange processes to the d wave amplitudes (after Donnachie et al., 1964): (a) Re $F_{2-}^{\frac{1}{2}}(s)$; (b) Re $F_{2+}^{\frac{1}{2}}(s)$; (c) Re $F_{2-}^{\frac{3}{2}}(s)$; (d) Re $F_{2+}^{\frac{3}{2}}(s)$ where $F_{\ell\pm}^{I}(s) = q^{-2\ell} t_{\ell\pm}^{I}(s)$, see page 388. The following symbolism is used: —·—·— for Born term; ----- for N^* exchange; -ı-ı-ı for $I = 0$ π-π exchange; -x-x-x for $I = 1$ π-π exchange; ———— for sum of contributions.

From this discussion the largest amplitudes are expected to be the p_{33} (dominant N and ρ exchange), and the d_{13} and f_{15} (dominant ρ exchange), and the f_{37} (dominant N exchange). Because of the breakdown of the N/D method for these large amplitudes, the phase shifts cannot be calculated in these cases [although the variational method of Donnachie and Hamilton (1965b) has been used with success for p_{33}]. As:

9.2. THE ANALYSIS OF LOW ENERGY π-N SCATTERING

$$q^{2\ell+1} \operatorname{Re} F^I_{\ell\pm}(s) = \frac{1}{2} \eta^I_{\ell\pm} \sin 2\delta^I_{\ell\pm} \qquad (9\text{-}40)$$

if $Q^I_{\ell\pm}(s) = q^{2\ell+1} \mathcal{F}^I_{\ell\pm}(s)$ exceeds $\frac{1}{2}$, the integrals over the physical cut:

$$g(s) = \frac{1}{\pi} \int \frac{\operatorname{Im} F^I_{\ell\pm}(s')}{(s'-s)} ds' \qquad (9\text{-}41)$$

must become negative. However, $g(s)$ is positive near threshold and passes through zero near a resonance. This is because near a narrow resonance Im $F_\ell(s')$ is of the form Im $F_\ell(s') \approx a\,\delta(s'-s_R)$ so that $g(s) \approx \frac{a}{\pi} \frac{1}{(s_R-s)}$. If $Q^I_{\ell\pm}$ becomes greater than 0.5, it then follows that the corresponding partial wave will resonate at a somewhat lower energy. Leaving aside the p_{11} wave (which is too inelastic for this treatment to be valid), $Q^I_{\ell\pm} = 0.5$ for the p_{33} at 380 MeV, for the d_{13} at 810 MeV, for the f_{15} at 1090 MeV, and for the f_{37} at 1500 MeV. All these waves (see Chapter 5) are actually resonant, as predicted, with resonance energies some 200 MeV below these values. In addition, there is a prominent resonance in the d_{15} wave at 900 MeV which is not predicted in this way. This resonance is very inelastic (although so is the f_{15}) and is presumably due to the interactions in the inelastic channel. Equally important is the prediction that none of the other partial waves should resonate in this energy region (up to 1000 MeV, say). This prediction has also been confirmed by the phase shift analyses.

Numerical predictions can be made for the small partial waves (p_{13}, p_{33}, d_{33}, d_{35}, d_{15}, f_{17}, and f_{35}). The dispersion relation, eq. (9-25), in this case can be solved by iteration. As a first approximation, the integral over the physical cut can be neglected, giving:

$$\operatorname{Re} F^I_{\ell\pm}(s) \simeq \mathcal{F}^I_{\ell\pm}(s). \qquad (9\text{-}42)$$

Unless the inelastic scattering is very large (for example, in the case of p_{11}), it can be verified that the integral over the physical provides only a small correction to the phase shift. Donnachie et al., (1964) made an addition correction for short range effects in the case of p waves, but this correction is only large for the p_{11} wave and can be safely ignored.

The phase shifts predicted are quite close to those given by Roper et al., (1965) in their phase shift analysis of the data below 300 MeV which was discussed in Chapter 4. To illustrate the agreement obtained, the predicted phase shifts at 300 MeV are shown in Table 9-4 compared with the phase shifts given by Rugge and Vick (1963). The particular Rugge and Vick solution known as $spdf$ II is shown, as this is in agreement with the energy dependent phase shift analysis.

The analysis of the π-N system by partial wave dispersion relations has been successful up to the point where inelastic processes are important. To treat the inelastic region, some multichannel formalism must be introduced. In the case of the p_{11} partial wave, attempts have been made to treat the system in terms of coupled $\pi+N$ and $\pi+N^*$ or $\pi+\sigma$ channels, but these have not proved very successful so far. An important point, which arises in cases where the scattering is strongly inelastic, concerns the presence or absence of CDD poles (Chapter 4). Bander, Coulter, and Shaw (1965) have shown (see also Atkinson et al., 1966; Hartle and Jones, 1965) that a resonance generated in a coupled channel problem, is not necessarily generated when the problem is treated as a single channel problem, even if the inelasticity parameter $R(s)$ is given correctly. A single channel treatment is, of course, valid, but in this case the resonance must be inserted as a CDD pole.

Table 9-4. The dispersion relation predictions of the small pion-nucleon phase shifts at 300 MeV.

	p_{13}	p_{37}	d_{33}	d_{35}	d_{13}	d_{15}
Predicted phase shifts	$-3.5^{+2.0}_{-3.4}$	$-13.0^{+1.3}_{-2.3}$	-1.3 ± 0.3	-2.1 ± 0.1	5.7 ± 1.6	0.7 ± 0.15
Experimental phase shifts	-3.6 ± 0.7	-11.8 ± 0.8	-3.1 ± 0.6	1.2 ± 0.8	5.9 ± 0.5	0.3 ± 0.6

9.3. DUALITY AND EXCHANGE-DEGENERACY

The relationship of the apparent (if limited) success of the particle exchange model, to the very successful L-excitation quark model, discussed in Chapter 6, is not well understood. If the nucleon resonances are in fact bound states of real quarks, then from the one channel viewpoint of this section the nucleon poles must presumably all be CDD poles. It appears that either the apparent success of the particle exchange model is illusory or perhaps the success may be used as evidence against the possibility that quarks have a real existence.

9-3. DUALITY, EXCHANGE-DEGENERACY, AND DUAL MODELS

In the last section, we saw how some of the low energy properties of the pion-nucleon system can be explained in terms of forces generated by the exchange of particles of low mass in channels II and III. The methods we discussed were not capable of handling the inelastic effects that become important as the energy of the system increases. We now turn to methods based on the description of the amplitude in terms of Regge poles, and the recognition that the parameters of the exchanged Regge poles can be related to the parameters of the low energy direct channel resonances through finite energy sum rules.

Consider a two-particle scattering amplitude $F(\nu,t)$, where $\nu = (s-u)/4M_N$ which for large ν has the asymptotic form $R(\nu,t)$. If the difference $[F(\nu,t)-R(\nu,t)]$ satisfies a dispersion relation in ν for a fixed value of the momentum transfer t and if $\nu[F(\nu,t)] \to 0$ as $|\nu| \to \infty$, then we know that $[F(\nu,t)-R(\nu,t)]$ satisfies the superconvergence relation (see Chapter 3):

$$\int_{-\infty}^{\infty} \text{Im}\,[F(\nu,t) - R(\nu,t)]\,d\nu = 0 \ . \tag{9-43}$$

This relation is identically satisfied for the part of the amplitude $F^+(\nu,t)$ even under the crossing $s \leftrightarrow u$, which corresponds to $\nu \leftrightarrow -\nu$, but for the part of the amplitude odd under crossing, $F^-(\nu,t)$, we find that:

$$\text{Im} \int_0^\infty [F^-(\nu,t) - R^-(\nu,t)] \, d\nu = 0 \qquad (9\text{-}44)$$

and if $F^-(\nu,t) \approx R^-(\nu,t)$ for $\nu > \nu_1$ we can write the finite energy sum rule:

$$\int_0^{\nu_1} \text{Im} \, F^-(\nu,t) \, d\nu = \int_0^{\nu_1} \text{Im} \, R^-(\nu,t) \, d\nu \quad. \qquad (9\text{-}45)$$

The amplitude $\nu F^+(\nu,t)$ is also odd under crossing, so we may also have the finite energy sum rule:

$$\int_0^{\nu_1} \nu \, \text{Im} \, F^+(\nu,t) \, d\nu = \int_0^{\nu_1} \nu \, \text{Im} \, R^+(\nu,t) \, d\nu \quad. \qquad (9\text{-}46)$$

If the asymptotic form of $F^-(\nu,t)$ is given in terms of a sum of Regge poles of the form:

$$\text{Im} \, R^-(\nu,t) = \sum_j \frac{\beta_j(t) \, \text{Im} \, \xi_j(t)}{\sin \pi a_j(t) \, \Gamma(1+a_j(t))} \nu^{a_j(t)} \equiv \sum_j C_j^-(t) \, \nu^{a_j(t)} \qquad (9\text{-}47)$$

where $\xi_j(t)$ is the signature factor:

$$\xi_j(t) = \pm 1 - \exp\{-i\pi a_j(t)\} \qquad (9\text{-}48)$$

the integral on the right-hand side of eq. (9-45) then becomes:

$$\text{Im} \, F^-(\nu,t) \, d\nu = \sum_j C_j^-(t) \left(\frac{\nu_1^{a_j(t)+1}}{a_j(t)+1} \right) \quad. \qquad (9\text{-}49)$$

A similar relation holds for the amplitude $\nu F^+(\nu,t)$.

If $\nu^{n+1}[F^-(\nu,t) - R^-(\nu,t)] \to 0$ as $\nu \to \infty$, we can obtain the finite energy sum rule:

9.3. DUALITY AND EXCHANGE-DEGENERACY

$$\int_0^{\nu_1} \operatorname{Im} F^-(\nu,t)\, \nu^n \, d\nu = \sum_j C_j(t) \frac{\nu_1^{a_j(t)+1+n}}{a_j(t)+1+n} \tag{9-50}$$

and a further generalization is possible by allowing n to be a noninteger.[5]

The finite energy sum rules relate the low energy parameters in the direct channel to the Regge trajectories which describe the crossed channels. In fact, because these relationships hold not only for the amplitude itself, but also for moments of the amplitude, it follows that the Regge asymptotic form will determine the amplitude $F^-(\nu,t)$ averaged over some finite region of ν at low energies. The important notion that the crossed channel Regge exchanges almost completely determine the whole of the direct channel amplitude is known as the principle of duality (Dolen, Horn, and Schmid, 1968), and the low energy resonances are said to be dual to the Regge trajectories in the crossed channel. A good example of how the Regge asymptotic amplitude can represent a local average of the low energy amplitude comes from the work of Igi and Matsuda (1967), who applied the sum rule, eq. (9-49) with only the ρ trajectory exchanged in the t-channel (channel II), (the subsequently discovered ρ' is weakly coupled).

To evaluate the integral, Igi and Matsuda set $t = 0$ and used the optical theorem to express $\operatorname{Im} F^-$ in terms of the difference of the π^- and π^+ total cross sections, which were taken from experiment:

$$\operatorname{Im} F^-(\nu,0) = \frac{q_L}{8\pi} \left(\sigma_{\text{tot}}(\pi^- - p) - \sigma_{\text{tot}}(\pi^+ + p) \right). \tag{9-51}$$

The upper limit of the integral was taken to be $\nu_1 = 6$ GeV. The sum rule was well satisfied with the values $a_\rho(0) = 0.54$, $C_\rho(0) = 5.72$, which fitted the high energy data.

---

[5] These generalized relationships are known as continuous momentum sum rules (Della Selva et al., 1968), and for noninteger n they depend on the real as well as the imaginary part of F.

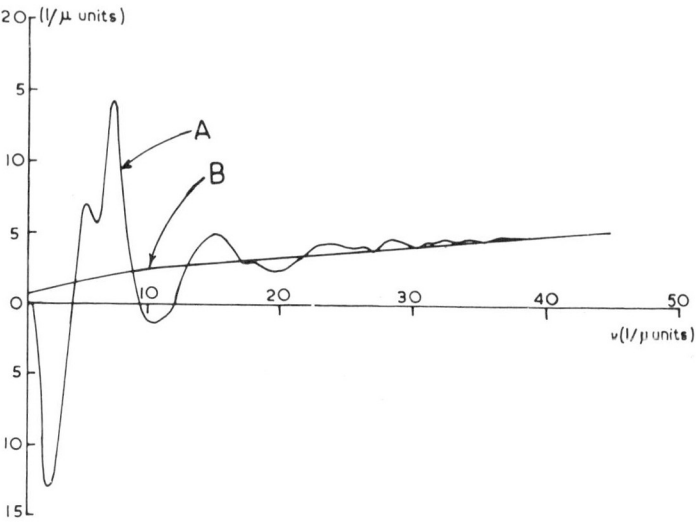

Fig. 9-3. The integral of the sum rule, eq. (9-45). Curve (a) represents $8\pi \operatorname{Im} F^-(\nu,0)$ calculated from the optical theorem, eq. (9-51). Curve (b) represents $8\pi \operatorname{Im} R^-(\nu,0)$ computed from a single ρ trajectory (after Igi and Matsuda, 1967).

In Fig. 9-3, the integrand, $8\pi \operatorname{Im} F^-(\nu,0)$ is plotted together with the Regge pole approximation to $8\pi \operatorname{Im} R^-(\nu,0)$. The Regge pole contribution is seen to average the amplitude in a remarkable manner. The dips and poles in the low energy amplitude are due to the resonances which occur at the positions shown on the figure.

Linear Trajectories and Bootstraps

If, in the low energy region, the amplitude $F(\nu,t)$ can be expressed entirely in terms of resonance contributions with no background, the finite energy sum rules express a connection between the resonance parameters and the parameters of the Regge trajectories in the crossed channels. Since the direct channel resonances also lie on Regge trajectories, this becomes a relationship between trajectories in the direct channel with those in the exchanged channels. If in a set of scattering processes only a finite number of trajectories take part, it then becomes possible to envisage calculating these trajectories by a self-consistent iteration or

9.3. DUALITY AND EXCHANGE-DEGENERACY

bootstrap procedure. Such a program has not been completely defined or carried out, but certain steps have been taken towards its realization.

The equations coupling the trajectories in different channels are complicated integral equations, but if the trajectories can be parametrized in a simple way, they reduce to algebraic equations between these parameters. Experimentally, the known trajectories are linear functions of s (or nearly so), and up to highest energies where resonances can be identified we may write:

$$\text{Re } \alpha(s) = a + bs \qquad (9\text{-}52)$$

where the two parameters a and b represent the intercept of the trajectory with the Re α axis at $s = 0$ and the slope, and we have $b > 0$. It is possible that this linear behavior persists to infinite values of s, providing a strong contrast to the corresponding behavior of trajectories in potential theory, where for large s the Re α decreases to a constant negative value. It may be noted that if the trajectories increase linearly and if $\alpha(s)$ is analytic with the exception of the unitarity or right-hand cut, the dispersion relation for $\alpha(s)$ requires two subtractions and can be written:

$$\alpha(s) = a + bs + \frac{1}{\pi} \int_{s_0}^{\infty} ds' \, \frac{\text{Im } \alpha(s')}{(s'-s)} \, . \qquad (9\text{-}53)$$

A condition for linearity is that Im α is small, which is the condition for the resonances on the trajectory to be narrow.

Exchange Degeneracy

One of the striking successes of the quark model, described in Chapter 6, is that it successfully predicts that in certain channels, such as the $K^+ + p$ or $\pi^+ + \pi^+$ channels, there are no resonances. Resonances forbidden by the quark model are known as "exotic" resonances, and no such resonances have been discovered experimentally. If such resonances do exist they are certainly weakly coupled to the channels that have been investigated. It is natural to ask how the presence or absence of resonances can explained by a Regge "bootstrap" model, where the averaged amplitude in the resonance region is related to the possibility of exchanging crossed channel Regge trajectories.

The explanation is connected with the idea of ordinary and exchange forces, which is familiar from nuclear physics. In scattering by an ordinary potential the Regge trajectories couple to each physical (integer) value of ℓ, but for scattering by exchange potential the amplitudes for odd and even value of ℓ are independent and we require two independent trajectories, of odd and even signature respectively, which couple only to odd or even values of ℓ. Experimentally all trajectories appear to have similar slopes ($\alpha' \sim 1$ GeV/c), but in addition, certain pairs of trajectories, such as the A_2 and ρ or the P' and ω, which for example, can all occur in the $K\bar{K} \to N\bar{N}$ channel, appear to be identical with each other. The quantum numbers of the A_2 and of the ρ trajectories are identical, apart from the signature factor. The ρ trajectory has $J^P = 1^-$ and signature $\sigma = -1$, while the A_2 has $J^P = 2^+$ and $\sigma = +1$. It follows that in the $K\bar{K} \to N\bar{N}$ channel, the degenerate A_2 and ρ trajectories can be considered as one trajectory coupled to all integer values of the orbital angular momentum and generated by ordinary forces, the exchange forces being absent. The same is true of the other pair of trajectories occurring in this channel, the P' and ω. In our usual notation, we label the $KN \to K^-N$, $K^+K^- \to N\bar{N}$ and $K^+N \to K^+N$ channels as channels I, II, and III:

$$\bar{K}^- + N \to \bar{K}^- + N \quad \text{I} \quad (s\text{-channel})$$
$$K^+ + K^- \to N + \bar{N} \quad \text{II} \quad (t\text{-channel}) \quad (9\text{-}54)$$
$$K^+ + N \to K^+ + N \quad \text{III} \quad (u\text{-channel})$$

The degeneracy of the pairs of trajectories in channel II is then completely consistent with the observed absence of resonances in channel III (weak scattering) and abundance of resonances in channel I (strong scattering).

In a self-consistent picture, we should also hope to explain the scattering in channels I and III, in terms of the exchange of channel II trajectories. The existence of a pair of trajectories with the same quantum

9.3. DUALITY AND EXCHANGE-DEGENERACY

numbers, such as the A_2 and ρ, will be responsible for forces in both channels I and III. From the finite energy sum rules, the average imaginary parts of the low energy amplitude in channels I and III are related to linear combination of the contributions from the A_2 and ρ. However as the A_2 and ρ trajectories have opposite signature, these contributions (which are of equal magnitude) add in channel I, but subtract in channel III. The same applies to the pair of trajectories P' and ω. The absence of strong forces in channel III and the presence of strong forces in channel I receives in this way an economical and convincing explanation.

By considering the π-π system, further conclusions can be reached about the degeneracy of the trajectories we have been considering. In this case the absence of resonances in the $\pi^+\pi^+(I=2)$ system, implies the degeneracy of the pairs of trajectories P' and ρ and of ω and A_2. It follows that all four trajectories P', ρ, ω, and A_2 must be, at least approximately, degenerate.

Duality Diagrams

In the examples discussed above the π^+-π^+ and K^+-p channels are "exotic," and the requirement that there should be no resonances in these channels gave constraints on the imaginary parts of the scattering amplitude and led to exchange degeneracy between certain trajectories. These constraints can be discussed in terms of the duality diagrams introduced by Harari (Harari, 1969; Rosner, 1969). It is assumed that every meson can be represented as a quark-anti-quark pair $(q\bar{q})$ and every baryon by a three quark configuration (qqq). We now connect the initial and final states in scattering processes by directed lines which represent the quark or anti-quark contact of initial or final particles. In a "proper" diagram no lines cross or double back on themselves. For example in Fig. 9-4, we show the diagrams drawn for π^-+p with n, p, or λ quark lines labeled accordingly.

In both diagrams (A) and (B), reading from left to right we obtain channel I. By reading the diagrams from bottom to top, diagram (A)

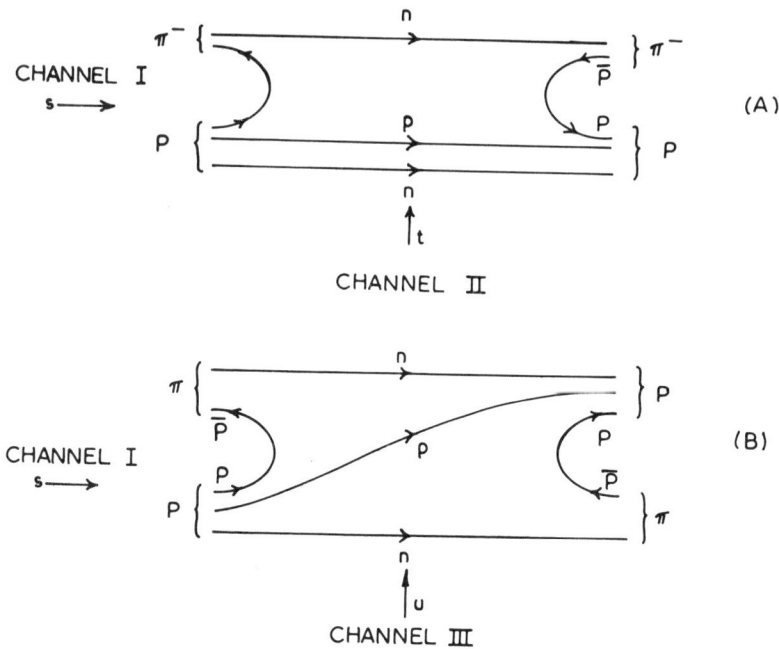

Fig. 9-4. Duality diagrams for π^-+p scattering.

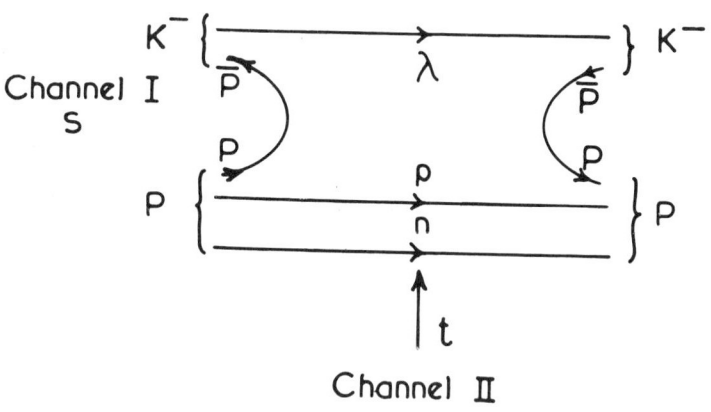

Fig. 9-5. Duality diagram for K^-+p scattering.

9.3. DUALITY AND EXCHANGE-DEGENERACY

corresponds to channel II, $p+p \to \pi+\pi$, and (B) to channel III, $\pi^++N \to \pi^++N$. Both diagrams are "proper" and exhibit duality between the s- and t- and s- and u-channels respectively, and there are no exchange degeneracy constraints on the trajectories, which are N and Δ in channel I, ρ and f in channel II, and Δ in channel III.

In contrast in Fig. 9-5 we show the $s \leftrightarrow t$ diagram for K^-+p scattering. No corresponding "proper" diagram can be drawn linking the s- and u-channels. For K^+p scattering no "proper" diagram can be drawn linking either the s- and t- or the s-and u-channels. It follows that the forces in the K^+p channel are weak and that the trajectories occuring in the corresponding t-channel (ρ, A_2, ω, P') must be exchange degenerate, and also those in the u-channel (Λ and Σ).

Background Scattering and the Pomeranchuk

In the preceding paragraph, we have discussed two-body scattering as if no scattering apart from resonant scattering occurs. This is far from correct. Scattering does occur in the nonresonant K^++p channel, and in π-N system, it is clear that considerable background scattering occurs in addition to scattering through the various resonant states. In a similar way, the trajectories that we have discussed have all been associated with particle (resonant) states and we have not mentioned the role of the Pomeranchuk or vacuum trajectory, which is apparently not associated with such states. It has been suggested that while the exchange of normal trajectories is associated with the generation of direct channel resonances, the Pomeranchuk trajectory is associated with background scattering (Harari, 1969). This separation of roles is possible because of the linear nature of the finite energy sum rules. If F^-_{Res} and $F^-{}_{Nonres}$ are the resonant and background parts of F^-, we have a relation of the form:

$$[\text{Im } F^-_{Res}(\nu,t) + \text{Im } F^-_{Nonres}(\nu,t)] = \sum_{\substack{\text{particle} \\ \text{trajectories}}} d_j(t)\, \nu^{a_j(t)} + d_p(t)\, \nu^{a_p(t)}. \tag{9-55}$$

This conjecture then suggests that this relation is satisfied separately for each of F^-_{Res} and F^-_{Nonres}:

$$\int_0^{\nu_1} \text{Im } F^-_{Res}(\nu,t) \, d\nu = \sum_j d_j(t) \, \nu^{\alpha_j(t)}$$

$$\int_0^{\nu_1} \text{Im } F^-_{Nonres}(\nu,t) \, d\nu = d_p \, \nu^{\alpha_p(t)} . \quad (9\text{-}56)$$

A test of this conjecture (Harari and Zarmi, 1969) is obtained by constructing from the $I = \frac{1}{2}$ and $I = \frac{3}{2}$, π-N partial wave amplitudes, the combinations that correspond to $I = 0$, or $I = 1$ in the t-channel ($\pi\pi \to N\bar{N}$). The Pomeranchuk can only contribute to the $I = 0$ amplitude, and it is expected that there should be significantly more background scattering in the $I = 0$ than in the $I = 1$ states. The results suggest that this is indeed the case. Naturally, the separation of resonant scattering from background is a highly nonunique process, and because of this it is difficult to assess the validity of these ideas.

A different split between background and resonance is given by the interference model (Barger and Cline, 1968; Alessandrini, Amati, and Squires, 1968). Here it is supposed that the amplitude F can be written as:

$$F = F_s + F_t + F_u \quad (9\text{-}57)$$

where F_s, F_t, F_u are amplitudes due to Regge poles in the s, t, and u channels respectively. For physical values of s, F will contain direct channel poles from F_s and crossed channel poles from F_t and F_u. For physical t, F will contain direct channel poles from F_t and crossed channel poles from F_s and F_u, and similarly for physical u. There appears to be no difficulty, in principle, in reconciling this model with duality and the finite energy sum rules, although the relationship between background and resonant scattering will be different from that in the model in which the background is linked to the Pomeranchuk trajectory in a simple way. A quite successful application of these ideas to π-N scattering has been described by Donnachie and Kirsopp (1969).

9.3. DUALITY AND EXCHANGE-DEGENERACY

Dual Models

Considerable impetus has been given to the study of dynamical ideas associated with duality and infinitely rising linear trajectories, by the construction of a model amplitude by Veneziano (1968), which exhibits complete crossing symmetry, and Regge asymptotic behavior in all channels. This work has led to a more general consideration of a whole class of 'dual models' which are crossing symmetric, Regge behaved and which are consistent with internal symmetries. We consider briefly the original Veneziano model, applied to $\pi\pi$ scattering.

We start by considering $\pi^+\pi^-$ scattering in channel I, the s-channel, in which case channel II also contains $\pi^+\pi^-$ scattering while channel III, the u channel, refers to $\pi^+\pi^+$ scattering:

$$
\begin{array}{lll}
\text{Channel I} & (s) & \pi^+ + \pi^- \to \pi^+ + \pi^- \\
\text{Channel II} & (t) & \pi^- + \pi^+ \to \pi^- + \pi^+ \\
\text{Channel III} & (u) & \pi^+ + \pi^+ \to \pi^+ + \pi^+ \ .
\end{array}
\qquad (9\text{-}58)
$$

The amplitude $F(s,t)$ must show resonances in both channels I and III (which are identical) and no resonances in channel III, which is exotic. The known trajectories in channel I are the ρ and the f, and because channel III is exotic these must be degenerate. The ρ-f degenerate trajectory rises linearly, and in a zero-width approximation where the trajectory function $a(s)$ is taken to be real, we can write:

$$a(s) = a + bs, \qquad b > 0 \ . \qquad (9\text{-}59)$$

The Veneziano prescription for $F(s,t)$ is then:

$$F(s,t) = -\lambda \frac{\Gamma(1-a(s))\,\Gamma(1-a(t))}{\Gamma(1-a(s)-a(t))} \ . \qquad (9\text{-}60)$$

This form displays obvious $s \leftrightarrow t$ crossing symmetry. The function $\Gamma(Z)$ is an analytic function of Z except for poles at negative real integer values of Z or zero. It follows that $F(s,t)$ possesses poles in s, $a(S_j) = J$, where J is a positive integer. These poles are equally spaced in s, since from eq. (9-59):

$$S_J = (J-a)/b \ . \tag{9-61}$$

At the same time for values of t such that $a(t_{J'}) = J$, $F(s,t)$ possesses an infinite series of poles in t. In order that such poles should represent particle exchange, each pole should be single and no double poles should occur. This is achieved by the presence of the denominator $\Gamma(1-a(s)-a(t))$, which cancels one of the poles when s and t are both at values s_J and $t_{J'}$.

The residue of the Jth pole in s is proportional to a polynomial in t of order J. This is interpreted by supposing that each pole corresponds not to a single particle, but to a multiplet of particles all of the same mass $\sqrt{s_J}$, and of spins $0, 1, 2, \ldots J$. The particles with spin less than J lie on trajectories which are known as "daughter" trajectories.

Because $\Gamma(Z) \propto Z^Z$ as $Z \to \infty$, the asymptotic form of $F(s,t)$ is:

$$F(s,t) \propto \Gamma(1-a(t))\,(-bs)^{a(t)} \qquad s \to \infty$$

and:

$$F(s,t) \propto \Gamma(1-a(s))\,(-bt)^{a(s)} \qquad t \to \infty \tag{9-62}$$

and so displays a Regge asymptotic form in both s and t.

The amplitude so constructed displays duality, in that the infinite series of resonances is shown to be equivalent to a Regge asymptotic form. It should be noted that this property can only be obtained if the trajectories rise infinitely, because a finite sum of resonances cannot by itself provide such an asymptotic form.

Because of these interesting properties, and because generalizations are possible to amplitudes with more than two particles in the final state,

9.4. PADÉ APPROXIMATES

the Veneziano amplitude has been studied extensively.[6] Although attempts have been made to analyze experimental data with the help of the model, this is difficult because all the poles occur on the real s axis so that unitarity is violated. Various methods can be devised for restoring unitarity and providing the amplitude with an imaginary part, but to date these attempts have succeeded only at the price of destroying some other feature satisfied by the model, such as crossing symmetry.

It is too early to attempt to summarize in this text the very extensive work in progress on the general class of dual models and of the interesting 'dual loop' calculus associated with it, but reference can be made to the reviews by Silvers and Yellin (1921) and by Alessandrini et al.,(1921).

9.4. PADÉ APPROXIMATES

Although the perturbation series for the S-matrix obtained from a conventional Lagrangian quantum field theory for the pion-nucleon system is highly divergent, some striking results have been obtained by re-expressing this series in terms of rational fractions. The perturbation series for a partial wave amplitude T can be written as:

$$T = g\,T_1 + g^2\,T_2 + \cdots$$

where g is the appropriate coupling constant. To a given order in g, this series can be re-expressed in terms of a rational function, called a Padé approximate. The Padé approximate $T^{N,m}$ contains a polynomial

---

[6] A review of Veneziano theory and its application to the analysis of experimental data has been given by Lovelace (1970).

in g of degree N in the numerator and m in the denominator. The sequence of Padé approximates as N and m increases is known to converge for potential scattering, and for problems like that of the anharmonic oscillator for which the perturbation series is known to diverge.

The convergence of the sequence of Padé approximates has not been proved in a field theoretical context, but numerical investigations based, for example, on the $\lambda\phi^4$ interaction suggest that the procedure converges. The diagonal approximates $T^{N,N}$ for the partial wave amplitudes are unitary, but not completely crossing symmetric, and if crossing symmetry is built into the model the unitarity property is lost.

The most extensive investigation based on this model [by Basdervant et al., (1969)] considers the coupled π-π, π-K and K-$\bar{K}$ system. In this case, three constants must be introduced, which determine the strengths of each of these couplings, but the masses of seven particles [the ρ, f^0, $K^*(890)$, $K^*(1420)$, ϕ, f^1 and A_2] are all obtained correctly in the output. This success is not entirely unclouded, as the s-waves in each channel are not given correctly. However the success so far obtained suggests that further interesting developments along these lines should be possible.

REFERENCES

A. Abashain, N. E. Booth and K. M. Crowe, Phys. Rev. 132, 2296, 2305 (1963).

V. A. Alessandrini, D. Amati and E. J. Squires, Phys. Letters 27B, 463 (1968).

V. A. Alessandrini, D. Amati, M. le Bellac and D. Olive, Phys. Rep. Phys. Letters C, 1C, 269 (1971).

D. Atkinson, K. Dietz and D. Morgan, Ann. Phys. (N.Y.) 37, 77 (1966).

REFERENCES

M. Bander, P. W. Coulter and G. L. Shaw, Phys. Rev. Letters 14, 230 (1965).

V. Barger and D. Cline, Phys. Rev. 155, 1792 (1962).

V. Barger and D. Cline, Phys. Rev. Letters 21, 392 (1968).

J. L. Basdervant, D. Bessis and J. Z. Justin, Nuovo Cimento 64A, 185 (1969).

L. M. Brown and P. Singer, Phys. Rev. Letters 8, 460 (1964).

G. F. Chew and F. E. Low, Phys. Rev. 101, 1570 (1956).

G. F. Chew and S. Mandelstam, Phys. Rev. 119, 467 (1960).

P. D. B. Collins and R. Johnson, Phys. Rev. 182, 1755 (1969).

S. Cuilli, Nuovo Cimento 61A, 787 (1969a).

S. Cuilli, Nuovo Cimento 62A, 301 (1969b).

R. Cutkosky and B. B. Deo, Phys. Rev. 174, 1859 (1968a).

R. Cutkosky and B. B. Deo, Phys. Rev. Letters 20, 1272 (1968b).

W. Deinet, et al., Phys. Letters 30B, 359 (1969).

A. Della Selva, L. Masperi and R. Odorico, Nuovo Cimento 54A, 979 (1968).

R. Dolen, D. Horn and C. Schmid, Phys. Rev. 166, 1768 (1968).

A. Donnachie and J. Hamilton, Phys. Rev. 133B, 1053 (1964).

A. Donnachie and J. Hamilton, Ann. Phys. (N.Y.) 31, 410 (1965a).

A. Donnachie and J. Hamilton, Phys. Rev. 138B, 678 (1965b).

A. Donnachie, J. Hamilton and A. T. Lea, Phys. Rev. 135B, 515 (1964).

A. Donnachie and R. G. Kirsopp, Nucl. Phys. B10, 433 (1969).

S. M. Flatte, M. Alston-Garnjost, A. Barbaro-Galtier, J. H. Friedman, G. R. Lynch, S. D. Protopopesan, M. S. Rubin and F. T. Solmitz, Phys. Letters 38B, 232 (1972).

W. R. Frazer and J. R. Fulco, Phys. Rev. 117, 1603, 1609 (1960).

J. Hamilton, *Elementary Particle Theory* (Oxford University Press, London, 1959).

J. Hamilton, *Springer Tracts in Modern Physics*, ed. G. Höhler (Springer-Verlag, Berlin, 1971), 57, p. 41.

J. Hamilton, T. D. Spearman and W. S. Woolcock, Ann. Phys. (N.Y.) 17, 1 (1962a).

J. Hamilton, P. Menotti, G. L. Oades and L. L. J. Vick, Phys. Rev. 128, 1881 (1962b).

H. Harari, Phys. Rev. Letters 22, 562 (1969).

H. Harari and Y. Zarmi, Phys. Rev. 187, 2230 (1969).

J. B. Hartle and C. E. Jones, Phys. Rev. 140, 390 (1965).

K. Igi and S. Matsuda, Phys. Rev. Letters 18, 625 (1967).

A. C. Lovelace, Proc. Roy. Soc. A 318, 321 (1970).

P. M. Morse and H. Feshbach, *Methods of Theoretical Physics* (McGraw-Hill, New York, 1953).

H. Nielsen, L. J. Petersen and E. Pietarinen, Nucl. Phys. B 22, 525 (1970).

L. J. Peterson, Nucl. Phys. B15, 549 (1970).

L. D. Roper, R. M. Wright and B. T. Feld, Phys. Rev. 138, B190 (1965).

J. L. Rosner, Phys. Rev. Letters 22, 689 (1969).

H. R. Rugge and O. T. Vik, Phys. Rev. 129, 2300 (1963).

REFERENCES

N. P. Samois, A. H. Bachmann, R. M. Lea, T. E. Kalogeropoulis and W. D. Shepard, Phys. Rev. Letters 9, 139 (1962).

J. R. Scharenguivel, et al., Phys. Rev. 18B, 1387 (1969).

D. Silvers and J. Yellin, Rev. Mod. Phys. 43, 125 (1971).

G. Veneziano, Nuovo Cimento 57A, 190 (1968).

CHAPTER 10

PION-NUCLEON INELASTIC SCATTERING

In earlier chapters, the analysis of the elastic scattering of pions by nucleons has been discussed in some detail, while some features of inelastic scattering at very high energies were outlined in Chapter 8. Neither the experimental information, nor its interpretation, is as complete for scattering into individual inelastic channels as for elastic scattering, but some progress has been made in understanding some of the principal features of these processes. The first inelastic threshold is at a laboratory pion kinetic energy of $T_\pi = 180$ MeV. Above this energy single pion production is possible:

$$\pi + N \to \pi + \pi + N \tag{10-1}$$

but the inelastic cross section remains small until $T_\pi = 300$ MeV; above that energy it increases, until by 600 MeV, it is comparable in size to the elastic scattering cross section, both for $\pi^- p$ and for $\pi^+ p$ scattering (see Figs. 10-1 and 10-2). Below 1 GeV, the only other inelastic channel of importance is that for η production:

$$\pi^- + p \to \eta + n \ . \tag{10-2}$$

The cross section rises steeply from threshold at $T_\pi = 561$ MeV to a peak near 650 MeV. By fitting the η production data simultaneously with the data on elastic scattering, Davies and Moorhouse (1967) have shown that the large cross section arises mainly from the decay of the S_{11} resonance ($W = 1535$ MeV, $T_\pi = 631$ MeV) with a d-wave background

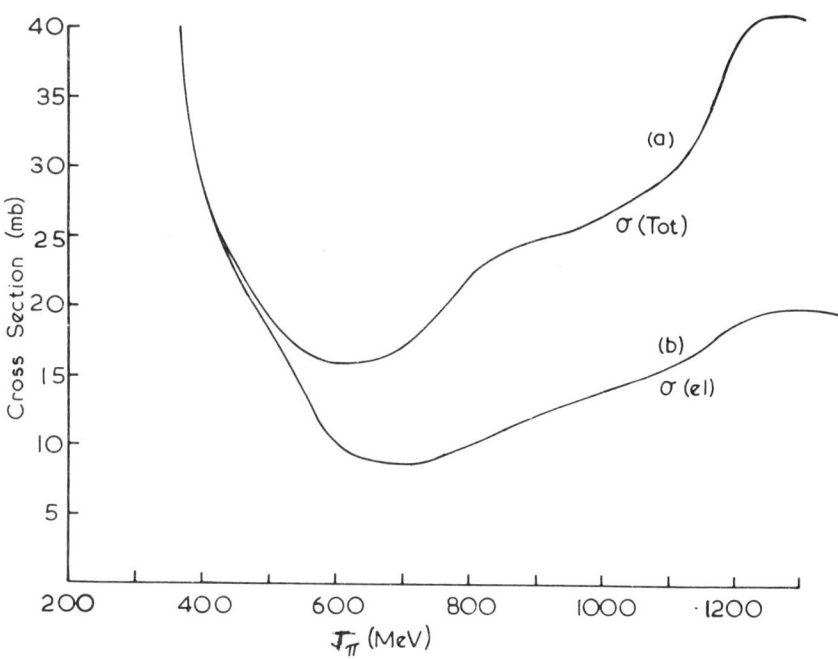

Fig. 10-1. The total cross section for $\pi^+ + p$ scattering (all processes) compared with the total elastic cross section.

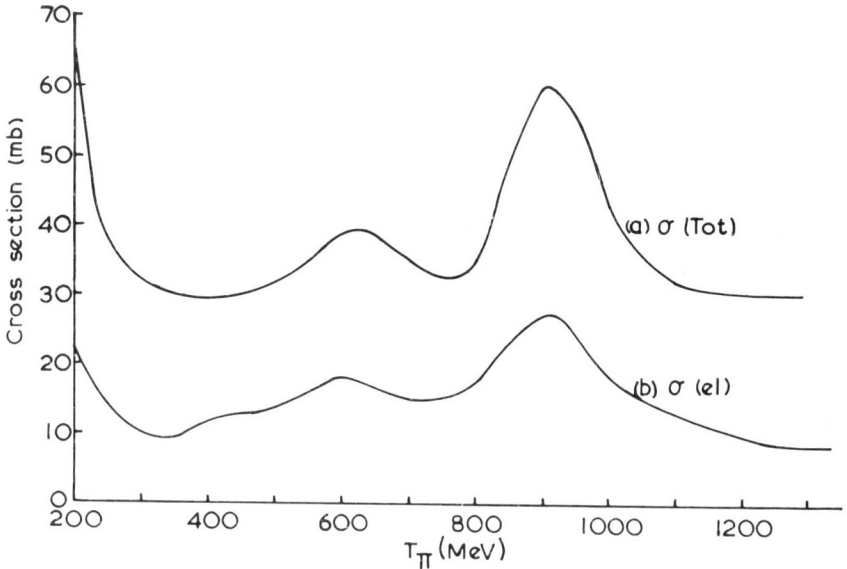

Fig. 10-2. The total cross section for $\pi^- + p$ scattering (all processes) compared with the total elastic cross section.

produced by the neighboring D_{13} resonance ($W = 1520$ MeV, $T_\pi = 616$ MeV), (see Chapter 5). Above $T_\pi = 1$ GeV, the cross section for the production of two pions becomes appreciable and as the incident energy increases, the chance of production of pions in the final state with high multiplicities steadily increases.

When sufficient energy is available, associated production of a K meson and hyperon can take place. The lowest threshold ($T_\pi = 771$ MeV) is for the reaction:

$$\pi^- + p \to \Lambda + K^0 \ . \tag{10-3}$$

At $T_\pi = 856$ MeV, Σ production can take place:

$$\pi^- + p \to \Sigma^- + K^+$$
$$\to \Sigma^0 + K^0 \ . \tag{10-4}$$

The cross sections rise above threshold to a peak near $T_\pi = 920$ MeV. At the maximum $\sigma(\Lambda + K^0) = 1.2$ mb, $\sigma(\overline{\Sigma^0} + K^{\overline{0}}) = 0.2$ mb, so that in this region the cross section for associated production is considerably less than that for pion production, which is of the order of 20 mb.

Much of the cross section for single pion production can be attributed to the inelastic decay of the resonances formed from the elastic channel:

$$\pi + N \to N^* \to \pi + \pi + N \ . \tag{10-5}$$

In the final state there is a strong tendency for a pair of particles to be associated in a resonant state giving the quasi-two-body reactions:

$$\pi + N \to N^* \to \pi + N^*$$
$$\to \rho + N \tag{10-6}$$
$$\to \sigma + N \ .$$

These reactions, proceeding largely through the formation and decay of an N^* resonance, can be distinguished from those involving the direct excitation of a nucleon:

$$\pi + N \to \pi + N^* \ . \tag{10-7}$$

These direct excitations do give rise to peaks in the inelastic cross sections and most of the N^* resonances can be detected in this way.[1] We shall now discuss the analysis of scattering experiments leading to three particle final states, and after this we shall describe the main features of single pion production in more detail.

10.1. ANALYSIS OF THREE PARTICLE FINAL STATES

The transition rate W_{fi} for a reaction in which there are two particles 1, 2 in the initial state i with momenta $\vec{p}_1, \vec{p}_2$ and particles 3, 4 ... n in the final state f, with momenta $\vec{p}_3, \vec{p}_4 ... \vec{p}_n$, is determined by a transition amplitude T_{fi}, where (for spinless particles):

$$T_{fi} = <\vec{p}_3, \vec{p}_4 ... \vec{p}_n, \text{ in } |T| \vec{p}_1, \vec{p}_2, \text{ in}> \ . \qquad (10\text{-}8)$$

If the single particle state vectors have the invariant normalization:

$$<\vec{p}_i | \vec{p}'_i> = E_i(p) \delta(\vec{p}_i - \vec{p}'_i) \qquad (10\text{-}9)$$

where $E_i(p) = \sqrt{M_i^2 + p_i^2}$, the transition amplitude T_{fi} will also be an invariant. The connection between ω_{fi} and T_{fi} is given by (see section 2.1):

$$W_{fi} = 4 \int d\vec{p}_3 \int d\vec{p}_4 ... \int d\vec{p}_n \frac{(2\pi)^{-4} |T_{fi}|^2}{E_3 E_4 ... E_n} \delta^3(\vec{p}_3 + \vec{p}_4 ... + \vec{p}_n)$$

$$\delta(W - \sum_{j=3}^{n} E_j) \qquad (10\text{-}10)$$

---

[1] For details the review articles by Rushbrooke (1968) and Plano (1969) should be consulted.

10.1. ANALYSIS OF THREE PARTICLE FINAL STATES

where we have employed the center of mass system in which $\vec{p}_1 + \vec{p}_2 = \sum_{j=3}^{n} \vec{p}_j = 0$ and where W is the total center of mass energy. The total cross section σ_{fi} for the reaction $1 + 2 \rightarrow 3 + 4 \ldots + n$ is then defined to be:

$$\sigma_{fi} = W_{fi}/F \tag{10-11}$$

where F is the incident flux defined by (2-17).

If we write the transition rate W_{fi}, in the form:

$$W_{fi} = |T_{fi}|^2 \, \rho(W) \tag{10-12}$$

the factor $\rho(W)$ is termed the phase space factor. For any particular differential cross section for scattering into a final state in which the particles be within certain ranges of momenta or energy, the variation of the cross section with the momentum or energy variables can in part be attributed to the phase space factor, which is kinematical in character, and in part to the variation of $|T_{fi}|^2$ which depends on the specific dynamical nature of the interaction.

Consider the particular case in which there are three particles 3, 4, and 5 in the final state. For each pair of particles (3 + 4), (4 + 5), or (3 + 5), an invariant mass M_{ij} can be defined by:

$$M_{ij}^2 = (p_i + p_j)^2 = (E_i + E_j)^2 - (\vec{p}_i + \vec{p}_j)^2 \;. \tag{10-13}$$

Since $W = E_3 + E_4 + E_5$, we see that:

$$M_{45}^2 = (W - E_3)^2 - p_3^2 \tag{10-14}$$

from which:

$$dM_{45}^2 = -2W dE_3 = -\frac{2W \, p_3 \, d p_3}{E_3} \;. \tag{10-15}$$

In the same way, we have:

$$dM_{35}^2 = -\frac{2W\, p_4\, dp_4}{E_4}. \qquad (10\text{-}16)$$

Using these results the transition rate can be expressed as a function of M_{34}, M_{35} and the polar angles (θ_3, ϕ_3) (θ_4, ϕ_4) of the vectors $\vec{p}_3$ and $\vec{p}_4$. After using the momentum conserving delta function to integrate over dp_5, we have:

$$W_{fi} = \int dM_{35}^2 \int dM_{45}^2 \int d\Omega_3 \int d\Omega_4 \frac{(2\pi)^{-4}|T_{fi}|^2\, p_3\, p_4}{W^2\, E_5}$$

$$\delta(W - E_3 - E_4 - E_5). \qquad (10\text{-}17)$$

Since $\vec{p}_5 = -(\vec{p}_3 + \vec{p}_4)$, E_5 is expressed in terms of $\vec{p}_3$ and $\vec{p}_4$ by:

$$E_5 = \sqrt{M_5^2 + p_3^2 + p_4^2 + p_3\, p_4 \cos\theta_{34}} \qquad (10\text{-}18)$$

where θ_{34} is the angle between $\vec{p}_3$ and $\vec{p}_4$. The energy conserving delta function can be expressed as:

$$\delta(W - E_3 - E_4 - E_5) = \delta(\cos\theta_{34})\frac{E_5}{p_3\, p_4} \qquad (10\text{-}19)$$

and taking p_3 as the polar axis for the integration over (θ_4, ϕ_4), we can integrate over $\cos\theta_{34}$) to obtain:

$$W_{fi} = \int dM_{35}^2 \int dM_{45}^2\, \frac{(2\pi)^{-4}}{W^2}\, D(W, M_{35}, M_{45}) \qquad (10\text{-}20)$$

where:

$$D(W, M_{35}^2, M_{45}^2) = \int d\Omega_3 \int d\phi_4\, |T_{fi}|^2. \qquad (10\text{-}21)$$

The differential cross section for scattering in which M_{35}^2 lies in the interval M_{35}^2 to $M_{35}^2 + dM_{35}^2$ and M_{45}^2 lies in the interval M_{45}^2 to $M_{45}^2 + dM_{45}^2$ is then:

10.1. ANALYSIS OF THREE PARTICLE FINAL STATES

$$\frac{d^2\sigma}{dM_{35}^2\, dM_{45}^2} = \frac{1}{F}\frac{(2\pi)^{-4}}{W^2}\, D(W, M_{35}, M_{45})\, . \qquad (10\text{-}22)$$

The Dalitz Plot

If the interaction between particles 3 and 5 or between 4 and 5 in the final state of the reactions is not very strong, then $D(W, M_{35}, M_{45})$ will vary with M_{35} or M_{45} slowly and smoothly. On the other hand, if the collision processes go via the production of a metastable state or resonance formed from a pair of particles, for instance 3 and 5:

$$\pi(1) + N^* + \pi(4)$$
$$\searrow$$
$$N(3) + \pi(5) \qquad (10\text{-}23)$$

the intensity $D(W, M_{35}, M_{45})$ will show a rapid variation with the invariant mass M_{35}, and for values of M_{35} close to the mass of the resonance M_5 will be of the general form:

$$D(W, M_{35}, M_{45}) = \left|\frac{A}{M_{35}^2 - M_R^2 + i\Gamma}\right|^2 \qquad (10\text{-}24)$$

where Γ is the width of the resonance.

The experimental data can be examined to decide whether a resonance is produced in a given two particle subsystem, by plotting each scattering event in a diagram, known as a Dalitz plot, with M_{35}^2 and M_{45}^2 as axes. Conservation of energy confines the possible events to within a certain contour, such as illustrated in Fig. 10-3. Such a diagram has the important property that if D is constant, that is if D does not vary with either M_{35} or with M_{45}, the density of events will be constant over the whole area enclosed by the boundary. A resonance in one, or both, the two particle subsystems $(3+5)$ or $(4+5)$ is then made manifest by a marked increase in the density of events in bands dated about the lines $M_{35}^2 = M_R^2$ or $M_{45}^2 = M_R'^2$.

A particular example is shown in Fig. 10-4.

To display the differential cross sections in one invariant mass $(d\sigma/dM_{45}^2)$ for example, we must integrate the cross section $d^2\sigma/dM_{45}^2 dM_{35}^2$ over the variable M_{35}^2. The phase space available for a fixed value of M_{45} can be obtained directly from the Dalitz plot. If EE' is the intersection of the line $M_{45}^2 = C$ with the boundaries of the Dalitz plot (see Fig. 10-3), then this is the phase space available at this value of M. A phase space curve, shown in Fig. 10-3b, can be obtained in this way, and if the intensity $D(W, M_{45}^2, M_{35}^2)$ is constant, this curve will be proportional to the differential cross section $d\sigma/dM_{45}^2$. If the interaction between particles 4 and 5 is strong, then the projection of the Dalitz plot on M_{45}^2 axis will appear as in curve A of Fig. 10-3b, which will show a marked departure from the phase space curve.

The differential cross section $d\sigma/dM_{45}^2$ for values of M_{45} near a resonance of mass M_R may be expected to have the form:

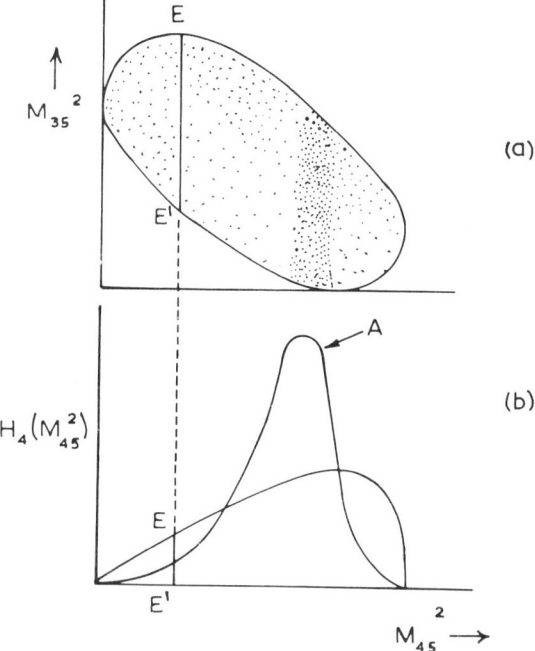

Fig. 10-3. (a) Illustrative diagram of a Dalitz plot; (b) the projection of the Dalitz plot onto the variable.

10.1. ANALYSIS OF THREE PARTICLE FINAL STATES

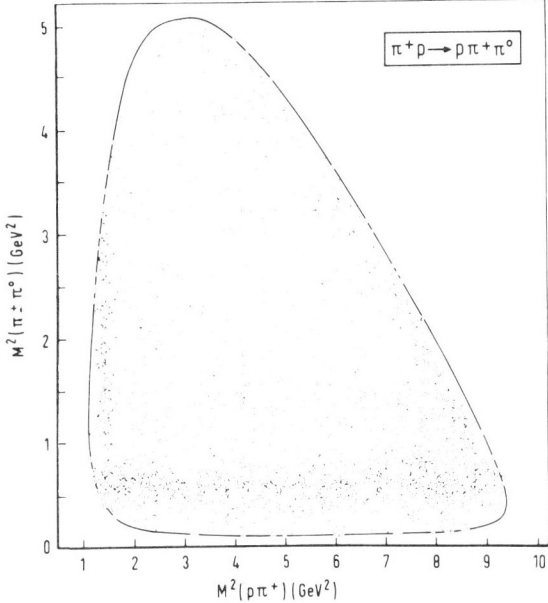

Fig. 10-4. The Dalitz plot for the reaction $\pi^+ + p \to p + \pi^+ + \pi^0$ at 5 GeV/c (Durham-Nijmegen-Paris-Turin Collaboration).

$$\frac{d\sigma}{dM_{45}^2} = \frac{F_1(M_{45})}{(M_{45}-M_R)^2 + \Gamma^2/4} + F_2(M_{45}) \qquad (10\text{-}25)$$

where F_1, F_2, and Γ are slowly varying functions of M_{45}. The factor F_1 is proportional to the cross sections for the production of the resonance, while F_2 represents that part of the scattering into the three particle final state that does not proceed via the production and decay of a metastable state.

10.2. SINGLE PION PRODUCTION BELOW 1 GeV AND THE ISOBAR MODEL

As we have already noted, single pion production is the most important inelastic reaction in pion nucleon collisions, at energies below 1 GeV. The reactions that can be studied in π^+ or π^- scattering by protons are:

$$\begin{aligned}
\pi^+ + p &\to \pi^+ + \pi^0 + p \quad &\text{(a)} \\
&\to \pi^+ + \pi^+ + n \quad &\text{(b)} \\
\pi^- + p &\to \pi^- + \pi^0 + p \quad &\text{(c)} \quad (10\text{-}26) \\
&\to \pi^- + \pi^+ + n \quad &\text{(d)} \\
&\to \pi^0 + \pi^0 + n \quad &\text{(e)} .
\end{aligned}$$

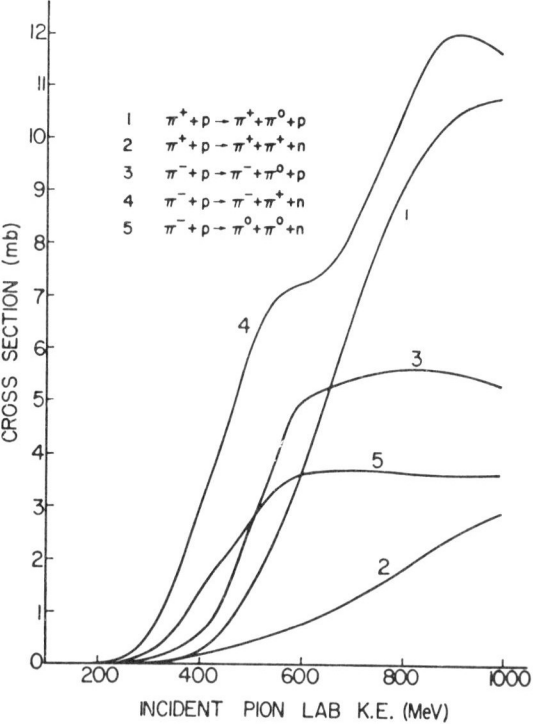

Fig. 10-5. Single pion production cross sections below 1 GeV (after Olsson and Yodh, 1966).

10.2. SINGLE PION PRODUCTION BELOW 1 GeV

The observed low energy behavior of each of these processes is shown in Fig. 10-5.

The total (elastic + inelastic) cross sections for $\pi^- p$ and $\pi^+ p$ scattering, $\sigma_{tot}(\pi^{\pm} p)$, to which the reactions (a) to (e) contribute, are related to the total cross sections for scattering in the $I = \frac{1}{2}$ and $I = \frac{3}{2}$ isospin states by (see Chapter 4):

$$\sigma_{tot}(\pi^+ + p) = \sigma\left(I = \frac{3}{2}\right)$$

$$\sigma_{tot}(\pi^- + p) = \frac{2}{3}\sigma\left(I = \frac{1}{2}\right) + \frac{1}{3}\sigma\left(I = \frac{3}{2}\right) .$$

(10-27)

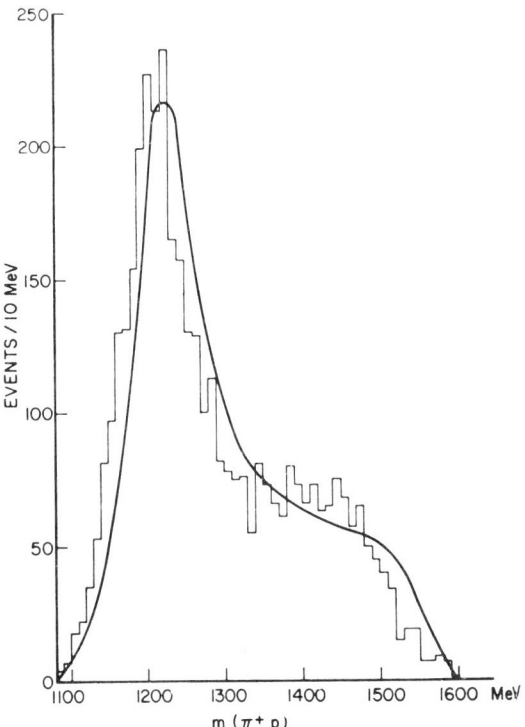

Fig. 10-6. $\pi^+ p$ mass distribution in the reaction $\pi^+ + p \rightarrow \pi^+ + \pi^0 + p$ at $T_\pi = 979$ MeV. The solid curve is the distribution expected, if the $\pi^+ p$ system is formed in a p_{33} isobar, produced with zero orbital angular momentum (after Olsson and Yodh, 1966).

It is then seen from Fig. 10-5, by comparing the sum of reactions (a) + (b) with those of (c) + (d) + (e), that pion production is considerably greater in the isospin state with $I = \frac{1}{2}$ than that with $I = \frac{3}{2}$.

The threshold energy for production of the $N^*(1236)$ resonance, in the reaction:

$$\pi + N \to N^*(1236) + \pi$$

is at a laboratory pion energy of $T_\pi = 400$ MeV. Above this energy the production of the $N^*(1236)$ becomes copious. This is illustrated in Fig. 10-6, which shows the distribution of events in the reaction $\pi^+ + p \to \pi^+ + \pi^0 + p$ as a function of the invariant mass of the $\pi^+ p$ combination in the final state, for an incident pion energy of $T_\pi = 979$ MeV. The production of the N^* isobar at 1236 MeV is shown very clearly. The isobar is produced nearly isotropically, showing that the N^* and π in the final state are mainly in a relative s state.

When the $N^*(1236)$ isobar is produced in an s state, the quantum numbers of the system must be $J = \frac{3}{2}$ $P = -1$, from which it follows at once that the system must have originated from the d_{33} partial wave of the initial $\pi^+ p$ system. When the isobar is produced in a p wave, the parity of the system is $P = +1$ and the angular momentum is either $J = \frac{1}{2}, \frac{3}{2}$ or $J = \frac{5}{2}$, so that the initial state was either $p_{3,1}$, $p_{3,3}$ or $f_{3,5}$.

When the initial state is a $\pi^- p$ state, both $I = \frac{1}{2}$ and $I = \frac{3}{2}$ isospin states contribute. Similar considerations apply to the analysis of the $I = \frac{1}{2}$ states. If the N^* originating from an initial $I = \frac{1}{2}$ πN state is produced in an s-wave, the initial partial wave was the d_{13}. Analysis of the $\pi^- p$ data shows (Olsson and Yodh, 1966) that the N^* production cross section from an initial $I = \frac{1}{2}$ state is strongly enhanced near $T_\pi = 600$ MeV, which indicates that the $N^*(1236)$ is associated with the decay of the d_{13} resonance near $T_\pi = 600$ MeV, predicted from elastic scatterers phase shift analysis.

Although the production of the $N^*(1236)$ isobar is of great importance, this process by itself does account for the complete pion production cross

10.2. SINGLE PION PRODUCTION BELOW 1 GeV

section. In particular, many events occur in which strongly interacting two pion states are produced. This can be seen by plotting the invariant mass distribution (d) of the two pion combination in the reactions (a) to (e). The two pion isobars of importance in the low energy region ($<$ 1 GeV) are the ρ-meson, at a mass of 750 MeV, with $J = I = 1$ and a strongly interacting s wave pion-pion state, with $J = I = 0$, the σ-meson. Clearly σ production can only take place in states with total $I = \frac{1}{2}$, while ρ production can take place in both isospin states.

The Isobar Model

To make a detailed analysis of the data using these ideas, a specific model for the production amplitude is required. The data are not sufficiently complete at present (particularly as polarization data are lacking) for an analysis based purely on expansions into partial waves to be feasible, although the necessary partial wave expansions have been developed.[2]

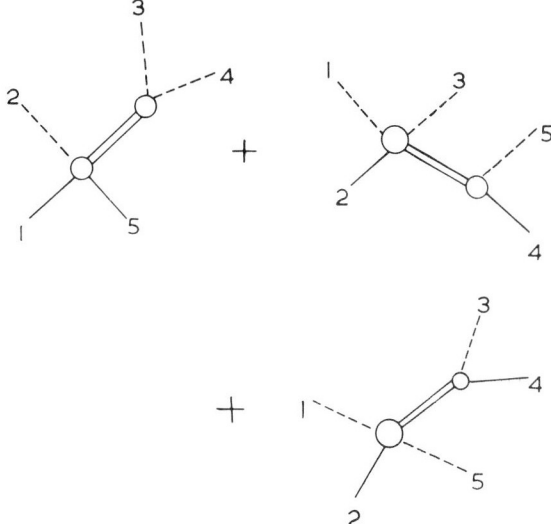

Fig. 10-7. Diagrams contributing to the isobar model.

---

[2] The partial wave formalism has been conveniently summarized by Morgan (1968).

In the isobar model[3] it is assumed that the production amplitude can be approximated by the sum of terms, each one of which represents the production of a strongly interacting pair of particles (or isobar) in the final state. This is represented diagramatically in Fig. 10-7. Each term in the partial wave amplitude is then in the form of a product of a factor representing the production of the isobar and a factor representing its decay. If the isobar is a resonance between particles 1 and 2, then the appropriate term T_{12}, would be of the form:

$$T_{12} = G(s,q_3) \frac{\Gamma/2}{(M_R-E_3) - i\Gamma/2} \tag{10-28}$$

where q_3 is the momentum in the center of mass system of particle 3 relative to the isobar (1+2). A more general form for T_{12} would be:

$$T_{12} = G(s,q_3) \left(\frac{e^{i\delta_{12}} \sin\delta_{12}}{q_{12}} \right) \tag{10-29}$$

where δ_{12} is the phase shift (real or complex) for the strong interaction between the pair of particles 1 and 2. In a typical application (Morgan, 1968) the production factor G was parametrized in the form:

$$G = B\, e^{i\Phi} \left[\frac{q_3}{(1+k\, q_3)^{\frac{1}{2}}} \right]^L \left(\frac{1}{q_{12}} \right)^{\ell} \tag{10-30}$$

where B and Φ are real functions of W and the last factor contains an effective range parameter k and provides the dependence on momentum appropriate to production of the isobar in an orbital state of angular momentum L relative to particle 3, with integral angular momentum ℓ.

---

[3] The progress of the isobar model can be traced from the following references: Sternheimer and Lindenbaum (1961), Freedman et al., (1966), Olsson and Yodh (1966), Morgan (1968), Namyslovski et al., (1967), Glauber (1959).

10.2. SINGLE PION PRODUCTION BELOW 1 GeV

The parametrization of δ_{12} is chosen to fit what is known about the subsystem (1+2). For example, if the (1+2) subsystem is a πN system in the P_{33} state, containing the $N^*(1236)$ resonance, an appropriate form would be (Olsson, 1965):

$$\tan\delta_{12} = \left[\frac{17.06\, q_{12}^2}{(1.236 - M_{23})(1 + 146.4\, q_{12}^2)} \right] \qquad (10\text{-}31)$$

where M_{23} is in GeV and q_{12} in GeV/c.

In the work of Olsson and Yodh (1966), the two isobars produced were:

(a) $N^*(1236)$ $\pi\text{-}N\ I = \frac{3}{2},\ J^P = \frac{3}{2}^+$

(b) N^1 $\pi\text{-}N\ I = \frac{1}{2},\ J^P = \frac{1}{2}^-$.

The N^1 isobar was taken to represent a strong final state interaction in the P_{11}, πN partial wave and was not resonant at the energies under consideration. The isobars were assumed to be produced in s waves only. The model was extended for the $I = \frac{1}{2}$ amplitudes by Morgan (1968), who also considered the production of the isobars:

(c) $\rho\ \pi\text{-}\pi\ I = 1\ J^P = 1^-$

(d) $\sigma\ \pi\text{-}\pi\ I = 0\ J^P = 0^+$

and who allowed for the possibility of production into p and as well as s states.

A more extensive analysis from centre of mass energy $\sim$ 1400 MeV to $\sim$ 2000 MeV is that of Herndon et al., (1973) who assumed the following four final state channels in appropriate change combinations: (i) $\pi + N$ (ii) $\pi + N^*(1236)$ (iii) $\rho + N$ (iv) $\sigma + N$. It is found that the production proceeds predominantly through resonance formation, for example, $\pi + N \to d_{13}(1520) \to \pi + N^*(1236)$. This is in accord with Harari-Freund duality (eqs. (8-82), (8-83)) which expects resonance dominance of the imaginary parts of inelastic amplitudes. One of the achievements of the analysis of

Herndon et al., was the discovery of the $d_{13}(1700)$, long required by the quark model as a 'missing' [8, 4] component of the $\{70\}$, $L = \Gamma$ multiplet (Chapters 5, 6). They also find the branching ratios of the previously known N^* resonances into $\pi + N^*(1236)$ and $\rho + N$ (and $\sigma + N$). The magnitudes and relative signs of such couplings, which are not predicted by SU3, could be important for theories of baryon structure such as the L-excitation quark model (similarly to the couplings to γN discussed on pages 276-279).

10.3. THE PERIPHERAL MODEL

As the incident momentum increases, the angular distribution of many inelastic processes becomes peaked in the forward direction. Typical examples, starting from an initial πN state are:

$$\begin{aligned} \pi + N &\to \rho + N \\ \pi + N &\to \rho + N^* \\ \pi + N &\to \omega + N \\ \pi + N &\to \omega + N^* \ . \end{aligned} \quad (10\text{-}32)$$

For small angles of scattering and at momentum in excess of 1 GeV/c, the scattering of the pion can be described semiclassically (see Chapter 8). In this case, the pion trajectory is to the first approximation the straight line $\vec{R} = \vec{b} + \hat{n}z$, where the impact parameter is b, $\hat{n}$ is a unit vector in the direction of incidence, and the direction of $\vec{b}$ is such that $\vec{b} \cdot \vec{n} = 0$ (see Fig. 10-8). Head-on collisions are associated with small impact parameters and large changes of momentum of the incident pions. Such collisions can be expected to lead to complicated many-particle final

10.3. THE PERIPHERAL MODEL

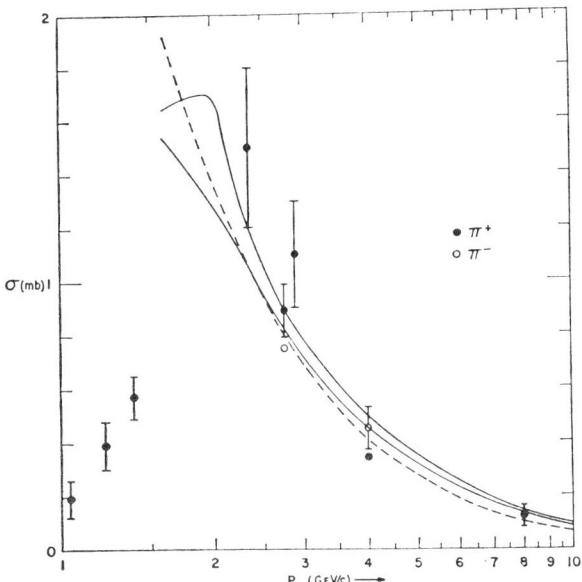

Fig. 10-8. The total cross section for the reaction $\pi^{\pm} + p \to \rho^{\pm} + p$. The theoretical curves are calculated assuming single pion exchange allowing for absorptive effects, the upper solid curve relates to π^+, and the lower solid curve to π^- scattering. The dashed curve is the cross section calculated with an empirical form factor (after Jackson et al., 1965).

states. On the other hand collisions that occur at large impact parameters will be less violent and will not involve large momentum transfers to the incident pion. As small angle scattering corresponds to scattering with small momentum transfers, $|t|$, it is expected that those inelastic collisions leading to two particle final states which are peaked in the forward direction are collisions in which the reaction takes place at large values of the impact parameter; that is, there are *peripheral collisions*.

Elementary considerations outlined in Chapters 3 and 4 suggest that the forces of largest range are due to the exchange of systems of small mass, and it follows that peripheral collisions are on the whole associated with the exchange of the lightest particle with suitable quantum numbers. For the reaction $\pi + N \to \rho + N$ the lightest particle that can be exchanged is a pion, so that the Feynman diagram involved is as in Fig. 10-9.

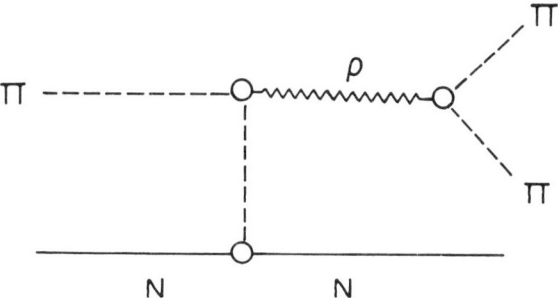

Fig. 10-9. Single pion exchange in ρ production.

The amplitude corresponding to this diagram will contain the pole term:

$$\frac{g_1(t)\, g_2(t)}{(t-M_\pi^2)} \qquad (10\text{-}33)$$

where $g_1(t)$ is the vertex function corresponding to the ρ-π coupling and $g_2(t)$ is that corresponding to the π-N coupling. This amplitude is not completely peripheral in character, since it provides substantial low order partial waves, but it can be expected to provide accurate high order partial waves. If this pole term dominates the cross section near the forward direction, the differential cross section is easily obtained. In the most simple version of the model, the t dependence of the vertex functions is ignored; g_1, g_2 are then taken to be respectively the ρ-π, π-N coupling constants. If this model provided an accurate description of forward scattering, the experimental data could be extrapolated to determine the residue of the pole in the scattering amplitude, that is, to obtain the product $g_1 g_2$. This procedure was originally suggested by Goebel (1958) and by Chew and Low (1959). In practice, such an extrapolation can be very difficult to carry out, since kinematical factors in certain cases can introduce a zero at $t = 0$ into the differential cross section.

The simple one pion exchange model disagrees with experiment both in the shape of the angular distribution and in the magnitude of the cross

10.3. THE PERIPHERAL MODEL

section. The model can be refined in a number of ways, of which the most important are the following: (a) allowance is made for the t variation of the form factors $g_1(t)$, $g_2(t)$; (b) since in any peripheral model the simple reactions should be associated with the higher partial waves, the lower partial waves being associated with complex reactions, steps can be taken to subtract or modify the lower partial waves, predicted by the model; (c) at very high energies, from the discussion of Chapter 8, we believe that the exchanged entity should be a complete Regge trajectory, rather than a single particle, and at lower energies corrections perhaps should be made to allow for this effect.

To allow for the t variation of the vertex functions Ferrari and Selleri (1962) explored the consequences of taking for the product $g_1(t) \, g_2(t)$, the phenomenological form:

$$g_1(t) \, g_2(t) = A + \frac{B}{t-C} \qquad (10\text{-}34)$$

where A, B, C are constants. A number of reactions taking place through one pion exchange could be fitted by the same values of these constants (for example, $\pi + N \to \rho + N + \pi + N \to \rho + N^*$, $K + N \to K^* + N^*$, $N + \overline{N} \to N^* + \overline{N}^*$). Unfortunately fits to processes proceeding via the exchange of vector mesons (for example, the ρ-meson) are not adequately given by this model, and even in the case of pion exchange the model fails to predict the observed angular correlation of the particles in the final state.

The Absorption Model

We now examine in a little more detail the notion that the lower partial waves are given incorrectly by the single particle exchange model because of absorption in these waves into final states of complex configuration. Projecting the one pion exchange amplitude into partial waves immediately reveals that the s-wave, in particular, often violates the unitarity limit and is thus manifestly too large. To correct this defect we proceed by rewriting the single particle exchange amplitude in terms of an impact parameter representation (see Chapter 8). A scattering amplitude $A(s,t)$ can be transformed into $B(s,b)$ where:

$$A(s,t) = \int_0^\infty bdb\, J_0((\sqrt{-t})b)\, B(s,b) \ . \qquad (10\text{-}35)$$

If $A(s,t)$ is given by the single pole term:

$$A(s,t) = \frac{1}{M^2 - t} \qquad (10\text{-}36)$$

the transform $B(s,b)$ turns out to be a modified Bessel function of the second kind:

$$B(s,b) = K_0\,(Mb) \ . \qquad (10\text{-}37)$$

Following Dar (1964) [see also Dar and Tobocman (1964)], the effect of absorption in the lower partial waves (small b) can be simulated by replacing the lower limit of the integral by a parameter b_0, in place of zero:

$$A(t) = \int_{b_0}^\infty bdb\, K_0\,(Mb)\, J_0((-\sqrt{t})b) \ . \qquad (10\text{-}38)$$

With a suitable value of b_0 (of the order 1 Fermi), the peaking of the angular distribution becomes much greater than in the single one pion exchange model, as suggested by experiment. The magnitude of the cross section is also reduced to near the experimental values.

Despite this success the use of a sharp cutoff introduces diffraction minima into the angular distribution, which are not observed in experiment. To overcome this difficulty the distorted wave Born approximation has been introduced (Durand and Chiu, 1964, 1965; Gottfried and Jackson, 1964; Jackson, 1965). To illustrate this approximation we may consider nonrelativistic potential scattering of particles of angular momentum ℓ. If a pair of channels are open at the energy considered, the coupled radial equations describing the reaction are of the form (Mott and Massey, 1965):

$$\left(\frac{d^2}{dr^2} - \frac{\ell(\ell+1)}{r^2} - V_{11} + k_1^2\right) f_1(r) = V_{12}\, f_2(r) \qquad (a)$$

$$\left(\frac{d^2}{dr^2} - \frac{\ell(\ell+1)}{r^2} - V_{22} + k_2^2\right) f_2(r) = V_{21}\, f_1(r) \qquad (b) \ .$$

10.3. THE PERIPHERAL MODEL

If the coupling between the channels is weak, the influence of the channel (b) on the scattering in channel (a) can be neglected; that is, the term $V_{12} f_2$ can be neglected in (a). Under these circumstances, the transition amplitude for scattering from channel (a) to (b), A_{21}, is given by:

$$A_{21} = -e^{i(\delta_{1\ell} + \delta_{2\ell})} \left| \int_0^\infty \bar{f}_2(r) V_{21} \bar{f}_1(r) \, dr \right| \frac{1}{k_1} \qquad (10\text{-}40)$$

where $\bar{f}_1, \bar{f}_2$ are the solution of the homogenous equation:

$$\left(\frac{d^2}{dr^2} - \frac{\ell(\ell+1)}{r^2} - V_{11} + k_1^2 \right) \bar{f}_1(r) = 0$$

$$\left(\frac{d^2}{dr^2} - \frac{\ell(\ell+1)}{r^2} - V_{22} + k_2^2 \right) \bar{f}_2(r) = 0 \qquad (10\text{-}41)$$

with the boundary condition:

$$\bar{f}_1(r) \sim e^{i\delta_{1\ell}} \sin\left(k_1 r - \frac{1}{2}\ell\pi + \delta_{1\ell}\right)$$

$$\bar{f}_2(r) \sim e^{i\delta_{2\ell}} \sin\left(k_2 r - \frac{1}{2}\ell\pi + \delta_{2\ell}\right) \qquad (10\text{-}42)$$

and $\delta_{1\ell}$ and $\delta_{2\ell}$ are the *phase* shifts for scattering in the uncoupled channels.

Making the further approximation that the functions $\bar{f}_i(r)$ are sufficiently well approximated by plane waves, we see:

$$A_{21} \simeq e^{i(\delta_{1\ell} + \delta_{2\ell})} B_{21} \qquad (10\text{-}43)$$

where B_{21} is the Born approximation matrix element. If absorption takes place from channels (a) and (b) into some other channels, this can be rerepresented by making the potentials V_{11} and V_{22} complex, with positive imaginary parts. Correspondingly the phase shifts $\delta_{i\ell}$ become complex also with positive imaginary parts:

$$\delta_{i\eta} \to \eta_{i\ell} + i\lambda_{i\ell} \tag{10-44}$$

so that A_{21} given by:

$$A_{21} = e^{i(\eta_{1\ell}+\eta_{2\ell})} e^{-(\lambda_{1\ell}+\lambda_{2\ell})} B_{21} \tag{10-45}$$

is decreased (through the exponential factor) by the effects of absorption into other channels.

It is now straightforward to translate this result into the semiclassical impact parameter formalism. The phase shifts $\eta_{i\ell}$, $\lambda_{i\ell}$ become functions of b, the impact parameter, and it is found that, in place of the expression eq. (10-35), we obtain:

$$A(s,t) = \int_0^\infty b\,db\, J_0((\sqrt{-t})b)\, B(s,b)\, e^{2i\eta(b)}\, e^{-2\lambda(b)} \tag{10-46}$$

where we have assumed for simplicity that the scattering in the initial and final states of the reaction is the same, so that:

$$\eta_1 = \eta_2 = \eta \quad \text{and} \quad \lambda_1 = \lambda_2 = \lambda \;. \tag{10-47}$$

The phase shift $\delta(b) = \eta(b) + i\lambda(b)$ is given, in the semiclassical approximation, by the expression (see eqs. (8-32), (8-35)):

$$2\delta(b) = -\frac{1}{v}\int_{-\infty}^\infty V_{11}(\sqrt{b^2+z^2})\, dz \;. \tag{10-48}$$

A derivation of these results can be consulted in the paper by Gottfried and Jackson (1964b).

In fitting experimental data, a parametric form is generally chosen to represent $\delta(b)$. The phase $\delta(b)$ is determined by the *elastic* scattering in the incident channel. At the energies under consideration (about 1 GeV) the elastic scattering amplitude is largely imaginary and the distribution in t is exponential in shape:

10.4. RESONANCE PRODUCTION AT HIGH ENERGIES

$$A(s,t)_{\text{elastic}} = i\left(\frac{q\,\sigma_{\text{tot}}(s)}{4\pi}\right) e^{+\frac{1}{2}\beta t} \tag{10-49}$$

where β is a constant. The phase reproducing this amplitude is given by:

$$e^{2i\delta(b)} = 1 - C\,e^{-\left(\frac{b^2}{2\beta}\right)} \tag{10-50}$$

with $C = \dfrac{\sigma_{\text{tot}}(s)}{4\pi\beta}$.

Experimental fits

As an example of the fit to the data that it is possible to obtain with this model, we can again consider the reaction $\pi + N \to \rho + N$ (see Jackson et al., 1965). The measurements extend to incident momenta of up to 8 GeV/c. The differential cross sections are well fitted by the model provided that complete absorption of the s-wave is assumed, corresponding to setting $C = 1$ in eq. (10-50). The energy dependence of the cross section is the same as that of the unmodified one pion exchange model, where $\sigma \propto 1/q^2$, and this agrees well with experiment for momenta greater than 2 GeV/c. This can be seen in Fig. 10-8.

Other processes in which single pion exchange dominates can equally well be fitted by the model, but processes in which the exchange of vector particles is important are less successfully treated.

10.4. RESONANCE PRODUCTION AT HIGH ENERGIES AND THE QUARK MODEL

In the previous paragraphs of this chapter, some methods of analysis of production processes at low or at intermediate energies have been outlined. At high incident momenta (for example at momenta in excess of 10 GeV/c) the methods based on the construction of Regge amplitudes were discussed in Chapter 8; however alternative procedure based on the quark model have been suggested and have met with some success.

In these models, the colliding particles are assumed to be composed of quarks or anti-quarks, and scattering takes place there in two-body quark-quark interactions. The situation is thus rather similar to the description of nuclear reactions in terms of assumed nucleon-nucleon forces. Several different forms of this general type of model have been put forward. As an illustration of one promising line of attack, we shall consider some work on nucleon isobar production by Hendry and Trefil (1969). In this work, an attempt is made to explain the major differences in the differential cross sections for the reaction:

$$\pi + p \to \pi + p \text{ (elastic scattering)}$$
$$\pi + p \to \pi + N^*(1400) \tag{10-51}$$
$$\pi + p \to \pi + N^*(1690) \ .$$

These cross sections for these reactions have been measured in the incident momentum interval of from 10 to 26 GeV/c (Foley et al., 1967). The qualitative features of the measurements are that for small values of (t), the inelastic differential cross sections are at least an order of magnitude smaller than the elastic cross section (see Fig. 10-9), and that the slopes of the forward peaks of the differential cross sections are very different for the different processes. For production of the $N^*(1690)$, which is in the F_{15} partial wave, the differential cross section is flat as a function of (t), while for the production of the $N^*(1400)$ (which is in the P_{11} partial wave) and for elastic scattering the forward cross section is fitted by the exponential form:

$$\frac{d\sigma}{d|t|} = A\, e^{-Bt} \tag{10-52}$$

where B is about 8 $(\text{GeV/c})^2$ for elastic scattering and 12-16 $(\text{GeV/c})^2$ for $N^*(1400)$ production.

10.4. RESONANCE PRODUCTION AT HIGH ENERGIES

In Hendry and Trefil's model, the nucleon and the nucleon isobars are considered to be composed of three quarks. The three-quark wave function was taken to be that of a harmonic oscillator (Faiman and Hendry, 1968, 1969). The nucleon is the ground state configuration which can be denoted as $(1s)$ $L = 0^+$, while the $N^*(1400)$ corresponds to a radial excitation with configuration $(1s)$ $(2s)$, $L = 0^+$, and the $N^*(1690)$ is an orbital excitation $(1s)$ $L = 2^+$. In principle the incident pion should be represented by a quark-anti-quark wave function; but as the pion is not excited in these collisions, the pion structure was not explicitly represented, but instead the forces responsible for the reaction were attributed to a pion-quark interaction. The pion quark scattering amplitude $f(\delta)$, where δ is the momentum transfer, as approximated by the form:

$$f(\delta) = \frac{\beta + i}{4\pi} (\sigma_{\pi Q}) e^{-\frac{1}{2} a \delta^2} \qquad (10\text{-}53)$$

where a, β, and σ are energy-independent constants.

At the energies under consideration semiclassical scattering theory is expected to be accurate and the scattering amplitude for any particular reaction can be written in terms of the amplitudes for pion-quark scattering. The explicit expression is:

$$T_{fi} = \frac{ip}{2\pi} <\psi_f | \int d\vec{b}\, e^{i\vec{\Delta}\cdot\vec{b}} (3 F_j - 3 C_2\, F_j F_k + C_3\, F_j F_k F_\ell) |\psi_i > \qquad (10\text{-}54)$$

with:

$$F_j = \frac{1}{2\pi i p_j} \int d\delta_j\, e^{i\delta_j \cdot (\vec{b} - \vec{s}_j + \vec{s})}\, e^{i \Delta_m (z_j - z)}\, f_j(\delta_j)\ .$$

In eq. (10-54), Δ is the transverse momentum transfer, and Δ_m is the longitudinal momentum transfer associated with the charge in mass of the baryon. The position of the jth quark is $r_j = s_j + z_j$ and $s = \sum s_j/A$, $z = \sum z_j/A$. For elastic scattering $C_2 = C_3 = 1$, and for inelastic scattering $C_2 = 2$, $C_3 = 3$. The formula expresses that the phase shift for scattering from the complex system is given by a sum of the phase shifts for scattering by the individual quarks.

The free parameters a, β, and σ appearing in the ansatz for the pion-quark scattering amplitudes were determined by fitting the elastic scattering data. The inelastic differential cross sections are then wholly determined, and predicted values are shown in Fig. 10-10. The shape of the differential cross sections is reasonably and accurately given by the

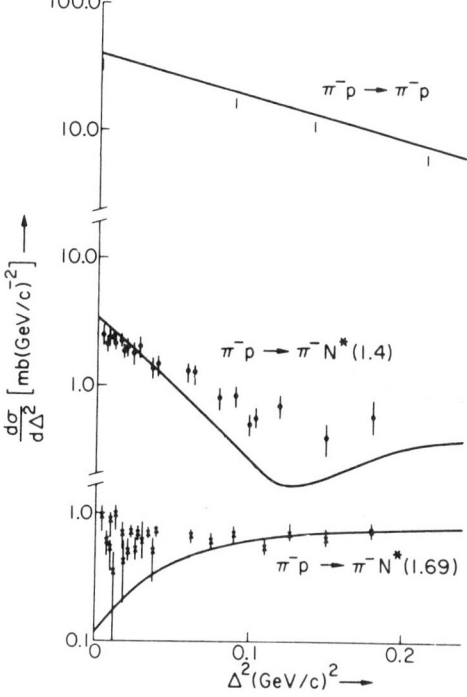

Fig. 10-10. A comparison of the predictions of the quark model of Hendry and Trefil (1969) with experimental data. The data which is energy-independent represents measurements taken at momenta between 14 GeV/c and 26 GeV/c. The quantity Δ is the transverse momentum transfer.

model, and the magnitude of the production cross section of the $N^*(1690)$ is correct as well. On the other hand, the absolute issue of the $N^*(1400)$ production cross section at $t = 0$ is about four times smaller than that measured, and the curve shown in Fig. 10-9 has been renormalized, so that the shape of the predicted and experimental cross sections can be compared.

It is remarkable that the success of the quark model persists when applied to collision phenomena. The predictions obtained are not considered to be sensitive to the precise form of isobar wave function chosen (harmonic oscillator in this case), but the difference between the $N^*(1400)$ and $N^*(1690)$ production cross section does depend on the former representing a radial and the latter an orbital excitation process.

10.5 MULTIPARTICLE PRODUCTION AT HIGH ENERGIES

The average number of particles produced in a two body collision increases with energy. These $2 \to n$ particle reactions have been the subject of much recent study,[†] stimulated by the introduction of the NAL accelerator, providing proton beams up to 500 GeV, and the CERN Intersecting Storage Rings (ISR), providing proton-proton collisions at energies up to 60 GeV in the centre of mass system, which is equivalent to an accelerator experiment in which protons with laboratory energies of up to 1900 GeV impinge on protons at rest. Some cosmic ray data also exists up to laboratory energies of 10^4 GeV.

Two important features of multiparticle production reactions are well established. The first is that the average number of particles produced (which are mainly pions) increases much more slowly with energy than

---

[†] General reviews of the subject, with full bibliographies have been given by Van Hove (1971), Frazer et al., (1972) and Jacob (1972).

would be expected from consideration of the phase space available. In pp collisions, the average observed multiplicity of charged particles produced $\bar{n}_c$, can be fitted very roughly to the expression

$$\bar{n}_c = A + B \log(W - 2M_p) \tag{10-55}$$

where W is the center of mass energy in GeV and $B \simeq 1.4$, $A \simeq 2.0$. This expression covers a wide laboratory momentum range from 10 to 10^3 GeV/c. Some typical values for pp collisions are $\bar{n}_c \simeq 4, 7.7, 13, 16$ at $P = 20, 200, 1500, 10{,}000$ GeV/c respectively.

The second interesting qualitative feature of these reactions is that nearly all the particles in the final state are produced at low values of the momentum transverse to the beam direction P_T. The mean value of P_T is of the order 300 MeV/c, for incident energies up to 30 GeV and a little larger for the highest ISR energies. This value is practically independent of the initial two body state (pp, $\pi^+ p$, $\pi^- p$ ...), and the multiplicity of the final state. For $P_T > 300$ MeV/c the momentum distribution appears to decrease exponentially at all energies. It follows that the available energy is concentrated in producing particles with large longitudinal momenta P_L, and the average value of P_L increases roughly as

$$\bar{P}_L \propto \left(\frac{W}{\log W}\right) \tag{10-56}$$

When a few particles are produced in a collision, it is possible to display the particle distributions in P_L in elegant multiparticle longitudinal phase space plots, generalizations of the Dalitz plot referred to earlier. An interesting feature shown by these displays, is that the heavily populated regions of the diagrams are associated with events that can be interpreted as being associated with the exchange of a pomeron (see Chapters 8 and 9). For example in Fig. 10-11 possible diagrams are shown which explain the observed momentum distribution of the final state particles in the reaction

$$\pi^- + p \rightarrow p + \pi^+ + \pi^- + \pi^- \tag{10-57}$$

at 11 and 16 GeV/c.

10.5 MULTIPARTICLE PRODUCTION AT HIGH ENERGIES

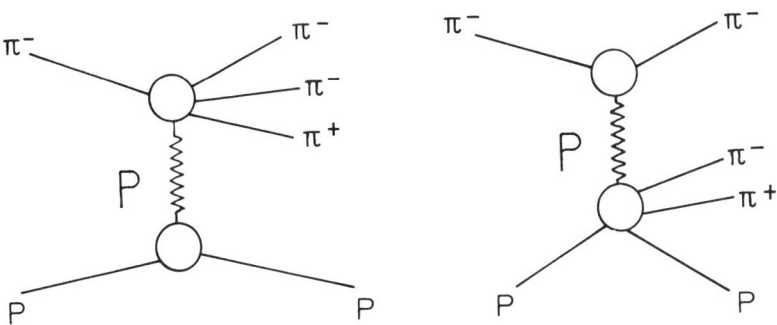

Fig. 10-11. Pomeron exchange mechanism explains the longitudinal momentum distributions in the reaction $\pi^- + p \to \pi^- + \pi^- + \pi^+ + p$ (Kittel et al., 1971).

Detailed examination and display of the distribution in momentum of all the particles in the final state becomes impractical when more than five particles are produced. Because of this, experimental effort has concentrated on measuring the cross-section as a function of the variables of a single particle (or sometimes two particles) in the final state, irrespective of the number of other particles produced. This type of experiment is termed an 'inclusive' experiment. For example in the reaction

$$A + B \to C + X \tag{10-58}$$

where X stands for any other particle or particle produced, distribution in momentum of the particle C is measured. The differential cross-section for an inclusive reaction depends on the (two dimensional) transverse momentum P_T and the longitudinal momentum P_L, and can be characterized by a function $f(P_T, P_L, W)$ defined as

$$f(P_T, P_L, W) = W \frac{d^2\sigma}{dP_T^2 \, dP_L} . \tag{10-59}$$

Certain theoretical models proposed by Feynman and by Yang, suggested that the function f might only depend on P_T and a scaled longitudinal momentum rather than on P_L and W separately. The experiments have confirmed this suggestion. In the Feynman form, in terms of center

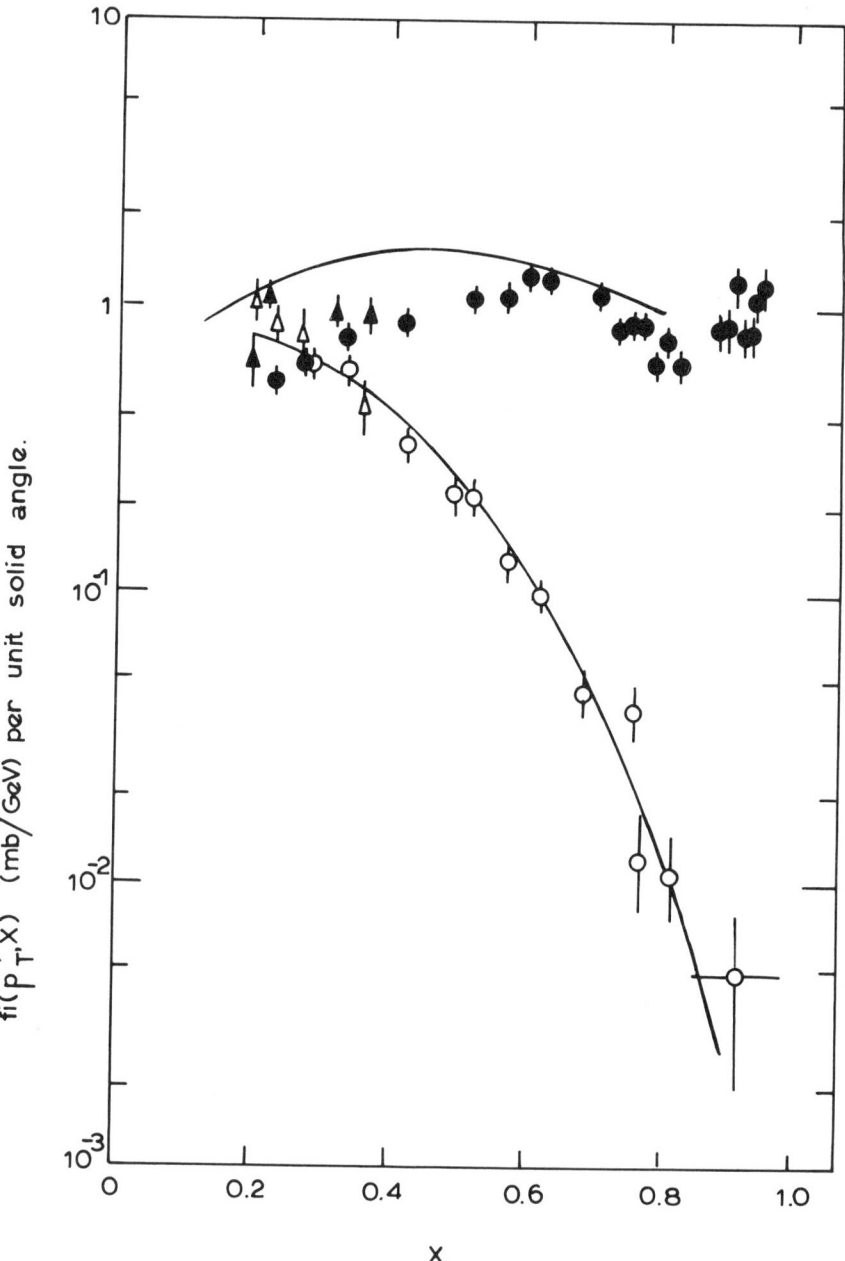

Fig. 10-12. The function $f(P_T, x)$ for the inclusive production of π^+ and p in pp collisions for $P_T = 0.8$ GeV/c. ●, ▲ P; 0, △ π^+ at 1995 (GeV)^2
——— data at 47 (GeV)^2.

10.5 MULTIPARTICLE PRODUCTION AT HIGH ENERGIES

of mass variables, the scaling law is that f is a function of P_T and x where x is defined as

$$x = (2 P_L/W) . \qquad (10\text{-}60)$$

The scaling law corresponds to the idea that at high energies the projectile and target particles break up independently. Viewed in the laboratory system, the target splits into fragments and at high energies the fragments produced approaches a limiting distribution irrespectively of the projectile. A fragment with a given momentum in the laboratory system has (in the high energy limit) a fixed value of x in the center of mass system. In just the same way, viewed from the rest frame of the projectile, the fragmentation of the projectile approaches a limiting distribution, also defined by the center of mass variable x.

The scaling law is well illustrated in Fig. 10-12 where the data at $s = 1995$ (GeV)^2 and 47 (GeV)^2 for the production of π^+ in pp collisions follows the same distribution in x, and the same is true for the proton distribution in its final state, also shown in the figure.

Several theoretical models have been proposed to explain multiparticle reactions. These are of a rather tentative nature and the reader is referred to the reviews cited for details. However an important framework in which the theory can be discussed has been proposed by Mueller (1970), in a generalization of the optical theorem (see Chapter 2). The optical theorem shows that the total cross-section for the reaction

$$A + B \to (\text{anything}) \qquad (10\text{-}61)$$

is proportional to the imaginary part of the forward elastic scattering amplitude, which is the discontinuity in s of the amplitude for $A + B \to A + B$. Similarly Mueller has shown that the total cross-section for

$$A + B \to C + (\text{anything}) \qquad (10\text{-}62)$$

is proportional to the discontinuity of the amplitude in the forward direction for the three body reaction

$$A + B + \overline{C} \to A + B + \overline{C} . \qquad (10\text{-}63)$$

This is illustrated in the diagram of Fig. 10-13. Models for the three body scattering amplitude can thus be explored, for example in Fig. 10-14 we show a Mueller diagram illustrating the double exchange of Regge poles, which may dominate the small x region of the cross-section.

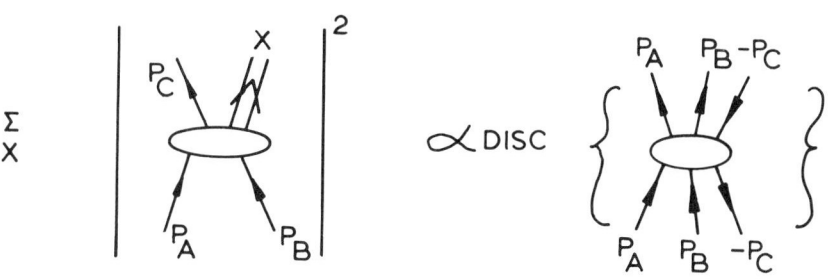

Fig. 10-13. The generalized optical theorem of Mueller.

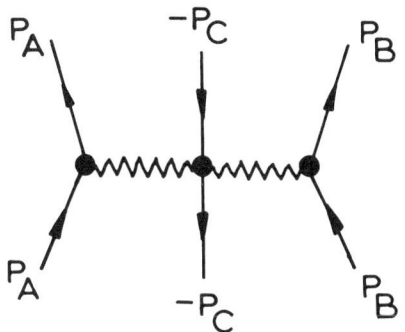

Fig. 10-14. Mueller diagram of double Regge exchange.

REFERENCES

G. F. Chew and F. E. Low, Phys. Rev. 113, 1640 (1959).

A. Dar, Phys. Rev. Letters 13, 91 (1964).

A. Dar and W. Tobocman, Phys. Rev. Letters 12, 511 (1964).

REFERENCES

A. T. Davies and R. G. Moorhouse, Nuovo Cimento 52A, 1112 (1967).

L. Durand and Y. T. Chiu, Phys. Rev. Letters 12, 399; 13, 45 (1964).

L. Durand and Y. T. Chiu, Phys. Rev. 137, B1530; 139, B646 (1965).

D. Faiman and A. W. Hendry, Phys. Rev. 173, 1720 (1968).

D. Faiman and A. W. Hendry, Phys. Rev. 180, 1572 (1969).

E. Ferrari and F. Selleri, Nuovo Cimento Supp. 24, 453 (1962).

K. J. Foley, R. S. Jones, S. J. Lindenbaum, W. A. Love, S. Ozaki, E. D. Platner, C. A. Quirles and E. H. Willen, Phys. Rev. Letters 19, 397 (1967).

W. R. Frazer, L. Inger, C. H. Melata, C. H. Poon, D. Silverman, K. Stowe, P. D. Tring and H. J. Yesian, Rev. Mod. Phys. 44, 284 (1972).

D. Z. Freedman, C. Lovelace and J. M. Namyslovski, Nuovo Cimento 43A, 258 (1966).

R. J. Glauber, in *Lectures in Theoretical Physics I* (Interscience Publishers, New York, 1959).

C. Goebel, Phys. Rev. Letters 1, 337 (1958).

K. Gottfried and J. D. Jackson, Nuovo Cimento 33, 309 (1964a).

K. Gottfried and J. D. Jackson, Nuovo Cimento 34, 735 (1964b).

A. W. Hendry and J. S. Trefil, Phys. Rev. 184, 1680 (1969).

D. Herndon, R. Longacre, R. Miller, A. H. Rosenfeld, G. Smadja, G. Yost, R. Cashmore, D. Leith, *Proc. XVI International Conf. on High Energy Phys.*, (ed. A. Roberts, National Accelerator Laboratory, 1973)

L. van Hove, Phys. Reports 1c, 347 (1971).

J. D. Jackson, Rev. Mod. Phys. 37, 484 (1965).

J. D. Jackson, J. T. Donohue, K. Gottfried, R. Keyser and B. E. V. Svensson, Phys. Rev. 139, B428 (1965).

M. Jacob, *Proc. XVI International Conf. on High Energy Physics*, Chicago (1972) (to be published).

W. M. Kittel, S. Ratti and L. van Hove, Nucl. Phys. B30, 333 (1971).

D. Morgan, Phys. Rev. 166, 1731 (1968).

N. F. Mott and H. S. W. Massey, *Theory of Atomic Collisions* (3rd Ed., Oxford Univ. Press, London, 1965).

A. H. Mueller, Phys. Rev. D3, 1486 (1971).

J. M. Namyslovski, M. S. K. Raznii and R. G. Roberts, Phys. Rev. 157, 1328 (1967).

M. G. Olsson, Phys. Rev. Letters 14, 118 (1965).

M. G. Olsson and G. B. Yodh, Phys. Rev. 145, 1309 (1966).

R. J. Plano, in *Proceedings of the Lund International Conference on Elementary Particles*, ed. G. von Daidel (Berlington Boktryckeriet, Lund, 1969), p. 321.

J. G. Rushbrooke, in *Proc. 14th Int. Conf. on High Energy Physics*, ed. J. Prentki and J. Steinberger (CERN, Geneva, 1968), p. 158.

R. M. Sternheimer and S. J. Lindenbaum, Phys. Rev. 123, 333 (1961).

CHAPTER 11

THE FORM FACTORS OF THE NUCLEON AND PION

11.1. INTRODUCTION

The structure of a composite particle, such as a nucleus, can be characterized by, amongst other things, the way in which the charge and magnetic moment of the particle is distributed. These distributions can be measured by allowing the composite particle to interact with an electromagnetic field. In particular, these distributions are connected with differential cross sections for the scattering of charged particles by the composite particle. The most suitable charged particles to use as probes in the experimental investigation of form factors are electrons, as these are structureless and do not take part in the strong nuclear interaction that would otherwise dominate the scattering cross sections. To take a simple example, the differential cross section for the scattering of electrons moving at a nonrelativistic velocity by a system with total charge Ze and with a charge density $\rho(r)$ is given, in Born's approximation by:

$$\frac{d\sigma}{d\Omega} = \frac{M_e^2 \, e^4 Z^2}{4k^4 \sin^4(\theta/2)} |F(q)|^2 \qquad (11\text{-}1)$$

where k is the momentum of the electron, θ is the angle of scattering, and $F(q)$ is a "form factor," which is proportional to the Fourier transform of the charge distribution:

$$F(q) = \frac{1}{Z} \int \rho(r) \, e^{i\vec{q}_1 \cdot \vec{r}} \, d\vec{r} = \frac{4\pi}{Zq} \int_0^\infty \rho(r) \sin(qr) \, r \, dr \quad . \tag{11-2}$$

The variable q represents the momentum transfer and is given by (in units with $\hbar = 1$):

$$q = 2k \sin(\theta/2) \quad .$$

Expanding $F(q)$ in powers of q, we find that:

$$F(q) = 1 - \frac{1}{6} q^2 \langle r^2 \rangle + \ldots \tag{11-3}$$

where $\langle r^2 \rangle$ is the mean square radius of the charge distribution. For small values of q, with $q^2 \ll 1$, the cross section is approximately that for scattering of an electron by a point particle of charge Z; for larger q, it is possible to measure $\langle r^2 \rangle$, and clearly as q increases further information can be obtained. In fact, to investigate nucleon structure, the values of q required will be of the order of 10^{12} (fermi)$^{-1}$. This requires the use of electrons with energies of several hundred MeV, and the simple formula eq. (11-1) must be suitably modified to take into account relativistic effects.[1]

Elementary particles that take part in the strong interactions necessarily possess a certain structure. For example the proton will spend part of the time in a (virtual) configuration, containing a neutron and a π^+-meson:

$$p \rightleftharpoons n + \pi^+ \quad .$$

---

[1] The relativistic Born approximation (see Mott and Massey, 1965, Chapter 9) is given by $\dfrac{d\sigma}{d\Omega} = \dfrac{e^4 Z^4 \cos^2\theta/2}{4E^2 \sin^4\theta/2} |F(q)|^2$ with $q = 2E \sin(\theta/2)$ (units $\hbar = c = 1$).
At large angles $(d\sigma/d\Omega)$ falls below the nonrelativistic expression because of the factor $(\cos^2\theta/2)$. The experiments on electron scattering by nuclei have been reviewed by Hofstadter (1957) and Hofstadter et al., (1965).

11.1. INTRODUCTION

As the wave function of the π^+-meson must possess a certain extension in space, this and other virtual processes will be effective in spreading the charge and magnetic moment of the proton over a finite region. Since this region is of the order 0.3 fermi in extent, particles of short wavelength are required as probes, capable of producing momentum transfers of 3 (fermi)$^{-1}$ or greater. Again, particles that do not take part in the strong interactions must be used, and most experiments have been performed with electron beams, although measurements of positron and μ-meson scattering are becoming available. It might be thought that the photon would also act as a suitable probe. This is not so, because (in units $\hbar = c = 1$) the momentum of a photon $|\vec{q}|$ is equal to its energy q^0, and this implies that either the photon will produce a high energy final state, as in photoproduction, or, as in Compton scattering, the intermediate state of the nucleon will be highly excited (see Fig. 11-1). In both cases the scattering cross sections are not very directly connected with the static properties of the unexcited system. Electron scattering takes place through the exchange of virtual photons, for which $q^2 = q^{0^2} - |\vec{q}|^2 \neq 0$, and does not possess this limitation.

The processes that have been measured are elastic scattering[2] of electrons and positrons by nucleons:

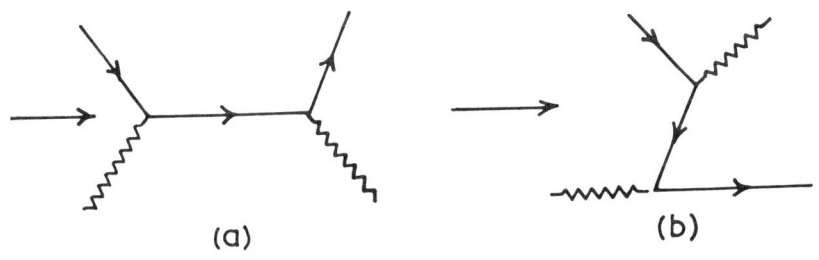

Fig. 11-1. The second order Feynman diagrams for Compton scattering of photons by nucleons.

---

[2] All accelerated charge particles radiate photons and, as in any collision experiment, scattered electrons over a small, but finite, range of energies are detected; the measured cross section contains a small contribution from scattering in which low energy photons have been produced. A correction can be made to eliminate this contribution and to obtain the true elastic cross section (Schwinger, 1949; Mo and Tsai, 1969).

$$e^{\pm} + N \to e^{\pm} + N$$

the electroproduction of pions, for example:

$$e^- + p \to e^- + n + \pi^+$$

and corresponding processes using μ^+- or μ^--mesons. We shall concentrate on the elastic scattering process. Electron-proton scattering can be measured directly, but electron-neutron scattering is known only indirectly through measurements of the scattering of electrons by deuterons. In this case, the electron-neutron cross section must be extracted by the use of the impulse approximations or by dispersion relation techniques.

A rather different class of experiments, which have received considerable attention, are those which measure the differential cross sections for the inelastic scattering of electrons by nucleons at high energies, observing only the scattered electron. The reaction, termed an "inclusive reaction," is:

$$e^- + p \to e^- + \text{"anything"}$$

and the cross section can be measured as a function of momentum transfer and incident energy. In the "deep inelastic" scattering region, at large values of the energy and of the momentum transfer, the cross section is larger than would be expected from a nucleon in which the charge distribution is continuous. This can be explained by a model in which the nucleon is composed of point-charge constituents called "partons." In different versions of this model the partons are variously identified with the quark components of the nucleon, or with the "bare" nucleons and pions that occur in a field theoretical formalism, and as yet the precise nature of the partons is unclear. In section 11.6 a brief discussion is given of the structure of the parton model and its application to deep inelastic scattering.

11.2. THE ELASTIC SCATTERING OF ELECTRONS BY PROTONS

The interaction between electrons and nucleons takes place through photon exchange. The strength of the coupling between the proton field A^μ and a current is given by $\sqrt{a}$ where a is the fine structure constant $(e^2/\hbar c)$, which has the value $(1/137)$. Because this is so small, only single-photon exchange, for which the cross section is proportional to a^2, need be considered. The corrections due to two-photon exchange are of the order a^3 and have not so far been identified.

Kinematics

The single-photon exchange diagram is shown in Fig. 11-2, where p_a and p_b denote the initial and final energy-momentum vectors of the electron, Q_a and Q_b denote the initial and final energy-momentum vectors of the nucleon, and q is the energy-momentum vector of the exchanged photon. Conservation of energy and momentum then requires:

$$p_a + Q_a = p_b + Q_b$$
$$q = p_a - p_b = Q_b - Q_a . \qquad (11\text{-}4)$$

As the external lines refer to real particles, we have:

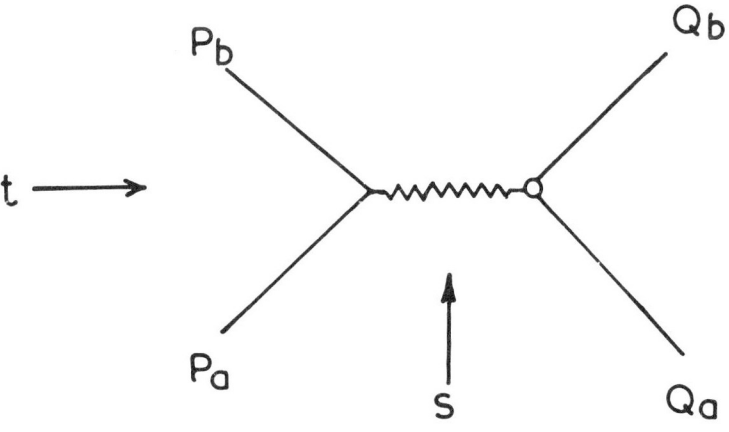

Fig. 11-2. The single-photon exchange diagram for scattering of electrons by nucleons.

$$p_a^2 = p_b^2 = M_e^2, \quad Q_a^2 = Q_b^2 = M_N^2 \qquad (11\text{-}5)$$

where M_e and M_N are the electron and nucleon masses. The internal line refers to a virtual photon, so that $q^2 \neq 0$.

The cross section is a function of the two invariants s and t introduced in Chapter 3. The square of the total energy in the center of mass system is equal to s, where:

$$s = (p_a + Q_a)^2 \qquad (a)$$

and t is defined by:

$$t = (p_a - p_b)^2 = -q^2 \qquad (b) . \qquad (11\text{-}6)$$

The results of electron scattering experiments are usually expressed in terms of the incident energy of the electron E and the scattering angle θ in the laboratory system, in which the nucleon is initially at rest. In terms of E and θ, s and t become:

$$s = M_N(M_N + 2E)$$

$$t = -(4E^2 M_N \sin^2(\theta/2))/[M_N + 2E \sin^2(\theta/2)] \qquad (11\text{-}7)$$

where the high energy approximation has been made in which $\dfrac{M_e}{|\vec{p}_a|}$ and $\dfrac{M_e}{M_N}$ are neglected compared with unity.

The Transition Matrix

The transition matrix corresponding to the diagram of Fig. 11-2 is given by:

$$2 T_{ba}(s,t) = \sum_{\mu,\nu} (2\pi)^4 <\vec{p}_b|j^\mu(0)|\vec{p}_a> \frac{g_{\mu\nu}}{(p_b - p_a)^2} <\vec{Q}_b|j^\nu(0)|\vec{Q}_a>. \qquad (11\text{-}8)$$

The first vertex function, $<\vec{p}_b|j^\mu(0)|\vec{p}_a>$, is the matrix element of the electromagnetic current operator with respect to single electron states.

11.2. THE ELASTIC SCATTERING OF ELECTRONS

As we are working to lowest order in α, and as the electron has only electromagnetic interactions, we can put $j^\mu(0) = \bar{\psi}_e(0) \gamma^\mu \psi_e(0)$, where $\psi_e(x)$ is the *free* field operator for the electron. This vertex in this approximation becomes:

$$<\vec{p}_b|j^\mu(0)|\vec{p}_a> = -e M_e (2\pi)^{-3} \bar{u}_e(\vec{p}_b) \gamma^\mu u_e(\vec{p}_a) \ . \tag{11-9}$$

A subscript e has been added to the spinors to indicate that they refer to the electrons.

If the proton were a structureless particle, the vertex function $<\vec{Q}_b|j^\mu(0)|\vec{Q}_a>$ could be written in a similar form as:

$$<\vec{Q}_b|j^\nu(0)|\vec{Q}_a> = e M_N (2\pi)^{-3} \bar{u}_N(\vec{Q}_b) \gamma^\nu u_N(\vec{Q}_a) \ . \tag{11-10}$$

The cross section, originally obtained by Mott (1929), is then found to be:

$$\left(\frac{d\sigma}{d\Omega}\right)_M = \frac{M_N\, e^4 \cos^2(\theta/2)}{4E^2 \sin^4(\theta/2)[M_N + 2E \sin^2(\theta/2)]} \ . \tag{11-11}$$

The exact matrix element of the electromagnetic current, taken between single proton states, can be expressed in the form:

$$<\vec{Q}_b|j^\nu(0)|\vec{Q}_a> = e M_N (2\pi)^{-3}[\bar{U}_p(\vec{Q}_b) O^\nu U_p(\vec{Q}_a)] \tag{11-12}$$

where O^ν is an operator that must transform like a four-vector under a Lorentz transformation. The current j^μ is an observable, and is therefore self-adjoint, which requires that:

$$[\bar{u}_N(\vec{Q}_b) O^\nu u_N(\vec{Q}_a)] = [\bar{u}_N(\vec{Q}_a) O^\nu u_N(\vec{Q}_b)]^* \ . \tag{11-13}$$

Remembering that $\bar{u}_N$ is defined as $(u_N^\dagger \gamma^0)$, it is easily seen that, with $\gamma_0^2 = 1$ and $\gamma^0 = \gamma^{0\dagger}$, for the condition eq. (11-13) to be satisfied we have:

$$O^\nu = \gamma^0 O^{\nu\dagger} \gamma^0 \ . \tag{11-14}$$

The most general form of O^ν will consist of linear combinations of all the possible four-vectors that can be constructed from the vectors Q_a^μ and Q_b^μ and from the γ-matrices. Each of these vectors can be multiplied by coefficients F_i^P, which, by definition, are the form factors of the proton. These form factors are functions of the invariants that can be built from the vectors Q_a^μ and Q_b^μ. These invariants are as follows: $Q_a^2 = Q_b^2 = M_p^2$ and $(Q_a \cdot Q_b)$ or, alternatively, $q^2 \equiv (Q_a - Q_b)^2$ and M_p^2. At present we are not interested in the dependence of the form factors on the mass M_p, and the form factors will be written as functions of q^2. Not all the vectors that can be written down (twelve in number) are independent. For example, as the matrix elements of O are required with respect to spinors representing free particles, the Dirac equation can be used to reduce multiplicative scalar factors such as $(\gamma \cdot Q_a)$. We have $(\gamma \cdot Q_a) u_N(\vec{Q}_a) = M_p u_N(\vec{Q}_a)$, and the factor M_p can be absorbed into one of the coefficients[3] F_i. Similarly the Dirac equation for $u_N(\vec{Q}_b)$ can be used to eliminate factors of $(\gamma \cdot Q_b)$. Keeping this in mind, the possible vectors can be reduced to three:

$$\gamma^\mu, \quad P = Q_a + Q_b, \quad q = Q_b - Q_a \ . \tag{11-15}$$

The law of current conservation $\partial_\mu j^\mu(x) = 0$ requires that:

$$\partial_\mu \langle Q_b | j^\mu(x) | Q_a \rangle$$
$$= \partial_\mu e^{i(Q_b - Q_a) \cdot x} \langle Q_b | j^\mu(0) | Q_a \rangle$$
$$= i q_\mu \langle Q_b | j^\mu(0) | Q_a \rangle e^{i(Q_b - Q_a) \cdot x}$$
$$= 0 \tag{11-16}$$

---

[3] Such factors can be eliminated only because the nucleons are real and on the "energy shell." This elimination cannot function for internal or "virtual" nucleon lines and the vertex function in that case contains many more terms.

11.2. THE ELASTIC SCATTERING OF ELECTRONS

from which it follows that the coefficient of the term in q must vanish. This leaves as the most general form:

$$O^\nu = \bar{F}_1^P(q^2)\, \gamma^\nu - \bar{F}_2^P(q^2)\, P^\nu\ . \tag{11-17}$$

From the condition eq. (11-13), it at once follows that the form factors $\bar{F}_1^P$ and $\bar{F}_2^P$ are both real for $q^2 \leq 0$.

Again, using the Dirac equation we have:

$$\bar{u}_N(\vec{Q}_b)\, P^\nu\, u_N(\vec{Q}_a) = \bar{u}_N(\vec{Q}_b)\, \{-2M_N\, \gamma^\nu + i\, \sigma^{\nu\mu} \cdot q_\mu\}\, u_N(\vec{Q}_a) \tag{11-18}$$

where $\sigma^{\mu\nu} = \frac{i}{2}[\gamma^\mu, \gamma^\nu]$, and it is common practice to write O^ν in the alternative form:

$$O^\nu = F_1^P(q^2)\, \gamma^\nu + F_2^P(q^2)\, i\, (\sigma^{\nu\mu} \cdot q_\mu) \tag{11-19}$$

where:

$$F_1^P(q^2) \equiv \bar{F}_1^P(q^2) - 2M_p\, F_2^P(q^2)\ . \tag{11-20}$$

The cross section is most easily calculated from the first of the forms we have given for O^ν, eq. (11-17). It is found that:

$$\frac{d\sigma}{d\Omega} = \left(\frac{d\sigma}{d\Omega}\right)_M \left[\{F_1^P(q^2)\}^2 - q^2\{F_2^P(q^2)\}^2 - \frac{q^2}{2M_p^2}\{F_1^P(q^2)\right.$$
$$\left. + 2M_p\, F_2^P(q^2)\}^2\, \tan^2\left(\frac{\theta}{2}\right)\right]\ . \tag{11-21}$$

This is known as the Rosenbluth formulae (Rosenbluth, 1950) and represents the most general form that the elastic electron-proton differential cross section can take, provided that only single photon exchange is important.

The ratio $\left(\frac{d\sigma}{d\Omega}\right)/\left(\frac{d\sigma}{d\Omega}\right)_M$, at a fixed value of q^2, is a linear function of $\tan^2(\theta/2)$. This linearity can be used to check that the assumption of single photon exchange is adequate, because two-photon exchange corrections can introduce terms with a different angular dependence.

The Low Momentum Transfer Limit

While the theoretical determination of the form factors requires a detailed knowledge of the structure of the photon-nucleon vertex, the limit of these functions as $q^2 \to 0$ is easily found, by studying the interaction of a proton moving at low velocities with an external electromagnetic field $A^\mu(x)$. The interaction energy density between the electromagnetic field and a current is of the form $A(x) \cdot j(x)$, and therefore for a single proton the interaction energy[4] is:

$$H' = e \int d\vec{x}\, A(x) \cdot <\vec{Q}_b|j(x)|\vec{Q}_a> \frac{(2\pi)^3}{\sqrt{E(|\vec{Q}_a|)\, E(|\vec{Q}_b|)}} \quad (11\text{-}22)$$

Making a space-time translation, to transform $j(x)$ to $j(0)$, the energy H' can be put in the form:

$$H' = e \int d\vec{x}\, A(x) \cdot <\vec{Q}_b|j(0)|\vec{Q}_a>\, e^{i(Q_b - Q_a)\cdot x} \frac{(2\pi)^3}{\sqrt{E(|\vec{Q}_a|)\, E(|\vec{Q}_b|)}} . \quad (11\text{-}23)$$

If we work in the frame of reference for which $\vec{Q}_b = -\vec{Q}_a$ (the Breit frame), we have that $E(|\vec{Q}_a|) = E(|\vec{Q}_b|)$ and:

$$H' = \frac{(2\pi)^3 e}{E(|\vec{Q}_a|)} \int d\vec{x}\, A(x) \cdot <-\vec{Q}_a|j(0)|\vec{Q}_a>\, e^{-i\vec{q}\cdot\vec{x}} . \quad (11\text{-}24)$$

In the limit $\vec{Q}_a$ and $\vec{q} \to 0$ using the explicit representation of the spinors, u_N in term of Pauli two component spinors, and working to first order in q, the following relations can be found:

---

[4] We wish to represent interactions of a single proton with the electromagnetic field and to take the volume of quantization as the unit volume. As the single particle state vectors $|Q>$ are normalized so that they correspond to $(2\pi)^{-3} E(|Q|)$ particles per unit volume, the factor $(2\pi)^3 \{E(Q_a)\, E(Q_b)\}^{-\frac{1}{2}}$ must be introduced in eq. (11-22). The interaction in (11-22) must be confined to unit volume and this avoids the apparent singularity of the final expression (11-28).

11.2. THE ELASTIC SCATTERING OF ELECTRONS

$$\bar{u}_N(\vec{Q}_b) u_N(\vec{Q}_a) \to \chi_b^\dagger \chi_a ; \quad \bar{u}_N(\vec{Q}_b) \gamma^0 u_N(\vec{Q}_a) \to \chi_b^\dagger \chi_a ;$$

$$\bar{u}_N(\vec{Q}_b) \vec{\gamma} \, u_N(\vec{Q}_a) \to \chi_b^\dagger \, \frac{i}{2}(\vec{\sigma} \times \vec{q}) \chi_a \tag{11-25}$$

where χ_b and $\chi_{a'}$ are two component Pauli spinors, and $\chi_b^\dagger \chi_a = \delta_{ba}$. Expressing $\langle \vec{Q}_b | j(0) | \vec{Q}_a \rangle$ in terms of 0^ν, by eq. (11-12) and using eq. (11-17) and the above relations, the low energy limit of H' is:

$$H' = \frac{e}{2M_p} \int d\vec{x} \, \vec{A}(x) \cdot \{2i \chi_b^\dagger (\vec{\sigma} \times \vec{q}) \chi_a\} \{F_1^P(0) + 2M_p F_2^P(0)\} \, e^{-i\vec{q} \cdot \vec{x}}$$

$$+ e \int d\vec{x} \, A^0(x) \, F_1^P(0) \, e^{-i\vec{q} \cdot \vec{x}} \delta_{ba} \, . \tag{11-26}$$

In an electrostatic field $(\vec{A} = 0)$, the second term indicates that the proton interacts in the limit $q^2 \to 0$ with an effective charge $e F_1^P(0)$. Since the proton has, in fact, a charge e, it follows that:

$$F_1^P(0) = 1 \, . \tag{11-27}$$

Now choosing $\vec{A}(x)$ to represent a magnetic field (with $A^0 = 0$), the interaction can be written in the form:

$$H' = \frac{e}{2M_p} \int d\vec{x} \, [F_1^P(0) + 2M_p F_2^P(0)] \, 2i \, \chi_b^\dagger (\vec{\sigma} \cdot \vec{B}) \chi_a \tag{11-28}$$

where $\vec{B} = \text{curl } \vec{A}$, and which represents the interaction energy of a magnetic dipole $\vec{\mu}$, with the magnetic field $\vec{B}$ where:

$$\mu = \frac{e}{2M_p} [F_1^P(0) + 2M_p F_2^P(0)] \equiv \mu_0 (1 + K_p) \, . \tag{11-29}$$

Since $F_1^P(0) = 1$, the first term in μ, $\mu_0 = \frac{e}{2M_p}$ represents the magnetic moment associated with a particle satisfying the Dirac equation. Assuming that the electromagnetic coupling is minimal [obtained by the replacement $p^\mu \to (p^\mu - \frac{e}{c} A^\mu)$], the second term represents the anomalous magnetic moment of the proton, K_p. If K_p is written in units of nuclear nagnetons[5] $\left(\frac{e}{2M_p}\right)$, we can set:

$$F_2^P(0) = K_p/2M_p \ . \tag{11-30}$$

For this reason $F_1^P(q^2)$ is often called the Dirac and $F_2^P(q^2)$ the Pauli form factor.

Electric and Magnetic Form Factors

In the analysis of the experimental data it is convenient to introduce form factors $G_E^P(q^2)$ and $G_M^P(q^2)$ which, in the Breit frame of the proton, are the distributions of charge and magnetic moment in the limit $q^2 \to 0$. These form factors are defined as (Sachs, 1962):

$$G_E^P(q^2) = F_1^P(q^2) + \frac{q^2}{2M_p} F_2^P(q^2)$$

$$G_M^P(q^2) = F_1^P(q^2) + 2M_p F_2^P(q^2) \ . \tag{11-31}$$

The important property of G_E^P and G_M^P is that they diagonalize the differential cross sections in the sense that no terms in (G_E^P, G_M^P) appear:

$$\left(\frac{d\sigma}{d\Omega}\right) = \left(\frac{d\sigma}{d\Omega}\right)_M \left[\frac{\{G_E^P(q^2)\}^2}{(1-q^2/4M_p^2)} + \frac{q^2}{4M_p^2}\{G_M^P(q^2)\}^2 \right.$$

$$\left. \times \left\{2\tan^2(\theta/2) - \frac{1}{(1-q^2/4M_p^2)}\right\}\right] \ . \tag{11-32}$$

It can be shown that the first term corresponds to scattering with no spin flip of the proton and the second corresponds to scattering with spin flip (Yennie et al., 1957).

---

[5] The observed value of K_p is $K_p \approx 1.79$. The anomalogous magnetic moment of the neutron K_n is found to be $K_n = -1.91$.

11.2. THE ELASTIC SCATTERING OF ELECTRONS

As large angles, near 180°, the term in G_M dominates the cross section and may be determined from the experimental data directly. Measurements at small angles (at the same value of q^2) then determine the combination $[\{G_E^P(q^2)\}^2 + (q^2/4M_p)\{G_M^P(q^2)\}^2]$.

Configuration Space

From our discussion of the nonrelativistic scattering of electrons by a charge distribution, it would at first seem as if we could identify the Fourier transforms of G_E^P and G_M^P with the spatial distributions of the charge and magnetic moment of the proton. This is not the case, because the Breit frame, which is obtained by applying a boost, depending on q, to the rest frame, is a different frame for each value of q, and it is difficult to attach a definite meaning to the Fourier transforms. Nevertheless these transforms exist and can be used to *define* mean square radii for the various form factors. From eq. (11-3) the mean square radius associated with any one of the form factors is seen to be, if $F(0) \neq 0$:

$$<r^2> = -6 \frac{1}{F(0)} \frac{\partial F(q^2)}{\partial |q^2|}\bigg|_{q^2=0} . \tag{11-33}$$

If $F(0) = 0$, the *definition* of $<r^2>$ is taken to be:

$$<r^2> = -6 \frac{\partial F(q^2)}{\partial |q^2|}\bigg|_{q^2=0} . \tag{11-34}$$

It should be noticed that $<r^2>$ is not necessarily a positive quantity, and it has the sign of $\{-F'(0)\}$.

The Experimental Data

The first measurements of electron scattering by protons in which the structure of the nucleon was observed were reported by Hofstadter and McAllister (1955), using electrons with energies up to 236 MeV. They found the mean square charge radius to be about $0.5\ f^2$ and obtained some evidence for scattering by the magnetic moment distribution.[6] In the ensuing period, electron beams of greater and greater energy have become available and independent measurements at several laboratories have resulted in determinations of the proton form factors for values of q up to 25 (GeV/c). The experimental methods and the methods of data analysis employed have been discussed by Wilson (1967) in a useful review, and we shall not enter into an account of these matters here.

The data obtained are summarized in Fig. 11-3. The form factors $G_E{}^p$ and $G_M{}^p$ decrease monotonically from their limiting values at $q^2 = 0$, and fit closely the following empirical formulae (Dunning et al., 1966):

$$G_E{}^p(q^2) = \left[\frac{1}{1-(|q^2|/0.71)}\right]^2$$

$$G_M{}^p(q^2) = (1+K_p)\, G_E{}^p(q^2) \tag{11-35}$$

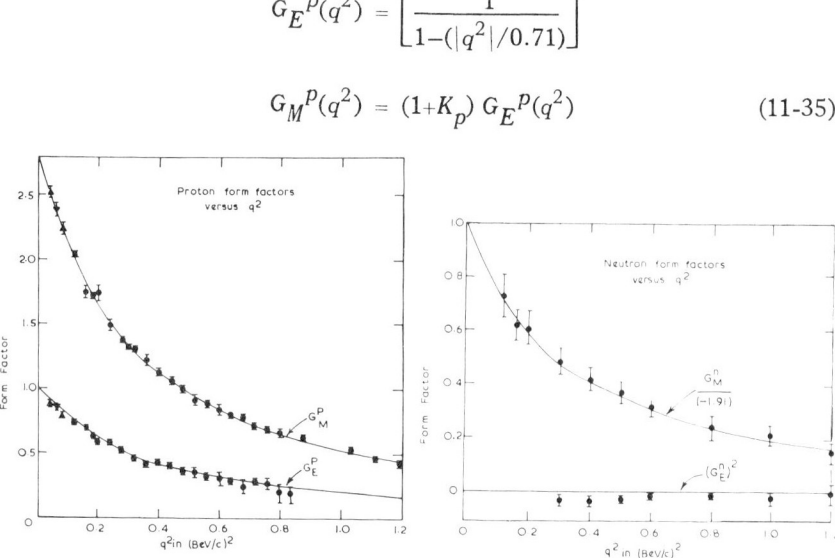

Fig. 11-3. (a) The proton form factors $G_E{}^p$ and $G_M{}^p$ as functions of $|q^2|$ in $(GeV/c)^2$; (b) the neutron form factors $G_E{}^n$ and $G_M{}^n$ as functions of $|q^2|$ in $(GeV/c)^2$ (after Hughes et al., 1965).

---

[6] Hofstadter (1963) has published a collection of all the important papers on the application of electron scattering to nucleon and nuclear structure problems up to 1962, with a valuable commentary that allows the development of the subject to be traced.

11.2. THE ELASTIC SCATTERING OF ELECTRONS

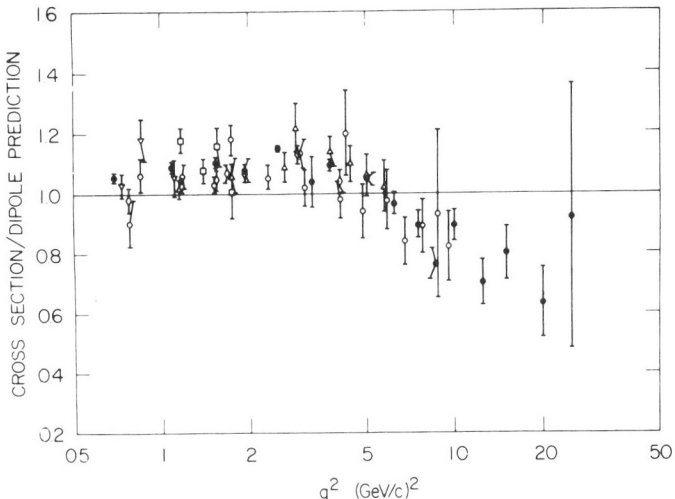

Fig. 11-4. The ratio of the electron-proton scattering cross section and that calculated from the dipole formulae of eq. (11-35) (after Coward et al., 1968).

where q^2 is in units of $(GeV/c)^2$. The mean square radius of the form factor G_E^p is then about $(0.66\ f^2)$. This simple fit is remarkably good (10-15%) over a region in which the form factors decrease by three orders of magnitude (see Fig. 11-4). However, above $10\ (GeV/c)^2$, G_E^p appears to decrease somewhat faster than $1/q^4$. The fit is of course not unique and an equally good fit (Drell et al., 1967), up to $15\ (GeV/c)^2$, can be obtained with the form:

$$G_E^p(q^2) = 27.8\ e^{-|q/0.040|^{\frac{1}{2}}}$$

$$G_M^p(q^2) = (1+K_p)\ G_E^M(q^2) \qquad (11\text{-}36)$$

although above $15\ (GeV/c)^2$, this form may decrease too rapidly.

Up to the highest electron energies so far employed, the ratio $\left(\frac{d\sigma}{d\Omega}\right)/\left(\frac{d\sigma}{d\Omega}\right)_M$ is a linear function of $\tan^2(\theta/2)$. This suggests that no contribution from two-photon exchange has been observed, but the test is inconclusive because under certain circumstances two-photon contributions can again give the Rosenbluth formulae. Details may be found in papers by Gourdin (1966), Flamm and Kummer (1963), and Drell and Fubini (1959). The contributions, if any, arising from two-photon exchange, can be studied by comparing the differential cross sections for the scattering of electrons and positrons of the same energy. If the contributions to the transition matrix from one- and two-photon exchange are $T^{(1)}$ and $T^{(2)}$, then $T^{(2)}$ has the same sign for both electron and positron scattering, but the sign of $T^{(1)}$ depends on the sign of the electronic charge. The ratio R, between the cross section σ_+ for positron and σ_- for electron scattering, is then:

$$R = \frac{\sigma_+}{\sigma_-} = \left|\frac{T^{(1)} + T^{(2)}}{-T^{(1)} + T^{(2)}}\right|^2 \approx 1 + \left|\frac{2\,\text{Re}\,T^{(2)}}{T^{(1)}}\right|. \qquad (11\text{-}37)$$

At present the data (Mar et al., 1968) which is available for $q^2 < 5$ (GeV/c) is consistent with $R = 1.0$.

Corresponding experiments have been carried out with μ meson beams, attaining momentum transfers up to $|q^2| = 0.9$ (GeV/c)^2. No difference has been found between electron and μ^+ or μ^- scattering in this momentum region (Ledermann et al., 1968).

11.3. THE NEUTRON FORM FACTORS AND THE ISOTOPIC SPIN DECOMPOSITION

The Neutron Form Factors

Form factors F_i^n and $G_{E,M}^n$ for the neutron can be defined in the same way as for the proton. Because the total charge of the neutron is zero, in the limit $q^2 \to 0$, G_E^n and F_1^n must vanish, while $F_2^n(0)$ and $G_M^n(0)$ are related to the magnetic dipole moment of the neutron ($K_n = -1.91$ nucleon magnetrons):

11.3. THE NEUTRON FORM FACTORS

$$F_1^n(0) = G_E^n(0) = 0$$

$$F_2^n(0) = \frac{1}{2M_n} \quad G_M^n(0) = \frac{K_n}{2M_n} \quad . \tag{11-38}$$

The electron-neutron differential cross section cannot be measured directly, but can be abstracted from measurements of the electron-deuteron scattering cross section. To do this it is necessary to assume the impulse approximation, which, in its simplest form, states that the electron-deuteron scattering amplitude is the sum of the amplitudes for the scattering of electrons by a free proton and free neutron, which however possess a momentum distribution determined by the wave-function of the deuteron. This treatment can be applied to both elastic electron-deuteron scattering and the electrodisintegration of the deuteron and, provided that the deuteron wave-function and the electron-proton scattering amplitudes are known, the electron-neutron form factors can be calculated. As far as the nucleon motion is concerned, this method is nonrelativistic, but corrections (which turn out to be small) can be applied to allow for relativistic effects. A detailed discussion of the impulse approximation applied to electron-deuteron scattering can be found in the book by Gourdin (1966). To attempt to provide a relativistic theory, dispersion relations have been developed for the electron-deuteron scattering amplitude (Durand, 1961; Gross, 1964), but these are not yet in a form suitable for a quantitative discussion of the data.

Although the study of the electron-deuteron interaction is the principle method of determining the neutron form factors, information for small q^2, can be obtained by studying the scattering of thermal neutrons by heavy elements. The cross section is determined by simultaneous scattering from the nucleus and from the atomic electrons, and it is possible to separate the two effects. The best measurement of this kind (Krohn and Ringo, 1966) shows that $\left.\frac{\partial G_E^n}{\partial |q^2|}\right|_{q^2=0} \approx K_n/4M_n^2$ and the square radius of the form factor G_E^n is:

$$\langle(r_E^n)^2\rangle = (-0.108 \pm 0.006)\, f^2 \quad . \tag{11-39}$$

Since:

$$\left.\frac{\partial F_1^n}{\partial |q^2|}\right|_{q^2=0} = \left.\left(\frac{\partial G_E}{\partial |q^2|} - \frac{1}{4M_n^2} G_M\right)\right|_{q^2=0} \tag{11-40}$$

this shows that:

$$\frac{\partial F_1^n}{\partial |q^2|} \approx 0, \text{ at } q^2 = 0 \quad .$$

The data obtained from electron-deuteron scattering can be summarized by the empirical fit:

$$G_E^n(q^2) = -\frac{q^2 K_n}{4M_n^2} G_E^p(q^2)$$

$$G_M^n(q^2) = K_n G_E^p(q^2) = \frac{K_n}{1+K_p} G_M^p(q^2) \tag{11-41}$$

where $G_E^p(q^2)$ is given by eq. (11-35). This fit indicates that the Dirac form factor of the neutron vanishes for all q^2, $F_1^n(q^2) = 0$, and shows that the electric form factor G_E^n is small for all momenta transfers subject to experiment. In fact, the data would be compatible with $G_E^n = 0$, were it not for the fact that the thermal neutron scattering data shows that $\left(\frac{\partial G_E^n}{\partial |q^2|}\right)$ is finite at $q^2 = 0$. However in the region of large q^2 [up to 7 $(\text{GeV}/c)^2$] a better fit is obtained with (Budnitz et al., 1968):

$$G_E^n(q^2) = -\left(\frac{q^2/4M_n^2}{1+q^2/M_n^2}\right) K_n G_E^p(q^2) \tag{11-42}$$

which seems more reasonable, in that the ratio (G_E^n/G_E^p) reaches a limiting value as $q^2 \to \infty$. With this fit $F_1^n(q^2)$ is small, but nonzero.

11.3. THE NEUTRON FORM FACTORS

Isotopic Spin Decomposition

When discussing the photoproduction (Chapter 2) of π mesons, we saw that the electromagnetic current could be divided into two terms, one of which, j_S^μ, transformed as an isotopic spin scalar and one of which, $j_{V_3}^\mu$, transformed like the third component of an isotopic spin vector:

$$j^\mu(x) = j_S^\mu(x) + j_{V_3}^\mu(x) \ . \tag{11-43}$$

The matrix elements of the current between single-proton or single-neutron states M^p, M^n are then related to matrix elements of the scalar and vector parts of the current M^S, M^{V_3}, by:

$$M^p = M^S + M^{\frac{1}{3}}$$

$$M^n = M^S - M^{\frac{1}{3}} \ . \tag{11-44}$$

From these relations we can define scalar and vector form factors F_i^S, $F_i^{V_3}$ such that:

$$F_i^S = \frac{1}{2}\{F_i^n + F_i^p\}$$

$$F_i^{V_3} = \frac{1}{2}\{F_i^p - F_i^n\} \tag{11-45}$$

and functions $G_{E,M}^{S,V_3}$ can be defined similarly. The zero momentum limits for the new form factors are:

$$F_1^{V_3}(0) = F_1^S(0) = \frac{1}{2}$$

$$F_2^S(0) = (K_n + K_p)/4M_N = 0.06$$

$$F_2^{V_3}(0) = (K_p - K_n)/4M_N = 1.85 \ . \tag{11-46}$$

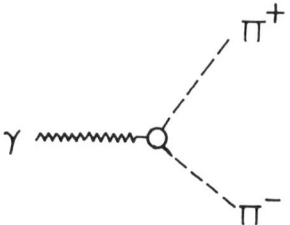

Fig. 11-5. The $\gamma \to 2\pi$ vertex.

It is important to notice that because the isotopic spin of the deuteron is $I = 0$, the experiments on elastic scattering of electrons by deuterons determine the isotopic scalar quantities $F_i^S(q^2)$ only. The electrodisintegration cross sections, on the other hand, depend on both F_i^S and $F_i^{V_3}$.

11.4. THE THEORY OF FORM FACTORS

The vertex function, eq. (11-12), representing the interaction of a virtual photon with a proton, is related by crossing symmetry with the vertex for the interaction $\gamma \to p + \bar{p}$, which is obtained by interchanging the photon and nucleon lines in the Feynman diagram of Fig. 11-5. This vertex is:

$$<\bar{Q}_a, Q_b \text{ (out)}|j^\mu(0)|0> \qquad (11\text{-}47)$$

where $\bar{Q}_a$ and Q_b are the energy-momentum vectors of the antiproton and proton respectively.

Energy-momentum conservation requires that:

$$q = \bar{Q}_a + Q_b$$

where q is the energy-momentum vector of the photon, so that in terms of our previous variables, $\bar{Q}_a = -Q_a$. The invariant t is then given by:

$$t = (\bar{Q}_a + Q_b)^2 = q^2 .$$

11.4. THE THEORY OF FORM FACTORS

The state vector $|0>$ on the right-hand side of eq. (11-47) represents the vacuum. Since $<\bar{Q}_a, Q_b \text{ (out)}|$ is not a free particle state, the boundary conditions must be specified, and in this case we choose the "out" state vector, which is specified by requiring that as $t \to +\infty$, $<\bar{Q}_a, Q_b \text{(out)}|$ must coincide with the free particle state $<\bar{Q}_a, Q_b|$.

For the time reversed process:

$$p + \bar{p} \to \gamma$$

the appropriate wave-function for the proton-antiproton state is specified by boundary conditions at $t = -\infty$, so that the "in" or $|\psi^+>$ state must be used and the appropriate matrix element is:

$$<0|j^\mu(0)|Q_a, \bar{Q}_b \text{ (in)}> \ . \tag{11-48}$$

Starting from eq. (11-47), and using the same invariance principles that enabled us to obtain eq. (11-19) from eq. (11-12), we find that:

$$<\bar{Q}_a, Q_b \text{(out)}|j^\mu(0)|0> = (2\pi)^{-3} M_N e\, \bar{u}(\vec{Q}_b) [F_1^{\,P}(q^2)\gamma^\mu + i\sigma^{\mu\nu} q_\nu F_2^{\,P}(q^2)] v(\vec{Q}_a) \ . \tag{11-49}$$

Crossing symmetry (see Chapter 4) tells us that the form factors $F_i^{\,P}$ are the same analytic functions as those previously introduced in eq. (11-19); but whereas for the vertex $\gamma + N \to N$ we were concerned with the region $q^2 < 0$, for the vertex $\gamma \to N + \bar{N}$, we are interested in the region $q^2 > -(2M_N)^2$, where a real two-nucleon state can be produced. For $q^2 < 0$, we saw that the functions $F_i^{\,P}$ were real, but this is not the case for $q^2 \geq -4M_N^2$. The point $q^2 = 4M_N^2$ is the threshold for the production of a real $(N+\bar{N})$ state and is therefore, from the general arguments given in Chapters 3 and 4, a branch point of the functions $F_i^{\,P}(q^2)$. This threshold is clearly not the lowest threshold involving strongly interacting particles. The lightest strong interacting particle is the pion and the two lowest thresholds are at $q^2 = (2M_\pi)^2$ and $q^2 = (3M_\pi)^2$, corresponding to the processes $\gamma \to 2\pi$ and $\gamma \to 3\pi$. It follows that the functions $F_i^{\,P}(q^2)$ develop imaginary parts for $q^2 \geq -4M_\pi^2$.

Perturbation theory can be used to expand the vertex function for the process $\gamma \to p + \bar{p}$ in the usual way, and the intermediate states of lowest mass, which may be expected to be most important, are the 2π and 3π states:

$$\gamma \to 2\pi \to p + \bar{p}$$

$$\gamma \to 3\pi \to p + \bar{p} \quad . \tag{11-50}$$

Such an expansion is not very useful in itself, as it fails to converge for strong couplings; however, by studying each term in the series, it can be shown to all orders that the form factors are real analytic functions of $t(= q^2)$ apart from the unitarity cut along the real axis $4M_\pi^2 \leq t \leq \infty$, that is required because of the branch points at $4M_\pi^2, 9M_\pi^2, \ldots, 4M_N^2$ associated with the thresholds for the production of 2π's, 3π's ..., $N + \bar{N}$, ... Assuming that $F_i{}^P(t) \to 0$ as $|t| \to \infty$, we can write the dispersion relations:

$$F_i{}^P(t) = \frac{1}{\pi} \int_{4M_\pi^2}^{\infty} \frac{\operatorname{Im} F_i{}^P(t') \, dt'}{(t'-t)} \quad . \tag{11-51}$$

Alternatively, if $F_i{}^P(t)$ does not vanish at infinity, but approaches a constant value, we can make a "subtraction" at $t = 0$ and write:

$$F_i{}^P(t) = F_i{}^P(0) + \frac{1}{\pi} t \int_{4M_\pi^2}^{\infty} \frac{\operatorname{Im} F_i{}^P(t') \, dt'}{(t'-t) t'} \quad . \tag{11-52}$$

The asymptotic properties of the form factors $F_i{}^P(t)$ are not known, but perturbation theory suggests that $F_1{}^{n,P}(t)$ satisfies the subtracted form of the dispersion relation eq. (11-52), while $F_1{}^{n,P}(t)$ satisfies the unsubtracted form, eq. (11-51).

Calculation of the Absorptive Parts of the Form Factors

The absorptive parts of the form factors, $\operatorname{Im} F_i{}^P(t)$, can be computed by examining the imaginary part of the vertex function $<\bar{Q}_a, Q_b \, (\text{out})| j^\mu(0) | 0 >$, since:

11.4. THE THEORY OF FORM FACTORS

$$\text{Im} <\bar{Q}_a, Q_b \text{ (out)}|j^\mu(0)|0> = (2\pi)^{-3} M_N e \, \bar{u}(\vec{Q}_b) [\text{Im}\{F_i{}^P(t) + 2M_N F_2{}^P(t)\}\gamma^\mu$$
$$+ \text{Im} \, F_2{}^P(t) \, (\bar{Q}_a^\mu - Q_b^\mu)] \, v(\vec{Q}_a) \; . \quad (11\text{-}53)$$

It is possible to proceed by using the reduction formula on the left-hand side of eq. (11-53), in the spirit of our discussion of πN forward scattering in Chapter 3; or the following argument, based on time reversal invariance, can be given. Under time reversal the electromagnetic current changes sign $\vec{j} \to -\vec{j}$, while the charge density remains unaltered $j^0 \to j^0$. The time reversal state vector to $<\bar{Q}_a Q_b; m_a, m_b; \text{(out)}|$ is $|-\bar{Q}_a, -Q_b; -m_a, -m_b; \text{(in)}>$, where the spins of the proton a and antiproton b, m_a, m_b, have been displayed. Accordingly we have:

$$<\bar{Q}_a, Q_b; m_a, m_b; \text{(out)}|\vec{j}|0> = - <0|\vec{j}|-\bar{Q}_a, -Q_b; -m_a, -m_b; \text{(in)}>$$

$$<\bar{Q}_a, Q_b; m_a, m_b; \text{(out)}|j^0|0> = <0|j^0|-\bar{Q}_a, -Q_b; -m_a, -m_b; \text{(in)}> . \quad (11\text{-}54)$$

Under a rotation combined with a space inversion (parity operation), the state vector $|-Q_a, -\bar{Q}_b; -m_a, -m_b; \text{(in)}>$ can be transformed to $|Q_a, \bar{Q}_b; m_a, m_b; \text{(in)}>$. At the same time, under the parity transformation we have that $\vec{j} \to -\vec{j}$ and $j^0 \to j^0$, and under the rotation $\vec{j}$ transforms like a (three-dimensional) vector and j^0 as a scalar. This allows us to write, using eq. (11-54), that:

$$<\bar{Q}_a, Q_b \text{ (out)}|j^\mu|0> = <0|j^\mu|\bar{Q}_a, Q_b \text{ (in)}>$$
$$= <\bar{Q}_a, Q_b \text{ (in)}|j^\mu|0>^* \quad (11\text{-}55)$$

where in the second step the hermiticity of j^μ has been used, and where the labels indicating the dependence on m_a, m_b have been dropped. The imaginary parts of the vertex $<\bar{Q}_a, Q_b \text{ (out)}|j^\mu|0>$ can be written, using this result, in the form:

$$2i \, \text{Im} <\bar{Q}_a, Q_b \text{ (out)}|j^\mu|0> = <\bar{Q}_a, Q_b \text{ (out)}|j^\mu|0> - <\bar{Q}_a, Q_b \text{ (in)}|j^\mu|0>$$
$$= \sum_n [<\bar{Q}_a, Q_b \text{ (out)}|n> - <\bar{Q}_a, Q_b \text{ (in)}|n>] <n|j^\mu|0>$$
$$(11\text{-}56)$$

where in the second line a complete set of states of state $|n>$ has been introduced. Choosing the states $|n>$ to be "in" or "+" states, we see that:

$$<\bar{Q}_a, Q_b \text{ (out)}|n> = S_{ab;n}$$

$$<\bar{Q}_a, Q_b \text{ (in)}|n> = \delta_{ab;n} \quad (11\text{-}57)$$

and since $S_{ab;n} - \delta_{ab;n} = 2i\, T_{ab;n}\, \delta^4(\bar{Q}_a + Q_b - P_n)$ where P_n is the energy-momentum vector of the state n we find that:

$$\text{Im} <\bar{Q}_a, Q_b\text{(out)}|j^\mu|0> = \sum_n T_{ab;n} <n|j^\mu|0> \times \delta^4(\bar{Q}_a + Q_b - P_n). \quad (11\text{-}58)$$

We have already observed that the intermediate state of lowest energy is the 2-pion state, and below the threshold at $t = 4M_\pi^2$, the imaginary part of the form factor vanishes, since the argument of the energy-conserving delta function cannot then vanish. The transition matrix $T_{ab;n}$ that occurs in eq. (11-58) represents a scattering in which a proton and an antiproton are produced from the state n. In the case of the lowest intermediate state, this is the transition amplitude for the process:

$$\pi + \pi \to N + \bar{N} .$$

For $4M_\pi^2 < t < 4M_N^2$, the term $T_{ab;2\pi}$ is an analytic continuation of the amplitude for the actual physical process, since the physical threshold is at $t = 4M_N^2$. The term $T_{ab;2\pi}$ can be expressed in terms of the invariant amplitudes $A^\pm(s,t)$ and $B^\pm(s,t)$ introduced into Chapter 3, as the process $\pi + \pi \to N + \bar{N}'$ is related to $\pi + N \to \pi + N$ by crossing symmetry. The factor $<n|j^\mu|0>$ that occurs in eq. (11-58) is vertex for the photon interaction with the state n; for example if n is the 2-pion state, then $<2\pi|j^\mu|0>$ is the vertex shown in Fig. 11-5, which is just the electromagnetic form factor of the pion.

Selection Rules

Selection rules which classify the possible intermediate states n can easily be derived by examining the vertex $<n|j^\mu|0>$. This matrix element

11.4. THE THEORY OF FORM FACTORS

will vanish unless the state $<n|$ has the same quantum numbers as the state $j^\mu|0>$, which represents a single photon state. These quantum numbers are: (a) angular momentum $J = 1$, with negative parity; (b) negative charge conjugation parity, $C = -1$; (c) zero charge; (d) total isotopic spin either $I = 0$, or $I = 1$. (States $|n>$ with $I = 0$ contribute to the isoscalar form factors F_i^S and those with $I = 1$ contribute to the isovector form factors $F_i^{V_3}$. Further information is obtained by using the concept of G parity introduced in Chapter 1, where it was shown that a state of n pions with total charge zero has G parity $G = (-1)^n$. The vacuum has positive G parity, while since the charge conjugation parity of the photon is $C = -1$ and the isoscalar part of j^μ does not change the isotopic spin, we must have, if j_S^μ and $j_{V_3}^\mu$ are the scalar and vector parts of j^μ, that:

$$G j_S^\mu G^{-1} = -j_S^\mu . \qquad (11\text{-}59a)$$

On the other hand as $j_{V_3}^\mu$ transforms like the third component of an isovector, we can show that:

$$G j_{V_3}^\mu G^{-1} = j_{V_3}^\mu . \qquad (11\text{-}59b)$$

It follows that multipion states with an odd number of pions contribute to the isoscalar part of the form factor, and states with an even number to the isovector part.

The lowest mass states that contribute to the isoscalar form factor are therefore 3-pion states, with $I = 0$, $J = 1$, odd parity, and zero charge, and the lower limit of the dispersion integral for $F_{1,2}^S$ is $t = 9M_\pi^2$. For the isovector form factor the lowest state is 2-pion state with $I = 1$, $J = 1$, odd parity, and zero charge, with a threshold of $t = 4M_\pi^2$, with the next contribution arising from a 4-pion state at a threshold of $16M_\pi^2$.

Vector Dominance of Form Factors

In the physical region for electron scattering ($t < 0$), we have seen that a natural assumption is that the smaller values of t' dominate the integrands of the dispersion integrals, eqs. (11-51) and (11-52), because it is in this region that the denominators $(t'-t)$ are smallest. From this, it follows that in the calculation of Im $F_1(t)$, the most important intermediate states in the sum eq. (11-58) are the lowest mass states, which are the 2-state for the isovector part of the form factor and the 3-pion state for the isoscalar part. Early calculations showed that such assumptions led to mean square radii of the charge and magnetic momentum distributions that were much too small, unless the pion-pion interaction was extremely strong, and later a detailed calculation of the isovector form factors led Frazer and Fulco (1960) to postulate the existence of a resonance in the $J = 1, I = 1$, 2-pion system. This resonance corresponds to a spin-one, negative parity particle with $J = 1$, and has been observed as the ρ meson in other processes, with a mass of $M_\rho = 770$ MeV and a width of ~ 110 MeV.

Nambu (1957) and Chew (1960), from different points of view, postulated the existence of a 3-spin resonant state with $J = 1, I = 0$, and odd parity, which would dominate the intermediate states in the calculation of the isoscalar form factor. Subsequently two such mesons have been discovered the ω, with mass $M_\omega = 783$ MeV and width $\Gamma_\omega \sim 16$ MeV, and the ϕ with mass $M_\phi = 1020$ MeV and width $\Gamma_\phi \sim 4$ MeV. The hypothesis of vector dominance is that the only important intermediate states are those due to the vector mesons and that the "background" of other states can be neglected. The situation is illustrated in Fig. 11-6, where the double line indicates a vector meson in the intermediate state.

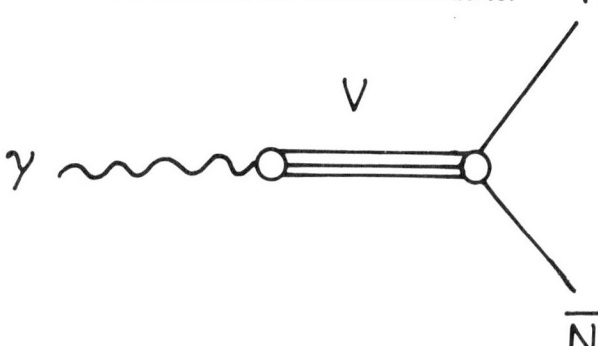

Fig. 11-6. Intermediate vector meson state in pion photoproduction.

11.4. THE THEORY OF FORM FACTORS

The predicted shape of the form factors is found easily in the extreme approximation in which the widths of the resonant vector particles are neglected. In this case from eq. (11-58) we see that if the intermediate state n is discrete (stable), and is of mass M_V, then $\text{Im} \langle \bar{Q}_a, Q_b | j^\mu | 0 \rangle$ is proportional to $\delta(t-M_V^2)$; accordingly from eq. (11-53) the contributions to the function $\text{Im } F_i(t)$ are of the form:

$$\text{Im } F_i(t) = C_i^V \delta(t-M_V^2) \tag{11-60}$$

where the C_i^V are real constants, which are proportional to the strength of the V-$N\bar{N}$ and V-γ couplings.

If we assume that F_1^{S,V_3} satisfies the subtracted dispersion relation, eq. (11-52), with $F_1^{S,V_3}(0) = \frac{1}{2}$ and F_2 the unsubtracted form, eq. (11-51), we have immediately that:

$$F_1^{V_3}(t) = \frac{1}{2} + \frac{1}{\pi} \frac{t\, C_1^\rho}{M_\rho^2(M_\rho^2-t)}$$

$$F_2^{V_3}(t) = \frac{1}{\pi} \frac{C_2^\rho}{(M_\rho^2-t)}$$

$$F_1^S(t) = \frac{1}{2} + \frac{t}{\pi}\left\{\frac{C_1^\omega}{M_\omega^2(M_\omega^2-t)} + \frac{C_1^\phi}{M_\phi^2(M_\phi^2-t)}\right\}$$

$$F_2^S(t) = \frac{t}{\pi}\left\{\frac{C_2^\omega}{(M_\omega^2-t)} + \frac{C_2^\phi}{(M_\phi^2-t)}\right\}. \tag{11-61}$$

The constant C_2^ρ, and one of the constants C_2^ω, C_2^ϕ can be eliminated if the conditions of eq. (11-46) are imposed that $F_2^S(0) = -0.06$, $F_2^{V_3}(0) = 1.85$. It is then possible to attempt to fit the experimental data by varying the remaining four constants. An acceptable fit at lower energies $|t| < 1.2$ (GeV/c)^2 can be found in the case of the isoscalar functions, F_i^S (or $G_{E,m}^S$); but the isovector form factors $F_i^{V_3}$ (or $G_{E,m}^{V_3}$) can only be fitted by these forms if the mass of the ρ resonance is left as a free parameter (Hughes et al., 1965). The value of M_ρ obtained is 580 MeV considerably lower than the observed value (760 MeV). This difficulty is certainly in part due to the zero width approximation, which is expected to be poor in the case of the ρ resonance, which has the substantial width of $\sim$ 100 MeV. Instead of treating M_ρ as a parameter, it is possible to add a second pole term to the expressions for $F_i^{V_3}$ proportional to $(M_X^2-t)^{-1}$. In this case a good fit can be found with M_X = 1200 MeV. This could be taken as evidence for the existence of a second vector meson with $I = 1$, and this mass, but no other evidence has been found for such a particle. Although the zero width model is an oversimplification, and it would be unwise to press the interpretation of the values of the constants C_i^V in terms of coupling constants for the V-$N\bar{N}$ vertices, its success does indicate the general reasonableness of the idea that the vector meson exchange is of great importance in mediating the photon-hadron interaction.

11.5. THE PION FORM FACTOR

From eq. (11-58) we have seen that if we retain only the 2-pion intermediate state in the calculation, the absorptive part of the isovector form factor is a product of the $\gamma \to 2\pi$ vertex function $<\pi^+ \pi^- (\text{out})|j^\mu|0>$ and the matrix element $T^*_{ab,2\pi}$ which represents the $\pi^+ + \pi^- \to N + \bar{N}$ interaction. The vertex function $<\pi^+ \pi^- (\text{out})|j^\mu|0>$ is proportional to the electromagnetic form factor of the pion. If k_1 and k_2 are four-momenta of the two pions, the only independent 4-vector quantities on which this vertex function can depend are (k_1-k_2) and (k_1+k_2), and so (omitting isotopic spin labels):

$$<\pi^+ \pi^- (\text{out})|j^\mu|0> = [(k_1+k_2)^\mu G_\pi(t) + (k_1-k_2)^\mu F_\pi(t)] e \quad (11\text{-}62)$$

where $t = (k_1+k_2)^2$.

11.5. THE PION FORM FACTOR

The first term in eq. (11-62) can be shown to vanish by using the requirement that the current is conserved:

$$\frac{\partial}{\partial x^\mu} j^\mu = 0 . \tag{11-63}$$

If P is the energy-momentum operator, this condition can be expressed as $[P_\mu, j^\mu] = 0$, from which we have:

$$\langle k_1, k_2 \text{ (out)} | [P_\mu, j^\mu] | 0 \rangle = (k_1+k_2)_\mu \langle k_1, k_2 \text{ (out)} | j^\mu | 0 \rangle$$

$$= e \left[(k_1+k_2)^2 G_\pi(t) + (k_1^2 - k_2^2) F_\pi(t) \right] = 0 .$$

The second term vanishes automatically since $k_1^2 = k_2^2 = M_\pi^2$, but the first only vanishes if we set $G_\pi(t)$ equal to zero; $F_\pi(t)$ is then the pion form factor. The pion must interact with a weak field like a point charge of magnitude e, and by following arguments similar to those leading to eqs. (11-26) and (11-27), we find that:

$$F_\pi(0) = 1 . \tag{11-64}$$

The form factor $F_\pi(t)$ will also be analytic in the t plane cut along the real axis from the 2-pion threshold at $4M_\pi^2$, and will obey a dispersion relation like that of eq. (11-51).

By inserting a set of intermediate states n we find the unitarity condition that:

$$\langle k_1 k_2 \text{ (out)} | j^\mu | 0 \rangle = \sum_n \langle k_1, k_2 \text{ (out)} | n \text{ (in)} \rangle \langle n \text{ (in)} | j^\mu | 0 \rangle . \tag{11-65}$$

For $4M_\pi^2 < t < 16M_\pi^2$, the only possible intermediate state is a 2-pion state, $|k_1' k_2' \text{ (in)}\rangle$, and this of course must have the quantum numbers $I = 1, J = 1$. As the pions have zero spin, J is equal to the orbital angular momentum and the π-π system is characterized in this region by a real phase shift $\delta_{J=1}$, describing elastic scattering.

The term $<k_1 k_2 (\text{out})| k_1' k_2' (\text{in})>$ is just the S-matrix for $\pi\pi$ scattering in the $J = 1, I = 1$ state and is equal to $\exp(2i\delta_1)$. We then have from eq. (11-65) that:

$$F_\pi(t) = e^{2i\delta_1} F_\pi^*(t) . \qquad (11\text{-}66)$$

Although this equation is exact only for $4M_\pi^2 < t < 16M_\pi^2$, it may be adopted as an approximation over the whole of the cut from $4M_\pi^2$ to infinity. It is then possible by combining the dispersion relation with eq. (11-66), to express $F_\pi(t)$ in terms of an integral depending on $\delta_1(t)$ only. Assuming that $\pi\pi$ scattering in this channel is dominated by the ρ resonance, a Breit-Wigner resonance form can be assumed for the $\pi\pi$ elastic scattering amplitude $T_{\pi\pi \to \pi\pi}$, which gives:

$$T_{\pi\pi \to \pi\pi} = \frac{e^{i\delta} \sin\delta}{k^3} = \frac{\Gamma_\rho M_\rho}{k^3 [(M_\rho^2 - t) - i \Gamma_\rho M_\rho]} \qquad (11\text{-}67)$$

where k is the momentum in the 2-pion center of mass system:

$$k = \left(\frac{t}{4} - M_\pi^2\right)^{\frac{1}{2}} \qquad (11\text{-}68)$$

and Γ_ρ is the width of the resonance. To take account of the p-wave threshold behavior, we may write $\Gamma = \gamma k^3/k_\rho^3$, where $k_\rho = (M_\rho^2/4 - M_\pi^2)^{\frac{1}{2}}$, and γ is the reduced width.

Then writing $F_\pi(t) = A\, T_{\pi\pi \to \pi\pi}$, where A is taken to be a constant, fixed by the condition $F_\pi(0) = 1$, we obtain:

$$F_\pi(t) = \left[\frac{M_\rho^2 + M_\rho \gamma_\rho (M_\pi^3/k_\rho^3)}{(M_\rho^2 - t) - i \gamma_\rho M_\rho (k^3/k_\rho^3)}\right] . \qquad (11\text{-}69)$$

Below the threshold at $t = 4M_\pi^2$, k becomes imaginary and $F_\pi(t)$ is real.

11.5. THE PION FORM FACTOR

Just as the nucleon form factor can be measured through the observation of electron scattering by nucleons, in principle, the pion form factor can be measured by scattering charged pions by atomic electrons. This experiment is only feasible for small negative values of t, but fortunately experiments involving colliding beams of electrons and positrons have been performed which are capable of reaching energies appropriate to the observation of the ρ resonance, and which allow the form factor to be measured by observing the reaction:

$$e^+ + e^- \to \rho^0 \to \pi^+ + \pi^- \ .$$

Virtually all the ρ^0 mesons decay into the π^+-π^- channel, and the branching ratio of the decay into the e^+-e^- channel is $B = 6.6 \times 10^{-5}$. Treating, as usual, the electron-photon interaction in first order as in eq. (11-9), the cross section for this reaction is found to be:

$$(e^+ e^- \to \pi^+ \pi^-) = \frac{8\pi a^2}{3} k^3(t)^{-\frac{5}{2}} |F_\pi(t)|^2 . \qquad (11\text{-}70)$$

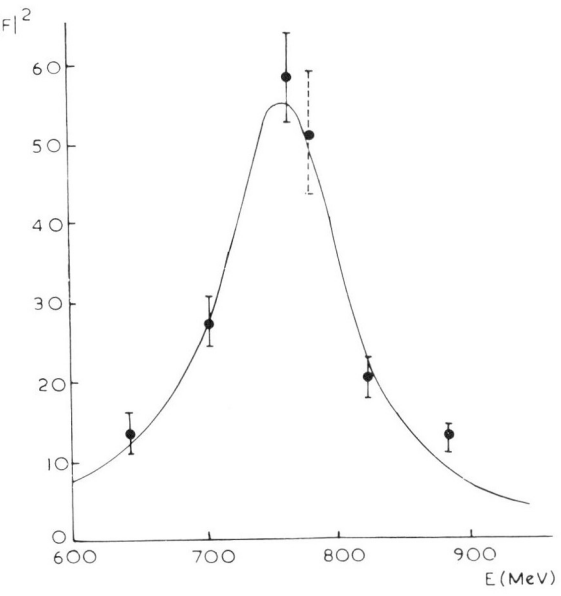

Fig. 11-7. The cross section for ρ^0 production in e^+-e^- scattering.

Experiments over the energy range from 600 to 900 MeV have been performed by Augustin et al., (1969a, b, c) and the results of the latter experiments are shown in Fig. (11-7), which beautifully demonstrates a resonance shape due to the production of the ρ^0-meson.

On fitting the expression eq. (11-70) to the experimental curve the resonance position and width are found to be approximately M_ρ = 760 MeV and γ = 112 MeV. Very slightly different values are obtained if different versions of the Breit-Wigner formula are employed.

Since the π^+-π^- system is in a p-wave, and if it is assumed that scattering is entirely through the ρ^0 state with no background, the cross-section at the peak must be represented by (see Chapter 3):

$$\sigma(\text{MAX}) = \frac{3\pi}{k^2} \frac{\Gamma_\rho \Gamma_\rho'}{\Gamma^2_{\text{tot}}} \approx \frac{3\pi}{k^2} B \qquad (11\text{-}71)$$

where Γ_ρ is the partial width for the decay $\rho^0 \to \pi^+ + \pi^-$, Γ_ρ' is the width for $\rho^0 \to e^+ + e^-$ and $\Gamma_{\text{tot}} = \Gamma_\rho + \Gamma_\rho'$. The value of Γ_ρ' obtained by measuring the peak cross section is found to be 7.4 MeV.

ω and ϕ Production

The isoscalar vector mesons can also be produced in a colliding beam experiment between electrons and positrons. The important decay channels for the ω-meson are the $\rho\pi$ channel (91%) and the $\pi\gamma$ channel (9%):

$$e^+ + e^- \to \omega \to \rho + \pi \to 3\pi$$
$$\searrow$$
$$\pi + \gamma \; . \qquad (11\text{-}72)$$

The ϕ meson is strongly coupled to the $K\bar{K}$ system, which accounts for 79% of the decays, while the remaining 21% are to the 3π channel:

$$e^+ + e^- \to \phi \to K + \bar{K}$$
$$\searrow$$
$$\pi + \pi + \pi \; . \qquad (11\text{-}73)$$

11.5. THE PION FORM FACTOR

The experiments of Augustin et al., (1969b,c) can be fitted by a Breit-Wigner resonance formula and both the total widths Γ_ω, Γ_ϕ and the partial widths Γ_ω', Γ_ϕ' for the decay back into the e^+-e^- channel, can be measured. The results are:

$$\Gamma_\omega = 12 \text{ MeV} \qquad \Gamma_\omega' = 0.94 \text{ keV}$$

$$\Gamma_\phi = 4 \text{ MeV} \qquad \Gamma_\phi' = 1.64 \text{ keV}$$

Phenomenological Couplings

As we have indicated, many features of the coupling between the electromagnetic field and the strong interacting particles (hadrons) can be explained if it is assumed that the interaction proceeds entirely through vector meson states (V) of zero charge $(\rho^0, \omega^0, \phi^0)$.

Phenomenological couplings can then be written for the γ-V vertex, and for the vertex coupling the vector mesons to the strongly interacting particles. There is some evidence that the coupling between a vector meson and the hadrons is "universal," in that, for example, the couplings for $\rho^0 \to 2\pi$ and $\rho^0 \to N + \bar{N}$ are the same.[7]

Let us start by considering the coupling between the photon and the neutral ρ^0-meson. We may write:

$$<\rho|j^\mu|0> = e[\epsilon^\mu(\rho) M_\rho^2 f_{\rho\gamma}^{(1)}(t) + q^\mu f_{\rho\gamma}^{(2)}(t)] \qquad (11\text{-}74)$$

where j^μ is the electromagnetic current, $\epsilon^\mu(\rho)$ is the polarization vector of the ρ and q^μ the energy momentum vector of the ρ-meson, and the functions $f_{\rho\gamma}^{(1)}(t)$ are form factors. To satisfy the current conservation condition $\frac{\partial}{\partial x_\mu} j^\mu = 0$, we have that $q_\mu <\rho|j^\mu|0> = 0$, which immediately shows that $f_{\rho\gamma}^{(2)}(t) \equiv 0$. The ρ-γ coupling constant is then *defined* as:

---

[7] Important discussions of the theory of vector dominance have been given by Sakurai (1960) and Kroll et al., (1967).

$$f_{\rho\gamma} = f_{\rho\gamma}^{(1)}(t=M_\rho^2) \ . \tag{11-75}$$

The form factor $f_{\rho\gamma}^{(1)}$ is not a constant, because the γ-ρ vertex must vanish at $t = 0$, to ensure that the γ-ρ interaction does not provide a finite contribution to the mass of the photon. In the same way we can examine the coupling at the $\rho \to 2\pi$ vertex. If η is the current (source) of the π^0 field, we define the coupling $f_{\rho\pi\pi}$:

$$\langle \pi^+ | \eta | \rho^0 \rangle = \epsilon^\mu(\rho) (k_1 - k_2)_\mu \, f_{\rho\pi\pi} \tag{11-76}$$

where k_1 and k_2 are the energy momentum vectors of the pions and the matrix element is evaluated at the point $(q-k_1)^2 = M_\pi^2$. This vertex can be thought of as arising from an interaction of the form:

$$H_{\rho\pi\pi} = f_{\rho\pi\pi} \, \rho^\mu \cdot \varphi \times \frac{\partial \varphi}{\partial x_\mu} \tag{11-77}$$

where ρ_μ is the ρ-meson vector field.

As we are now treating the ρ^0 as a stable particle, rather than as a 2π state of finite width, the pion form factor is given by [eq. (11-67)]:

$$F_\pi(t) = \frac{M_\rho^2}{M_\rho^2 - t} \ . \tag{11-78}$$

In terms of the couplings $f_{\rho\gamma}$ and $f_{\rho\pi\pi}$, we can also write:

$$F_\pi(t) = \frac{f_{\rho\gamma} \, f_{\rho\pi\pi}}{M_\rho^2 - t} \tag{11-79}$$

and on comparison we see that:

$$f_{\rho\gamma} = \frac{M_\rho^2}{f_{\rho\pi\pi}} \ . \tag{11-80}$$

11.5. THE PION FORM FACTOR

This is equivalent to the requirement that $F_\pi(0) = 1$, which ensures that the pion behaves like a point particle of charge e in a weak electromagnetic field.[8]

The constants $f_{\rho\pi\pi}$ can now be determined from the decay width of the ρ into (2π)'s. From perturbation theory we find that:

$$\gamma_\rho = \frac{1}{12} \left(\frac{f^2_{\rho\pi\pi}}{4\pi}\right) M_\rho \left(1 - \frac{4M_\pi^2}{M_\rho^2}\right)^{\frac{3}{2}} \qquad (11\text{-}81)$$

and using the observed width (111 MeV), we find that $\frac{f^2_{\rho\pi\pi}}{4\pi} = 2.10 \pm 0.11$. Alternatively, the width for the $\rho \to e^+ - e^-$ decay, can be written using the relation eq. (11-80):

$$\gamma_\rho' = \frac{1}{3} \alpha^2 \left(\frac{f_{\rho\pi\pi}}{4\pi}\right)^{-2} M_\rho \;. \qquad (11\text{-}82)$$

Using the value $\gamma_\rho' = 4.7$ MeV, we find that:

$$\frac{f^2_{\rho\pi\pi}}{4\pi} = 1.86 \pm 0.18 \;. \qquad (11\text{-}83)$$

Turning now to the nucleon form factors, we can again discuss the isovector parts in terms of the ρ-$N\bar{N}$ and ρ-γ couplings. The vertex functions must be of the form:

$$<N|\bar{\eta}(0)|\rho^0> v(Q) = \bar{u}(N') \; v(Q) \left\{\frac{\gamma^\mu}{2} f^1_{\rho NN}(t) + i \sigma^{\mu\lambda} (Q+\bar{Q})_\lambda \; f^2_{\rho NN}(t)\right\}\epsilon_\mu$$

$$(11\text{-}84)$$

---

[8] In the gauge invariance theory of Kroll et al., (1967), which also leads to the identification (80), we have "field-current identities" in which the isovector vector part of the electromagnetic current is identified with the ρ field:

$$j_{V_3}^\mu = \frac{M_\rho^2}{f_{\rho\pi\pi}} \rho_\mu^0 \;.$$

The masses and coupling constants in these relations are always renormalized quantities.

and this is exactly similar to the coupling between the photon (which is also a vector particle) and the nucleon. In the calculation of the nucleon isovector form factors, we require $\text{Im} < N\bar{N}|j^\mu|0>$ for $t = M_\rho^2$, and we easily find that:

$$\text{Im } F_1^{V_3} = \frac{\pi}{2} f_{\rho\gamma} f^{(1)}_{\rho N\bar{N}} \delta(t-M_\rho^2)$$

$$\text{Im } F_2^{V_3} = \pi f_{\rho\gamma} f^{(2)}_{\rho N\bar{N}} \delta(t-M_\rho^2) \qquad (11\text{-}85)$$

and in terms of the constants C_i^V introduced in eq. (11-60):

$$C_1^\rho = \frac{1}{2} \pi f_{\rho\gamma} f^{(1)}_{\rho N\bar{N}}, \quad C_2^\rho = \pi f_{\rho\gamma} f^{(2)}_{\rho N\bar{N}} . \qquad (11\text{-}86)$$

To fit the data on the isovector form factors accurately, we saw that a form involving two poles was required, but a reasonable fit to the data, for small t only, can be found with:

$$C_1^\rho \approx \frac{3\pi}{2} M_\rho^2 .$$

Combining this with the relation eq. (11-80) for $f_{\rho\gamma}$ we find that:

$$\frac{f^{(1)}_{\rho N\bar{N}}}{f_{\rho\pi\pi}} \approx 1.5 . \qquad (11\text{-}87)$$

An alternative calculation, in which the nucleon form factors are related to the $N + \bar{N} \to \pi + \pi$ amplitude, which in turn is related by crossing to the $\pi + N \to \pi + N$ amplitude, suggests that (Donnachie and Hamilton, 1962):

$$\frac{f^{(1)}_{\rho N\bar{N}}}{f_{\rho\pi\pi}} \simeq 1.3 . \qquad (11\text{-}88)$$

It has been suggested (Sakurai 1960) that a less approximate calculation would yield the result $f_{\rho\pi\pi} = f^{(1)}_{\rho N\bar{N}} = f_\rho$. This could imply that the vector part of the coupling between the ρ-mesons and hadrons is given by:

11.5. THE PION FORM FACTOR

$$H_\rho = f_\rho\, P_\mu \cdot J^\mu \tag{11-89}$$

where J^μ is the conserved isospin current:

$$J^\mu = \bar\psi\, \gamma_\mu \frac{\tau}{2} \psi - \left(\varphi \times \frac{\partial}{\partial x_\mu}\varphi\right) + \ldots \ . \tag{11-90}$$

The contributions from the nucleons and pions are shown on the right-hand side, and further terms giving the contribution from hyperons, K mesons, and so on, must be added as indicated.

The Isoscalar Charges

The couplings of the ω and ϕ mesons to the photon field and the isoscalar parts of the nuclear form factors can be treated along similar lines. The coupling of the electromagnetic field to these particles can be analyzed using the ideas of SU3 invariance. It will be remembered from Chapter 6, that the physical particles ω and ϕ (which have the identical quantum numbers $Y = I_3 = 0$), can be considered as mixtures of an SU3 singlet ω_0 and the central component of an octet ϕ_0. If a mixing angle θ_V is defined, the wave-function of the physical particles can be written in terms of those for ϕ_0 and ω_0 as:

$$\omega = \omega_0 \cos\theta_V + \phi_0 \sin\theta_V$$
$$\phi = \omega_0 \sin\theta_V - \phi_0 \cos\theta_V \ . \tag{11-91}$$

From the relation:

$$Q = I_3 + \tfrac{1}{2} Y \tag{11-92}$$

it is a natural assumption that the electromagnetic current transforms in SU3 space like the combinations $\left(F^3 + \frac{1}{\sqrt{3}} F^8\right)$ (where the matrices $F^{(i)}$ were defined in Chapter 6), which is a $U = 0$ member of an octet. It is of course, possible that part of the electromagnetic current may transform like an SU3 singlet, but if we ignore this possibility, we see that the photon is coupled to the ϕ_0 only and not to the ω_0. This implies that if λ is the strength of the γ-ϕ_0 coupling, the physical particles ω and ϕ couple with strengths λ_ω and λ_ϕ, where:

$$\lambda_\omega = \lambda \sin\theta_V$$

$$\lambda_\phi = \lambda \cos\theta_V \ . \tag{11-93}$$

The mixing angle can be determined by comparing the decay rate γ_ω' for $\omega \to e^+ + e^-$, with γ_ϕ' for $\phi \to e^+ + e^-$. It is found that:

$$\frac{\gamma_\omega'}{\gamma_\phi'} = \tan^2\theta_V \left(\frac{M_\omega}{M_\phi}\right). \tag{11-94}$$

From the experimental results already quoted (Augustin et al., 1969a,b,c), it is found that $\theta_V \approx 40°$.

As the electromagnetic current transforms like a U-spin scalar, $U = 0$, it is coupled to the combination $(\sqrt{3}\,\rho^0 + \phi_0)$ but not to the combination $(\sqrt{3}\,\phi_0 - \rho^0)$ which is a member of a U-spin triplet. This suggests that the coupling constant λ is given by:

$$\lambda = \frac{1}{\sqrt{3}}\, f_{\rho\gamma} = \frac{1}{\sqrt{3}}\left(\frac{M_\rho^2}{f_\rho}\right). \tag{11-95}$$

The vertex between the ω- and ϕ-mesons and the strongly interacting particles can be discussed by supposing that just as the ρ is coupled to a conserved isospin current, the ϕ_0 is coupled to a conserved hypercharge current and the ω_0 to a conserved baryon current. The constants appearing in the isoscalar parts of the nucleon form factors, eq. (11-45), can then be related to these fundamental couplings. The details of the model have been given by Massam and Zichichi (1966a,b). A vector meson has both vector (γ^μ) and tensor $(\sigma^{\mu\lambda})$ couplings to the nucleon. For each of these is a mixture of F and D SU3 couplings for the ϕ_0 which is member of an octet, while for the ω_0 which is an SU3 singlet there

11.5. THE PION FORM FACTOR

is just one coupling. Overall there are six coupling constants to be determined, apart from f_ρ and the mixing angle θ_V, which also appear. Clearly not all these constants can be determined from the data without imposing additional assumptions, and not very definite results can be obtained.

Form Factors at Large Momentum Transfers

Although the idea of the vector dominance provides some explanation of the behavior of form factors for small values of the momentum transfer, the model fails to explain the rapid rate of decrease of the form factors at large negative values of t, which experimentally appear to be of the form $F(t) \propto [1 + a|t|]^{-2}$. Such a rate of decrease is clearly at variance with a single pole model, which requires a behavior like $F(t) \propto [1 + a|t|]^{-1}$. If $F(t)$ decreases as fast as $|t|^{-2}$, then a superconvergence relation:

$$\int_{M_\pi}^\infty \text{Im } F(t) \, dt = 0$$

must be satisfied by the spectral functions Im $F(t)$. If there are a number of vector mesons, Im $F(t)$ is of the form:

$$\text{Im } F(t) = \sum_i C_i \, (t - M_i)$$

so that the residues C_i must satisfy a relation such as:

$$\sum_i C_i = 0$$

and even if there are contributions from as yet undiscovered particles, such as the hypothetical ρ', there is nothing in the theory to explain this constraint on the residues.

In looking elsewhere for an explanation of the behavior at large q^2 (large $-|t|$), models have been put forward in which the nucleon is represented as a composite particle. It has been shown (Drell et al., 1967) that the form factor of a composite particle in a nonrelativistic theory has a behavior at large momentum transfers which is determined by the behavior of the interaction potential at the origin. With a potential that behaves like $1/r$ at the origin, the form factors are asymptotic to $1/q^{2\ell+2}$ where ℓ is angular momentum of the state concerned, while potentials that are more singular at the origin can lead to an exponential decrease of the form factor. The problem is then how to construct a relativistic version of such a theory. A possible line of attack is (see Amati et al., 1968; Ball and Zacharisan, 1968; Ciafaloni and Menotti, 1968) to use the Bethe-Salpeter equation to represent the structure of the composite particle, and it appears that it is possible to obtain rapidly decreasing form factors in this way.

11.6. DEEP INELASTIC SCATTERING AND THE PARTON MODEL

So far we have explored at some length, the information that can be obtained from elastic scattering of electrons by nucleons. We now turn to a consideration of the very interesting measurements of inelastic scattering of electrons by protons obtained principally at the Stanford Linear Accelerator. In these experiments only the scattered electron is observed in the final state, so the process involved is:

$$e^- + p \rightarrow e^- + \text{``anything''} \qquad (11\text{-}96)$$

where "anything" stands for some hadronic state with baryon number $B = 1$ and charge $Q = +e$. It will be assumed that the reaction takes place through single photon exchange as in the Feynman diagram shown in Fig. 11-8.

Kinematics and the Transition Matrix

As earlier in this chapter, we denote the initial and final energy-momentum vectors of electron by P_a and P_b, Q_a is the initial energy-momentum vector of proton target, and Q_b is the total energy-momentum vector of the final hadronic state. Conservation of energy and momentum requires:

11.6. DEEP INELASTIC SCATTERING

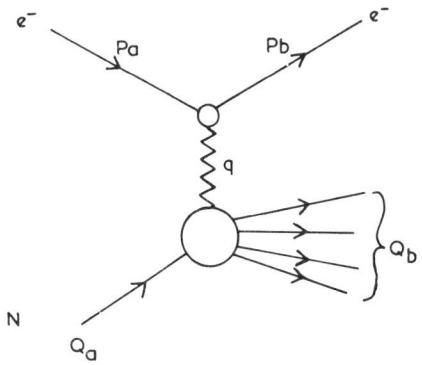

Fig. 11-8. Inelastic scattering of electrons by protons through single photon exchange.

$$P_a + Q_a = P_b + Q_b$$
$$q = P_a - P_b = Q_b - Q_a \tag{11-97}$$

where q is the energy-momentum vector of the exchanged photon.

In addition we have the mass constraints:

$$P_a^2 = P_b^2 = M_e^2 \qquad Q_a^2 = M_N^2$$
$$Q_b^2 = W^2 \tag{11-98}$$

where W is the mass of the final hadronic state. At a given laboratory energy of the incident electron E, if we do not observe the hadronic final states, the differential cross section is a function of the energy E' and angle of scattering θ of the electron in the final state only. It is useful to express the experimental results in terms of two Lorentz scalars in place of E' and θ, and these can be chosen to be q^2 and ν where $\nu = (q \cdot Q_a/M_N)$. In the laboratory system $Q_a = (M_N, 0, 0, 0)$ so that:

$$\nu = q_0 \tag{11-99}$$

the energy of the exchanged virtual photon, or in terms of E and E' and θ:

$$\nu = (E-E')$$

$$q^2 = -4E\,E'\sin^2\frac{\theta}{2} \tag{11-100}$$

where we have used the high energy approximation in which M_e^2 has been neglected compared with E or E'.

Just as in the case of elastic scattering [see eqs. (11-8) and (11-9)] the cross section matrix will contain as a factor the matrix element of the electromagnetic current operator with respect to single electron states (which can be calculated to lowest order in α), and also a further factor $<Q_b(n)|j^\nu(0)|Q_a>$, the matrix element of the current with respect to the initial and final hadronic states. Since we are summing over all possible final hadronic states, n, consisted with conservation of energy and momentum, the cross section will depend on the quantity $W_{\mu\nu}$ where:

$$W_{\mu\nu} = \frac{1}{4\pi a}\sum_n <Q_a|j^\mu(0)|Q_b(n)><Q_b(n)|j^\nu(0)|Q_a> \times (2\pi)^3\,\delta^4(Q_b(n)-Q_a-q). \tag{11-101}$$

Following a procedure analogous to that outlined in (section 11-2) for elastic scattering, it can be shown, using Lorentz and gauge invariance, that $W_{\mu\nu}$ depends on two functions W_1 and W_2 of ν and q^2 where:

$$W_{\mu\nu} = W_1(\nu,q^2)\left(\delta_{\mu\nu} - \frac{q_\mu q_\nu}{q^2}\right)$$

$$+ \frac{W_2(\nu,q^2)}{M_N^2}\left(Q_{a_\mu} - \left(\frac{Q_a\cdot q}{q^2}\right)q_\mu\right)\left(Q_{a_\nu} - \left(\frac{Q_a\cdot q}{q^2}\right)q_\nu\right). \tag{11-102}$$

(For details see de Forest and Walecka, 1966.)

The measured differential cross section can now be expressed as:

$$\frac{d^2\sigma}{dE'd\Omega} = \frac{4a^2\,E'^2}{q^4}\left[2\,W_1(\nu,q^2)\sin^2\left(\frac{\theta}{2}\right) + W_2(\nu,q^2)\cos^2\left(\frac{\theta}{2}\right)\right]. \tag{11-103}$$

11.6. DEEP INELASTIC SCATTERING

Unfortunately this differential cross section is not directly measured by the experiments, since the accelerated electrons radiate photons and this contribution to the experimental cross section must be removed before a comparison can be made with the formula given in eq. (11-103). These radiative corrections, although important, are well understood and can be calculated accurately. Details can be consulted in papers by Maximan (1969) and Mo and Tsai (1969). Having performed the radiative corrections, one can obtain series of spectra at different incident energies and angles of scattering.

Experimental Spectra and Scaling

The general characteristics of the observed spectra can be summarized as follows (Gilman, 1969). For small values of $t \equiv -q^2$, (t less than ~ 2 GeV^2) and at fixed W, the spectra exhibit prominent resonance peaks, which disappear rapidly as larger values of t are reached. At these larger values of t, the cross section is of the same order of magnitude as the Mott cross section for the scattering of an electron by a point charge. This is a much greater cross section than would be expected from scattering from a continuous charge distribution. If the cross section is plotted against t for a fixed value of W which is greater than 2 GeV, it decreases slowly like $1/t$, rather than the rapid $(1/t)^4$ behavior of the nucleon form factors.

For values of $t > 1$ GeV^2 and $W > 2$ GeV, the cross section data exhibit the property of scale invariance. This property, originally suggested to hold in the limit $\nu \to \infty$, $t \to \infty$ by Bjorken (1969), is that W_1 and νW_2 should depend only on the value of the dimensionless ratio $\omega = (2M_N \nu/t)$ and not on ν or t separately. This behavior is illustrated for the function νW_2 in Fig. 11-9, where for fixed $\omega = 4$, νW_2 is shown to be independent of t for $t > 1$.

The measurements as functions of E' and θ can be used to separate the contributions to the cross section from W_1 and W_2. It is usual in doing this to employ the two related quantities σ_T and σ_S where:

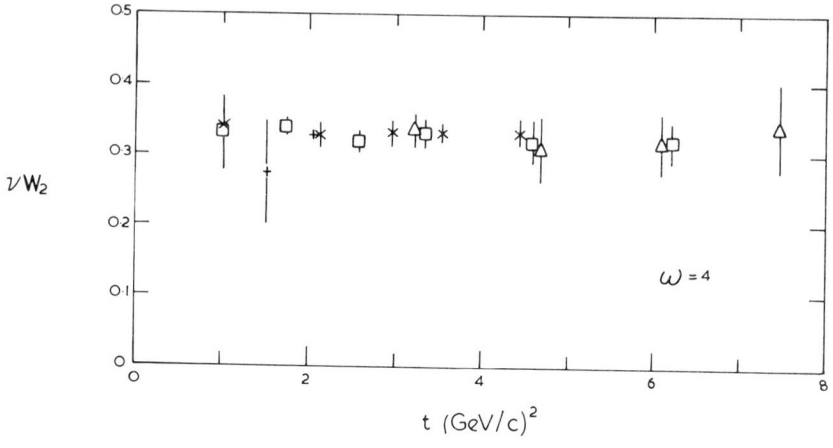

Fig. 11-9. The scale invariance of the function νW_2 (see text).

$$\sigma_T = \frac{4\pi^2 \alpha}{K} W_1 \qquad (11\text{-}104)$$

$$\sigma_S = \frac{4\pi^2 \alpha}{K}\left(\frac{t+\nu^2}{t}\right) W_2 - \sigma_T \qquad (11\text{-}105)$$

where $K = \left(\nu + \frac{t}{2M_N}\right) = \frac{W^2 - M_N^2}{2M_N}$. The quantities σ_T and σ_S can be interpreted (Gilman, 1968) as the total cross sections for the scattering of virtual photon ($mass^2 = q^2$) by nucleons, σ_T corresponding to transversely and σ_S to longitudinally polarized photons. Since a real photon has zero mass, at $q^2 = 0$, σ_T becomes equal to the total photoabsorption cross section (which has been measured up to 20 GeV) while σ_S vanishes. The experimental situation (summarized by Gilman, 1969) is that $R = \sigma_S/\sigma_T$ is small; $R = 0.18 \pm 0.05$ for $1 < \omega < 10$, and does not depend strongly on Ω or on ν or q^2.

Some data is also available on the inelastic scattering of electrons by deuterium, from which the electron-neutron cross section can be deduced. The data show clearly that $W_{2n} < W_{2p}$ over the range $1.5 < \omega < 6$, and that W_{2n} exhibits the scaling property. The ratio W_{2n}/W_{2p} varies slowly with ω, taking the value $\frac{2}{3}$ between $\omega = 3$ and $\omega = 4$. This shows

11.6. DEEP INELASTIC SCATTERING

clearly that the scattering process is not purely diffractive; in other words is not only due to a process involving the exchange of a Pomeron.

The Parton Model

The large magnitude of the cross section and the scaling behavior both can be explained if it is assumed that the nucleon is composed of a number of pointlike particles or "partons." In the deep inelastic region it is supposed that the conditions for the impulse approximation to be accurate apply and that the electron scatters from individual partons which can be treated as free particles during the collision. More precisely the model will apply in the infinite momentum frame of the nucleon $\vec{Q}_a \to \infty$, $(\vec{Q}_a) \gg M_N$ in which by time dilatation the motion of the partons is "frozen" (Feynman, 1969; Bjorken and Paschos, 1969). The partons are assumed to have momentum components in a direction normal to $\vec{Q}_a$ which are small enough to be neglected and in this case if the ith parton has a fraction X_i of the nucleons momenta $\vec{Q}_a$, then the energy momentum vector of the parton (neglecting the mass of the parton) is:

$$P_\mu^{(i)} = X_i Q_{a_\mu}. \tag{11-106}$$

An elementary calculation shows that for a *free* parton of unit charge and any spin:

$$W_2(\nu, q^2) = \delta\left(\nu - \frac{t}{2M_N}\right)$$
$$= M_N \delta\left(q \cdot P - \frac{1}{2} t\right) \tag{11-107}$$

while since $\sigma_t = 0$ for a spin 0 target and $\sigma_\ell = 0$ for a spin $\frac{1}{2}$ target,

$$W_1 = 0 \text{ for spin } 0$$
$$= \left(1 + \frac{\nu^2}{t}\right) W_2 \text{ for spin } \frac{1}{2}. \tag{11-108}$$

Making use of eq. (11-107), the contribution to W_2 from the tth parton with charge Q_i is:

$$W_2^{(i)} = X_i\, Q_i^2\, M_N\, \delta\!\left(X_i q \cdot Q_a - \tfrac{1}{2} t\right)$$

$$= Q_i^2\, \delta\!\left(\nu - \frac{t}{2 M_N X_i}\right) \tag{11-109}$$

where the factor X_i is necessary to insure that:

$$\lim_{E \to \infty} \frac{d\,\sigma^{(i)}}{d\,q^2} = Q_i^2 \left(\frac{4\pi a^2}{q^4}\right) \tag{11-110}$$

the Rutherford scattering cross section for scattering from a point charge. If $P(N)$ is the probability of N partons occurring and $f_N^{(i)}(X_i)$ is the distribution of the fraction of momentum carried by each parton:

$$\sum_N P(N) = 1$$

$$\int_0^1 f_N^{(i)}(X_i)\, dX_i = 1 \tag{11-111}$$

then since it is assumed that the contributions to W_2 add incoherently we find:

$$\nu W_2(\nu, q^2) = \sum_N P(N) \sum_{i=1}^{N} \int_0^1 dX_i\, f_N^{(i)}(X_i)\, \nu\, W_2^{(i)}(\nu, q^2; X_i)$$

$$= \sum_N P(N) \sum_{i=1}^{N} Q_i^2 \times f_N^{(i)}(X)\Bigg|_{X = -\frac{q^2}{2 M_N \nu}} \tag{11-112}$$

where in the last line the expression of eq. (11-109) for $W_2^{(i)}$ has been employed. This expression for νW shows the desired scaling behavior, being a function of $\frac{1}{\omega} = \frac{q^2}{2 M_N \nu}$ only, as well as having the magnitude expected from the point-charge assumption.

Various suggestions have been made as to the nature of the partons. Since the ratio R is small, the number of partons of spin 0 must be much less than those with spin $\frac{1}{2}$. This fact can be used to reject certain

11.6. DEEP INELASTIC SCATTERING

possibilities — for example, that the partons are bare pions and nucleons. An attractive possibility is that the partons are to be identified with quarks or a mixture of quarks and anti-quarks.

Sum Rules

If each parton has the same momentum distribution, then:

$$\int_0^1 X_i f_N^{(i)}(X) \, dX_i = \frac{1}{N} \qquad (11\text{-}113)$$

and from eq. (11-112), defining $F_2(X) = \nu W_2(\nu, q^2)$, $X = \frac{1}{\omega}$, we have:

$$\int_0^1 dX \, F_2(X) = \sum_N P(N) \sum_{i=1}^N \left(\frac{Q_i^2}{N}\right) = \overline{Q}^2 \qquad (11\text{-}114)$$

where Q is in the mean parton charge. If the partons are quarks and the parton is composed of three quarks:

$$\sum_{i=1}^3 Q_i^2 = \frac{4}{9} + \frac{4}{9} + \frac{1}{9} = 1 \qquad (11\text{-}115)$$

and:

$$\overline{Q}^2 = \frac{1}{3} \sum_{i=1}^3 Q_i^2 = \frac{1}{3}$$

Experimentally for protons:

$$\int_{0.1}^1 dx \, F_2(X) = 0.14 \, (\pm 15\%) \, . \qquad (11\text{-}116)$$

The contribution to the integral from $0 < X < 0.1$ (corresponding to $10 < \omega < \infty$) can be estimated by extrapolation and appears to be less than ~ 0.03. It follows that the integral is too small to be consistent with a single quark model. The model can be elaborated by supposing that configurations with neutral quarks or from an infinite set of quark-anti-quark pairs contribute to the structure functions. We shall not enter into these implications here, but reference can be made to the review by Gilman (1969) for further details.

REFERENCES

D. Amati et al., Abstract Vienna Conference (1968).

J. E. Augustin, D. Benaksas, J. C. Bigot, J. Buon, B. Delcourt, V. Gracio, J. Haissinski, J. Jeanjean, D. Lalanne, F. Laplande, J. Lefrancois, P. Lehmann, P. Marin, H. Nguyen Ngoc, J. Perez-y-Jorba, F. Richard, F. Rumpf, E. Silva, S. Tarvenier and D. Treille, Phys. Letters 28B, 503 (1969a).

J. E. Augustin, J. C. Bigot, J. Buon, J. Haissinski, D. Lalanne, P. Marin, H. Nguyen Ngoc, J. Perez-y-Jorba, F. Rumpf, E. Silva and S. Tarvenier, Phys. Letters 28B, 508 (1969b).

J. E. Augustin, J. C. Bigot, J. Buon, B. Delcourt, J. Haissinski, J. Jeanjean, D. Lalanne, P. Marin, H. Nguyen Ngoc, J. Perez-y-Jorba, F. Richard, F. Rumpf and D. Treille, Phys. Letters 28B, 513 (1969c).

J. S. Ball and F. Zacharisan, Phys. Rev. 170, 1541 (1968).

J. D. Bjorken, Phys. Rev. 179, 1547 (1969).

J. D. Bjorken and E. A. Paschos, Phys. Rev. 185, 1975 (1969).

R. J. Budnitz, J. Appel, L. Carroll, J. Chen, J. R. Dunning, M. Goitein, K. Hanson, D. Imrie, C. Mistretan, J. K. Walter and R. Wilson, Phys. Rev. 173, 1357 (1968).

G. F. Chew, Phys. Rev. Letters 4, 142 (1960).

M. Ciafaloni and P. Menotti, Nuovo Cimento 46, 162 (1966).

D. H. Coward, H. De Staebler, R. A. Early, J. Litt, A. Minten, L. W. Mo, K. H. Panofsky, R. E. Taylor, M. Breidenbach, J. I. Friedman, H. W. Kendall, N. Kirk, B. C. Barish, J. Mar and J. Pire, Phys. Rev. Letters 20, 292 (1968).

A. Donnachie and T. Hamilton, Phys. Rev. 133B, 678 (1964).

REFERENCES

S. D. Drell and S. Fubini, Phys. Rev. 113, 741 (1959).

S. D. Drell, A. C. Finn and M. H. Goldhaber, Phys. Rev. 157, 1402 (1967).

J. R. Dunning, K. W. Chen, A. A. Cone, G. Hartwig, N. G. Ramsey and J. K. Walker, Phys. Rev. 141, 1286 (1966).

L. Durand, Phys. Rev. 123, 1393 (1961).

R. P. Feynman, Phys. Rev. Letters 23, 1415.

D. Flamm and W. Kummer, Nuovo Cimento 28, 33 (1963).

T. de Forest and J. Walecka, Ann. Phys. 15, 1 (1966).

W. R. Fraser and J. R. Fulco, Phys. Rev. 117, 1603 (1960).

F. J. Gilman, Proc. 4th International Symposium on Electron and Photon Interactions at High Energies 1969, p. 177 (1969).

F. J. Gilman, Phys. Rev. 167, 1365 (1968).

M. Gourdin, *Diffusion des Electron de Haute Energie* (Mason et Cie, Paris 1966).

F. Gross, Phys. Rev. 134B, 405; 136B, 140 (1964).

R. H. Hofstadter and R. W. McAllister, Phys. Rev. 98, 47 (1955).

R. Hofstadter, Ann. Rev. of Nuclear Science 7, 231 (1957), in Springer Tracts in Modern Physics, Vol. 39, ed. G. Hohler (Springer-Verlag, Berlin, 1965).

R. H. Hofstadter, Nuclear and Nucleon Structure (W. A. Benjamin, New York, 1963).

E. B. Hughes, T. A. Griffy, M. R. Yeorian and R. Hofstadter, Phys. Rev. 139, B458 (1965).

V. E. Krohn and G. R. Ringo, Phys. Letters 18, 297 (1966).

N. M. Kroll, T. D. Lee and B. Sumino, Phys. Rev. 157, 1376 (1967).

L. M. Ledermann et al., Paper submitted to the Vienna Conference (1968).

J. Mar, B. Barish, J. Pire, D. H. Coward, H. Staebler, J. Litt, A. Minten, R. E. Taylor and M. Breidenbach, Phys. Rev. Letters 21, 482 (1968).

T. Massam and A. Zichichi, Nuovo Cimento 43, 1137; 44, 369 (1966).

L. C. Maximan, Rev. Mod. Phys. 41, 193 (1969).

L. W. Mo and Y. S. Tsai, Rev. Mod. Phys. 41, 405 (1969).

N. F. Mott, Proc. Roy. Soc. A 124, 425 (1929).

N. F. Mott and H. S. W. Massey, *Theory of Atomic Collisions* 3rd Ed., (Oxford Univ. Press, London, 1965).

Y. Nambu, Phys. Rev. 106, 1366 (1957).

M. N. Rosenbluth, Phys. Rev. 76, 615 (1950).

R. G. Sachs, Phys. Rev. 126, 2256 (1962).

J. J. Sakurai, Annals of Physics 11, 1 (1960).

J. Schwinger, Phys. Rev. 7b, 790 (1949).

R. Wilson, in *Particle Interactions at High Energies*, ed. T. W. Priest and L. L. J. Vick (Oliver and Boyd, Edinburgh, 1967).

D. R. Yennie, M. M. Levy and D. G. Ravenhall, Rev. Mod. Phys. (1957); and R. Wilson, Phys. Rev. 141, 1286, 29, 144 (1966).

APPENDIX A

ANGULAR MOMENTUM

In this appendix, some useful formulae and tables of quantities related to angular momentum and rotations are collected together for reference.

A.1. THE LEGENDRE POLYNOMIALS AND SPHERICAL HARMONICS

The Legendre polynomials $P_\ell(\mu)$ are defined by:

$$P_\ell(\mu) = \frac{1}{2^\ell \, \ell!} \frac{d^\ell}{d\mu^\ell} (\mu^2-1)^\ell \qquad \ell = 0, 1, 2, \ldots \qquad (A-1)$$
$$-1 \leq \mu \leq +1$$

with the recurrence relation:

$$(\ell+1) P_{\ell+1}(\mu) = (2\ell+1)\mu \, P_\ell(\mu) - \ell \, P_{\ell-1}(\mu) \ . \qquad (A-2)$$

They satisfy the orthonormality conditions:

$$\int_{-1}^{+1} P_\ell(\mu) \, P_{\ell'}(\mu) \, d\mu = \frac{2}{2\ell+1} \, \delta_{\ell\ell'} \qquad (A-3)$$

and special values are:

$$P_0(\mu) = 1, \quad P_1(\mu) = \mu, \quad P_2(\mu) = \frac{1}{2}(3\mu^2-1) \ . \qquad (A-4)$$

The associated Legendre functions $P_\ell(\mu)$ are defined by:

$$P_\ell^m(\mu) = (1-\mu^2)^{m/2} \frac{d^m}{d\mu^m} P_\ell(\mu), \quad m = 0, 1, 2, \ldots \ell \tag{A-5}$$

with:

$$P_\ell^0(\mu) \equiv P_\ell(\mu) \tag{A-6}$$

and the recurrence relation:

$$(\ell-1-m) P_{\ell+1}^m(\mu) = (2\ell+1)\mu P_\ell^m(\mu) - (\ell+m) P_{\ell-1}^m(\mu) . \tag{A-7}$$

They satisfy the orthonomality relations:

$$\int_{-1}^{+1} P_\ell^m(\mu) P_{\ell'}^m(\mu) \, d\mu = \left(\frac{2}{2\ell+1}\right) \frac{(\ell+m)!}{(\ell-m)!} \delta_{\ell\ell'} . \tag{A-8}$$

It is on occasion useful to employ the derivatives of the associated Legendre function, $\frac{d}{d\mu}[P_\ell^m(\mu)]$, which satisfy the recurrence relation:

$$(1-\mu^2) \frac{d}{d\mu}[P_\ell^m(\mu)] = (\ell+1)\mu P_\ell^m(\mu) - (\ell+1-m) P_{\ell+1}^m(\mu) . \tag{A-9}$$

The spherical harmonics $Y_{\ell,m}(\theta,\phi)$ are defined by:

$$Y_{\ell,m}(\theta,\phi) = (-1)^m \left[\frac{2\ell+1}{4\pi} \frac{(\ell-m)!}{(\ell+m)!}\right]^{\frac{1}{2}} P_\ell^m(\cos\theta) e^{im\phi} \quad \begin{array}{l} \ell = 0, 1, 2 \ldots \\ m = 0, 1, 2, \ldots \ell \end{array}$$

and:
$$\tag{A-10}$$

$$Y_{\ell,-m}(\theta,\phi) \equiv (-1)^m Y_{\ell,m}^*(\theta,\phi) . \tag{A-11}$$

They satisfy the orthonomality conditions:

$$\int_{-1}^{+1} \int_0^{2\pi} Y_{\ell',m'}^*(\theta,\phi) Y_{\ell,m}(\theta,\phi) \, d\phi \, d(\cos\theta) = \delta_{\ell\ell'} \delta_{mm'} . \tag{A-12}$$

A.2. THE CLEBSCH-GORDAN COEFFICIENTS

The spherical harmonics are eigenfunctions of L^2 and L_z, where $\vec{L}$ is the orbital angular momentum operator, belonging to the eigenvalues $\ell(\ell+1)$ and m (where the units are such that $\hbar = 1$). The phase convention is observed that:

$$L_\pm Y_{\ell,m}(\theta,\phi) \equiv (L_x \pm iL_y) Y_{\ell,m}(\theta,\phi)$$

$$= + [\ell(\ell+1) - m(m\pm 1)]^{\frac{1}{2}} Y_{\ell,m\pm 1}(\theta,\phi) . \quad \text{(A-13)}$$

The Legendre polynomials obey the following addition theorem. If (θ_1,ϕ_1) and (θ_2,ϕ_2) define two directions in space, such that $\theta_1 + \theta_2 < \pi$ and γ is the angle between these directions, then:

$$\cos\gamma = \cos\theta_1 \cos\theta_2 + \sin\theta_1 \sin\theta_2 \cos(\phi_1 - \phi_2) .$$

The addition theorem then reads that:

$$P_\ell(\cos\gamma) = \sqrt{\frac{4\pi}{2\ell+1}} \sum_{m=-\ell}^{\ell} Y_{\ell,m}(\theta_1,\phi_1) Y_{\ell,-m}(\theta_2,\phi_2) . \quad \text{(A-14)}$$

A.2. THE CLEBSCH-GORDON COEFFICIENTS

The Clebsch-Gordon coefficients are defined as the transformation matrix:

$$< j_A, j_B, m_A, m_B \mid j, j_A, j_B, m >$$

where $\vec{J}_A, \vec{J}_B$ are two commuting angular momentum vectors and $\vec{J} = \vec{J}_A + \vec{J}_B$, and the eigenvalues of $J^2, J_A^2, J_B^2; J_z, J_{A_z}, J_{B_z}$ are $j(j+1)$, $j_A(j_A+1)$, $j_B(j_B+1)$; m, m_A, m_B respectively. We employ the notation that:

$$C_{j_A j_B}(j, m; m_A, m_B) \equiv < j_A, j_B; m_A, m_B \mid j; j_A, j_B; m > . \quad \text{(A-15)}$$

The simultaneous eigenfunctions of J^2, J_A^2, J_B^2 and J_z are then given in terms of the simultaneous eigenfunctions of J_A^2, J_B^2, J_{A_z} and J_{B_z} by:

$$|j, j_A, j_B; m\rangle = \sum_{m_A, m_B} C_{j_A j_B}(j, m; m_A, m_B) |j_A, j_B; m_A, m_B\rangle \quad \text{(A-16)}$$

with the inverse:

$$|j_A, j_B; m_A, m_B\rangle = \sum_{j=|j_A-j_B|}^{j_A+j_B} \sum_m C_{j_A j_B}(j, m; m_A, m_B) |j, j_A, j_B; m\rangle. \quad \text{(A-17)}$$

The coefficient $C(j, m; m_1, m_2)$ vanishes unless:

(a) $m = m_1 + m_2$

(b) $-j \leq m \leq +j$; $-j_1 \leq m_1 \leq +j_1$; $-j_2 \leq m_2 \leq +j_2$ \quad (A-18)

(c) $|j_1 - j_2| \leq j \leq (j_1 + j_2)$.

The basic symmetry relations are:

$$C_{j_1 j_2}(j, m; m_1, m_2) = (-1)^{j_1+j_2-j} C_{j_1 j_2}(j, -m; -m_1, -m_2)$$

$$= (-1)^{j_1+j_2-j} C_{j_2 j_1}(j, m; m_2, m_1)$$

$$= (-1)^{j_1-m_1} \sqrt{\frac{2j+1}{2j_2+1}} C_{j_1 j}(j_2, -m_2, m_1, -m). \quad \text{(A-19)}$$

For integer ℓ, ℓ_1, ℓ_2, $C_{\ell_1 \ell_2}(\ell, 000)$ vanishes unless $(\ell_1 + \ell_2 + \ell)$ is even. When one of the angular momenta vanishes, say j_2, we have:

$$C_{j_1, 0}(j, m; m_1, 0) = \delta_{jj_1} \delta_{mm_1}. \quad \text{(A-20)}$$

A.2. THE CLEBSCH-GORDAN COEFFICIENTS

Tables of Clebsch-Gordon coefficients are given in Condon and Shortley (1935) for $j \leq 2$ and by Simon (1954)* for $j \leq \frac{9}{2}$. Tables A-1 and A-2 give the values of the coefficients for the common cases in which one of the angular momenta has the value $\frac{1}{2}$ or 1.

Table A-1 $C_{j_1 \frac{1}{2}}(j, m; m-m_2, m_2)$

j	$m_2 = \frac{1}{2}$	$-\frac{1}{2}$
$j_1 + \frac{1}{2}$	$\left[\dfrac{j_1 + m + \frac{1}{2}}{2j_1 + 1}\right]^{\frac{1}{2}}$	$\left[\dfrac{j_1 - m + \frac{1}{2}}{2j_1 + 1}\right]^{\frac{1}{2}}$
$j_1 - \frac{1}{2}$	$-\left[\dfrac{j_1 - m + \frac{1}{2}}{2j_1 + 1}\right]^{\frac{1}{2}}$	$\left[\dfrac{j_1 + m + \frac{1}{2}}{2j_1 + 1}\right]^{\frac{1}{2}}$

Table A-2 $C_{j_1 1}(j, m; m-m_2, m_2)$

j	$m_2 = 1$	0	-1
$j_1 + 1$	$\left[\dfrac{(j_1 + m)(j_1 + m + 1)}{(2j_1 + 1)(2j_1 + 2)}\right]^{\frac{1}{2}}$	$\left[\dfrac{(j_1 - m + 1)(j_1 + m + 1)}{(2j_1 + 1)(j_1 + 1)}\right]^{\frac{1}{2}}$	$\left[\dfrac{(j_1 - m)(j_1 - m + 1)}{(2j_1 + 1)(2j_1 + 2)}\right]^{\frac{1}{2}}$
j_1	$-\left[\dfrac{(j_1 + m)(j_1 - m + 1)}{2j_1(j_1 + 1)}\right]^{\frac{1}{2}}$	$\left[\dfrac{m^2}{j_1(j_1 + 1)}\right]^{\frac{1}{2}}$	$\left[\dfrac{(j_1 - m)(j_1 + m + 1)}{2j_1(j_1 + 1)}\right]^{\frac{1}{2}}$
$j_1 - 1$	$\left[\dfrac{(j_1 - m)(j_1 - m + 1)}{2j_1(j_1 + 1)}\right]^{\frac{1}{2}}$	$-\left[\dfrac{(j_1 - m)(j_1 + m)}{j_1(2j_1 + 1)}\right]^{\frac{1}{2}}$	$\left[\dfrac{(j_1 + m + 1)(j_1 + m)}{2j_1(2j_1 + 1)}\right]^{\frac{1}{2}}$

---

* E. U. Condon and G. H. Shortley, *The Theory of Atomic Spectra* (Cambridge Univ. Press, Cambridge, 1935).
  A. Simon, Oak Ridge Nat. Lab. Report 1718 (1954).

A.3. THE ROTATION MATRICES

The rotation matrix $\mathcal{D}^j_{m'm}(\alpha,\beta,\gamma)$ for a rotation defined by Euler angles (α, β, γ) is defined as:

$$\mathcal{D}^j_{m'm}(\alpha,\beta,\gamma) = \langle j, m'| e^{-i\alpha J_x} e^{-i\beta J_y} e^{-i\gamma J_z} |j, m\rangle . \quad (A-21)$$

It may be written in reduced form as:

$$\mathcal{D}^j_{m'm}(\alpha,\beta,\gamma) = e^{+i(m'\alpha-m\gamma)} d^j_{m'm}(\beta) \quad (A-22)$$

where:

$$d^j_{m'm}(\beta) = \langle j,m'|e^{-i\beta J_y}|j,m\rangle = [(j+m)!\,(j-m)!\,(j+m')!\,(j-m')!]^{\frac{1}{2}}$$

$$\sum_\lambda \frac{(-1)^\lambda}{(j-m'-\lambda)!(j+m-\lambda)!(\lambda+m'-m)!\lambda!} \left(\cos\frac{\beta}{2}\right)^{2j+m-m'-2\lambda} \left(-\sin\frac{\beta}{2}\right)^{m'-m+2\lambda}$$

where the sum is over all integer values of λ for which the arguments of the factorials are non-negative.

The symmetry properties of the $d^j_{m'm}(\beta)$ are:

$$d^j_{m'm}(\beta) = (-1)^{m'-m} d^j_{mm'}(\beta) = d^j_{-m,-m'}(\beta) = d^j_{mm'}(-\beta) \quad (A-23)$$

and:

$$d^j_{m'm}(\beta) = (-1)^{j-m'} d^j_{m',-m}(\pi-\beta) = (-1)^{j+m} d^j_{m',-m}(\pi+\beta) .$$

If the rotation (α,β,γ) is the result of rotating first through $(\alpha_1,\beta_1,\gamma_1)$ and then through $(\alpha_2,\beta_2,\gamma_2)$, we have:

$$d^j_{m'm}(\beta) = \sum_{m''} d^j_{m'm''}(\beta_2)\, d^j_{m''m}(\beta_1) . \quad (A-24)$$

Special values for ℓ integer are:

A.3. THE ROTATION MATRICES

$$d_{00}^\ell(\beta) = P_\ell(\cos\beta)$$

$$d_{m0}^\ell(\beta) = (-1)^m \left[\frac{(\ell-m)!}{(\ell+m)!}\right]^{\frac{1}{2}} P_\ell^m(\cos\beta) \qquad \text{(A-25)}$$

and:

$$d_{m1}^\ell(\beta) = \frac{-m}{\sqrt{\ell(\ell+1)}} \frac{(1+\cos\theta)}{\sin\theta} d_{m0}^\ell(\beta) - [(\ell-m)(\ell+m+1)]^{\frac{1}{2}} d_{m+1,0}^\ell(\beta) \qquad \text{(A-26)}$$

for j a half integer, with $j = \ell + \frac{1}{2}$, special values are:

$$d_{\frac{1}{2}\frac{1}{2}}^j(\beta) = \frac{1}{(\ell+1)} \cos\left(\frac{\beta}{2}\right) (P_{\ell+1}'(\cos\beta) - P_\ell'(\cos\beta))$$

$$d_{\frac{1}{2}\frac{1}{2}}^j(\beta) = \frac{1}{(\ell+1)} \sin\left(\frac{\beta}{2}\right) (P_{\ell+1}'(\cos\beta) + P_\ell'(\cos\beta)) \qquad \text{(A-27)}$$

$$d_{\frac{1}{2}\frac{3}{2}}^j(\beta) = \frac{1}{(\ell+1)} \sin\left(\frac{\beta}{2}\right) \left[\sqrt{\frac{\ell}{\ell+2}} P_{\ell+1}'(\cos\beta) + \sqrt{\frac{\ell+2}{\ell}} P_\ell'(\cos\beta)\right]$$

$$d_{-\frac{1}{2}\frac{3}{2}}^j(\beta) = \frac{1}{(\ell+1)} \cos\left(\frac{\beta}{2}\right) \left[-\sqrt{\frac{\ell}{\ell+2}} P_{\ell+1}'(\cos\beta) + \sqrt{\frac{\ell+2}{\ell}} P_\ell'(\cos\beta)\right].$$

The rotation matrices have the following symmetry properties:

$$\mathcal{D}_{m'm}^{j*}(\alpha,\beta,\gamma) = (-1)^{m'-m} \mathcal{D}_{-m',-m}^j(\alpha,\beta,\gamma)$$

$$= \mathcal{D}_{mm'}^j(-\gamma,-\beta,-\alpha) . \qquad \text{(A-28)}$$

They satisfy the Clebsch-Gordon series

$$\mathcal{D}_{\mu_1 m_1}^{j_1}(\alpha,\beta,\gamma) \mathcal{D}_{\mu_2 m_2}^{j_2}(\alpha,\beta,\gamma) = \sum_{j,\mu,m} C_{j_1 j_2}(j,\mu;\mu_1,\mu_2)$$

$$= C_{j_1 j_2}(j,\mu_2; m_1, m_2) \mathcal{D}_{\mu_1 \mu_2}^j(\alpha,\beta,\gamma)$$

$$\text{(A-29)}$$

with the inverse:

$$\mathcal{D}^j_{\mu,m}(\alpha,\beta,\gamma) = \sum_{\mu_1,m_1} C_{j_1 j_2}(j,\mu;\mu_1\mu_2) \, C_{j_1 j_2}(j,m;\mu_1,\mu_2) \, \mathcal{D}^{j_1}_{\mu_1 m_1}(\alpha,\beta,\gamma)$$

$$\mathcal{D}^{j_2}_{\mu_2 m_2}(\alpha,\beta,\gamma) \qquad \text{(A-30)}$$

where in both eqs. (A-29) and (A-30) all the arguments of the rotation matrices are the same.

The orthonomality properties of the rotation matrices are:

$$\sum_m \mathcal{D}^{j*}_{m_1 m}(\alpha,\beta,\gamma) \, \mathcal{D}^j_{m_2 m}(\alpha,\beta,\gamma) = \delta_{m_1 m_2} \qquad \text{(A-31)}$$

$$\sum_m \mathcal{D}^j_{m m_1}(\alpha,\beta,\gamma) \, \mathcal{D}^j_{m m_2}(\alpha,\beta,\gamma) = \delta_{m_1 m_2} \qquad \text{(A-32)}$$

If $\int d\Omega = \int_0^{2\pi} d\alpha \int_0^{\pi} d\cos\beta \int_0^{2\pi} d\gamma$, then:

$$\int d\Omega \, \mathcal{D}^{j_1}_{\mu_1 m_1}(\alpha,\beta,\gamma) \, \mathcal{D}^{j_2}_{\mu_2 m_2}(\alpha,\beta,\gamma) = \frac{8\pi^2}{2j_1+1} \delta_{\mu_1 \mu_2} \delta_{m_1 m_2} \delta_{j_1 j_2} . \qquad \text{(A-33)}$$

Special values of $\mathcal{D}^j_{mm'}$ are:

$$\mathcal{D}^\ell_{m 0}(\alpha,\beta,\gamma) = \sqrt{\frac{4\pi}{2\ell+1}} \, Y^*_{\ell,m}(\beta,\alpha) \quad \text{with} \quad \ell = 0, 1, 2 \ldots$$

APPENDIX B

FORMALISM FOR PION PHOTOPRODUCTION

Equation (2-136) can be written:

$$f_{\mu\lambda}(\theta,\phi) = \sum_j f^j_{\mu\lambda}(2j+1)\, d^j_{\lambda\mu}(\theta)\, e^{i(\lambda-\mu)\phi} \tag{B-1}$$

where:

$$\lambda = \lambda_1 - \lambda_2 = \text{(photon helicity)} - \text{(initial nucleon helicity)}$$

$$\mu = -\mu_2 = -\text{(final nucleon helicity)}.$$

We here use a normalization conventional in photoproduction for the amplitudes (B-1). They are a factor $\sqrt{\omega_b/\omega_a}$ (see eq. (2-151)) times the helicity amplitudes of Chapter 2.

Define parity-conserving helicity amplitudes by:[1]

$$A_{\ell+} = -\frac{1}{\sqrt{2}} \left(f^j_{\frac{1}{2},\frac{1}{2}} + f^j_{-\frac{1}{2},\frac{1}{2}} \right)$$

$$A_{(\ell+1)-} = \frac{1}{\sqrt{2}} \left(f^j_{\frac{1}{2},\frac{1}{2}} - f^j_{-\frac{1}{2},\frac{1}{2}} \right)$$

$$B_{\ell+} = \frac{1}{\sqrt{2}} \left(f^j_{\frac{1}{2},\frac{3}{2}} + f^j_{-\frac{1}{2},\frac{3}{2}} \right), \quad \ell \geq 1$$

$$B_{(\ell+1)-} = -\frac{1}{\sqrt{2}} \left(f^j_{\frac{1}{2},\frac{3}{2}} - f^j_{-\frac{1}{2},\frac{3}{2}} \right), \quad \ell \geq 1 \tag{B-2}$$

---

[1] Note that the definition of the B amplitudes differs by a factor of $\frac{1}{2}\sqrt{\ell(\ell+1)}$ from those defined by Hebb and used by R. L. Walker, Phys. Rev. **182**, 1729 (1969).

where $j = \ell + \frac{1}{2}$ is the angular momentum, and the integer suffix on A or B denotes the orbital angular momentum of the pion; e.g., $A_{(\ell+1)-}$ is an amplitude corresponding to angular momentum $j = (\ell+1) - \frac{1}{2} = \ell + \frac{1}{2}$ and pion orbital angular momentum $\ell + 1$. We define a mnemonic notation for helicity amplitudes without the ϕ dependence as follows:

$$H_0(\theta) \equiv f_{\frac{1}{2},\frac{1}{2}}(\theta,\phi) \quad ; \quad H_1(\theta) \equiv f_{\frac{1}{2},\frac{3}{2}}(\theta,\phi)\,e^{-i\phi}$$

$$H_{-1}(\theta) \equiv f_{-\frac{1}{2},\frac{1}{2}}(\theta,\phi)\,e^{-i\phi}; \quad H_2(\theta) \equiv f_{-\frac{1}{2},\frac{3}{2}}(\theta,\phi)\,e^{-2i\phi} \quad \text{(B-3)}$$

so that H_0 is a zero helicity flip amplitude, H_1 and H_{-1} are unit helicity flip amplitudes; and H_2 is a double helicity flip amplitude.

$$H_0(\theta) = \sum_{\ell=0}^{\infty} (A_{\ell+} - A_{(\ell+1)-})\,\sqrt{2}(\ell+1)\,d_{\frac{1}{2},\frac{1}{2}}^{\ell+\frac{1}{2}}(\theta)$$

$$= \sum_{\ell=0}^{\infty} (A_{\ell+} - A_{(\ell+1)-})\,\sqrt{2}\,\cos\tfrac{1}{2}\theta\,(P_\ell'(\cos\theta) - P_{\ell+1}'(\cos\theta))$$

$$H_{-1}(\theta) = \sum_{\ell=0}^{\infty} -(A_{\ell+} + A_{(\ell+1)-})\,\sqrt{2}(\ell+1)\,d_{\frac{1}{2},-\frac{1}{2}}^{\ell+\frac{1}{2}}(\theta)$$

$$= \sum_{\ell=0}^{\infty} (A_{\ell+} + A_{(\ell+1)-})\,\sqrt{2}\,\sin\tfrac{1}{2}\theta\,(P_\ell'(\cos\theta) + P_{\ell+1}'(\cos\theta))$$

$$H_1(\theta) = \sum_{\ell=1}^{\infty} -(B_{\ell+} - B_{(\ell+1)-})\,\sqrt{2}(\ell+1)\,d_{\frac{3}{2},\frac{1}{2}}^{\ell+\frac{1}{2}}(\theta)$$

$$= \sum_{\ell=1}^{\infty} (B_{\ell+} - B_{(\ell+1)-})\,\frac{\sqrt{2}\,\sin\theta\,\cos\tfrac{1}{2}\theta}{\sqrt{\ell(\ell+2)}}\,(P_\ell''(\cos\theta) - P_{\ell+1}''(\cos\theta))$$

$$H_2(\theta) = \sum_{\ell=1}^{\infty} (B_{\ell+} + B_{(\ell+1)-})\,\sqrt{2}(\ell+1)\,d_{\frac{3}{2},-\frac{1}{2}}^{\ell+\frac{1}{2}}(\theta)$$

$$= \sum_{\ell=1}^{\infty} (B_{\ell+} + B_{(\ell+1)-})\,\frac{\sqrt{2}\,\sin\theta\,\sin\tfrac{1}{2}\theta}{\sqrt{\ell(\ell+2)}}\,(P_\ell''(\cos\theta) + P_{\ell+1}''(\cos\theta))\,.$$

(B-4)

FORMALISM FOR PION PHOTOPRODUCTION

The above equations express the helicity amplitudes as a sum of partial wave amplitudes of definite J, P, and λ; λ is the relative photon-nucleon helicity and has the value $\frac{1}{2}$ (amplitudes A) or $\frac{3}{2}$ (amplitudes B), the corresponding amplitudes of negative λ being given through (2-137).

Some experimental quantities in terms of the $H_i(\theta)$ are as follows, where q, k are the center of mass momenta of the pion and the photon respectively: (a) differential cross section in the center of mass:

$$\frac{d\sigma(\theta)}{d\Omega} = \frac{1}{2}\frac{q}{k}\sum_{i=-1}^{2}|H_i(\theta)|^2 \tag{B-5}$$

(b) differential cross section for photons polarized perpendicular and parallel to the production plane:

$$\left(\frac{d\sigma(\theta)}{d\Omega}\right)_\perp = \frac{1}{2}\frac{q}{k}[|H_1+H_{-1}|^2 + |H_0-H_2|^2]$$

$$\left(\frac{d\sigma(\theta)}{d\Omega}\right)_\parallel = \frac{1}{2}\frac{q}{k}[|H_1-H_{-1}|^2 + |H_0+H_2|^2] . \tag{B-6}$$

From eq. (B-6) the assymetry from polarized photons is:

$$\sum(\theta) \equiv \frac{\sigma_\perp - \sigma_\parallel}{\sigma_\perp + \sigma_\parallel} = \frac{2\,\mathrm{Re}\,(H_1 H^*_{-1} - H_0 H^*_2)}{\sum_{i=-1}^{2}|H_i|^2} . \tag{B-7}$$

The polarization of the recoil nucleon in the direction $\vec{k}\times\vec{q}$ is, for unpolarized incident photons:

$$P(\theta) = -2\,\mathrm{Im}\,(H_1 H^*_2 + H_0 H^*_{-1})/\sum_{i=-1}^{2}|H_i|^2 . \tag{B-8}$$

If σ_+ and σ_- are the differential cross sections for target nucleons polarized "up" and "down" (that is, parallel and antiparallel to $\vec{k}\times\vec{q}$) respectively, for unpolarized photons, then the polarized target asymmetry is defined as:

$$T(\theta) \equiv \frac{\sigma_+ - \sigma_-}{\sigma_+ + \sigma_-} = 2 \text{ Im } (H_1 H_0^* + H_2 H_{-1}^*) / \sum_{i=-1}^{2} |H_i|^2. \quad \text{(B-9)}$$

In terms of the partial wave amplitudes, the total cross section is:

$$\sigma_T = 4\pi \frac{q}{k} \sum (\ell+1) \left[|A_{\ell+}|^2 + |A_{(\ell+1)-}|^2 + |B_{\ell+}|^2 + |B_{(\ell+1)-}|^2 \right]. \quad \text{(B-10)}$$

The partial wave amplitudes $A_{\ell\pm}(W)$, $B_{\ell\pm}(W)$ are functions of the center of mass energy W. If we assume a simple resonance dominance situation in a partial wave (without any background phase) such as is assumed in deriving eq. (6-117) we can relate the partial wave amplitudes for pion photoproduction, at $W = M_R$, to the resonance photoexcitation amplitudes $A_{\frac{3}{2}}$ and $A_{\frac{1}{2}}$. Comparison of eq. (6-117) and eq. (B-10) gives (for example, for the case of a resonance of pion orbital angular momentum ℓ, $j = \ell + \frac{1}{2}$):

$$A_{\ell+}(W = M_R) = \left[\frac{1}{\pi} \frac{\Gamma_\pi}{\Gamma^2} \frac{k}{q} \frac{1}{2j+1} \frac{M_N}{M_R} \right]^{\frac{1}{2}} A_{\frac{1}{2}} \quad \text{(B-11a)}$$

$$B_{\ell+}(W = M_R) = \left[\frac{1}{\pi} \frac{\Gamma_\pi}{\Gamma^2} \frac{k}{q} \frac{1}{2j+1} \frac{M_N}{M_R} \right]^{\frac{1}{2}} A_{\frac{3}{2}} \quad \text{(B-11b)}$$

and similarly for a resonance of pion orbital angular momentum $\ell + 1$, $j = (\ell+1) - \frac{1}{2}$.

Equivalently to eq. (2-150) we can express the CGLN multipole amplitudes for pion photoproduction in terms of the $A_{\ell\pm}$, $B_{\ell\pm}$ as:

$$(\ell+1)\, E_{\ell_+}(W) = A_{\ell_+}(W) + \sqrt{\frac{\ell}{\ell+2}}\, B_{\ell_+}(W)$$

$$(\ell+1)\, M_{\ell_+}(W) = A_{\ell_+}(W) - \sqrt{\frac{\ell+2}{\ell}}\, B_{\ell_+}(W)$$

$$(\ell+1)\, E_{(\ell+1)_-}(W) = - A_{(\ell+1)_-}(W) + \sqrt{\frac{\ell+2}{\ell}}\, B_{(\ell+1)_-}(W)$$

$$(\ell+1)\, M_{(\ell+1)_-}(W) = A_{(\ell+1)_-}(W) + \sqrt{\frac{\ell}{\ell+2}}\, B_{(\ell+1)_-}(W) \ . \qquad (B\text{-}12)$$

Evidently by substituting $A_{\frac{1}{2}}$ for $A_{\ell_+}(W)$ or $A_{(\ell+1)_-}(W)$ and $A_{\frac{3}{2}}$ for $B_{\ell_+}(W)$ or $B_{(\ell+1)_-}(W)$ in eq. (B-12) gives the equivalent electric and magnetic resonance photoexcitation amplitudes.

In pion photoproduction a photon of isospin 1 can produce final states of isospin $\frac{1}{2}$ or $\frac{3}{2}$ for which the corresponding amplitudes are conventionally designated by a superscript 1 or 3 respectively. For example, for an amplitude called A these would be written $A^{(1)}$ and $A^{(3)}$. A photon of isospin 0 can only produce a state of isospin $\frac{1}{2}$ and the corresponding amplitude is conventionally denoted by $A^{(0)}$. In writing the amplitudes for π^+, π^0 and π^- photoproduction in terms of $A^{(1)}$, $A^{(3)}$, $A^{(0)}$, there is a sign convention entering. Using one convention the expressions are:

$$\pi^+ : A^+ = \sqrt{\tfrac{1}{3}}\, A^{(3)} - \sqrt{\tfrac{2}{3}}\, (A^{(1)} - A^{(0)})$$

$$\pi^0 : A^0 = \sqrt{\tfrac{2}{3}}\, A^{(3)} + \sqrt{\tfrac{1}{3}}\, (A^{(1)} - A^{(0)})$$

$$\pi^- : A^- = \sqrt{\tfrac{1}{3}}\, A^{(3)} - \sqrt{\tfrac{2}{3}}\, (A^{(1)} + A^{(0)})$$

$$\pi^0 : A^{n_0} = \sqrt{\tfrac{2}{3}}\, A^{(3)} + \sqrt{\tfrac{1}{3}}\, (A^{(1)} + A^{(0)}) \ . \qquad (B\text{-}13)$$

The formulae (B-11) refer to pure isospin amplitudes (0) or (1) or (3). Equation (B-13) has to be used to find the charged or neutral pion photoproduction amplitudes.

AUTHOR INDEX

Italicized page numbers indicate full book or journal reference(s) which are given at the chapter end.

Abashian, A., 399, *420*
Adair, R. K., 181, *212*
Adler, S. L., 286, 301, 303, 305, 306, 310, *317*
Akimov, Yu. K., 16, *19*
Alcock, J. W., 198, *212*
Alei, E. F., *213*
Alessandrini, V. A., 416, 419, *420*
d'Alfaro, A., 128, 138, *164*
Allaby, J. V., *378*
Alston, M., 168, *212*
Alston-Garnjost, M., *422*
Alvarez, L. W., *212*
Amati, D., 337, *377*, 416, *420*, 498, *506*
Amblard, B., 118, *123*
Anderson, C. D., 14, *19*
Anderson, E. W., 206, *212*
Anderson, H. L., 55, *85*, 115, *123*
Arndt, R. A., 55, *85*, 340, *379*
Arnold, R. C., 330, 331, 360, *377*
Atkinson, D., 406, *420*
Augustin, J. E., 490, 491, 496, *506*
Auvil, P., 201, *212*

Baacke, J., *124*
Bachmann, A. H., *423*
Ball, J. S., 498, *506*
Ballam, J., *377*
Bander, M., 406, *421*
Barash-Schmidt, N., 8, 9, *19*, *87*
Barbaro-Galtieri, A., *19*, *87*, *422*

Barbour, I., 358, *377*
Bareyre, P., 201, 208, 211, *212*
Barger, V. D., 372, *377*, 416, *421*
Barnes, V. E., 268, *281*
Barut, A. O., 109, *123*
Basdervant, J. L., 420, *421*
Baud, R., 187, *212*
Becchi, C., 275, *281*
Beg, M. A., 274, *281*
Bellac, M. Le, *420*
Berends, F. A., 85, *85*
Berger, E. L., 373, *377*
Bernstein, J., 294, *317*
Bertocchi, L., 376, *377*
Bessis, D., *421*
Bethe, H. A., *213*
Bingham, H., 348, *377*
Bjorken, J. D., 92, *123*, 147, *164*, 501, 503, *506*
Blackmon, M. L., 360, *377*
Blankenbecler, R., 133, *164*, 328, *377*
Blatt, J. M., 53, *85*
Bogoluibov, N. N., 96, *123*
Bonamy, P., 350, *377*
Booth, N. E., *420*
Borgeaud, P., *123*, *377*
Bowcock, J., 197, *213*
Bransden, B. H., 65, *85*, 123, *123*, 202, 203, 204, *212*, *213*, *215*
Bremmerman, H. J., 96, *124*
Bricman, C., *87*, 208, 211, *212*
Bros, J., 127, 144, *164*

523

Brown, L. M., 400, *421*
Budnitz, R. J., 476, *506*
Bugg, D. V., 54, *86*
Burckhardt, H., 198, *213*
Bussey, P. J., *86*
Byers, N., 335, *377*
Calucci, G., 190, *213*
Carter, A. A., 54, *86*
Cashmore, R., *457*
Castillejo, L., 163, *164*
Caverzasio, C., *377*
Chadwick, G. B., 206, *213*, *377*
Chamberlain, O., *86*
Chau, A. Y., 272, *281*
Chew, G. F., 109, 117, *124*, 128, 142, 146, *164*, 383, 392, *421*, 442, *456*, 484, *506*
Chiu, Y. T., 444, *457*
Chou, T. T., 335, *377*
Chu, S. Y., 374, *377*
Ciafaloni, M., 498, *506*
Ciulli, S., 401, *421*
Cline, D., 372, *377*, 416, *421*
Coleman, S., 271, *281*
Collins, P. D. B., 345, 353, 360, *377*, 383, *421*
Conforto, G., *19*
Copley, L. A., 273, 276, 278, 279, *281*
Cottingham, W. N., 197, 198, *213*, 198, *212*
Coulter, P. W., 406, *421*
Coward, D. H., 473, *506*, *508*
Cronin, J. W., 181, *213*
Crowe, K. M., *420*
Cutkosky, R., 198, *213*, 401, *421*

Dalitz, R. H., 163, *164*, 179, 180, 184, 193, 211, *213*, 250, 257, 267, 272, 275, *281*, 429, 430
Dance, D. R., *86*
Dar, A., 326, 379, 444, *456*
Darriulat, P., 187, *213*
Dashen, R. F., 269, *281*, 286, *317*

Davidson, W. C., *85*, *123*
Davier, M., 364, *377*
Davies, A. T., 185, 203, 206, 208, 209, 210, 211, *213*, 425, *457*
Dean, N. W., 334, *378*
Deans, S. R., 54, *86*
Deinet, W., 399, *421*
Della Selva, A., 409, *421*
Denisov, S. P., 339, *378*
Deo, B. B., 198, *213*, 401, *421*
Dietz, K., *124*, *420*
Dolen, R., 364, *378*, 409, *421*
Dombey, N., 80, *86*, 272, *281*, 358, *378*
Donald, R. A., 65, *86*
Donnachie, A., 69, 83, 84, *86*, 201, 208, 209, 210, 211, *212*, *213*, 384, 388, 393, 395, 402, 404, 405, 416, *421*, 494, *506*
Donohue, J. T., *450*
Donskov, S. V., *378*
Dorfan, D., 251, *282*
Drell, S. D., 92, *123*, 147, *164*, 473, 474, 497, *506*
Ducros, Y., *123*
Dunning, J. R., 472, *506*, *507*
Durand, L., 444, *457*, 475
Dyson, F. J., 163, *164*

Eades, J., *282*
Ebel, G., *214*
Eberhard, P., *212*
Eden, R. J., 128, 142, *164*, 185, *214*
Edmonds, A. R., 80, *86*
Eisenberg, Y., *377*
Eisenbud, L., 192, *214*
Epstein, H., *164*
Erwin, A. R., 167, *214*
Evans, W. H., *86*

Faiman, D., 250, 251, 254, 266, 279, *282*, 449, *457*
Falk-Vairant, P., *123*, *377*

AUTHOR INDEX

Feld, B. T., 59, 61, 62, *87*, 202, 204, *215*, *422*
Ferrari, E., 443, *457*
Feshbach, H., 139, *165*, 389, *422*
Feynman, R. P., 273, *282*, 296, *317*, 503, 507
Fierz, M., *377*
Flamm, D., 474, *507*
Flatte, S. M., 400, 402, *422*
Focacci, M. N., 187, *214*
Foley, K. J., 113, 119, 120, *124*, 323, 328, *378*, 448, *457*
Foley, K. T., 206, *214*
Fonda, L., 190, *213*
Foote, J., 49, *86*
de Forest, T., 500, *507*
Fox, G. C., 373, *377*
Frautschi, S. C., 185, *214*
Frazer, W. R., 168, 183, *214*, 397, 398, *422*, 450, *457*, 484, *507*
Freedman, D. Z., 438, *457*
Fretter, W. B., *377*
Freund, P. G. O., 362, *378*
Friedman, J. H., *422*
Froissart, 337, *378*
Fubini, S., 343, *378*, 474, *506*
Fulco, J. R., 168, *214*, 397, 398, *422*, 484, *507*

Gajdicar, T. J., 121, *124*
Gasiorowitz, S., 328, *378*
Gellert, E., 210, *214*
Gell-Mann, M., 96, *124*, 230, 232, 260, 267, 268, *282*, 285, 296, *317*
Ghirardi, G. C., 190, *213*
Giacomelli, G., *378*
Gickaman, M., *85*
Giesecke, J., *124*
Gilman, F. J., 363, *378*, 501, 502, 505, *507*
Gilmore, R. S., *124*, *378*
Glashow, S. L., 269, 271, *281*, *282*
Glasser, V., 164, *377*

Glauber, R. G., 330, *378*, 438, *457*
Goebel, C. J., 179, *214*, 442, *457*
Goldberger, M. L., 21, 25, *86*, 109, 114, *124*, 187, *214*, 328, *377*
Goldhaber, G., *86*
Goldhaber, S., 66, *86*
Good, M. L., *212*
Gorin, Yu. P., *378*
Gottfried, K., 444, 446, *457*
Gourdin, M., 272, *282*, 474, 475, *507*
Graziano, W., *212*
Green, H. S., 252, *282*
Greenberg, O. W., 250, 252, *282*
Gross, F., 475, *507*
Guillaud, J. P., *377*
Guisan, O., *123*, *377*
Gursey, F., 274, *282*

Hamilton, J., 49, 64, 65, *86*, 115, 117, *124*, 316, *317*, 341, *379*, 381, 384, 387, 393, 396, 400, 401, 402, 404, *421*, *422*, 494, *506*
Hammermesh, M., 219, 226, 241, *282*
Hara, Y., 153, *165*
Harari, H., 231, *282*, 358, 362, 363, 364, 376, *377*, *379*, 413, 415, 416, *422*
Hart, W., *86*
Hartle, J. B., 406, *422*
Hartley, H. O., 55, *87*
Hendry, A. W., 183, 206, *214*, 250, 251, 254, 266, 279, *282*, 374, *377*, 448, 449, 450, *457*
Henyey, F. S., 360, 361, 366, *379*, *380*
Hepp, K., 103, *124*
Herndon, D., 439, *457*
Herzberg, G., 187, *214*
Hoffmann, F. de, 167, *213*
Hofstadter, R. H., 460, 471, 472, *507*
Höhler, G., 117, 118, *124*, 181, *214*

Holliday, G., 54, *86*
Holngren, H. D., *213*
Horn, D., 364, *378*, 409, *421*
Hughes, E. B., 472, 486, *507*
Hull, M. M., 59, 63, *86*, 204, *214*

Igi, K., 409, 410, *422*
Igo, G., *213*
Inger, G., *4'7*

Jackson, J. D., 441, 444, 446, 447, *457, 458*
Jacob, M., 71, *86*, 450, *458*
Jakob, H. P., 159, *165*
Jauch, J. M., 79, *87*
Johnson, C. H., 201, *214*
Johnson, R., 383, *421*
Jones, C. E., 406, *422*
Jones, H. F., 109, *125*
Jones, R. S., *124, 457*
Justin, J. Z., *421*

Kabir, P. K., 80, *86*
Källen, G., *7, 19*
Kalogeropoulis, T. E., *423*
Kane, G. L., 203, 205, *215*, 361, 366, *379, 380*
Karl, G., 273, 276, 278, 279, *281*
Kelly, R. L., 361, *379*
Kemmer, N., 14, *19*
Keyser, R., *450*
Khuri, N. N., 130, *164, 165*, 368, *379*
Kinoshita, T., 161, *165*, 368, *379*
Kirsopp, R. G., 208, 209, 210, 211, *213*, 416, *421*
Kislinger, M., 273, *282*
Kittel, W. M., 450, *458*
Klein, A., 130, *165*
Kozlowsky, B., 326, *379*
Krohn, V. E., 475, *507*
Kroll, N. M., 491, 492, *508*
Kruse, E. U., *85, 123*
Kummer, W., 474, *507*

Landshoff, P. V., *164*
Lasinski, T., *87*
Laskar, W., *123*
Lattes, C. M. G., 14, *19*
Layson, W. M., 61, *87*
Lea, A. T., *421*
Lea, R. M., *423*
Leader, E., 357, *379*
Lederman, L., *282*, 474, *508*
Lee, B. W., 274, *281*
Lee, T. D., *508*
Lee, W., *86, 282*
Lehmann, H., 96, 104, *124*
Leith, D., *457*
Len, A. T., *212*
Levi-Setti, R., 266, *282*
Lin, F., 59, 63, *86*, 204, *214*
Lindenbaum, S. J., *124, 378*, 438, *457, 458*
Lipkin, H., 264, *282*
Logan, R. K., 206, *215*
Longacre, R., *457*
Love, W. A., *124, 378, 457*
Lovelace, A. C., 208, 209, 210, 211, *212, 213, 215*, 419, *422, 457*
Low, F. E., *124*, 392, *421, 442, 456*
Lynch, G. R., *422*

MacDowell, S. W., 156, 160, *165*
MacGregor, M. H., 55, *85*, 340, *379*
Malone, W., 358, *377*
Mandelstam, S., 359, *379*, 383, *421*
Manneli, I., *377*
Mar, J., 474, *508*
March, R., *214*
Marshak, R. E., 294, *317*
Martin, A., 103, *125*, 127, 128, 153, *165*, 337, *380*
Martin, A. D., 66, 71, *87*, 123, *125*
Mason, P., *86*
Masperi, L., *421*

AUTHOR INDEX

Massam, T., 496, *508*
Massey, H. S. W., 21, 46, 48, *87*, 444, *458*, 460, *508*
Matsuda, S., 409, 410, *422*
Maximan, L. C., 501, *508*
McAllister, R. W., 471, *507*
McVoy, K. W., 179, *214*
Melater, C. H., *457*
Menotti, P., *422*, 498, *506*
Meshkov, S., 264, *282*
Messiah, A., 46, *87*, 106, *125*, 129, *165*, 252, *282*
Metropolis, N., *213*
Miller, R., *457*
Miller, R. C., 198, *215*
Mirjazaura, H., *124*
Mitra, A. N., 250, 251, *282*
Mo, L. W., 461, 501, *508*
Moffat, J. W., 121, *124*
Moffeit, K. C., 377
Moorhouse, R. G., 85, *87*, 179, 185, 193, 202, 206, 211, *212*, *213*, *214*, 272, 276, *281*, *283*, 358, 377, 425, *457*
Morgan, D., 210, *215*, *420*, 437, 438, 439, *458*
Morpurgo, G., 275, *281*
Morrison, D., 340, *380*
Morse, P. M., 139, *165*, 389, *422*
Mott, N. F., 21, 46, 48, *87*, 444, *458*, 465, *508*
Mueller, A. H., 450, *458*
Muirhead, H., *19*
Muzenich, I., *123*

Nambu, Y., *124*, 484, *508*
Namyslovski, J. M., 438, *457*, *458*
Nauenberg, M., 185, *215*
Nearing, J. C., 185, *215*
Neddermeyer, S. H., 14, *19*
Ne'eman, Y., 230, 232, 267, *282*
Newton, R. G., 21, *87*
Ng, P., 197, 198, *213*
Nielsen, H., 401, 402, *422*

Nishijuma, K., 7, *19*
Noble, B., 161, *165*
Noelle, P., 85, *87*

Oades, G. L., *422*
Obryk, E., 273, 276, 278, 279, *281*
Occhialini, G. P. S., *19*
O'Donnell, P. J., 202, *212*, *213*
Odorico, R., *421*
Oehme, R., *124*
Ogden, P. J., 203, *212*
O'Halloran, T., *86*
Okubo, S., 268, *283*
Olive, D. I., *164*, *420*
Ollson, L. E., 200, *215*
Olsson, M. G., 434, 435, 436, 438, *458*
Ozaki, S., *124*, 378, *457*

Pagnamenta, A., 334, *380*
Pais, A., 274, *281*
Paschos, E. A., 503, *506*
Pearson, E. S., 55, *87*
Perkins, D. H., 336, *380*
Perrin, R., 66, *87*, 203, *215*
Petersen, L. J., *422*
Peterson, J. L., 159, *165*
Petrukhin, A. I., 378
Pfeil, W., 85, *87*
Phillips, R. J. N., 356, *380*
Pietarinen, E., *422*
Pjerrou, G. M., *86*
Plane, D. E., *86*
Plano, R. J., 428, *458*
Platner, E. D., *124*, *457*
Podolsky, W. J., 377
Poirier, J. A., 16, *19*, 65, *87*
Polkinghorne, J. C., *164*
Pomeranchuk, I. Ia, 337, *380*
Poon, C. H., *457*
Powell, C. F., *19*
Price, L. R., *19*
Pripstein, M., 16, *19*, 65, *87*

Prokoshkin, Yu. D., *378*
Protopopesan, S. D., *422*
Pugh, H. G., *213*
Pumplin, J., 363, *378, 379*

Quarles, C. A., 124, *457*
Queen, N. M., 123, *125*

Rabin, M. S., *377*
Radicati, L. A., 274, *282*
Rajasekaran, G., 184, *213*
Rankin, W. A., 85, *87*
Rarita, W., 356, *380*
Ratti, S., *458*
Ravndal, F., 273, *282, 283*
Raznii, M. S. K., *458*
Reader, J. C., *86*
Regge, T., 128, 138, *164*
Resnikoff, S., 250, *282*
Restignoli, M., *125*
Riazzudin, 294, *317*
Richards, D. R., *379*
Richter, B., 358, *380*
Ringo, G. R., 475, *507*
Rittenberg, A., 54, *87*
Roberts, R. G., *458*
Rodgers, E., *86*
Rohrlich, F., 79, *87*
Roos, M., *19, 87*
Roper, L. D., 59, 61, 62, *87*, 202, 204, *215*, 406, *422*
Rosenbluth, M. N., 467, *508*
Rosenfeld, A. H., *19, 87*, 377, *457*
Rosner, J. L., 413, *422*
Ross, M., 250, 251, *282*, 361, 366, *379, 380*
Roychoudhury, R. K., 203, *215*
Rubin, M. S., *422*
Rugge, H. R., 406, *422*
Rushbrooke, J. G., 428, *458*
Russell, J. J., *378*
Ryan, C. P., 294, *317*

Sachs, R. G., 470, *508*
Sakata, S., 232, *283*
Sakita, B., 274, *283*
Sakurai, J. J., 7, 16, *19*, 107, *125*, 491, *508*
Salin, Ph, 272, *282*
Samaranayake, U. K., 119, *125*
Samois, N. P., 399, *423*
Savchenko, O. V., *19*
Scadron, M. D., 109, *125*
Scharenguivel, J. R., 399, 402, *423*
Schmid, C., 364, *378*, 409, *421*
Schneider, J., *377*
Schweber, S. S., 293, *317*
Schwela, D., *87*
Schwimmer, A., 363, *378*
Schwinger, J., 310, *317*, 461, *508*
Scotti, A., 341, *380*
Selleri, F., 443, *457*
Serber, R., 328, *380*
Sergiampetri, F., *377*
Seyboth, P., *377*
Shaw, G., 69, 83, 84, *86*
Shaw, G. L., 406, *421*
Shepard, W. D., *423*
Shinowsky, W., *86*
Silvers, D., 419, *423*
Singer, P., 400, *421*
Siverman, D., *457*
Skillicorn, I. O., *377*
Smadja, G., *377, 457*
Smith, A. R., *86*
Smith, F. T., 187, 188, 191, *215*
Söding, P., *19, 87*
Solmitz, F. T., 49, *87, 422*
Somers, G., 153, *165*
Sonderegger, P., *123, 377*
Sovoko, L. M., *19*
Spearman, T. D., 71, *87*, 203, 205, *215, 422*
Spitzer, H., *377*
Squires, E. J., 185, *215*, 345, 353, 360, *377, 380*, 416, *420*

Steine, F., 159, *165*
Steiner, H. M., 201, *214*
Steinger, H., *86*
Sternheimer, R. M., 438, *458*
Stirling, A. V., *123, 212*
Stodolsky, L., 363, *378*
Stork, D. H., *86*
Stowe, K., *457*
Stoyanova, D. A., *378*
Stubbs, T. F., *86*
Sturge, K., *449*
Sutherland, D. G., 272, 275, *281*
Svensson, B. E. V., *450*
de Swart, J. J., 262, *281*

Takeda, G., 168, *215*
Taylor, J. G., *124*
Taylor, J. R., 185, *214*
Thirring, W., *124*
Ticho, H., *212*
Ticho, K. K., *86*
Ting, C., *282*
Titchmarsh, E. C., 90, *125*
Tobocman, W., 444, *456*
Tran Ha, A., *123*
Trefil, J. S., 448, 450, *457*
Tremain, S. B., *164*
Tring, P. D., *457*
Tripp, R. D., 266, *283*
Trippe, T. G., *87*
Trower, W. P., 200, *215*
Tsai, Y. S., 461, 501, *508*

Uchiyama-Campbell, F., 206, *215*

Van Hove, L., 49, *86*, 451, *457, 458*
Veneziano, G., 417, *423*
Vick, L. L. J., *422*
Vik, O. T., 406, *422*
Villet, G., 208, 211, *212*
Vincelli, L., *377*
Violini, G., *125*

Walecka, J., 500, *507*
Walker, R. L., 85, *87*, 272, 277, *283, 517*
Walzer, W., *214*
Warshaw, S. D., *123*
Watson, G. N., 325, *380*
Watson, K. M., 21, 25, *86, 165*, 187, *214*
Weaver, D. L., 85, *85*
Weinberg, S., 316, *317*, 337, *380*
Weisberger, W., 310, *317*
Weisskopf, V. F., 53, *85*
West, E., *214*
Weyl, H., 219, 226, *283*
Wick, G. C., 71, *86*
Wigner, E. P., 188, *215*
Willen, E. H., *124, 449*
Williams, D. N., *123*
Williams, J. R., *86*
Williamson, G., *379*
Willen, E. H., *378, 457*
Wilson, R., 472, *506, 508*,
Wohl, C. G., *19, 87*
Wojeicki, S. G., *212*
Wolf, G., *377*
Wong, D. Y., 341, *380*
Woolcock, W. S., 49, *86*, 115, 117, *124, 125, 422*
Wright, R. M., 59, 61, 62, *87*, 202, 204, *215*, 340, *379, 422*
Wu, T. T., 335, *380*

Yamada, R., *378*
Yamaka, R., *124*
Yank, C. N.,'335, *377, 380*
Yellin, J., 419, *423*
Yennie, D. R., 470, *508*
Yessian, H. J., *457*
Yodh, G. B., 434, 435, 436, 438, *458*
Yoert, M., *123*
Yost, G. P., *377, 457*
Yuan, L. C. L., *124, 378*
Yukawa, H., 14, *19*
Yvert, M., *377*

Zacharisan, F., 498, *506*
Zarmi, Y., 363, *379*, 416, *422*
Zeemach, C., 130, *165*

Zichichi, A., 496, *508*
Zovko, N., *124*
Zwingenberger, J., *214*

SUBJECT INDEX

A and B amplitudes, 344
absorption model for inelastic scattering, 443-447
absorption parameter (see inelasticity parameter)
Adler consistency condition, 293, 303-306
Adler-Weisberger sum rule, 310
adjoint representation, 239
analytic behavior of scattering amplitudes for forward scattering, 101-102
  at fixed momentum transfer, 102-105
  for potential scattering, 131-37
  for complex angular momenta, 137-42
  for πN scattering, 153-61
  for πN partial waves, 155-60
angular distributions (see differential cross-section)
  polynomial fit to, 199-200
anti-bound state, 184
anti-quarks, 239-40
Argand diagram for resonance, 172, 173, 176, 177
associated production, 425
axial vector current, 294, 307

backward πN scattering, 369-76
Baryon decays and SU3, 266
  electromagnetic, 271-79
Baryon multiplets, 231, 244-48
black sphere scattering, 323-26
"bootstrap" hypothesis, 383
  and duality, 410-13
Born approximation for pion nucleon scattering, 286-93

bound state poles, 131, 144, 183
box diagram, 146
Breit-Wigner formula, 52, 53, 141, 170, 171
  for photoproduction, 69
  and the N/D method, 164
  and Regge trajectories, 185-86

Cabibbo angle, 295
causality, 94-97
C.D.D. poles, 163, 406
charge conjugate quarks (see anti-quarks)
charge conjugation, 16, 17
  in the quark model, 249, 257
charge exchange-scattering, 342, 360
  and duality, 409-10
charge independence, 44, 65, 217
Chew-Low plot, 392
chiral algebra, 309
  chirality operator, 309
circle cut in πN scattering, 396-400
Clebsch-Gordon coefficients, 503-505
conserved vector current, 296
conspiracy, 357
continuous moment sum rules, 409
continuum model for nucleus, (see optical model), 334-35
Coulomb corrections to scattering, 46-49
coupling constants
  pseudoscalar, 100, 111, 287
  pseudo-vector, 114, 291-92
  determination for πN system, 115, 116

531

coupling constants (*cont'd*)
   determination for KN system, 123
   and SU3, 250-66
   pion-quark, 266
   $\rho N\bar{N}$, 494-96
   $\rho\pi\pi$, 493-94
crossing symmetry in forward scattering, 97
   for $\pi^{\pm}N$ scattering, 109, 110, 393
   for two-body scattering, 142-44
cross-over effect, 356-366
cross-section defined, 25
   expression for $2 \to n$ particle reaction, 26
   expression for $2 \to 2$ particle reactions, 27
   inequalities, 33
   πN at high energies, 118-21
   total at high energies, 336-40
crossing matrix for πN system, 393
current algebra, 285, 306-11

Dalitz plot, 167, 431-33
daughter trajectories, 375-76, 418
decay widths
   electromagnetic, 272-79
   of the $N^*(1260)$, 274
decuplet representation
   of SU3, 231
   and quark wave-functions, 233-34
   and SU6, 246
   mass splitting of, 267-68
deep inelastic scattering, 462, 498-505
derivative pion-nucleon coupling, 291-92
differential cross-section
   for 2 particle scattering, 28, 36
   for scattering from polarized targets, 57, 58
   for photoproduction (see angular distribution), 78
diffraction scattering, 319, 325
   in elastic πN collisions, 357-61

dips in forward scattering, 350, 357-59, 366-67
   in inelastic scattering, 359-361, 367
dispersion relations, 89-92
   for forward scattering, 92-102
   at fixed momentum transfer, 102-105
   for πN scattering, 110, 113-21
   for KN scattering, 121-23
   validity at higher energies, 117-19
   and partial wave analysis, 115
   in potential scattering, 131
   for partial waves, 160
   and the Pomeranchuk theorem, 338
distorted wave approximation, 444-46
double scattering experiments, 41, 43, 58, 59
double spectral function, 132
dual models, 417-19
duality diagrams, 413-15
duality, 361-363, 409

effective range expansion (see also scattering length), 32, 66
eigenphases
   defined, 178
   no-crossing theorem, 179
eightfold way, 231
eikonal approximation in scattering, 329-344, 360
electromagnetic interactions
   and SU3, 269-70
   and mass splitting, 270, 271
   e.m. decays, 272-279
electromagnetic form factors, 443-44, 466-67, 469-74
   experimental data for protons, 470-74
   for neutrons, 474-478
   isospin decomposition, 477
   and dispersion theory, 478-83
   absorptive parts, 480-82
   and vector dominance, 483-86

SUBJECT INDEX

of pions, 486-90
  behavior at large momentum
    transfers, 497-98
electron scattering by nucleons,
    461-62
  elastic scattering, 462-74
  and two photon exchange,
    473-74
  electron scattering by deuterons,
    475
  inelastic differential cross-
    section, 500
equal time commutators, 309
error matrix, 55
exchange degeneracy, 360, 411-13
exchange forces, 412
exchange mechanism (see particle
  exchange, Regge pole exchange)
exotic channels, 358, 361, 411

F and D couplings, 260-66
factorization of pole residues,
  185, 353
Fermi-Yang ambiguity, 57
Feynman integration formula, 147,
  148
finite energy sum rules, 361,
  407-409
field-current identities, 485
forward cross-section (see diffraction scattering) and Regge poles,
  347
Froissart bound, 337

ghost-killing mechanism, 353
Glauber approximations (see
  eikonal approximation)
Goldberger-Treiman relation, 298
G-parity, 18, 19
  in the quark model, 257
Green's functions, 129
groups (see symmetry groups)
  generators of, 221, 230

helicity
  defined, 71
  and partial wave expansions,
    72-74

helicity flip, 365
Hilbert transforms, 90

impact parameter, 32
impact parameter representation,
  329
incident flux
  defined, 25, 26
  invariance of, 27
inclusive reactions, 453, 462
inelasticity parameter
  definition, 33
infinitesimal transformations, 222
interference model, 416-17
intersecting storage rings, 449
invariant amplitudes
  for πN scattering, 105-107
irreducible representations (see
  representations)
isobar model, 437-500
isospin (see isotopic spin)
isotopic spin, 11-16
  and the electromagnetic inter-
    action, 79, 80
  and $SU2$, 217-18

J-plane cuts, 358-61
Jost functions, 136

K-matrix
  defined, 178
  poles in, 181
  analytic properties, 182-85
kaon-nucleon coupling, 123
  in the quark model, 250-66
kaon-nucleon scattering, 65-68
  at high energies, 358, 363, 364,
    368
  and exchange degeneracy,
    412-13
kinematical singularities, 159

laboratory system variables, 51
Landau equations, 149
Legendre polynomials, 509-10
Lehmann ellipse, 104, 132
L-excitation quark model, 250-57
  and baryon multiplets, 252-53

L-excitation quark model (cont'd)
  and meson multiplets, 256-57
  and strange baryon decays, 266
  and e.m. baryon decays, 273-79
Lie algebra, 221, 307
longitudinal momentum distributions, 449
low energy πN scattering and partial wave dispersion relations, 384-407

McDowell symmetry, 160
magnetic moment of protons, 470
  of neutrons, 470, 474
magnetic moments
  in quark model, 274
Mandelstam representation, 132
Mandelstam variables s, t, u, 94
mass splitting
  within isospin multiplets, 218, 231
  within SU3 multiplets, 231
  in the L-excitation quark model, 251
  in SU3, 267-69
  electromagnetic, 270
maximum analyticity, 146
meson multiplets, 231, 248-49
metastable states (see resonance)
Minami ambiguity, 57, 201
minimal electromagnetic coupling, 828
Mott scattering cross-section, 465
Mueller diagram, 456
multiparticle production, 451-56
multipole expansion, 75-77
multiplicities in particle production, 452

$N^*(1238)$
  energy and width, 53, 54
  in photoproduction, 69, 167, 168, 206
  radiative decay, 274-76
  production, 434

N_1^* resonances, 206, 277, 278
  and (Δ) exchange in low energy πN scattering, 392-96
  and inelastic processes, 425
N/D method, 161-64, 390
natural units, 3
nonets of mesons, 242-43
normal thresholds, 144-46
normalization
  single particle states, 22
  two particle states, 30
nucleon exchange in low energy πN scattering, 388-92

octet representation
  of SU3, 231
  and quark wave-functions, 236-38, 242, 252
  and SU6, 246
  and F and D couplings, 260-66
  mass splitting in, 268-69
one pion exchange model, 442-43
optical model (see black sphere scattering), 326, 335
optical theorem, 29, 321
  Mueller generalization, 455-56
ordinary forces, 412

pp scattering, 363
Padé approximates, 382, 419-20
Panofsky ratio, 70
parastatistics, 252
parity, 7, 35, 36
  and scattering, 72
  in the quark model, 257
partial cross-sections, 32, 33
partial wave expansion
  for spinless particles, 30-32
  for particles with spin, 34-36, 195-97
particle exchange, 340-43, 381-82
partons, 462, 503-505
PCAC-partially conserved axial current, 285, 299

SUBJECT INDEX

PDDAVC - pole dominance of the divergence of the axial vector current, 285
peripheral method for partial waves, 385-88, 402-407
peripheral model of quasi-two body reactions, 438-41
permutation symmetry, 224-226
phase rule, 350
phase shift
  definition, 32
  low energy πN, 62, 402-407
phase shift analysis, 54-55
  with correlated errors, 55
  for low energy πN scattering, 56, 57, 61-63
  ambiguities in, 57, 201
  at discrete energies, 202
  energy dependent, 203
photoexcitation, 279-80
photo-nucleon vertex function, 465-67
  low momentum transfer limit, 468-500
photoproduction of pions
  at low energies, 68-85
  in helicity representation, 74, 75
  multipole expansion for, 75-77
  and isospin, 80-82
  and the quark model, 279-80
  and the Regge pole model, 358
  and formalism for, 517-20
Phragmen-Lindelöff theorem, 368
physical cut in πN partial waves, 157
pion-nucleon scattering
  elastic, partial wave expansion, 35-36
  isospin decomposition, 43, 44
  Coulomb corrections to, 46-49
  p_{33} resonance, 53, 54
  low energy data and analysis, 59-65
  forward dispersion relation analysis, 114-21
  and the Mandelstam representation, 153-5
  and partial wave dispersion relations, 160-61

  and higher resonances, 205-12
  Born approximation for, 286-93
  empirical fit to high energy data, 328-29
  and Regge pole model, 347-61
  in the backward direction, 369-76
  analysis of low energy s waves, 400-407
  analysis of low energy p waves, 396, 402-407
  analysis of higher low energy partial waves, 402-407
pion-pion scattering, 396-402
pion-quark coupling, 266
pion production in collisions, 434-40
polarization, in scattering, 37-43
  polarization vectors, 39
  in πN elastic scattering, 57-60
  in πN charge exchange, 360
polynomial fitting, 199-200
Pomeranchuk theorem, 337-40
Pomeranchuk trajectory, 354-6, 362, 415
Pomeron (see Pomeranchuk trajectory)
positron scattering by protons, 474
potential scattering, 128-42

quark model for inelastic scattering, 447-51
quarks, 233
  and SU3 wave-functions, 234-43
  with spin, 244-49
  and the L-excitation model, 256-57
  and high energy scattering, 333
  quantum numbers of, (see antiquarks), 235
quasi-two body reactions, 427

radius of proton, 471
range of interaction, 385-86
reaction matrix, 81, 82
reduced energies, 28
reduced width, 54

reducibility, 219
Regge cuts (see J-plane cuts)
reggeons, 185, 187
Regge pole
  in potential scattering, 140-41
  in high energy scattering, 343
  and forward scattering, 344-47
  and ρ exchange (see Regge trajectory, Residue functions), 347-51
Regge pole exchange, 343
Regge trajectory (see Regge pole), 343
  ρ trajectory, 349, 350, 409, 412, 415
  P trajectory, 354-55, 415-17
  P' trajectory, 356, 412
  ω trajectory, 356-57, 412, 415
  N_α, nucleon trajectory, 370, 415
  and nucleon resonances, 371
  Δ trajectory, 370, 415
  N_j trajectory, 371
  A_2^j trajectory, 412, 415
  Λ trajectory, 415
  Σ trajectory, 415
  f trajectory, 417
representations
  of SU2, 219
  irreducible, 219
  equivalent, 229
  of SU3, 231-40
  of SU6, 244-49
residue functions, 351-54
  and factorization, 185
rescattering corrections, 341
resonance, 52, 167, 168
  width, 52, 53, 173
  elastic, 169-73
  inelastic, 173-77
  and time delay, 170, 171, 187-94
  energy, 173
  poles, 181-84
  criterion for, 194-95
  and partial wave analysis, 204-212
Rosenbluth formula, 459
rotation matrices, 514-16

scale invariance, 501-503
scaling law, 454-55
scattering amplitude
  defined, 28
  partial wave expansions, 31, 34-36
scattering lengths, 32, 64-65, 117, 291-93
  and current algebra, 311-16
s-channel helicity, 363
Schwinger terms, 310
selection rules
  for e.m. decays, 274-79
singlet representation of SU3, 231
  and quark wave-functions, 236, 242
scattering amplitude
  defined, 28
  in laboratory system, 112
sense and nonsense Regge vertices, 351-52, 354, 373
S-matrix
  definition, 22, 174
  partial wave expansions, 31, 34, 72
  helicity representation, 72-74
  poles in, 181-85
  in eikonal approximation, 331
shrinkage of forward peak in πN scattering, 349
signature factors, 352, 353, 354
single spectral function, 132
soft pions, 300
Sommerfeld-Watson transform, 139, 346
spin-flip amplitude, 35
spectral function (see single spectral function; double spectral function)
spherical harmonics, 510-11
spin rotation coefficients, 57, 58
S operator (see S-matrix)
spinors
  Dirac, notation, 4
  2-component, 218

SUBJECT INDEX

structure constants, 221, 307
subtraction constants, 92, 115, 132
sum rules in the parton model, 505
super-convergence relations, 91, 285
symmetry groups
  SU_2, 218-29
  SU_3, 230-32
  and quark model, 232-43
  and particle classification, 232
  SU_6, 244-49
  and the L-excitation model, 252-57
  and couplings, 258-66
  SU_{6W}, 264, 265
  and F/D ratios, 260-66
  and electromagnetic current, 495-96

Tamm-Dancoff method, 382
t-channel exchange, 370
time-delay, 187-94
time reversal invariance in photoproduction, 69
three particle states, 428-33
threshold behavior,
total cross-section
  high energy behavior (see Froissart bound, Pomeranchuk theorem), 336
transition matrix (T-matrix), 24, 174
  poles in, 181
  analytic properties, 181-85
  for $e^- - p$ scattering, 464
transition operator, 23
transition rate
  defined, 25
transverse momentum distributions, 452-53

u channel amplitudes for πN scattering, 372
u channel exchange, 369-76
U-spin, 270

unitarity
  of S-matrix, 23, 31
  relation for transition matrix, 24
  in potential scattering, 133
unitary matrices, 219

vector current, 294
Veneziano model, 417-19
virtual bound state, 184

weak currents, 294
weight diagrams, 235-36
width of resonance, 52, 53
Wiener-Hopf technique, 161
Wigner's causality condition, 188
Wolfenstein parameters (see spin rotation coefficients), 59

$Y_1^*(1385)$, 68
$Y_0^*(1405)$, 68
Young tableaux, 226-29
  partition numbers, 231
  for SU_3, 232
  for SU_6, 244

Δ baryon resonance (see $N^*(1238)$)

η particle
  production in πN collisions, 403, 423

μ meson scattering by protons, 474

ρ exchange in low energy πN scattering, 399
ρ meson, 167, 168, 399
  and πN charge exchange, 342, 347-51
  and $\pi\pi$ bootstrap, 383
  and nucleon form factors, 484
  production in $e^- - e^+$ collisions, 489-90
  and the vector dominance model, 491-95
ρ' meson, 348

σ exchange in low energy πN scattering, 399-400

ϕ meson, 484, 490-91, 495

ω meson, 484, 490-91, 495
$\omega-\phi$ mixing, 243, 269, 495-96